Pollination Biology and Ecology

Pollination Biology and Ecology

Edited by Blake Hughes

SYRAWOOD
PUBLISHING HOUSE

New York

Published by Syrawood Publishing House,
750 Third Avenue, 9th Floor,
New York, NY 10017, USA
www.syrawoodpublishinghouse.com

Pollination Biology and Ecology
Edited by Blake Hughes

© 2022 Syrawood Publishing House

International Standard Book Number: 978-1-64740-069-9 (Hardback)

Cataloging-in-Publication Data

Pollination biology and ecology / edited by Blake Hughes.
    p. cm.
Includes bibliographical references and index.
ISBN 978-1-64740-069-9
1. Pollination. 2. Plant ecology. 3. Fertilization of plants. 4. Fertilization of plants by insects. I. Hughes, Blake.
QK926 .P65 2022
571.864 2--dc23

TABLE OF CONTENTS

Chapter 23   **Pollen transfer efficiency of** *apocynum cannabinum* **(apocynaceae)**..192
          Tatyana Livshultz, Sonja Hochleitner and Elizabeth Lakata

          Permissions

          List of Contributors

          Index

PREFACE

This book was inspired by the evolution of our times; to answer the curiosity of inquisitive minds. Many developments have occurred across the globe in the recent past which has transformed the progress in the field.

Pollination biology or anthecology is the science concerned with the study of pollination and the association between flowers and pollinators. Pollination is an important process in plant reproduction that is essential to the production of fruits and seeds. The process is mediated through an interaction between flowers and pollen vectors. Pollinators (or pollen vectors) are insects, birds or mammals that move pollen from the anther to the stigma of a flower. This allows the process of fertilization to occur. The study of pollination ecology encompasses pollination studies at many levels, such as the effectiveness of specific pollinators of a plant species, properties of interactions occurring within communities of plants and floral visitors, etc. This book unravels the recent studies in the fields of pollination biology and ecology. Also included in it is a detailed explanation of the various interactions between pollinators and plants. This book is a vital tool for all researching or studying pollination ecology as it gives incredible insights into emerging trends and concepts.

This book was developed from a mere concept to drafts to chapters and finally compiled together as a complete text to benefit the readers across all nations. To ensure the quality of the content we instilled two significant steps in our procedure. The first was to appoint an editorial team that would verify the data and statistics provided in the book and also select the most appropriate and valuable contributions from the plentiful contributions we received from authors worldwide. The next step was to appoint an expert of the topic as the Editor-in-Chief, who would head the project and finally make the necessary amendments and modifications to make the text reader-friendly. I was then commissioned to examine all the material to present the topics in the most comprehensible and productive format.

I would like to take this opportunity to thank all the contributing authors who were supportive enough to contribute their time and knowledge to this project. I also wish to convey my regards to my family who have been extremely supportive during the entire project.

Editor

Resource abundance and distribution drive bee visitation within developing tropical urban landscapes

Victoria Wojcik[1,2]*

[1]Pollinator Partnership, 423 Washington Street 5th Floor, San Francisco, CA 94111
[2]University of California, Berkeley, Department of Environmental Science, Policy, and Management, Berkeley, CA 94720

Abstract—Urban landscapes include a mix of biotic and anthropogenic elements that can interact with and influence species occurrence and behaviour. In order to outline the drivers of bee (Hymenoptera: Apoidea) occurrence in tropical urban landscapes, foraging patterns and community characteristics were examined at a common and broadly attractive food resource, *Tecoma stans* (Bignoniaceae). Bee visitation was monitored at 120 individual resources in three cities from June 2007 to March 2009. Resource characteristics, spatial distribution, and other local and regional landscape variables were assessed and then used to develop descriptive regression models of forager visitation. The results indicated that increased bee abundance and taxon richness consistently correlated with increased floral abundance. Resource distribution was also influential, with more spatially aggregated resources receiving more foragers. Individual bee guilds had differential responses to the variables tested, but the significant impact of increased floral abundance was generally conserved. Smaller bodied bee species responded to floral abundance, resource structure, and proximity to natural habitats, suggesting that size-related dispersal abilities structure occurrence patterns in this guild. Larger bees favoured spatially aggregated resources in addition to increased floral abundance, suggesting an optimization of foraging energetics. The impact of the urban matrix was minimal and was only seen in generalist feeders (African honey bees). The strongly resource-driven foraging dynamics described in this study can be used to inform conservation and management practices in urban landscapes.

Key Words: urban ecology, foraging, resource characteristics, Tecoma stans, bees (Hymenoptera), Costa Rica

Introduction

Bees (Hymenoptera: Apoidea) and other pollinators are facing global survival challenges that are attributed in part to local anthropogenic change (NRC 2007). Land use intensification, particularly in the form of agricultural and urban development, has been shown to diminish the nesting and foraging habitats of native bees, resulting in decreases in abundance and changes in pollinator function (Kremen et al. 2002; Buchmann & Ascher 2005). In recent years, Northwestern Costa Rica has experienced an increase in urban development linked to growth in tourism, international business, manufacturing, and agriculture. Changes in bee community composition that correspond to local fragmentation have already been recorded (Frankie et al. 1997) and are particularly evident at the edge of urban development (Frankie et al. 2009a). Investigating how bees respond to urbanization is imperative given the ever-increasing rate of landscape conversion and their importance as pollinators to nearly 90% of tropical floral species (Ollerton et al. 2011) and over 30% of the crops that make up the human diet (Buchmann & Nabhan 1996).

Recent studies focusing on urban bees have documented species occurrences (Owen 1991; Frankie et al. 2002), examined community composition with respect to general land use patterns (Saure 1996; McIntyre & Hostetler 2001; Cane et al. 2006; Nates-Parra et al. 2006; Matteson et al. 2008), and evaluated trends in floral host usage and preferences (Hismatsu & Yamane 2006; Antonini et al. 2006; Frankie et al. 2005). Although it is clear that bees are resident and abundant in some urban landscapes, field observations and published works show their distribution to vary. Nearby cities in the same regional landscape can have significantly different resident bee faunas (Frankie et al. 2009b). Within individual cities variability in bee community structure has been noted in different land use types, such as residential areas (Winfree 2007) and community gardens (Matteson et al. 2008), and with respect to the size and age of habitat remnants (Cane et al. 2006). Changes in bee community structure (Lui & Koptur 2003) and absolute bee numbers have also been recorded along urban-natural gradients (Ahrne et al. 2009), as have seasonal transitions driven by shifts in resource availability (Cane et al. 2005).

Both landscape context and local site variables can influence bee occurrences and foraging patterns. In the highly modified landscapes of central Europe the overall landscape composition was found to be more influential in determining honey bee occurrences than site-specific variables (Steffan-Dewenter & Kuhn 2003). A similar trend was reported by Heard et al. (2007) for bumble bees foraging within different agro-environments. Nearness to intact forest habitat was also noted to be more important than local land management in determining the number of bees expected at local coffee plantations in the tropical dry forest of Costa Rica (Brosi et

*Corresponding author;
email: vw@pollinator.org;
victoriawojcik@hotmail.com

FIG I: A common street view showing two *T. stans* resources within the urban matrix (Cañas, Costa Rica).

al. 2007). In contrast, Collevatti et al. (2000) showed that the quality and quantity of resources present at a patch determined foraging patterns, irrespective of the local landscape. Bee assemblages are functionally diverse and factors that impact the community might not necessarily have analogous effects on individual groups and species. For example, Jha and Vandermeer (2009) recorded contrasting responses to resource variables and local habitat variables between native bees and non-native honey bees foraging in coffee plantations in central Mexico. Foraging patterns might also be driven by different variables in natural and anthropogenic systems, but they have yet to be investigated in sufficient detail to make generalized statements.

The goal of this study is to achieve a better understanding of how resource characteristics and local landscape variables structure the foraging patterns of bees in urban ecosystems. The following analysis aims to: (I) describe and model the factors that are influencing bee visitation to food resources in a tropical dry forest urban system and (2) to examine these factors at the community level and in terms of individual bee guilds to see if congruent or contrasting patterns emerge. The system of bees visiting *Tecoma stans*, a common and preferred food resource in this region, is used as a model. Distance to wildland habitats, proximity to other landscape features (riparian areas and open space), land use type (residential versus commercial), local resource characteristics (floral abundance and life form), and resource distribution are examined for their effect on abundance and richness. Within the mix of variables that might be influencing resource usage in cities are those that are unique to urban landscapes, such as residential and commercial land use and the separation of a site from the wildland habitat by the urban matrix. A parallel focus of this analysis is to determine to what extent these 'urban' factors impact bee resource usage.

MATERIALS AND METHODS

Site description

The cities of Liberia (10°37'47"N, 85°26'18"W), Bagaces (10°31'34"N, 85°15'18"W), and Cañas (10°25'36"N, 85°05'28"W) lie in a row along Central America Highway I in the Guanacaste province of Costa Rica. These three cities are characteristic of urban development in the region, displaying a trend of increasing peripheral growth due to increased commercial and residential demands. Liberia covers 7.97 km2, Bagaces covers 0.76 km2, and Cañas covers 2.98 km2. The most recent census data puts the urban population of Liberia, the provincial capital, at 34,469; Bagaces at 3,645; and Cañas at 16,512 (INEC 2000).

Regional landscape classification

Tropical dry forest was the dominant habitat in this region prior to urbanization and agricultural development. Today, habitat has been fragmented to varying degrees by cattle ranching, agriculture, and commercial development, however there are remnant areas of savannah. Ranching was dominant around both Bagaces and Liberia, but these two cities were also in closer proximity to a network of biological reserves and conservation areas. Cañas was surrounded by more intensive irrigated agriculture.

Urban landscape classification

A combination of satellite images, GIS data, and ground observations were used to code urban land use and landscape characteristics. Appendix I provides the summary statistics and abbreviations for the variables used. All spatial calculations were preformed in ArcMap 9.3 © (ESRI Inc. 2009, Chicago IL). The *Near Feature* tool was used to calculate the distance of each *T. stans* to the wildland-urban interface (D WUI) and nearest riparian area (D RIP). The *Nearest Neighbour* tool was used to find the closest conspecific *T. stans* individual. The number *T. stans* individuals within radii of 10 m, 50 m, and 100 m were calculated by constructing corresponding buffer polygons and using the *Points in Polygon* tool in the *Hawth's Tools* package. Proximity to open space (OPEN) was coded as present when the floral resource was on or adjacent to a city block that contained unmanaged open spaces (e.g., large yards, vacant lots) and public parks or fields. A residential classification (RES) was given when the dominant use on the block was either single or multifamily housing. A commercial designation (COM) was given when the dominant use consisted of either public or private business, shopping, government offices, banks, and parking lots.

Floral resource characterization

The yellow trumpet tree, *Tecoma stans* Kunth (Bignoniaceae) is a common mass-flowering woody perennial found across Costa Rica, and much of the tropics. The native range in the Americas extends from the south central and southeastern United States (Arizona to Florida) down into Argentina (Hammel 2005; Zuchowski 2007). *T. stans* was ubiquitous and well distributed in the three landscape studied here and multiple morphologies were common due to variation in pruning and management. Although *T. stans* produces flowers year-round, the most intense flowering occurs between November and March. Fig. 1 depicts a common residential street view where multiple *T. stans* resources are present within the urban matrix.

Previous work has indicated that *T. stans* attracts a wide range of bee species from multiple feeding, social, and size guilds that account for approximately 10% of the local wildland bee species diversity (Wojcik 2009). The majority of these bee species are in the genus *Centris* and have been observed to collect both pollen and nectar (see Wojcik 2011). The pollination system of *T. stans* in Central America has yet to be studied in detail. Work from South America identified *Centris tarsata* Smith and *Exomalopsis fulvofasciata* Smith as effective pollinators in Brazil (Silva et al. 2007). The size and behaviour of the most abundant visitor to *T. stans* in Northwestern Costa Rica, *C. eurypantana* Snelling, suggest that it could be an effective pollinator (pers. obs.). Other large bodied species such as *Ephicharis elegans* and *Eulema* sp. could be potential pollinators, but are not common visitors. *Halictus* and *Lasioglossum* species that visit this resource do not appear to be effective pollinators based on their size and behaviour (pers. obs.).

A survey of each city catalogued all visible *T. stans* individuals using a hand-held GARMIN Vista C etrex GPS unit. In total, 147 individuals were marked, of which 120 were easily accessible from public land and were used in the study. Each individual resource was photographed digitally for visual reference, character coding, and resource metric estimation. Height (HEIGHT) and crown width (CROWN) at the widest part of the crown were estimated from the referenced photographs to the nearest 25 cm. Each individual

was also coded into three life form classes (SHRUB, HEDGE, or TREE) depending on shape and growth habit.

A stratified sub-sample of the canopy was used to estimate the number of flowers (FLW) per individual *T. stans*. The canopy of each resource was fractioned into a manageable sector based on overall size (2, 4, 8, or 10 sectors); this fraction was then used as an expansion factor (k). The total number of inflorescences (i) in one randomly selected sector was counted. A subsample (n = 5) of inflorescences was used to calculate the average number of flowers per inflorescence (f). The calculated average number of flowers per inflorescence was then multiplied by the number of inflorescences and the expansion factor as follows: FLW = f x i x k. In the case of some individual resources with very few flowers the total number of flowers was enumerated using a census.

TAB I: Bee species comprising the taxonomic categories used for analysis and model development. The relative occurrence of each species in the cities of Bagaces, Cañas, and Liberia is indicated as common (C), rare (R), or known from a single record (S) based on collections made over the study period.

Species	Location		
	Bagaces	Cañas	Liberia
Solitary Large			
Ancyloscelis sp.	R	–	R
Augochlora nigrocyanea Cockerell	C	C	C
Centris aethyctera Snelling	C	C	–
Centris eurypatana Snelling	C	C	C
Centris heithausi Snelling	C	C	–
Epicharis elegans Smith	–	–	C
Euglossa viridissima Friese	C	C	C
Eulema sp.	S	–	C
Xylocopa fimbriata Fabricius	R	C	C
Xylocopa subvirescens Cresson	–	–	C
Solitary Small			
Agapostemon nasutus	–	–	S
Halictus lutescens Friese	C	C	C
Lasioglossum Dialictus sp.1	–	C	–
Lasioglossum Dialictus sp.2	–	C	–
Native Eusocial			
Nannotrigona perilampoides Cresson	C	C	C
Plebeia frontais Friese	R	–	R
Tetragonisca angustata Lep.	C	–	C
Trigona fulviventris Guérin-Méneville	C	C	C
Non-native Eusocial			
Apis mellifera scutellata Latreille	C	C	C

Bee monitoring and identification

Bee visitation was documented from June 2007 to March 2009. Visitor abundance and richness were measured at each resource within a standardized one by one metre square observation frame that was visually projected onto an easily observable and unobstructed area of the floral resource. The visitation rates of bees to all of the flowers within this visual frame were recorded for the duration of three minutes. Eight evenly spaced counts occurred throughout the day between 0800h and 1400h in order to account for temporal variability in bee occurrence (see Wojcik 2011). Samples were repeated across multiple days (at least four) and in both the wet and dry seasons. Bees were identified on-the-wing to one of four categories: large solitary, small solitary, native eusocial, or non-native eusocial bees. A list of the species that comprise each taxonomic category and their documented occurrences in each city is presented in Tab. 1. Bee identification was assisted by Laurence Packer and Jason Gibbs of York University, Toronto, Canada and by Ricardo Ayala of Universidad Nacional Autónoma de México.

Statistical analysis

The average number of bees and the average number of different taxon groups recorded per three-minute count were calculated for each resource from all of the counts taken during the entirety of the study period. The visitation rates of each bee group, the abundance, and the taxon richness were compared between the three cities using multivariate ANOVA followed by pairwise comparisons with Tukey's HSD in order to determine if parallel patterns existed in each landscape. A Bonferroni adjusted α of 0.016 (0.05/3) was used to reduce Type I error.

The averaged count data were checked for normality using both normal-quartile plots and the Shapiro-Wilk test. The untransformed data plotted within a normal distribution, with some wobble at the ends of the distribution, but the results from the Shaprio-Wilk test were less then p = 0.05, indicating a deviation from the normal distribution. Subsequent standard data transformations were performed (log, square root, arcsine, etc.), but neither the plotted fit nor the Shaprio-Wilk statistic changed significantly. Given the large size of the data set (n > 30), a decision was made to use the untransformed data for the analysis. An analysis of covariance was used to assess the independence of the variables in their description of bee visitation patterns. The variable CROWN was found to be highly correlated with FLW and was removed from the regression analysis. The remaining thirteen variables were independent in their explanation of the data. Regional and city specific descriptive models were generated for abundance, taxon richness, solitary bees (large and small) and eusocial bees (native and non-native), and for each individual bee group (large solitary bees, small solitary bees, native eusocial bees, and non-native eusocial bees) using backward stepwise regression with p = 0.10 as the entrance criterion and p = 0.05 as significant. A total of 32 models were generated in this manner. Model selection was based on significant improvement in fit ($r2$ value) and ΔAIC, favouring simpler models with fewer variables when significant changes were not present in the analytical metrics. All of the statistical analyses were performed in SPSS 16.0 (© Chicago IL 2007).

RESULTS

General and site-specific foraging patterns

Patterns of overall abundance and richness varied significantly between the three cities ($F_{2,2083}$ = 4.94, p = 0.007 and $F_{2,2083}$ = 19.50, p = 0.000, respectively). Abundance ranged from 0 to 32 bees observed per count, with a mean of 3.55 ± 1.01 (mean ± S.E) and was highest in Liberia. Richness ranged from 0 to 4 bee types per count, with a mean of 1.35 ± 0.73 and was again higher in Liberia. Pairwise comparison with Tukey's HSD indicated that large native solitary bees (F_2 = 18.68, p = 0.000), small native solitary bees (F_2 = 24.52, p = 0.000), and native eusocial bees (F_2 = 35.91, p = 0.000) exhibited significantly different mean foraging rates in at least one city comparison, often with Cañas differing from both Liberia and Bagaces (see Tab. 2). Non-native eusocial bees (African honey bees) had significantly different occurrences in each city (F_2 = 38.50, p = 0.000); they were highest in Liberia, intermediate in Bagaces, and lowest in Cañas (Tab. 2). Variability between the three cities validated the development of site- and guild-specific foraging models.

Community foraging models

Increasing resource abundance positively influenced the abundance and richness of foragers attracted to individual *T. stans* resources across the entire (all city) sample. The number of flowers per *T. stans* individual (FLW: t = 5.48, p = 0.000) and the number of other *T. stans* resources within a 10 metre radius (CON-SP 10: t = 1.99, p = 0.049) were globally significant drivers of bee abundance at a site. Overall taxon richness was influenced only by the total number of flowers per resource (FLW: t = 6.92, p = 0.000). A city-specific analysis of the drivers of bee visitation revealed further variability between the study sites. The summary of the multivariate regression models constructed for the bee community (total abundance and taxon richness) is displayed in Tab. 3, with more detailed models available in Appendix II.

In each city, abundance patterns were driven by a mix of resource characteristics and land use variables, however, in each case, the number of flowers per individual resource remained significant (FLW: Bagaces, t = 2.82, p = 0.010; Cañas, t = 4.03, p = 0.000; Liberia, t = 4.10, p = 0.000). In Bagaces residential land use (RES; t = 2.27, p = 0.032) indicated increased bee visitor abundance; per count, on average 2.64 (± 1.16) more bees were attracted to resources that were on residential land compared to commercial land. In Cañas, proximity to open space (OPEN: t = 2.00, p = 0.054) and an increasing distance from the wildland-urban interface (D WUI: t = 2.00, p = 0.054) positively correlated with increased bee abundance. In Liberia a shrub life form (SHRUB: t = 2.26, p = 0.028) was positively associated with increasing bee visitation, while open space was negatively associated (OPEN: t = -2.14, p = 0.037).

	City Comparison			Diff.	Std. Error	p	95% CI Upper	Lower
Abundance	B	vs	C	-0.609	0.202	0.003*	-1.006	-0.212
	B	vs	L	-0.184	0.193	0.339	-0.562	0.194
	C	vs	L	0.425	0.184	0.021	0.063	0.786
Richness	B	vs	C	-0.218	0.056	0.000*	-0.327	-0.109
	B	vs	L	-0.329	0.053	0.000*	-0.433	-0.226
	C	vs	L	-0.111	0.051	0.028	-0.210	-0.012
Large Solitary	B	vs	C	-0.773	0.148	0.000*	-1.063	-0.482
	B	vs	L	-0.047	0.141	0.741	-0.323	0.230
	C	vs	L	0.726	0.135	0.000*	0.462	0.990
Small Solitary	B	vs	C	0.538	0.082	0.000*	0.377	0.698
	B	vs	L	0.133	0.078	0.088	-0.020	0.286
	C	vs	L	-0.405	0.075	0.000*	-0.551	-0.258
Native Eusocial	B	vs	C	-0.521	0.076	0.000*	-0.670	-0.373
	B	vs	L	0.012	0.072	0.870	-0.129	0.153
	C	vs	L	0.533	0.069	0.000*	0.398	0.668
Non-native Eusocial	B	vs	C	0.110	0.039	0.004*	0.035	0.186
	B	vs	L	-0.191	0.037	0.000*	-0.263	-0.119
	C	vs	L	-0.302	0.035	0.000*	-0.370	-0.233

Tab 2: The results of multivariate ANOVA comparing the mean number of bee foragers per count at *T. stans* resources in the three sample landscapes: Bagaces (B), Cañas (C), and Liberia (L). Significant differences (p < 0.016, Bonferroni adjusted) are indicated by the * symbol.

The number of flowers per individual *T. stans* resource consistently correlated with increased forager richness. This was the singular factor that influenced patterns in Liberia (FLW: t = 4.49, p = 0.000) and Cañas (FLW: t = 3.64, p = 0.001). Other variables including residential land use (RES: t = 3.76, p = 0.001) and distance to the wildland-urban interface (D-WUI: t = -2.33, p = 0.032) were important in Bagaces. Residential land use predicted almost one additional taxon to be present per observation compared to commercial land use. Forager richness was also higher on resources closer to the wildland-urban edge.

Guild specific resource usage

Solitary bees—Floral abundance was consistently positively correlated to increased solitary bee visitation in each city (FLW: Bagaces, t = 2.40, p = 0.025; Cañas, t = 4.67, p = 0.000; Liberia, t = 2.97, p = 0.004), but was not the only factor that drove visitation patterns. A summary of the foraging models developed for each guild is presented in Tab. 3, with more detailed models presented in Appendix III. In Bagaces, residential land use significantly increased the number of solitary bees by 2.36 (±0.92) individuals per count compared to commercial land use (RES: t = 2.58, p = 0.016). In Cañas, an increasing distance from the wildland-urban interface (D WUI: t = 2.80, p = 0.001) and a tree life form (TREE: t = 2.68, p = 0.012) significantly positively increased solitary bee numbers. In Liberia, there was a negative association with solitary bee foraging rates and proximity to open space (OPEN: t = -2.10, p = 0.041) and a positive correlation with resources that were considered shrubs (SHRUB: t = 2.44, p = 0.018). When solitary bees were examined by functional size guilds more specific patterns became evident.

Large bees—The overall occurrence of large solitary bees was significantly positively correlated with the number of flowers that a resource had (FLW: t = 3.84, p = 0.000) and with the number of con-specific resources within 10 metres (CON-SP 10: t = 2.23, p = 0.015); both are measures of resource aggregation. In Bagaces, the foraging rates of large solitary bees were significantly positively influenced by the total number of flowers per individual resource (FLW: t = 2.62, p = 0.018), residential land use (RES: t = 2.89, p = 0.002), and decreasing distance to riparian areas (D RIP: t = -3.44, p = 0.000). In Cañas, large solitary bee foraging rates were significantly positively correlated with resource characteristics (FLW: t = 3.32, p = 0.002 and TREE: t = 2.51, p = 0.018) and distance to the wildland-urban interface (D WUI: t = 4.10, p = 0.000). In Liberia, only the number of flowers per individual resource (FLW: t = 2.00, p = 0.050) was significant, with more large solitary bees visiting resources with greater floral abundance.

Small bees—The overall visitation rates of small solitary bees correlated significantly and positively with the number of flowers per individual resource (FLW: t = 3.06, p = 0.003) and with resource characteristics. Shrubs were preferred (SHRUB: t = 3.07, p = 0.003), with 0.63 (± 0.21) more visitors per observation noted on shrubs as compared to hedge- and tree-like growth forms. The influence of floral abundance on this guild was not consistent across the three cities. Cañas was the only city in which floral abundance per resource significantly increased small bee visitation (FLW: t = 4.27, p = 0.000); decreasing resource height (HEIGHT: t = -2.74, p = 0.010) was also a driving factor. In Liberia, a shrub life form (SHRUB: t = 3.10, p = 0.003) significantly positively correlated with small solitary bee visitation. An increasing distance from riparian areas (D RIP: t = 3.21, p = 0.000) was the only significant descriptor of small solitary bee visitation in Bagaces; small bee numbers were inversely related to nearness to riparian habitats.

Eusocial bees—The local community of eusocial bees contains many native stingless bees (*Nannotrigona perilampoides*, *Plebeia* sp., *Plebeia frontais*, *Tetragonisca angustata*, *Trigona fulviventris*, and *Trigona* sp.), but also the non-native African honey bee (*Apis mellifera scutellata*). In the pooled city analysis, eusocial bees were positively influenced by the number of flowers per resource (FLW: t = 5.97, p = 0.000) and by the distribution of con-specific resources within a 10 metre radius (CON-SP 10: t = 2.65, p = 0.009). Flowers per resource remained an important influence on visitation rates of eusocial bees in all of the study cities (FLW: Bagaces, t = 3.41, p = 0.002; Cañas, t = 2.72, p = 0.001; Liberia, t = 5.28, p = 0.000). In Bagaces the life form was significant, with hedges attracting more visitors than shrubs or trees (HEDGE: t = 1.86, p = 0.075), while in Cañas resource distribution (CON-SP 10: t = 1.94, p = 0.062) was important, with more aggregated resources attracting more eusocial bees. In Liberia the location of the resource within the urban landscape correlated with increasing visitation (D WUI: t = 2.50, p = 0.015), with resources located closer to the perimeter receiving more visits.

Native eusocial bees—Native eusocial bees were influenced exclusively by resource quantity (FLW: t = 3.25, p = 0.00) and distribution (CON-SP 10: t = 3.32, p = 0.049) in the pooled city sample. Floral abundance per resource remained significantly positively influential in Bagaces and Liberia (FLW: t = 2.35, p = 0.000 and t = 1.96, p = 0.055, respectively), but not in Cañas where proximity to open space (OPEN: t = 3.11, p = 0.004) and life form (HEDGE: t = 2.02, p = 0.052) were the significant drivers of foraging rates.

Non-native eusocial bees—Floral abundance per resource was a clear and consistent factor in describing African honey bee occurrence in each city (FLW: Bagaces, t = 2.63, p = 0.015; Cañas, t = 1.82, p = 0.072; Liberia, t = 6.77, p = 0.000), but landscape variables also correlated significantly in some cities. Residential land use had a particular significant positive association with increased occurrence in Liberia (RES: t = 3.21, p = 0.002), as did an increasing distance from the wildland (D WUI: t = 4.39, p = 0.000); neither of these variables impacted local native bee species.

TAB 3: Summary of the pooled and individual city bee visitation models for the total community (abundance and taxon richness) and the individual bee guilds that are using *T. stans* as a pollen and nectar resource in the small and developing urban landscapes of Guanacaste, Costa Rica. All of the variables shown are significant at the α = 0.05 level unless indicated by (‡) where they are significant at α = 0.08. Variable abbreviations are as follows: FLW = total flowers, CON-SP 10 = the number of other *T. Stans* within a 10 metre radius of the sampled resource, RES = residential land use, OPEN = proximity to open space, D WUI = distance to the wildland-urban interface, and D RIP = distance to riparian areas. The life form of the resource is indicated as a TREE, SHRUB, or HEDGE. The + or − sign next to the variable in parenthesis indicates the direction of correlation. The explicit models for each taxon group are presented in Appendices II and III.

	All Cities	Bagaces	Cañas	Liberia
Community:				
Abundance	(+) FLW (+) CON-SP 10	(+) FLW (+) RES	(+) FLW (+) OPEN‡ (+) D WUI‡	(+) FLW (-) OPEN (+) SHRUB
Richness	(+) FLW	(+) FLW (+) RES (-) CON-SP 10 (-) D WUI (-) SHRUB	(+) FLW	(+) FLW
Guilds:				
Solitary	(+) FLW	(+) FLW (+) RES	(+) FLW (+) D WUI (+) TREE	(+) FLW (-) OPEN (+) SHRUB
Large	(+) FLW (+) CON-SP 10	(+) FLW (+) RES (-) D WUI (-) D RIP	(+) FLW (+) D WUI (+) TREE	(+) FLW (-) OPEN
Small	(+) FLW (-) OPEN (+) SHRUB	(-) D RIP	(+) FLW (-) HEIGHT	(+) SHRUB
Eusocial	(+) FLW (+) CON-SP 10	(+) FLW (+) HEDGE‡	(+) FLW (+) CON-SP 10‡	(+) FLW (+) D WUI
Native	(+) FLW (+) CON-SP	(+) FLW	(+) OPEN (+) HEDGE‡	(+) FLW‡ (-) RES
Non-native	(+) FLW (+) RES (+) D WUI	(+) FLW (+) D RIP	(+) FLW‡	(+) FLW (+) RES (+) CON-SP 10 (+) D WUI

DISCUSSION

Bees using *T. stans* as a pollen and nectar source in the urban landscapes of Northwestern Costa Rica preferentially visited resources with more abundant flowers, irrespective of their location in the urban landscape. The importance of characteristically anthropogenic variables was minimal, indicating that bees were generally not influenced by the urban matrix. Similar trends highlighting the greater importance of floral resources are emerging from studies conducted elsewhere (Colla et al. 2009; Werrell et al. 2009; Matteson & Langellotto 2010), further supporting the idea that the constructed elements of the urban landscape generally do not interfere with the occurrence and foraging of bees. The preferences of individual functional bee groups examined in this study are generally consistent with this trend, but also indicated that organism size and other life history characteristics are responsible for some unique responses to modified habitats.

Solitary bees are the dominant and most diverse assemblage of bees in the tropical dry forest (Frankie & Vinson 2004). Larger solitary bees have been documented to preferentially visit flowering trees, while smaller bodied species visit more herbaceous forms (Frankie & Vinson 2004). Within urban environments both size guides can be found visiting similar resources (Wojcik 2009). In this study, large bees responded with increased foraging at *T. stans* individuals with more flowers and at those that were clustered with other con-specifics – both are measures of resource aggregation at different scales. The occurrence of large bees in these landscapes may be influenced by the principals of energetics and optimal foraging. Larger species with increased metabolic needs have been shown to pattern their resource visitation to favour nearby patches of flowers, optimizing the energy required for flight with that acquired from nectar (Heinrich 1979).

Small bees in this study were more likely to visit *T. stans* resources with abundant flowers, but were also influenced by the life form of the resource, foraging in larger numbers on lower-growing resources that were shrubs. Vertical stratification was recorded by Roubik (1993) in the canopy of tropical forests near this study region and is suggested as a strategy to reduce direct competition. The observed preference for lower-growing resources could indicate that smaller bees are employing vertical stratification in an effort to reduce competition and exploit resources that larger species are not visiting. The proximity of a food resource to open space, riparian areas, and the wildland-urban interface (edge of the city) also correlated significantly with the number of small solitary bees recorded. Distance-based foraging responses in small bees in these landscapes could be due to dispersal limitations. The dispersal range from nest to food sites of small bees is thought to be 250 metres (Greenleaf et al. 2007), but this is likely a maximum capability and not an optimum range as indicated by Zurbuchen et al. (2010) who found reduced fitness effects as nest distance increased from food sources within this range. Areas of nearby natural habitat present more nesting opportunities for species such as *Agapostemon nasutus* and *Halictus lutescens* that commonly nest in the ground (Michener 2000), and these species would preferentially forage on more proximal *T. stans* resources to improve fitness.

The importance of urban variables

Certain landscape and habitat characteristics are unique to cities. Urban fauna must interact with features that wildland species do not experience, which leads urban ecologists to question if these unique *urban* habitat elements have a significant or corresponding unique effect on resident species. Although the results of this study generally indicated little to no influence from anthropogenic variables, residential land use was shown to have a significant and large impact on one particular group of bees, African honey bees, nearly doubling their occurrence. Residential areas have been noted by many authors as sites of increased bee richness due to the diversity of the floral resource base located in ornamental gardens (Tommasi et al. 2004; Frankie et al. 2005; Winfree et al. 2009). Others have found that floral richness *per se* does not strictly correlate with individual species occurrence patterns in gardens (Werrel et al. 2009) as most bees possess narrow feeding ranges (Michener 2000) or restrict daily foraging to specific plants (Heinrich 1979). In this study, only generalist species displayed significantly increased abundance rates at residential garden sites, supporting findings that indicate floral diversity patterns do not universally increase bee occurrence.

Characteristics of the urban matrix and peri-urban landscape may have a unique or combined influence on bee visitation within guilds that respond to different spatial scales. Steffan-Dewenter et al.'s (2002) examination of bees in complex landscapes indicated heterogeneous and equally complex responses that were scale depended and guild-specific; similar responses might be occurring in the urban systems studied here. Bagaces and Liberia have the most similar patterns of per count bee visitation and species occurrences and both cities are situated within a similar regional land use dominated by cattle grazing and punctuated by native dry forest remnants. Cañas is different, sitting in the centre of agricultural intensification. On a finer scale, Bagaces and Liberia are interspersed with more streams and riparian areas while Cañas has a higher density of housing units, but also has more managed public parks. Given the city-specific variability seen in this study, it is the author's opinion that generalizing the responses of bees to 'urban' areas is premature. Few studies have compared multiple urban landscapes, and fewer still have focused on meso- and micro-scale landscape characteristics, especially in tropical systems. The trends outlined here provide a starting point to further studies of comparative bee ecology and behaviour in a rapidly growing landscape type.

ACKNOWLEDGEMENTS

Thanks to Meaghan Jastrebski and Laura Fine for their assistance with data collection in the field. Financial support for this work was provided by the Schwabacher Memorial Scholarship in Forestry, and made trips into the field possible. The staff of the GIF (ESPM, University of California Berkeley) provided technical support and advice for GIS analysis. Thanks to Greg Biging for continued advice on statistical modelling. Thanks also to Patina Mendez, Maria Wojcik, Jennifer Tsang, Laurie Davies Adams, and

Sean Fine for early and final reviews of the manuscript text. Constructive comments from the two reviewers improved the manuscript and are much appreciated. Special thanks to Joe R. McBride for supporting this work and pushing to see it complete.

APPENDICES

Additional supporting information may be found in the online version of this article.

APPENDIX I: The resource and landscape variables used in backward stepwise regression model development.

APPENDIX II: The backward stepwise regression models describing bee visitor abundance and taxon richness at *T. stans* resources and across all of the and at each individual study site.

APPENDIX III: The backward stepwise regression models describing the occurrence individual guilds at *T. stans* resources and across all of the cities and at each individual study site.

REFERENCES

Ahrne K, Bengtsson J, and Elmqvist T (2009) Bumble Bees (*Bombus* spp.) along a gradient of increasing urbanization. PLoS 4(5): e5574:1-9.

Antonini Y, Costa RG, and Martins RP (2006) Floral preferences of a neotropical stingless bee, *Melipona quadrifasciata* Lepeletier (Apidad: Meliponia) in an urban forest fragment. Brazilian Journal of Biology 66(2A): 463-471.

Brosi BJ, Daily GC, and Ehrlich PR (2007) Bee Community Shifts With Landscape Context In A Tropical Countryside. Ecological Applications 17(2): 418-430.

Buchmann S, and Nabhan GP (1996) The Forgotten Pollinators. Island Press, New York.

Buchmann SL, and Ascher JS (2005) The plight of pollinating bees. Bee World 86: 71-74.

Cane J, Minckley R, Kervin L, and Roulston T (2005) Temporally persistent patterns of incidence and abundance in a pollinator guild at annual and decadal scales: the bees of *Larrea tridentata*. Biological Journal of the Linnean Society 85: 319-329.

Cane J, Minckley R, Kervin L, Roulston T, and Williams N (2006) Complex responses within a desert bee guild (Hymenoptera: Apiformes) to urban habitat fragmentation. Ecological Applications 16(2): 632-644.

Colla SR, Willis E, and Packer L (2009) Can green roofs provide habitat for urban bees (Hymenoptera: Apidae)? Cities and the Environment 2(1): 12 [online] URL: http://escholarship.bc.edu/cate/vol2/iss1/4/.

Collevatti RG, Schoereder JH, and Campos LAO (2000) Foraging behavior of bee pollinators on the tropical weed *Triumfetta semitriloba*: Flight distance and directionality. Revista Brasileira de Biologia 60(1): 29-37.

ESRI (1999) ArcGIS v 9.3 [online] URL: www.esri.com.

Frankie GW, and Vinson BS (2004) Conservation and Environmental Education in Rural Northwestern Costa Rica. In: Frankie GW, Mata A, and Vinson SB (eds) Biodiversity Conservation in Costa Rica, University of California Press, Berkeley, pp 247-256.

Frankie GW, Rizzardi M, Vinson BS, and Griswold T (2009a) Decline in bee diversity and abundance from 1972-2004 on a flowering leguminous tree, *Andira inermis* in Costa Rica at the interface of disturbed dry forest and the urban environment. Journal of the Kansas Entomological Society 82(1): 1-20.

Frankie GW, Thorp RW, Hernandez JL, Rizzardi M, Ertter B, Pawelek JC, Witt SL, Schindler M, Coville R, and Wojcik VA

(2009b) Native bees are a rich natural resource in urban California gardens. California Agriculture 63(3): 113-120.

Frankie GW, Thorp RW, Schindler M, Hernandez JL, Ertter B, and Rizzardi M (2005) Ecological patterns of bees and their host ornamental flowers in two northern California cities. Journal of the Kansas Entomological Society 78: 227-246.

Frankie GW, Thorp RW, Schindler MH, Ertter B, and Przybylski M (2002) Bees in Berkeley? Fremontia 30(3-4): 50-58.

Frankie GW, Vinson BS, Rizzardi M, Griswold T, O'Keefe S, and Snelling RR (1997) Diversity and abundance of bees visiting a mass flowering tree in disturbed seasonal dry forest, Costa Rica. Journal of the Kansas Entomological Society 70: 281-296.

Greenleaf SS, Williams N, Winfree R, and Kremen C (2007) Bee foraging ranges and their relationship to body size. Oecologia 153: 589-596.

Hammel B (2005) Plantas ornamentales nativas de Costa Rica Native Ornamental Plants. Instituto Nacional de Biodiversidad (INBio), San Jose.

Heard MS, Carvell C, Carreck NL, Rothery P, Osborne JL, and Bourke AFG (2007) Landscape context not patch size determines bumble-bee density on flower mixtures sown for agri-environmental schemes. Biology Letters 3: 638-641.

Heinrich B (1979) Bumble bee Economics. Harvard University Press, Cambridge.

Hisamatsu M, and Yamane S (2006) Fuanal makeup of wild bees and their flower utilization in a semi-urbanized area in central Japan. Entomological Science 9: 137-145.

INEC (Instituto Nacional de Estadistica y Censos) Costa Rica (2000) del V Censo Nacional de Vivienda [online] URL: http://www.inec.go.cr/.

Jha, Shalene and J Vandermeer (2009) Contrasting bee foraging response to resource scale and local habitat management. Oikos 118: 1174-1180.

Kremen C, Williams N, and Thorp RW (2002) Crop pollination from native bees at risk from agricultural intensification. PNAS 99(26): 16812-16816.

Liu H, and Koptur S (2003) Breeding system and pollination of a narrowly endemic herb of the Lower Florida Keys: impacts of the urban-wildland interface. American Journal of Botany 90(8): 1180-1187.

Matteson KC, and Langellotto GA (2010) Determinates of inner city butterfly and bee species richness. Urban Ecosystems 13: 333-347.

Matteson KC, Ascher JS, and Langellotto GA (2008) Bee Richness and Abundance in New York City Urban Gardens. Annals of the Entomological Society of America 101(1): 140-150.

McIntyer NE, and Hostetler ME (2001) Effects of urban land use on pollinator (Hymenoptera: Apoidea) communities in a desert metropolis. Basic and Applied Ecology 2: 209-218.

Michener CD (2000) The Bees of the World. Johns Hopkins University Press, Baltimore.

Nates-Parra G, Parra A, Rodrigues A, Baquero P, and Velez D (2006) Wild bees (Hymenoptera: Apoidea) in urban ecosystems: preliminary survey in the city of Bogota and its surroundings. Revista Colombiana de Entomologia 32: 77-84.

National Research Council (NRC) – National Academies of Science: Committee on the Status of Pollinators in North America NRC (2007) Status of Pollinators in North America. The National Academies Press, Washington, D.C..

Ollerton J, Winfree R, and Tarrant S (2011) How many flowering plants are pollinated by animals? Oikos 120: 321-326.

Owen J (1991) Ecology of a Garden, The First Fifteen Years. Cambridge University Press, Cambridge.

Roubik DW (1993) Tropical Pollinators in the Canopy and Understory: Field data and theory for stratum "preferences". Journal of Insect Behavior 6(6): 659-673.

Saure C (1996) Urban habitats for bees: the example of the City of Berlin. In: Matheson A, Buchmann SL, O'Toole C, Westrich P, and Williams IH (eds) The conservation of bees, Academic Press, New York, pp 47-54.

Silva CI, Augusto SC, Sofia SH, and Moscheta IS (2007) Bee diversity in *Tecoma stans* (L.) Kunth (Bignoniaceae): Importance for pollination and fruit production. Neotropical Entomology 36(6): 331-341.

SPSS (2007) SPSS v 6 [online] URL: www.spss.com.

Steffan-Dewenter I, and Kuhn A (2003) Honey bee Foraging in Differentially Structured Landscapes. Proceedings: Biological Sciences 270(1515): 569-575.

Steffan-Dewenter I, Munzenberg U, Burger C, Thies C, and Tscharntke T (2002) Scale-dependent effects of landscape context on three pollinator guilds. Ecology 83(5): 1421-1432.

Tommasi D, Miro A, Higo HA, and Winston ML (2004) Bee diversity and abundance in an urban setting. Canadian Entomologist 136(6): 851-869.

Werrell PA, Langellotto GA, Morath SU, and Matteson KC (2009) The influence of garden size and floral cover on pollen deposition in urban community gardens. Cities and the Environment 2(1): 16 [online] URL: http://escholarship.bc.edu/cate/vol2/iss1/6.

Winfree R, Aguilar R, Vazquez DP, LeBuhn G, and Aizen MA (2009) A meta-analysis of bees' responses to anthropogenic disturbance. Ecology 90(8): 2068-2076.

Winfree R, Griswold T, and Kremen C (2007) Effect of Human Disturbance on Bee Communities in a Forested Ecosystem. Conservation Biology 21(1): 213-223.

Wojcik VA (2009) Bees in urban landscapes: An investigation of habitat utilization. Ph.D. Dissertation, University of California, Berkeley Department of Environmental Science, Policy, & Management: i-204.

Wojcik, VA (2011) The urban bees (Hymenoptera: Apoidea) of *Tecoma stans* (Bignoniaceae): Phenology and community composition in three tropical dry forest cities. Journal of the Kansas Entomological Society *In press*.

Zuchowski W (2007) Tropical Plants of Costa Rica. Zona Tropicala, Ithaca.

Zurbuchen A, Chessman S, Klaiber J, Muller A, Hein S, and Dorn S (2010) Long foraging distances impose high costs on offspring production in solitary bees. Journal of Animal Ecology 79: 674-681.

POLLINATION ECOLOGY OF *DESMODIUM SETIGERUM* (FABACEAE) IN UGANDA; DO BIG BEES DO IT BETTER?

Dara A. Stanley[1*], Mark Otieno[2], Karin Steijven[3,4], Emma Sandler Berlin[5], Tiina Piiroinen[6], Pat Willmer[7] & Clive Nuttman[8]

[1]*Botany and Plant Science, School of Natural Sciences and Ryan Institute, National University of Ireland, Galway, Ireland*

[2]*Department of Agricultural Resource Management, Embu University College, P.O. Box 6-60100, Embu, Kenya.*

[3]*Department of Animal Ecology and Tropical Biology, Biocentre - University of Würzburg, Am Hubland, 97074 Würzburg, Germany*

[4]*Department of Bee Health, Van Hall Larenstein – University of Applied Sciences, Agora 1, P.O. Box 1528, 8901 BV Leeuwarden, the Netherlands*

[5]*Department of Biology, Lund University, Sölvegatan 37 223, 62 Lund, Sweden*

[6]*Faculty of Science and Forestry, Department of Environmental and Biological Sciences, University of Eastern Finland, P. O. Box 111, FI-80101 Joensuu, Finland*

[7]*School of Biology, Harold Mitchell Building, University of St Andrews, St Andrews, Fife, KY16 9TH, UK.*

[8]*Tropical Biology Association, The David Attenborough Building, Pembroke Street, Cambridge CB2 3QZ, UK.*

Abstract—Explosive pollen release is documented in many plant families, including the Fabaceae. *Desmodium setigerum* E. Mey (Fabaceae) is a perennial herb with single trip explosive pollen release found in eastern Africa, and the unique ability to reverse floral colour change if insufficient pollination has occurred. However, little else is known about the pollination ecology of this species, what visitors can trigger explosive pollen release, and whether bee body size is related to pollination efficiency. We investigated: 1) the breeding system of *D. setigerum,* and whether it is pollen limited; 2) whether flowers are visited early in the day allowing sufficient time for a second opportunity for pollination; and 3) what insect species visit *D. setigerum* and the relative efficacy of different flower visitors in relation to visitor size and pollination success. We found that although self-compatible, *D. setigerum* requires insect visitation to set seed as explosive pollen release is needed even for selfing. Most flowers are initially visited before 1400h, and by 1800h nearly all flowers have been tripped. Flowers were not pollen limited in this study, and were visited primarily by bees. We observed 16 visiting species, and there was a wide variation (0-404 grains) in the amount of pollen deposited on stigmas. Although almost all bees deposited some pollen, the mean number of pollen grains deposited in a single visit per species was negatively related to body size. However, one particular megachilid species deposited significantly more pollen grains than any other visitor and so is likely an important pollinator of this species. This provides insights into the pollination biology of this unique plant species, and adds to increasing literature on the relationships between bee body size, explosive pollen release and pollination effectiveness.

Keywords: explosive release, Fabaceae, Leguminosae, pollen deposition, single visit, size matching

INTRODUCTION

Explosive pollen release, where pollen is rapidly expelled from a flower often following visitation, is presumed as an adaption to promote pollination by the "perfect" pollinator (Aluri & Reddi 1995). This trait has been recorded in over 17 plant families (Aluri & Reddi 1995), including the Fabaceae which contains 727 genera and 19,327 species (Lewis et al. 2005). Many Fabaceae display a tripping mechanism, where the 'mechanical' handling by the visitor is critical to release the pollen (Yeo 1993). When landing on a flower, visitors use the wing-keel arrangement as a landing platform, and probe for nectar using their proboscis beneath the base of the keel petals whilst forcing against the immoveable flag (Westerkamp 1997). The wing-keel complex is the only floral component that moves; it moves in a relative lowering movement away from the flag that is related to force applied as opposed to actual weight of the visitor, before returning to its original position (Westerkamp 1997). In some instances the mechanism is explosive with no return of the wing-keel complex to the original position – an 'all or nothing' response. Without this complicated manipulation of the flower resulting in the triggering mechanism, visitors cannot access flower reproductive parts. Although several studies have addressed the issue of how pollen release is triggered in papilionate flowers (e.g. Cordoba & Cocucci 2011; Stout 2000; Vivarelli et al. 2011), detailed information is still required on how different visitors and visitor size might affect various aspects of plant reproductive fitness in flowers of this type.

*Corresponding author: darastanley@gmail.com

Desmodium setigerum E. Mey. is a common, scrambling perennial found in disturbed areas throughout eastern Africa (Fig. I). It has typical papilionate legume flowers that last a single day and 'trip' explosively when visited, with the keel petals remaining open and the filaments and gynoecium remaining uncovered. Unusually for species with flowers that only last a day (Van Doorn 1997), *D. setigerum* displays rapid floral colour change following visitation. The flowers are initially lilac but rapidly change colour becoming paler before turning white and eventually turquoise (Willmer et al. 2009). In addition to this, when tripped, flowers retain their tripped form, thus providing a morphological signal to signify prior visitation as well as the ensuing colour change. Perhaps the most striking feature of *D. setigerum* is that some flowers, if they have not been successfully pollinated, have the ability to regain some of their former colour to allow a second opportunity for pollination (Willmer et al. 2009). The ability to change colour and then reverse this change is so far unique to this species, and occurs over the short time frame of a single day. As a result we expect that *D. setigerum* has evolved this ability as it benefits from insect pollination, and that most flowers are visited early in the day so that there is ample time for a second chance at pollination following colour reversal; however, the reproductive ecology of this species has not been studied previously.

We investigate the pollination ecology of *D. setigerum* and use it as a model to determine what insect visitors elicit explosive pollen release, and whether there is any relationship between pollen deposition and visitor size. Although *D. setigerum* is plentiful and has been noted to receive visits by a wide range of different sized visitors (Willmer et al. 2009), the effectiveness of these visitors as pollinators has not been assessed. Specifically we asked the following questions:

1) What is the breeding system of *D. setigerum* and is it pollen limited? Are visits necessary for pollination success?

2) How does flower availability vary throughout the day?

3) Which visitors are most effective (i.e. can all visitors 'trip' flowers to cause explosive pollen release), and is their ability to deliver pollen to the floral stigma determined by pollen placement and/or body size?

Materials and Methods

The study was conducted in Kibale Forest National Park, Uganda (0° 13' to 0° 41' N and 30° 19' to 30° 32' E) which lies north of the equator in the foothills of the Rwenzori Mountains. The area comprises 796 km2 of mid-altitude (1,590 m asl in the north to 1,110 m asl in the south) tropical moist forest. Rainfall pattern is bimodal with two rainy seasons, but with considerable variation between years (Struhsaker 1997). *D. setigerum* is plentiful in the northern part of the national park at forest edges and data were collected along trails, in forest gaps and in open areas in the vicinity of the Makerere University Biological Field Station (MUBFS) close to the village of Kanyawara (0° 35' N, 30° 20' E).

Breeding system and pollen limitation

To examine the breeding system of *D. setigerum*, and whether the species is pollen limited in our study sites, a manipulation experiment was conducted in August 2008. Nine sites were selected around MUBFS, and five pollination treatments were applied to six flowers in each site ($N = 54$ flowers per treatment), with each flower selected on a different spike. The treatments comprised: 1) **self-pollinated**: manual tripping with addition of self-pollen; 2) **cross-pollinated**: manual tripping with addition of pollen from a neighbouring conspecific flower; 3) **natural pollination**: tripping and open pollination by insect visitors; 4) **artificial tripping with no pollen added**: pressure was applied to the base of the wing-keel complex with a dissecting needle to mimic explosive pollen release without visitation, and; 5) **control**: flowers were bagged (using fine netting) and remained untripped. The treated flowers were left for 5 days after which the developing seed pods were then collected and seeds counted in the field station laboratory. Where a flower had abscised without a seed pod being produced, seed set was recorded as zero.

Flower availability and tripping rates over time

To determine the rate at which flowers were visited, two 1 km transects were walked hourly through areas rich in *D. setigerum* multiple times throughout one day in August 2008. From 0800 h to 1800 h, untripped lilac flowers on the transect line were counted (transect walks lasted approximately 45 minutes each hour) to determine how many untripped flowers were available throughout the day.

Pollinator effectiveness

To examine which pollinators delivered the most pollen to the stigmas of *D. setigerum*, data were collected between the 13th and 18th July 2009. Our tripping rate observations and an earlier study (Willmer et al. 2009) had indicated that anthesis occurred between 0800 h and 0900 h with highest bee visitation from the latter time through to noon. Hence, sampling took place from 0830 h to 1100 h to capture peak pollinator activity. Individual bees were noted arriving on and foraging within randomly selected patches (ca. 5 × 5 m) of flowers (> 25 m apart) and followed during foraging bouts (up to 2 minutes) whilst tripping flowers. Preliminary observations indicated that not all bees were able to effect explosive pollen release; some small *Lasioglossum/Pseudapis* solitary bees (ca. 6-8 mm estimated body length from head to tip of abdomen) foraged for pollen at the tip of the keel petals rather than attempting to access nectar at the base of the flag (Fig.I; these bees were classed as "illegitimate visitors" and are not included in further analyses). To determine pollen deposition, visited and legitimately tripped flowers were immediately removed from the plants and fixed in 70% ethanol in small glass vials, ensuring that pollen deposition was the result of a single visit by a single bee, and the bee species recorded. We also collected stigmas from untripped flowers that had received visits from illegitimate

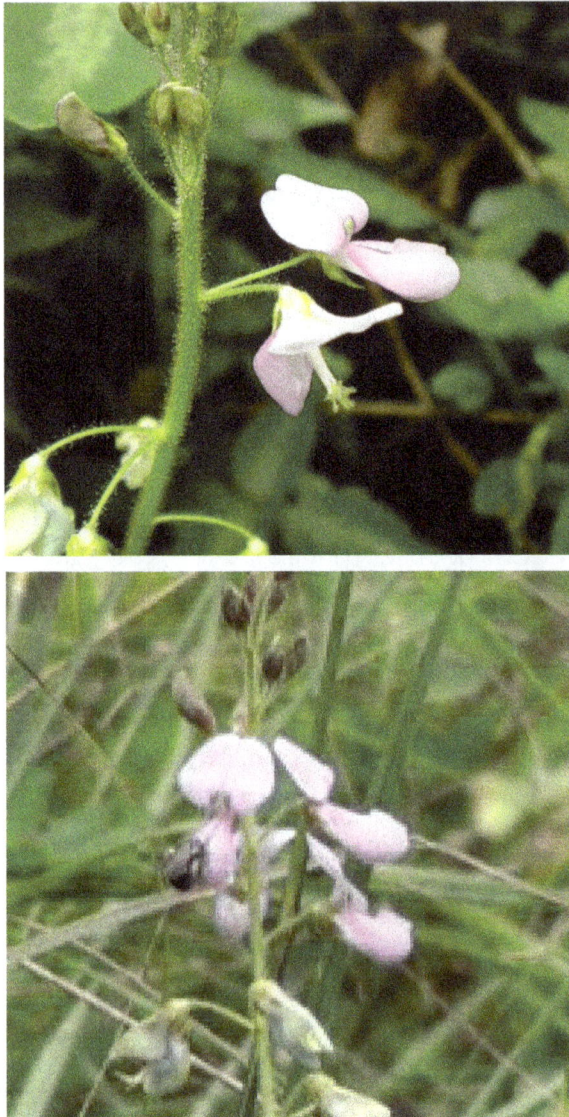

FIGURE I.   A tripped and untripped flower of *D. setigerum* (above, photo DS), and a small solitary bee (*Lasioglossum/Pseudapis* sp.) foraging at tip of keel complex in untripped *D. setigerum*; an illegitimate visit that does not trigger explosive pollen release (below, photo CN).

visitors that foraged at the tip of the keel, attempting to prise apart the petals to access pollen. As we had found artificially tripped flowers did set some seed (and for comparison with pollen deposition by both legitimate and illegitimate visitors), we also artificially tripped (see treatment 4 above) 30 flowers and collected their stigmas to assess the background number of pollen grains deposited in this way. In the laboratory, we removed the stigma of each collected flower, made a squash preparation with glycerin jelly stained with fuchsin red (Brunel Micro Ltd. UK) and counted the number of pollen grains adhered to the stigma under a light microscope.

Single specimens of each visiting bee species were collected, and sent for identification (see acknowledgements). To quantify size of each species, morphometric measurements were made for any of the visitor species that were available in collections at the Natural History Museum, London, UK (9 species out of 14 species, Tab. I). Digital calipers (Vernier) were used to take measurements (mm) of mounted specimens ($N = 6 - 10$ individuals per species). Measurements made included head width (from a view of the lateral aspect of the head of the specimen from eye to eye at the widest point) and wing length (using the measurement from the base to the apex of the wing). We also measured dimensions of 50 *D. setigerum* flowers in the field.

Data analysis

To determine breeding system and pollen limitation in *D. setigerum*, we tested for differences in pollination treatments applied to flowers using generalized linear mixed effects models in the lme4 package (Bates et al. 2014) in R version 3.3.0 (R Core Team 2014). Pollination treatment was a fixed categorical effect with seed set as a response variable. Site was included as a random factor to control for spatial variance in the model, and a Poisson distribution specified as data were counts. Where statistically significant results were found between treatments, we used a Tukey HSD post-hoc test procedure to separate them by making pairwise comparisons between each treatment using the multcomp package (Hothorn et al. 2008).

We analysed pollinator effectiveness in two ways. Firstly, we used analysis of variance (ANOVA) to test for differences in the mean number of pollen grains deposited per species, and per pollen placement area. Secondly, we investigated linear relationships between morphometric measurements (for the 9 species where this was possible) and amount of pollen deposited. Head width and wing length were highly correlated (Pearson Product Moment Correlation; t = 49.67, df = 109, $P < 0.001$); therefore only head width was used in subsequent analyses as this variable resulted in models with a lower AIC value. As size and pollen deposition data were from different individuals, we calculated means of both measures per species. We then used a linear model to investigate the relationship between mean number of pollen grains deposited and mean head width per species. All models were validated by plotting standardized residuals versus fitted values, normal qq-plots, and histograms of residuals.

RESULTS

Breeding system and pollen limitation

There was a significant impact of treatment on the number of seeds produced (Generalized linear mixed effects model: $\chi^2 = 140.38$, df = 4, $P < 0.001$, Fig. 2). There was no difference in the number of seeds produced between naturally visited flowers and flowers manually tripped with cross pollen added, indicating that this species is currently not pollen limited in our study sites. The origin of pollen did not seem to be important; flowers set similar levels of seed when self-pollen was applied in comparison to cross pollen from another individual indicating that this species is self-compatible. However, no seed was set in flowers that remained untripped (apart from one isolated case), and very little in manually tripped flowers where no stigmatic contact occurred. This indicates that although *D. setigerum* is self-

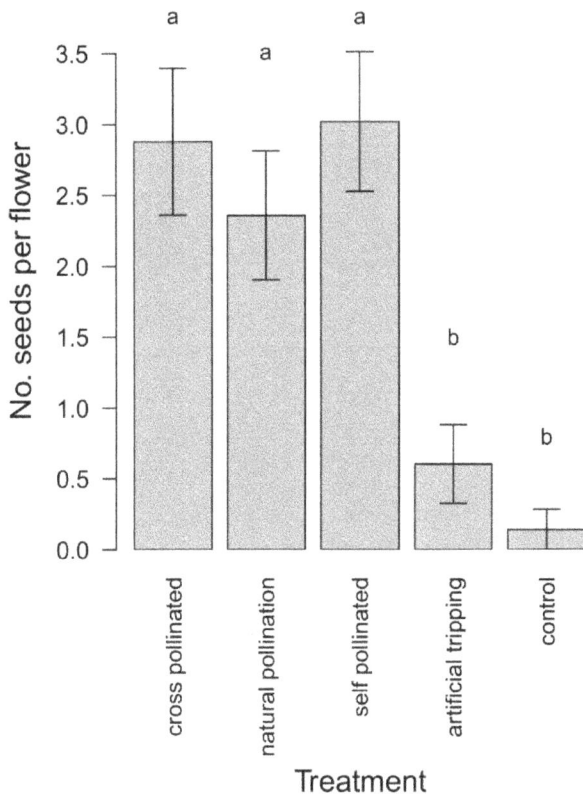

FIGURE 2. Mean number of seeds produced per flower of *D. setigerum* following pollination treatments. Error bars indicate standard error, and letters indicate significant differences ($P <$ 0.05). $N = 270$ flowers (54 flowers per treatment).

compatible it is visitor dependent, as without insect visitation to cause the tripping mechanism fertilization is unlikely to occur.

Flower availability and tripping rates over time

Flowers of *D. setigerum* were available throughout the day, with an increase in numbers as flowers emerged through the morning until 0900 h (transect 2) and 1100 h (transect 1). After this point, there was a rapid decline in the availability of untripped flowers (as tripping rates exceeded emergence rates). By 1400 h most flowers had been tripped, and by 1800 h almost all flowers were tripped (Fig. 3).

Pollinator effectiveness – visitor identity, size and behaviour

D. setigerum flowers were relatively small; the length of the keel was 9.45 mm ($\pm$ 0.50; range 8.40 - 10.30 mm; $N = 50$ flowers) from the base of the keel to the tip, and 6.67 mm ($\pm$ 0.45; range 5.50 - 7.75 mm; $N = 50$ flowers) from base to stigma tip. *D. setigerum* was only visited by bees, and 16 distinct bee species were recorded (although post-tripping visits by other taxa have been recorded at low frequency later in the day; Willmer et al. 2009). Of the 9 species that we could measure in the NHM collections, visitors ranged in body length (head to tip of abdomen) from 6.4 mm (*Anthidiellum* sp.) to 25.4 mm (*Xylocopa flavorufa*). In total, we counted 14,417 pollen grains attached to the stigmas of 184 flowers that were

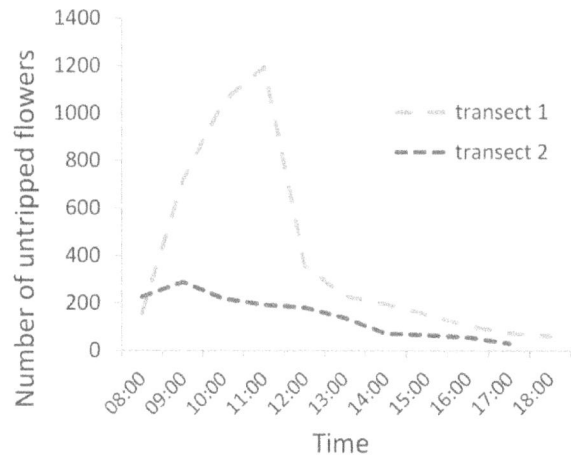

FIGURE 3. The number of untripped *D. setigerum* flowers available across the day in two transects. Numbers of flowers increase initially as flowers open in the morning, but steadily decrease during the day as nearly all flowers become tripped by 1800 h.

on stigmas from 30 further flowers that had been artificially tripped (Tab. 1). The number of pollen grains that adhered to stigmas ranged from 0 (hand-triggered; illegitimate visitors *Lasioglossum* sp. 2/*Pseudapis* sp.); to 404 (*Megachile* sp. 1), but pollen was deposited on at least some flowers by each visitor bee species represented in the analysis. Hand-triggered flowers had extremely low pollen deposition (6.8 $\pm$ 3.32 grains; $N = 30$ flowers), as did flowers receiving illegitimate visitation (37.4 $\pm$ 6.03; $N = 27$) (Tab. 1).

We found a highly significant effect of species on the number of pollen grains deposited (ANOVA: $F_{13,138} = 7.85$, $P < 0.001$, Fig. 4); "*Megachile* sp. 1" deposited a larger number of pollen grains than most other species. We also found an impact of pollen placement on the number of pollen grains deposited (ANOVA: $F_{4,147} = 7.31$, $P < 0.001$); bees that collected pollen on their abdomen deposited significantly more pollen grains than bees with pollen placed on other body parts (Fig. 5; although this was most likely driven by *Megachile* sp. 1). There was a significant negative relationship between head width and mean number of pollen grains deposited per species on floral stigmas (linear model: $F_{1,7} = 9.87$, $P = 0.02$, Fig. 6).

DISCUSSION

Our work has shed further light on the relationships between flower visitors, explosive pollen release, and single visit pollen deposition. We found that although *D. setigerum* is self-compatible and not pollen limited, insect pollination is required for substantial seed set. Although the species is visited and tripped by a diverse assemblage of bee species that all deposit pollen, bee size was negatively associated with pollen deposition and a particular megachilid bee species deposited more pollen than others.

That *D. setigerum* has evolved the trait of reversible floral colour change to allow a second opportunity for pollination (Willmer et al. 2009) suggests that insect pollination is of significant reproductive benefit to the

TABLE 1. The 15 bee species collected visiting *D. setigerum*. Legend for Family: A = Apidae, H = Halictidae, M = Megachilidae, Legend for 'pollen placement: C = corbiculae; A = abdomen; HL = hind legs; T = thorax; S = scattered (i.e. noted on all four preceding parts). For analysis of pollen placement, the most important location was used for any species with more than one recorded, which is highlighted in bold. Size parameters (wing length and head with) are ranges recorded from approximately 10 specimens per species in the Natural History Museum, London.

Species	Family	Wing length range (mm)	Head width range (mm)	Pollen counts Mean ± SE (no. of flowers)	Min-Max	Pollen placement
Anthidiellum sp.	M			56.3 ± 18.4 (4)	24-103	A
Illegitimate visitors*	H			37.4 ± 6 (27)	0-102	C
Megachile frontalis Smith	M	5.08 - 5.60	3.11-4.06	84.2 ± 17.6 (9)	30-170	A
Lasioglossum sp. 1	H			39.9 ± 16.7 (10)	1-160	HL, **C**
Megachile semierma Vachal	M	7.60 - 10.07	3.79-4.52	106.3 ± 49.1 (3)	10-171	**A**, HL
Apis mellifera	A	7.43 - 9.17	3.64-4.22	84.1 ± 6.8 (31)	15-168	C
Amegilla fallax Smith	A	7.23 - 9.10	3.81-4.89	106.7 ± 39.3 (7)	29-316	S
Nomia sp. 1	H			50.6 ± 11.5 (7)	20-112	A, **C**
Megachile sp. 1	M			256.2 ± 38.4 (9)	95-404	A
Nomia sp. 2	H			69.1 ± 17.2 (11)	28-228	S
Xylocopa senior Vachal	A	14.05-15.39	5.74-7.34	46.3 ± 16.1 (7)	0-112	T
Xylocopa calens Lepeletier	A	15.04-17.71	6.37-7.73	78.4 ± 9.6 (38)	0-315	T
Megachile cincta Fabricius	M	14.12-15.94	5.35-6.16	53.3 ± 16.4 (10)	6-109	S
Xylocopa nigrita Fabricius	A	21.68-26.77	7.66-10.22	21.0 ± 4 (2)	17-25	HL
Xylocopa flavorufa DeGeer	A	24.08-28.37	7.66-9.02	62.1 ± 12.5 (9)	18-112	S
ARTIFICIAL 'Hand-triggered'				6.8 ± 3.3	0-75	-

Note: one additional bee species (*Pachyanthidium* sp.) was observed visiting *D. setigerum*, but no pollen deposition data was recorded, and two further bee species were recorded visiting and successfully tripping *D. setigerum* in 2009 that were not included in this study; *Megachile chrysopogon* Vachal (Megachilidae) and *Braunsapis* sp. (Apidae) (CN, unpublished data).
Lasioglossum sp. 2; *Pseudapis* sp.

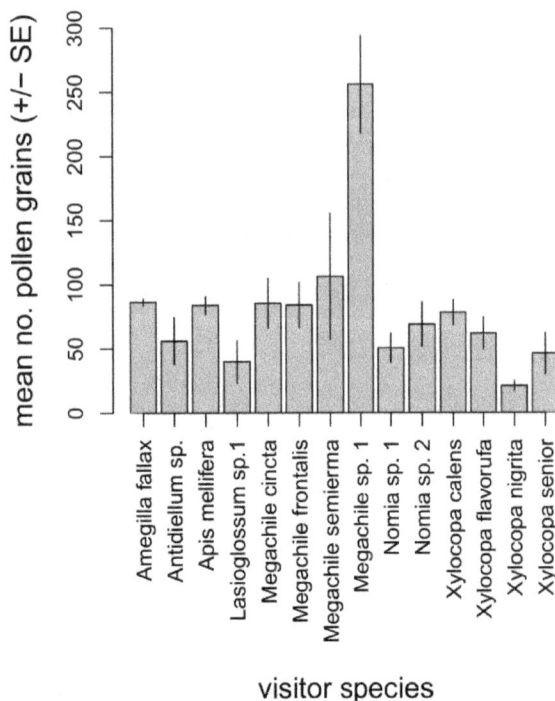

FIGURE 4. The mean (+/- standard error) number of pollen grains deposited on stigmas of *D. setigerum* by each visitor species. Significantly more pollen grains were deposited by *Megachile* sp. 1 than any of the other bee species. Error bars show standard error.

species, and that initial flower visitation happens early in the day to allow sufficient time for 'second round' pollination. Our findings show that *D. setigerum* is self-compatible; flowers artificially tripped and hand-pollinated with self-pollen set similar amounts of seed as those artificially tripped and hand-pollinated with cross-pollen. Self-compatibility has also been identified in other *Desmodium* species from South America (Alemán et al. 2014). In addition flowers that were artificially tripped but with no addition of pollen set some seed, suggesting that the act of tripping by hand leads to some deposition of self-pollen. However, although confirming self-compatibility, these circumstances are artificial and would not occur naturally. As flowers that were not visited by insects (and therefore not tripped) did not set any seed, and as those that were tripped and visited produced more seed than those tripped with no visits, this confirms that insect visitation is required for seed set. Notably, one outlier flower set seed without being tripped; as observations showed that small 'illegitimate' flower visitors can deposit pollen without tripping, this form of visitation may have occurred on this flower. Together, this shows that the process of insect visitation is fundamental to pollination and seed production. In addition, we found most flowers are initially visited prior to 1400 h. This gives substantial time before darkness (ca. 1800 h) and flower senescence for flowers that have not been successfully pollinated to have a second chance.

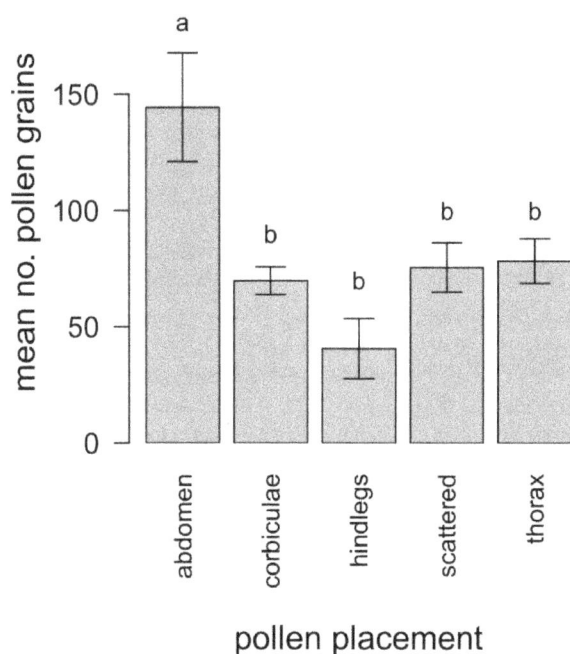

FIGURE 5. The mean (+/- standard error) number of pollen grains deposited by bees according to where pollen was predominantly placed on their bodies. N = 152 bees. Letters indicate significant differences (P < 0.05)

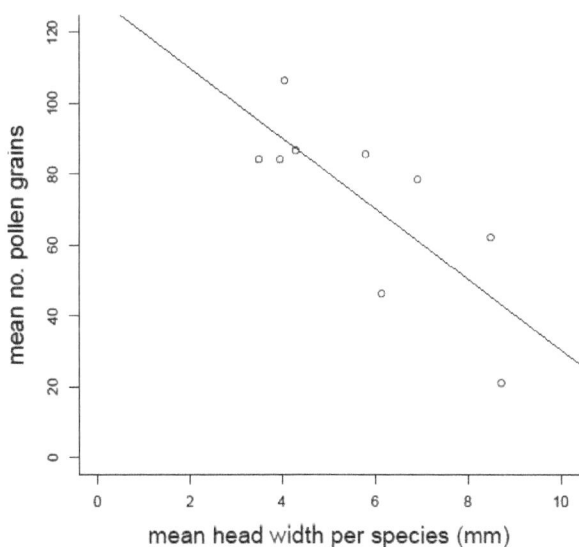

FIGURE 6. The relationship between the mean number of pollen grains deposited on stigmas and the head width (in mm) of the species visiting a flower. N = 9 bee species.

We found that all bees 'tripping' $D.$ $setigerum$ deposited some pollen, supporting the hypothesis that this explosive mechanism has evolved to 'filter' pollinators (Cordoba & Cocucci 2011). However, one species of megachilid ($Megachile$ sp. I, Tab. I) deposited more pollen per single visit than most other visitors. This suggests that this plant species may be more specialized in terms of its flower visitors than usually assumed for papillionate flowers, and visits by other bees may result in insufficient pollen delivery requiring a second pollination event (see Willmer et al. 2009). There

are a number of reasons why this megachilid species may be more successful in pollen deposition than others; firstly, this family of bees has pollen-carrying hairs on the underside of the abdomen, potentially making visitors more likely to deposit pollen when positioning themselves on the keel of $D.$ $setigerum.$ Indeed, bees that had pollen placed under their abdomen (Tab. I) deposited more pollen grains overall than bees with pollen in corbiculae, on the thorax or on the hind legs. Secondly, high deposition may also be due to morphological "fit" (or size matching) in that the size and shape of certain species is better for achieving contact with the reproductive parts of $D.$ $setigerum.$ Thirdly, it could be that this species behaves differently to others when handling flowers, thereby collecting and depositing more pollen.

We found a linear relationship between bee body size and pollen deposition, with smaller bees depositing more pollen than larger ones. Other work has also shown differently sized bees to be optimal in terms of pollen transfer/seed set in other Fabaceae species. Small bees (no measurement parameter documented) were found to have the lowest number of pollen grains on their bodies following non-tripping visits to $Pongamia$ $pinnata,$ which releases pollen explosively (Raju & Rao 2006). Using body length of bumblebees ($Bombus$ spp.), Stout (2000) found that smaller bees were more successful in tripping flowers of the legume $Cytisus$ $scoparius;$ large queens were too big to physically effect tripping and this may have had impacts on plant reproduction. Vivarelli et al. (2011) reported lower seed set in $Ononis$ $masquillierii$ when visitation was mainly by smaller bees (halictids and megachilids) as measured by dry weight. $Ononis$ $masquillierii$ released pollen through a pump action as opposed to explosive release and a higher level of 'selfing' was apparent when small bees visited and pollen was released several times in a single visit (Vivarelli et al. 2011). Cordoba & Cocucci (2011) demonstrated that $Apis$ $mellifera$ was relatively weak compared to smaller (by weight) and stronger (measured force (mN)) megachilids which could trip flowers of high operative strength such as $Spartium$ $junceum$ that $A.$ $mellifera$ could not access. Investigating intra-specific variation in size in bumblebees (scored as large, medium or small) and pollination effectiveness in a non-leguminous plant, $Vinca$ $minor$ (Apocynaceae), Willmer & Finlayson (2014) found that pollen deposition to stigmas in single visits varied between individuals with larger bees depositing more grains.

In this work we only measured pollen deposition, a single component of pollinator "performance". Ne'eman et al (2010) reviewed the issue of pollinator performance, noting that plant reproductive success is not solely dependent on female fitness through pollen deposition on the stigma. They put forward a model that takes into account pollen deposition, visitation rates, and the necessity to continue any study to seed set to enable complete evaluation of reproductive success (Ne'eman et al 2010). Single visit deposition (SVD) of pollen has been put forward as the most practical measure of pollinator effectiveness (PE) in a range of tropical and temperate plant species (King et al. 2013). Pollinator effectiveness has been measured for relatively few individual plant species but is critical to the understanding of the relative importance of flower visitors where many and varied visitors are involved. Only recently has the necessity to examine PE been highlighted in community-level studies of pollination; Ballantyne, Baldock & Willmer (2015) provided the first plant-pollinator network based on pollinator evidence rather than just visitation or pollen transport. Although we measured just a single component of the Ne'eman et al (2010) model, pollen deposition, *D. setigerum* is an ideal plant for further studies of this type. Pollen deposition by visitors can be assessed at species level through SVD (as visit frequency is not a factor here), and seed counts subsequently confirm overall individual pollinator performance. In addition, the structure of the bee assemblages and relative abundances of bees of different sizes could be taken into account in future studies.

D. setigerum has evolved a variety of cues to attract and direct pollinators. These include the papillionate legume tripping mechanism, but also floral colour change and ability to reverse this change. Our work shows that these mechanisms may have evolved as, although self-compatible, *D. setigerum* needs insect visitation to set seed. The large assemblage of bees that visit this bee-adapted flower are not equally important in one aspect of pollinator performance; one species appears to be significantly more successful in pollen deposition than others. As the majority of pollination studies focus on flower visitation and not single visit deposition (King et al. 2013), our work adds to the literature as it shows that not all pollinators are equal in terms of pollination efficiency, and that even individuals within the same genus (e.g. *Megachile*) can vary hugely in their pollen deposition. These findings give more insights into the pollination ecology of this papillionate legume, but also into explosive pollen release and the relationship between bee body size and pollination effectiveness.

ACKNOWLEDGEMENTS

We would like to give thanks to the Tropical Biology Association for organizing the field course on which these data were collected, and giving us the opportunity to carry out this study at Kibale Forest. In addition, we would also like to thank Makerere University for the use of their Biological Field Station and facilities, the Uganda Wildlife Authority for permission to work in Kibale National Park, and both the British Ecological Society and BAT Biodiversity Partnership for financial assistance. Special thanks to Connal Eardley for all bee identifications and David Notton at the Natural History Museum, London for access to bee collections.

AUTHOR CONTRIBUTIONS

CN, DAS, ESB, KS, PW and TP designed and carried out the study; DAS and MO analysed the data; CN, DAS, KS and MO wrote the manuscript; all authors provided comments on subsequent manuscript versions.

REFERENCES

Alemán M, Figueroa-Fleming T, Etcheverry Á, Sühring S, Ortega-Baes P (2014) The explosive pollination mechanism in Papilionoideae (Leguminosae): an analysis with three *Desmodium* species. Plant Systematics and Evolution 300:177-186.

Aluri RJS, Reddi CS (1995) Explosive pollen release and pollination in flowering plants. Proceedings of the Indian National Science Academy Part B Biological Sciences 61:323-332.

Ballantyne G, Baldock KCR, Willmer PG (2015) Constructing more informative plant–pollinator networks: visitation and pollen deposition networks in a heathland plant community. Proceedings of the Royal Society of London B: Biological Sciences 282.

Bates D, Maechler M, Bolker BM, Walker S (2014) lme4: Linear mixed-effects models using Eigen and s4. R package version 1.1-6 http://CRAN.R-project.org/package=lme4.

Cordoba SA, Cocucci AA (2011) Flower power: its association with bee power and floral functional morphology in papilionate legumes. Annals of Botany 108:919-931.

Hothorn T, Bretz F, Westfall P (2008) Simultaneous inference in general parametric models. Biometrical Journal 50:346-363.

King C, Ballantyne G, Willmer PG (2013) Why flower visitation is a poor proxy for pollination: measuring single-visit pollen deposition, with implications for pollination networks and conservation. Methods in Ecology and Evolution 4:811-818.

Lewis G, Schrire B, MacKinder B, Lock M (2005) Legumes of the world. Royal Botanical Gardens, Kew, UK.

Ne'eman G, Jurgens A, Newstrom-Lloyd LE, Potts SG, Dafni A (2010) A framework for comparing pollinator performance: effectiveness and efficiency. Biological Reviews 85:435-451.

R Core Team (2014) R: A language and environment for statistical computing. R Foundation for Statistical Computing, Vienna, Austria. URL http://www.R-project.org.

Raju AJS, Rao SP (2006) Explosive pollen release and pollination as a function of nectar-feeding activity of certain bees in the biodiesel plant, *Pongamia pinnata* (L.) Pierre (Fabaceae). Current Science 90:960-967.

Stout JC (2000) Does size matter? Bumblebee behaviour and the pollination of *Cytisus scoparius* L. (Fabaceae). Apidologie 31:129-139.

Struhsaker TT (1997) Ecology of an African rain forest: logging in Kibale and the conflict between conservation and exploitation. University Press of Florida, USA.

Van Doorn WG (1997) Effects of pollination on floral attraction and longevity. Journal of Experimental Botany 48:1615-1622.

Vivarelli D, Petanidou T, Nielsen A, Cristofolini G (2011) Small-size bees reduce male fitness of the flowers of *Ononis masquillierii* (Fabaceae), a rare endemic plant in the northern Apennines. Botanical Journal of the Linnean Society 165:267-277.

Westerkamp C (1997) Keel blossoms: Bee flowers with adaptations against bees. Flora 192:125-132.

Willmer P, Stanley DA, Steijven K, Matthews IM, Nuttman CV (2009) Bidirectional flower color and shape changes allow a second opportunity for pollination. Current Biology 19:919-923.

Willmer PG, Finlayson K (2014) Big bees do a better job: intraspecific size variation influences pollination effectiveness. Journal of Pollination Ecology 14:244-254.

Yeo PF (1993) Secondary pollen presentation. Form, Function and Evolution. Springer, Vienna, Australia.

3

A TIGHT RELATIONSHIP BETWEEN THE SOLITARY BEE *CALLIOPSIS* (*CEROLIOPOEUM*) *LAETA* (ANDRENIDAE, PANURGINAE) AND *PROSOPIS* POLLEN HOSTS (FABACEAE, MIMOSOIDEAE) IN XERIC SOUTH AMERICAN WOODLANDS

Favio Gerardo Vossler*

Laboratorio de Actuopalinología, CICyTTP–CONICET / FCyT-UADER, Dr. Materi y España, E3105BWA, Diamante, Entre Ríos, Argentina

Abstract—The large genus *Calliopsis* (Andrenidae, Panurginae) is composed of ten subgenera with polylectic and presumably oligolectic species. These categories have been mainly developed from floral visits of female bees collecting pollen. In the present study, pollen analyses of nest provisions and scopal loads from museum specimens of the monotypic subgenus *Ceroliopoeum* were carried out to assess its degree of specialization to pollen host-plants. Despite the great variety of floral resources close to two active nest aggregations in the Chaco sites (83 and 44 melittophilous taxa from 36 and 17 families, respectively), the only host-plant recorded in all nest pollen samples was *Prosopis*. This genus was represented by six species and their hybrids, all having similar pollen morphology. The nesting sites in Monte scrub also contained several *Prosopis* species, some of which had different pollen morphology from those of the Chaco forest. Two different *Prosopis* pollen types were identified in all samples. Since the whole geographic distribution of *C. laeta* matches with the range of *Prosopis*, its strong association with this pollen host seems to be well supported. However, the low number of study populations (four) could erroneously indicate oligolectism. A broader sampling is necessary to ensure the character of specialization. Most *Calliopsis* species have been identified as oligolectic. Yet, this categorization has mainly been based on floral visits and a large diversity of floral hosts has been recorded for each bee species. Further analyses are necessary to confirm the relationship of this genus with its pollen hosts. Moreover, as most of them have short to medium phenologies (up to 4 months) their presumably oligolecty can be due to a local specialization (i.e. variable according to location) typical of polylecty.

Keywords: Calliopsini, Chaco forest, emergence, narrow oligolecty, pollen specialization, specialist bee

INTRODUCTION

The legume family (Fabaceae *sensu lato* or Leguminosae) is a mainly bee-pollinated plant group that constitutes a major food source for the entire taxonomic spectrum of bees (Arroyo 1981). The South American Chaco is a forest characterized by the abundance of woody and herbaceous Fabaceae, mainly the woody mimosoids *Prosopis* and *Acacia*, Zygophyllaceae, Anacardiaceae, Celastraceae, Rhamnaceae, Capparidaceae, Santalaceae, Ulmaceae s.l., Cactaceae and Bromeliaceae (Cabrera & Willink 1973). *Prosopis* is a species-rich plant genus and most species are quite uniform in floral and inflorescence phenotypes (Burkart 1937; Palacios & Bravo 1981). The flowers are open with exposed nectar and pollen resources (Arroyo 1981). *Prosopis* contains 45 species and is mainly distributed in arid and semi-arid regions of the world (Burkart 1976; Palacios & Brizuela 2005). However, recent studies showed that it is not a natural group, and that Old world species are not true

Prosopis (Catalano et al. 2008). In the Americas it is distributed from southwestern USA to central Chile and Argentina, mainly in warm and dry regions. The most important centre of differentiation of the genus is found in Argentina (27 species and 19 varieties) (Burkart 1976; Palacios & Brizuela 2005).

Bees are the most important group of pollinators of *Prosopis* and highly attracted to its flowers for both pollen and nectar resources (Moldenke and Neff 1974; Simpson et al. 1977; Keys et al. 1995). The most common bee genera visiting the flowers of *Prosopis* are *Colletes*, *Pygopasiphae*, *Chilicola*, *Calliopsis*, *Megachile*, *Centris*, *Eremapis*, *Exomalopsis*, *Svastrides* and *Xylocopa* in South American deserts and *Lasioglossum*, *Nomia*, *Perdita*, *Megachile*, *Centris* and *Melissodes* in North American deserts (Simpson et al. 1977). Oligolectic bees of *Prosopis* have been recorded for the South American *Colletes*, *Pygopasiphae*, *Chilicola*, one *Megachile* species and *Eremapis parvula* Ogloblin (Neff 1984; Simpson et al. 1977; Vossler 2013). In North America, *Prosopis* specialists include a complex of *Perdita* species, several *Colletes* and *Ashmeadiella prosopidis* (Simpson et al. 1977).

---

*Corresponding author: favossler@yahoo.com.ar; 0054-343-498-3086

In South America, the relationship between the bee fauna and *Prosopis* has been studied using flower visits (Simpson et al. 1977; Genise et al. 1990, 1991; Michelette & Camargo 2000). Flower visits allow identifying the association of bees with particular flowering plants, but may miss other host plants. The analysis of pollen from nest provisions is a reliable method to reveal the degree of pollen specialization (as shown by Neff (1984) and Vossler (2013) for *Eremapis parvula*, an oligolectic bee on *Prosopis*). Further advantages of pollen analysis from nest provisions compared to field observations is the unbiased representation of inaccessible or unanticipated pollen hosts, such as forest canopy species or alternative hosts of presumed oligolectic species (Cane & Sipes 2006). However, stenopalynous plant taxa (Erdtman 1952), that show only marginal or no morphological pollen variation can be identified only to a certain taxonomic level (i.e. genus, tribe, subfamily or family). In such cases, field observations will be necessary to identify the specialization status of a presumably oligolectic bee (Cane & Sipes 2006; Vossler 2013).

The genus *Calliopsis* consists of 10 subgenera and approximately 80 species and is found in temperate zones of Western Hemisphere, mainly in xeric areas (Michener 2007). It seems that each subgenus of *Calliopsis* has species specialized in different taxa of pollen host-plants, mostly Fabaceae (Mimosoideae and Papilionoideae, Tab. 1). For the three South American subgenera of *Calliopsis* (i.e. *Ceroliopoeum*, *Liopoeodes* and *Liopoeum*), floral hosts of only three *Liopoeum* species have been identified (Tab. 1).

Calliopsis (Ceroliopoeum) laeta (Vachal) is endemic to Argentina, where it has been recorded in dry areas of La Rioja, Santiago del Estero and Chaco provinces (Jörgensen 1912; Ruz 1991; Michener 2007; Moure & Dal Molin 2012). Biological data of this monotypic subgenus has not yet been documented. The objectives of this survey were to identify the botanical origin of pollen samples from nests and museum specimens of the solitary bee *Calliopsis laeta* and reveal its degree of pollen specialization, using mainly the pollen analysis method. This study further aimed at documenting its phenology and its global geographic distribution.

MATERIALS AND METHODS

Study sites of field observations and museum specimen

To identify pollen specialization of *Calliopsis laeta*, pollen samples were taken from two nesting sites (1-Villa Río Bermejito, 25° 37' S, 60° 15' W and 2-Juan José Castelli, 25° 56' S, 60° 37' W) in Chaco forest (Fig.1), from 22nd to 26th of September 2008 (for 1) and from 19th to 20th of September 2011 (for 2). Notes on nesting were taken from these same sites and dates. Nests aggregated in horizontal hard packed soils along dirt roads. Cells occurred up to a depth of 10 cm from the soil surface where the soil texture was sandy loam. The presence of clay probably prevented deeper nests. The soil was moist during the nesting period. Nest initiation occurred during a short period of two days after copious rain (96 mm), and before

the soil dried out and hardened. Neither nest building nor flight activity around nest entrances was observed after this period. However, five females and eight males were recorded foraging on *P. alba* on a site 3 km away from this nesting area and five days after the rainfalls. Similar to the statement of Rozen (1967) for many panurgines, there was no indication that a female of *Calliopsis laeta* uses water in building her nest.

Further, two museum specimens from different sites in Monte scrub (3-Amaichá del Valle (Tucumán) and 4-San Fernando del Valle de Catamarca (Catamarca)), captured in early November 2004 (for 3) and 1989 (for 4) could be examined for scopal pollen loads.

Pollen analysis of nest samples and museum specimens

Pollen samples from nests included brood provisions (N = 3 from 2008 and 2 from 2011), stomach contents of larvae (N = 10 from 2008), feces of post-defecating larvae (N = 11 from 2008), and scopal pollen from adults caught returning to nest entrances (N = 3 from 2008). Bees were caught at nests by hand and in nets, identified by Arturo Roig-Alsina and deposited in the Entomology collection of the Museo Argentino de Ciencias Naturales "Bernardino Rivadavia" (MACN). Nest pollen samples were dissolved in distilled water at 80-90 °C for 10-15 minutes, pressed when necessary using a glass rod, stirred by hand or, when necessary, by a magnetic stirrer for 5-10 minutes, and filtered. Finally, to obtain pollen sediment, samples were centrifuged at 472 × g for 5 minutes. Processing included Wodehouse (1935) and acetolysis methods (Erdtman 1960; Lieux 1980).

Sixtysix museum specimens of female *C. laeta* were examined but only two bore pollen loads. These are very moist and may get lost during handling (netting, pan trapping). The discovered samples stemmed from two localities in western Argentina (Tucumán and Catamarca provinces). One leg per individual was mounted on a slide, immersed in acetolysis fluid and heated directly over a flame for 20 seconds. This sediment was mounted using a glycerine-jelly mixture. A cover glass was added and sealed with paraffin. After this short process, pollen grains acquired a brownish colour similar to that obtained via the acetolysis method of Erdtman (1960). Pollen types were identified using a Nikon Eclipse E200 light microscope at 1000 × magnification. Pollen grains from flowers of herbarium plants collected in the study area and deposited in the Herbaria of La Plata (LP) and of the Museo Argentino de Ciencias Naturales "Bernardino Rivadavia" (BA), Argentina served as reference collection. The classification of host-plant specialization by bees follows Cane & Sipes (2006) and Müller & Kuhlmann (2008).

Additional data on the phenology and geographic distribution of *Calliopsis laeta* were taken from specimens examined from MACN collection (see Appendix 1).

Availability of floral resources

The availability of floral resources next to the two nest aggregations was recorded, as stressed by Cane & Sipes

TABLE 1.   Host-plant associations of the ten *Calliopsis* subgenera and specialization degree as suggested by different authors. Literature references: 1_ this article; 2_ Michelette & Camargo (2000); 3_ Simpson et al. (1977); 4_ Dumesh & Packer (2011); 5_ page 309 in Michener (2007); 6_ Shinn (1967); 7_ page 264 in Wcislo & Cane (1996); 8_ Rozen (1970); 9_ Danforth (1990); 10_ Rozen (2008); 11_ Rozen (1963); 12_ Wcislo (1999); 13_ Vossler *(in prep.)*; 14_ Michener (1954); 15_Robertson (1929). References: Pollen collection (P); nectar collection (N).

No. of species	Bee species	Host-plant genus (plant family)	Pollen samples (PS)/ floral visits (FV) and study sites	Specialization suggested by authors
South American subgenus				
Ceroliopoeum (1 species endemic to Argentina)	C. (Ceroliopoeum) laeta (Vachal)	*Prosopis* (Fabaceae, Mimosoideae)[1]	29 PS (Dry Chaco forest) and 2 PS (Monte desert)[1]	Possibly narrowly oligolectic[1]
Liopoeodes (1 species endemic to Argentina)	C. (Liopoeodes) xenopous Ruz	Unknown	No data	
Liopoeum (5 species from Chile and Argentina[4,5])	C. (Liopoeum) argentina (Jörgensen)	*Larrea* and *Bulnesia* (Zygophyllaceae)[2]	FV (two Argentinean sites from Catamarca and La Rioja provinces, Monte desert)[2,3]	Polylectic[3]
	C. (Liopoeum) rigormortis Dumesh & Packer	*Adesmia* (Fabaceae, Papilionoideae)[4]		Apparently specialist[4]
	C. (Liopoeum) mendocina (Jörgensen)	Prosopis, Solanum (Solanaceae) and Brassicaceae[13]	31 PS (Dry Chaco forest), 3 PS (Monte desert) and 2 PS (Wet Chaco forest)[13]	Polylectic with strong preference for Prosopis[13]
North and Central American subgenus				
Calliopsis s. str. (12 species, North and Central America to Panama)				Subgenus narrowly polylectic[5]. Subgenus widely polylectic, mostly on Fabaceae, especially the small-flowered clovers *Trifolium* and *Melilotus*[6]
	C. (Calliopsis) andreniformis Smith	*Trifolium* and *Melilotus* (Fabaceae, Papilionoideae)[5,6], *Malva* (Malvaceae) (P)[6]. Fabaceae, Asteraceae, Verbenaceae and Malvaceae (P and N)[6]	17 PS (pure Fabaceae, pure Malvaceae or mixed Asteraceae and Malvaceae)[6] and 98 FV[6]	
	C. (Calliopsis) hondurasica Cockerell	*Aeschynomene americana* (Fabaceae, Papilionoideae) in Panamá[12]	PS (a nest population near Veracruz, Panamá, open field regularly mowed surrounded by deciduous tropical forest)[12]; FV (Panamá)[14]	Possibly pollen specialist[12]
	Other species	Apocynaceae, Convolvulaceae, Asteraceae, Fabaceae, Oxalidaceae, Lamiaceae, Verbenaceae, etc[6]	FV[6]	
Calliopsima (15 species, Canada, USA and Mexico)	Most species	*Heterotheca, Gutierrezia, Baileya, Senecio, Solidago, Bidens, Boltonia, Coreopsis, Rudbeckia, Cirsium, Encelia, Hemizonia, Haplopappus, Grindelia, Verbesina*, etc. (Asteraceae), *Melilotus albus* and *Medicago sativa* (Fabaceae)[6]	FV[6]	Primarily on the Asteraceae, particularly tribes Heliantheae and Astereae[6]
	C. (Calliopsima) coloradensis Cresson	*Boltonia, Solidago, Bidens, Rudbeckia* and *Coreopsis* (P)[6,15]	FV (Carlinville, Illinois, USA)[15]	
	C. (Calliopsima) rozeni Shinn	Primarily on *Heterotheca subaxillaris*[6]	FV[6]	

TABLE I. continued.

No. of species	Bee species	Host-plant genus (plant family)	Pollen samples (PS)/ floral visits (FV) and study sites	Specialization suggested by authors
Hypomacrotera (3 species, SW USA to Mexico)	*C. (Hypomacrotera) persimilis* (Cockerell)	*Physalis* (Solanaceae) (P)[8,9]	PS (Animas, New Mexico, USA, mixed grassland adjacent to a cotton field)[9]	
	C. (Hypomacrotera) subalpina (Cockerell)	*Sphaeralcea* (Malvaceae)[8,5]	FV (Douglas, Arizona, USA)[8]	
Micronomadopsis (20 species, Western North America)				Subgenus oligolectic of *Trifolium*[7]. Many species are oligolectic[5]
	C. (Micronomadopsis) snellingi (Rozen)	*Salvia* (Lamiaceae)[11]	FV[11]	
	C. beamerorum (Rozen)	*Prosopis*[11]	FV[11]	
	C. fracta (Rozen)	*Eriodictyon* (Boraginaceae)[11]	FV[11]	
Nomadopsis s.s. (8 species) (the 5 species of *Macronomadopsis* were separately analyzed in this article) (Western North America)	*C. (Macronomadopsis) micheneri* (Rozen), *C. anthidia* Fowler, and *C. filiorum* (Rozen)	*Trifolium* (Fabaceae) (P)[11]	FV[11]	All *Macronomadopsis* species appear to be oligolectic of Fabaceae[10]
	C. zebrata Cresson	*Astragalus* (Fabaceae) (P)[11]	FV[11]	
	C. (Nomadopsis) trifolii (Timberlake)	*Mimulus* (Phrymaceae or Scrophulariaceae s.l.) (P)[11]	FV[11]	Species of *Nomadopsis s.s.* are oligolectic on a wide assortment of plant families, but none visits Fabaceae[10]
	C. zonalis sierrae Cresson	*Monardella* (Lamiaceae)[11]	FV[11]	
Perissander (7 species, (SW USA to NW Mexico)				Subgenus oligolectic of *Euphorbia*[7]
	C. (Perissander) anomoptera Michener	*Euphorbia* (Euphorbiaceae) (six species), *Cladothrix* (Amaranthaceae), *Eriogonum* (Polygonaceae), *Lepidium* (Brassicaceae) and *Tidestromia* (Amaranthaceae)[6]	2 PS (*Euphorbia*)[6]; most FV on *Euphorbia*[6]	Species principally on Euphorbiaceae[6]
	Other species	*Euphorbia, Tidestromia* and *Verbesina*[6]	FV[6]	
Verbenapis (4 species, USA and Mexico)				Subgenus oligolectic of *Verbena*[7]
	C. (Verbenapis) verbenae Cockerell and Porter	*Verbena* (Verbenaceae), *Sphaeralcea* (Malvaceae) and *Chamaesaracha* (Solanaceae)[6]	Most FV on *Verbena*[6]; PS (mixed Verbenaceae and Fabaceae)[6]	Species oligolectic of *Verbena*[6]
	C. (Verbenapis) hirsutifrons Cockerell	*Verbena, Vernonia* (Asteraceae), *Ambrosia* (Asteraceae), *Asclepias* (Apocynaceae) and *Medicago sativa* (Fabaceae)[6]	Most FV on *Verbena*[6]	

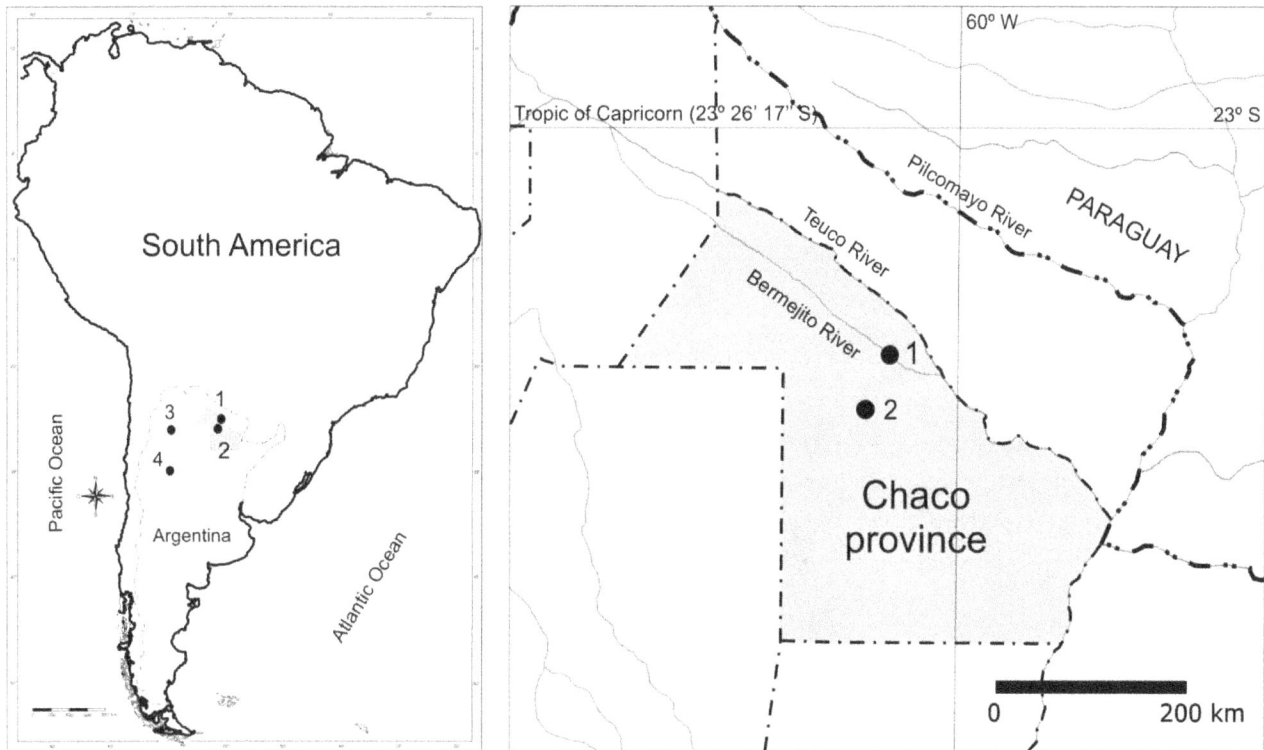

FIGURE 1.   Study sites where pollen samples from nests (1-Villa Río Bermejito and 2-Juan José Castelli, both Chaco) and bee specimens of museum collections were taken (3-Amaichá del Valle (Tucumán) and 4-San Fernando del Valle de Catamarca (Catamarca)).

(2006). The relative abundance of flowers and flowering individuals as a whole was estimated and simply classified as 1 = rare, 2 = common, and 3 = highly abundant (Appendix II) All the entomophilous plants flowering within a radius of up to 100 m of the two nesting aggregations from the Chaco forest were recorded (Appendix II). Plants around the nests mainly belonged to riparian (site 1) and xerophylous forest vegetation (site 2), accompanied by alien plants (Fig. 1; Appendix II).

RESULTS

A total of 31 pollen samples from two nest aggregations and two museum specimens were analyzed. All samples consisted only of *Prosopis* pollen. Using light microscopy, two pollen types belonging to *Prosopis* could be distinguished: *Prosopis* type 1 had an exine <2 (mainly from 1 to 1.5) µm thick and a smooth or slightly scabrate wall sculpture (all Chaco and Tucumán samples) while *Prosopis* type 2 had an exine 1.5-3 µm thick and a strongly scabrate wall sculpture (the Catamarca sample). These two pollen morphologies belong to different sections of the genus *Prosopis* according to Caccavari (1972): pollen type 1 to *Algarobia* and type 2 to sections *Strombocarpa* (*P. abbreviata* Benth., *P. reptans* Benth., *P. strombulifera* (Lam.) Benth. and *P. torquata* (Lag.) DC.), *Cavenicarpa* (*P. ferox* Gris.) and *Monilicarpa* (*P. argentina* Burk.) (following the taxonomical classification of Burkart (1976)).

The fecal pollen of post-defecating larvae, found at the end of brood cells, consisted of collapsed *Prosopis* pollen grains (Fig. 2a).

A total of 83 and 44 melittophilous plant taxa belonging to 36 and 17 families were recorded around nesting areas during September 2008 and 2011, respectively (Appendix II). In the riparian forest at study site 1, most blooming species belonged to the families Asteraceae (20 taxa, mainly from tribe Heliantheae), Fabaceae (8 taxa, mainly from subfamiliy Mimosoideae), Verbenaceae (6 taxa), Solanaceae (5 taxa) and Bignoniaceae (4 taxa). In the xerophilous forest at study site 2, the major families were Fabaceae (14 taxa), Asteraceae (7 taxa), Capparidaceae (3 taxa), Verbenaceae (3 taxa) and Bignoniaceae (3 taxa). The most frequent growth habits were herbs in locality 1 (47 taxa), while trees or shrub-trees (17) were equally abundant as herbs (16) in locality 2. During the nesting period, flowers of *Prosopis alba*, *Albizia inundata*, *Leucaena leucocephala*, *Ziziphus mistol*, *Cissampelos pareira* and *Clematis montevidensis* were highly abundant in locality 1 within 20 to 50 m from the nests (Appendix II). In the xerophilous forest of locality 2, flowers of all *Prosopis* species (*P. alba*, *P. nigra*, *P. ruscifolia*, *P. vinalillo*, *P. elata* and *P. kuntzei*) and hybrids, as well as of three *Capparis* species, two Celastraceae, *Cercidium praecox*, *Castela coccinea* and *Ziziphus mistol* were abundant in close proximity to the nests (5 to 10 m) (Fig. 2b). A crop of *Melilotus albus* at flowering peak was also found in 4 m distance to the nests of locality 2. The *Prosopis* species found close to the nesting sites belong to section *Algarobia* (Burkart 1976) and have pollen grains of similar morphology (Caccavari 1972). Floral visits were recorded on *P. ruscifolia* and *P. alba* during a whole day (22nd September 2011), but *Calliopsis laeta* was only observed on *Prosopis alba*.

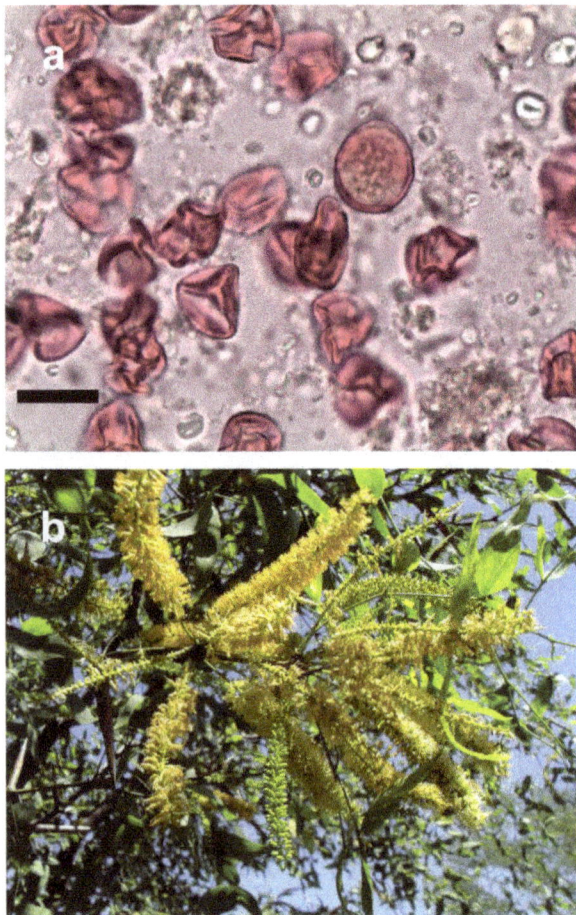

FIGURE 2. a)   Collapsed pollen grains of *Prosopis* from feces of bee larvae. Unacetolyzed pollen grains dyed using fuchsine, seen in light microscope at 40 × magnification. Scale bar = 10 μm. b)   Flowers of *Prosopis ruscifolia* at its flowering peak in early spring.

The examination of museum specimens revealed that *C. laeta* occurs in the dry Monte and Chaco regions of Argentina, from Salta and Formosa in the north to Mendoza and San Luis provinces in the south and that flight activity lasted less than a month. In Chaco sites, bees were active for two weeks in late September 2008 and only one week in late September 2011 (after 30 mm and 96 mm rain, respectively). From the dates recorded on the museum specimen from the Monte sites, it was concluded that bees were active from mid October to early November for females and up to 22nd November considering only males.

DISCUSSION

Prosopis *as the only pollen host of* Calliopsis laeta

Only *Prosopis* pollen was found in nest samples in spite of the presence of abundant alternative pollen hosts being available, such as *Melilotus, Capparis, Cercidium, Albizia* and others (Appendix II). Taking into account all examined samples throughout the geographic range of *Calliopsis laeta,* two pollen types belonging to *Prosopis* species were well distinguished using light microscopy. One of them belonged to *Prosopis* section *Algarobia* (Chaco and Tucumán samples) and the other to the sections *Strombocarpa,*

Cavenicarpa or *Monilicarpa* (the Catamarca sample). Therefore, pollen analysis of nests and museum specimens showed that *Calliopsis laeta* could be specialized in pollen collection from the genus *Prosopis* alone, suggesting that it is a *narrowly oligolectic* bee species. Although the low number of study populations (two museum specimens and two field studies) could indicate oligolectism, a broader sampling is necessary to ensure the character of specialization. *C. laeta* seems to be quite limited in its geographic distribution (only eight Argentine provinces), but occurs within the range of *Prosopis,* which could add to the hypothesis of oligolecty on this plant-host.

Most *Calliopsis* species have been identified as oligolectic (Tab. I). Yet, this categorization has mainly been based on floral visits. A large diversity of floral hosts has been recorded for each bee species demanding for further analyses to confirm the relationship of this genus with its pollen hosts. Moreover, as most of them have short to medium phenologies (up to 4 months) (Shinn 1967, see page 803) their presumably oligolecty can be due to a local specialization (i.e. variable according to location) typical of polylecty.

In a review of floral relationships of panurgine bees in Northeastern Brazil, Schlindwein (2003) highlighted that they have not been recorded visiting flowers of trees, but only herbs or small shrubs, and only in open areas (but not in tropical rainforest habitats). However, in other regions this bee group has also been associated to arboreal pollen hosts, such as *Prosopis* and *Salix* with *Perdita* in North America (Rozen 1967; Simpson et al. 1977), *Prosopis* with *Acamptopoeum* and *Calliopsis argentina* in South America (Simpson et al. 1977; Rozen & Yanega 1999).

After a 5-month autumn-winter unfavourable period of cold climate and drought, the first heavy rain commonly occurs between August and November (Appendix III). In the Chaco sites, emergence of the ground nesting *Calliopsis laeta* was observed after rain, presumably because of the rain softening the hard packed sandy loam soil making nest excavation possible. The shallow nests of *C. laeta* (of no more than 10 cm) appear to support this hypothesis. When the soil dried out, nest building (and provisioning) stopped even though *Prosopis* was still flowering. On the other hand, foraging activity of females and males was observed on *P. alba* when nesting had already stopped suggesting the presence of active nests outside the study area. This would indicate that soil moisture varies among microsites, such as shadowed areas in the forest vs sunny dirt roads, explaining the differences in duration of provisioning periods observed. Many *Prosopis* species are deep-rooted perennial phreatophytes which do not rely on rainfall but rather on changes in day length as the cue for floral initiation (Peacock & McMillan 1965; Simpson et al. 1977).

As different cues trigger both *Prosopis* blooming and the emergence of the ground nesting *Calliopsis*, synchronization between them might not occur. Nevertheless, even when triggers are different, the bee nesting and peak flowering of their pollen hosts overlap. The museum specimens examined from Tucumán and Catamarca, Western Argentina, had scopal loads composed only of *Prosopis.* These bees were collected in early November, co-occurring with the single blooming period of *Prosopis flexuosa* and *P. chilensis* and the first flowering of *P. torquata* in this region (Simpson et al. 1977). During five field samplings carried out in the Chaco region during rainy episodes (September 2008 and 2011, late October, late November and early December 2008, see Vossler 2013), nest aggregations of *C. laeta* were only found in September (late winter - early spring) in both years when all *Prosopis* species (including their hybrids) were in their flowering peak. In both areas, bee phenology is triggered by the first rains, which in Western Chaco occur in spring (Appendix III), while in northern Monte first rain falls towards the end of spring but more often in summer (Cabrera 1976). Since all examined specimens were from the whole geographic distribution, its flight activity could be narrowed to a short period of no more than one month (from mid October to mid November in the Monte scrub and during late September in the Chaco forest), indicating its univoltine nature.

ACKNOWLEDGEMENTS

I am especially grateful to Arturo Roig-Alsina for identifying the bees, providing museum specimens and for suggestions and comments on the manuscript. I would also like to thank Nora Brea for her help with the English translation. I would also like to thank the editor and reviewers for greatly improving the quality of the manuscript. This study was supported by CONICET (Consejo Nacional de Investigaciones Científicas y Técnicas).

APPENDICES

Additional supporting information may be found in the online version of this article:

APPENDIX I. Museum specimens examined from MACN entomological collection.

APPENDIX II. Available floral resources during the *Calliopsis* nesting period (September 2008 and 2011) in the Chaco forest.

APPENDIX III. Climate diagram from 2008 to 2010 for "Los Frentones" metheorological station (26° 22' 13" S, 61° 27' 34" W), Chaco province, Argentina.

REFERENCES

Arroyo MTK (1981) Breeding systems and pollination biology in Leguminosae. In: Polhill RM, Raven PH (eds) Advances in Legume Systematics. Royal Botanic Gardens, Kew, pp 723-769.

Burkart A (1937) Estudios morfológicos y etológicos en el género *Prosopis*. Darwiniana 3:27-47.

Burkart A (1976) A monograph of the genus *Prosopis* (Leguminosae Subfam. Mimosoideae). Journal of the Arnold Arboretum 57:219-249.

Cabrera AL (1976) Regiones fitogeográficas argentinas. In: Kugler WF (ed) Enciclopedia argentina de agricultura y jardinería. Tomo II, Acme, Buenos Aires, Argentina, pp 1-85.

Cabrera AL, Willink A (1973) Biogeografía de América Latina. Secretaría General de la Organización de los Estados Americanos, Washington D.C.

Caccavari MA (1972) Granos de polen de Leguminosas de la Argentina II. Subfam. Mimosoideae; Tribu Adenanthereae. Revista del Museo Argentino de Ciencias Naturales "Bernardino Rivadavia" 4:281-320.

Cane JH, Sipes S (2006) Characterizing floral specialization by bees: analytical methods and a revised lexicon for oligolecty. In: Waser NM, Ollerton J (eds) Plant-pollinator interactions from specialization to generalization. University of Chicago Press, Chicago, pp 99-122.

Catalano SA, Vilardi JC, Tosto D, Saidman BO (2008) Molecular phylogeny and diversification history of *Prosopis* (Fabaceae: Mimosoideae). Biological Journal of the Linnean Society 93:621-640.

Danforth BN (1990) Provisioning behavior and the estimation of investment ratios in a solitary bee, *Calliopsis (Hypomacrotera) persimilis* (Cockerell) (Hymenoptera: Andrenidae). Behavioral Ecology and Sociobiology 27:159-168.

Dumesh S, Packer L (2011) The *Calliopsis* (Hymenoptera; Andrenidae; Panurginae) of Chile with the description of a new species. Zootaxa 2908:64-68.

Erdtman G (1952) Pollen Morphology and Plant Taxonomy, Angiosperms. Almqvist and Wiksell, Stockholm.

Erdtman G (1960) The acetolysis method, a revised description. Svensk Botanisk Tidskrift 54:561-564.

Genise J, Palacios RA, Hoc PS, Carrizo R, Moffat L, Mom MP, Agulló MA, Picca P, Torregrosa S (1990) Observaciones sobre la biología floral de *Prosopis* (Leguminosae, Mimosoideae). II. Fases florales y visitantes en el Distrito Chaqueño Serrano. Darwiniana 30:71-85.

Genise J, Palacios RA, Hoc PS, Agulló MA, Picca P (1991) Some new aspects of the floral biology of *Prosopis* (Leguminosae, Mimosoideae): behaviour of the main insect visitors in West-Chaco (Argentina). Bulletin of the International Group for the Study of Mimosoideae 19:130-145.

Jörgensen P (1912) Revision der Apiden der Provinz Mendoza, Republica Argentina. Zoologische Jahrbücher, Abteilung für Systematik, Geographie und Biologie der Tiere 32:89-162.

Keys RN, Buchmann SL, Smith SE (1995) Pollination effectiveness and pollination efficiency of insects foraging *Prosopis velutina* in South-eastern Arizona. Journal of Applied Ecology 32:519-527.

Lieux MH (1980) Acetolysis applied to microscopical honey analysis. Grana 19:57-61.

Michelette ERF, Camargo JMF (2000) Bee-plant community in a xeric ecosystem in Argentina. Revista Brasileira de Zoologia 17:651-665.

Michener CD (1954) Bees of Panamá. Bulletin of the American Museum of Natural History 104:1-175.

Michener CD (2007) The bees of the world. The John Hopkins University Press, Baltimore.

Moldenke AR, Neff JL (1974) The bees of California, a catalogue with special reference to pollination and ecological research. Origin and Structure of Ecosystems, Technical Reports 74-1 to 74-6. Santa Cruz: University of California, p. -v + 245 pp.

Moure JS, Dal Molin A (2012) Calliopsini Robertson, 1922. In: Moure JS, Urban D, Melo GAR (orgs). Catalogue of Bees (Hymenoptera, Apoidea) in the Neotropical Region [online] URL: http://www.moure.cria.org.br/catalogue (accessed December 2013).

Müller A, Kuhlmann M (2008) Pollen hosts of western palaearctic bees of the genus *Colletes* (Hymenoptera: Colletidae): the Asteraceae paradox. Biological Journal of the Linnean Society 95:719-733.

Neff JL (1984) Observations on the biology of *Eremapis parvula* Ogloblin an anthophorid bee with a metasomal scopa (Hymenoptera: Anthophoridae). Pan-Pacific Entomologist 60:155-162.

Palacios R, Bravo L (1981) Hibridación natural en *Prosopis* (Leguminosae) en la región chaqueña argentina. Evidencias morfológicas y cromatográficas. Darwiniana 23:3-35.

Palacios R, Brizuela MM (2005) *Prosopis*. Flora Fanerogámica Argentina 92:1-25.

Peacock JT, McMillan C (1965) Ecotypic differentiation in *Prosopis* (mesquite). Ecology 46:35-51.

Robertson C (1929) Flowers and insects. Lists of visitors to four hundred and fifty-three flowers. Science Press Printing Company, Lancaster, Pennsylvania, USA.

Rozen JG Jr. (1963) Notes on the biology of *Nomadopsis*, with descriptions of four new species (Apoidea, Andrenidae). American Museum Novitates 2142:1-17.

Rozen JG Jr. (1967) Review of the biology of panurgine bees, with observations on North American forms (Hymenoptera, Andrenidae). American Museum Novitates 2297:1-44.

Rozen JG Jr. (1970) Biology and inmature stages of the panurgine bee genera *Hypomacrotera* and *Psaenythia* (Hymenoptera, Apoidea). American Museum Novitates 2416:1-16.

Rozen JG Jr. (2008) The solitary bee *Calliopsis zebrata*: Biological and distributional notes and description of its larva (Hymenoptera: Andrenidae: Panurginae). American Museum Novitates 3632:1-12.

Rozen JG Jr, Yanega D (1999) Nesting biology and inmature stages of the South American bee genus *Acamptopoeum* (Hymenoptera: Andrenidae: Panurginae). In: Byers GW, Hagen RH, Brooks RW (eds) Entomological contributions in memory of Byron A. Alexander. The University of Kansas Natural History Museum, special publication No. 24, Lawrence, Kansas, pp 59-67.

Ruz L (1991) Classification and phylogenetic relationships of the panurgine bees: The Calliopsini and allies. University of Kansas Science Bulletin 54:209-256.

Schlindwein C (2003) Panurginae (Hymenoptera, Andrenidae) in Northeastern Brazil. In: Melo GAR, Alves-dos-Santos I (eds) Apoidea Neotropica: Homenagem aos 90 Anos de Jesus Santiago Moure. Editora UNESC, Criciúma, pp 217-222.

Shinn AF (1967) A revision of the bee genus *Calliopsis* and the biology and ecology of *C. andreniformis* (Hymenoptera, Andrenidae). University of Kansas Science Bulletin 46:753-936

Simpson BB, Neff JL, Moldenke AR (1977) *Prosopis* flowers as a resource. In: Simpson BB (ed) Mesquite, its biology in two desert scrub ecosystems. Dowden, Hutchinson & Ross, Stroudsburg, Pennsylvania, pp 85-107.

Vossler FG (2013) The oligolecty status of a specialist bee of South American *Prosopis* (Fabaceae) supported by pollen analysis and floral visitation methods. Organisms Diversity and Evolution 13:513-519.

Wcislo WT (1999) Male territoriality and nesting behavior of *Calliopsis hondurasicus* Cockerell (Hymenoptera: Andrenidae). Journal of the Kansas Entomological Society 72:91-98.

Wcislo WT, Cane JH (1996) Floral resource utilization by solitary bees (Hymenoptera: Apoidea) and exploitation of their stored foods by natural enemies. Annual Review of Entomology 41:257-286.

Wodehouse RP (1935) Pollen grains. Mc. Graw-Hill, New York.

4

Feeding behaviour of the dawn bat (*Eonycteris spelaea*) promotes cross pollination of economically important plants in Southeast Asia

Pushpa Raj Acharya[1], Paul A Racey[2], Sunthorn Sotthibandhu[1], Sara Bumrungsri[1]

[1]*Department of Biology, Prince of Songkla University, Hat Yai Thailand*
[2]*Centre for Ecology and Conservation, University of Exeter in Cornwall, UK*

Abstract—*Eonycteris spelaea* is recognized as the principal pollinator of most chiropterophilous plants in SE Asia. The present study describes its feeding behaviour and clarifies its role in cross pollinating these highly self-incompatible plants. Ten individuals of *E. spelaea* were radio-tracked during the flowering period of durian (*Durio zibethinus*) and petai (*Parkia speciosa*) in an agricultural mosaic in southern Thailand. *Eonycteris spelaea* makes a mean of seven visits per hour to these trees and 80-86% of each feeding bout involves visits to multiple conspecific trees. During each visit, 93% of *D. zibethinus* stigmas and 50% of *P. speciosa* stigmas were loaded with conspecific pollen. *Eonycteris spelaea* was the most common bat visitor to the trees. High visitation frequency and conspecific pollen deposition by *E. spelaea* to *D. zibethinus* and *P. speciosa* indicates that this nectarivorous bat is an effective pollinator. Mixed planting of chiropterophilous trees in fruit orchards is recommended to ensure regular visits of *E. spelaea*. Protecting natural roost caves of *E. spelaea* is also essential in order to maintain the vital ecosystem service provided by these bats.

Keywords: *Eonycteris, effective pollination, Parkia spp., Durio zibethinus*

Introduction

Although bat pollination is relatively uncommon among existing pollination systems, which rely heavily on insects, over 528 species of 250 genera of plants from tropical and subtropical regions show various degrees of dependency on bats for pollination (Dobat & Peikert Holle 1985; Fleming & Muchhala 2008; Fleming et al. 2009). Some of these plants are either economically valuable for human use as food and timber or ecologically important members of arid and semi-arid ecosystems of America and Africa and mangrove forests of Southeast Asia (Fujita & Tuttle 1991; Mickleburgh et al. 1992; Yetman 2007; Fleming et al. 2009, Kunz et al 2011). In Southeast Asia, commercial crops such as durian (*Durio zibethinus*) and the canopy leguminous trees (*Parkia speciosa* and *P. timoriana*) generate millions of dollar each year through local and global markets (Fujita & Tuttle 1991; Lim & Luder 1997; Kingston 2010; Bumrungsri et al. 2013). A study of the pollination of *D. zibethinus* and *P. speciosa* by fruit bats in southern Thailand estimated crop values of 137 million USD in 2008 (Petchmunee 2008). Production of durian fruit is important for the livelihoods of farmers in several countries in South and Southeast Asia while durian production in some areas is low and erratic (Subhadrabandhu et al. 1991; Honsho et al. 2004) especially where nectarivorous bats have been exterminated.

Eonycteris spelaea is the largest nectarivorous bat in South and Southeast Asia. It was inferred from a diet study in west Malaysia that *E. spelaea* is a long distance forager (~38 km, Start 1974). More recently, radio-tracked bats were located up to 18 km from the roost cave in an agricultural mosaic (Acharya et al. in press). High mobility in bats is described as a potential factor in promoting gene flow between plants in fragmented habitats (Law & Lean 1996; Corlett 2004; Molina-Freaner et al. 2003; Fleming et al. 2009). The annual diet of *E. spelaea* consisted of over 31 plant species in the west Malaysian mainland (Start 1974) and eleven taxa in southern Thailand (Bumrungsri et al. 2013). *Durio zibethinus*, *Parkia* spp. and *Musa* spp. are major components of the diet of *E. spelaea* during their flowering season (Start 1974; Bumrungsri et al. 2013).

The pollination biology of *D. zibethinus* and two species of *Parkia* showed that they are highly or exclusively self-incompatible and require cross pollination for fruit set (Valmayor et al. 1965; Soepadmo & Eow 1971; Bumrungsri et al. 2008, 2009). Although insects were major visitors to inflorescences of *D. zibethinus* and a common visitor to *Parkia*, fruit bats, especially *E. spelaea*, have been identified as their principal or sole pollinators (Bumrungsri et al. 2008, 2009). The larger body surface of *E. spelaea* may facilitate more pollen transfer during each visit than insects. However, it is still unknown how the foraging behaviour of this bat promotes out-crossing between conspecific trees.

In pollination biology, the effectiveness of flower visitors as pollinators (Stebbins 1970), is measured by the number of compatible pollen grains transferred by a visit of the pollinator (Primack & Silander 1975; Fenster 1991; Mayfield et al. 2001) or the resultant fruit set after such a visit (Spears 1983; Schemske & Horvitz 1984). In addition, a legitimate pollinator is defined as a pollen vector that deposits conspecific pollen onto the stigma during a visit (Fleming & Sosa 1994). The objective of the present study was to describe the foraging behaviour of *E. spelaea* on *D. zibethinus* and *Parkia* spp. We hypothesized that this behaviour would promote out-crossing of these plants, and that *E. spelaea* is thus an effective and legitimate pollinator.

MATERIALS AND METHODS

Study species

Eonycteris spelaea

The dawn bat (*Eonycteris spelaea*) is a 40-80 g nectarivorous bat which normally forms large colonies inside large caves in the Oriental region, from the Indian subcontinent to Southeast Asia and Asia–Pacific as far as East Timor (Bates & Harrison 1997). The home-range size of *E. spelaea* in agricultural areas extends to almost 500 ha and the foraging area up to 60 ha (Acharya et al. in press) *Eonycteris spelaea* has a strong fidelity to its foraging areas, consistently using them for at least several months (Acharya et al. in press). It emerges from its roosts 43 ± 13 min (mean ± SD, $N = 64$ nights) after sunset and spends 6 h 17 min ± 1 h 56 min away from the cave. Mature females spent significantly longer foraging than mature males and immature individuals (Acharya et al. 2015).

The conservation value of *E. spelaea* across Southeast Asia is linked to its role as the pollinator of economically and ecologically important plants such as *Durio*, *Parkia* and coastal mangrove trees (Kingston 2010; Kunz et al. 2011; Bumrungsri et al. 2013). Although the IUCN Redlist categorises the species as "lower risk" (IUCN 2014), population declines have been recorded in *E. spelaea* colonies in Thailand (Bumrungsri et al. 2008). Cave tourism and harvesting for bushmeat are common threats to the bats in Southeast Asia (Mickleburgh et al. 2009).

Durio zibethinus and Parkia spp.

Durio zibethinus is a commercial species of the subfamily Bombacoideae (family Malvaceae), commonly planted in the countries of Southeast Asia (Brown 1997). Durian trees are characterized by ephemeral mass flowering as well as the classical syndrome of chiropterophilous flowers such as nocturnal anthesis, a musty smell and cauliflory (Faegri & van der Pijl 1979). The major durian flowering in our study areas occurred during April and May. The flowering period extended for only 2-3 weeks. Each night, durian flowers start to open in the late afternoon and anthesis occurs after dusk when the flowers have fully opened. The flowers are hermaphroditic and protogynous so that the stigma protrudes spatially beyond the anthers. Durian flowers begin nectar secretion in the late afternoon and cease when the floral parts including the corolla and androecia drop. Floral abscission occurs around 01h00 and all the flowers have dropped by 06h00. The nectar which accumulates up to 19h00 (the time when the bat activity begins) was measured as c. 0.37 ml and the secretion rate thereafter was 0.05 ml h⁻¹ until floral abscission (Bumrungsri et al. 2009). The sucrose concentration of durian nectar was highest (21.9%) during the evening and then decreased gradually later in the night (Sripaoraya 2005; Bumrungsri et al. 2009).

Two species of *Parkia* (*P. speciosa* and *P. timoriana*) are common in both wild and cultivated areas of Southeast Asia. *Parkia speciosa* produces flowers from mid-May to mid-September while *P. timoriana* does so from October to December in our study sites. The inflorescences of *Parkia* spp. consists of compact biglobose bell heads called capitula each of which possess thousands of small tubular flowers. The capitula open during the evening for only one night. *Parkia speciosa* produces a few (up to 20) capitula while *P. timoriana* produces a greater number of capitula per tree per night. The flowering period of a tree lasts for 4-5 weeks in both species. We radiotracked the bats to flowering patches of *P. speciosa*, which is the preferred species by the villagers for cultivation.

Each capitulum of *P. speciosa* consists of three types of flowers: fertile (hermaphrodite), nectar-secreting and staminoidal (Hopkins 1984). Each capitulum comprises 2-3 thousand flowers, of which 70 to 75% are fertile (Bumrungsri et al. 2008).The capitula of *P. speciosa* begin to flower around 18h00 and anthesis and nectar secretion starts at 19h30. Each capitulum produces a nectar volume of 12.4 ml night⁻¹, with the highest volume and sucrose concentration from 20h00 to 21h00, which decreases gradually until 02h00 when nectar secretion ceases (Sripaoraya 2005; Bumrungsri et al. 2008).

Study Areas

This study was carried out in agricultural areas in Rattaphum district, Songkhla Province, Southern Thailand. We identified two colonies of *E. spelaea* in caves 10 km apart (Srikesorn Cave: 07° 04' 29.3" N, 100° 10' 07.4" E and Khaosoidao Cave: 06° 58' 53.91" N, 100° 08' 25.21" E). The land-use pattern within a 20 km radius around Srikesorn Cave consists of 26% tropical lowland forest (inside Ton Nga Chang wildlife sanctuary) and 74% agricultural mosaic and human settlements (data source: Southern GIS centre, Prince of Songkla University). The agricultural mosaic is dominated by rubber plantations (67%) with the remaining land used for fruit orchards and human settlements (33%). The fruit orchards comprise mixed fruit crops with various tropical fruits including durian (*D. zibethinus*), rambutan (*Nephelium lappaceum*), longsat (*Lansium domesticum*), mangosteen (*Garcinia mangostana*), santol (*Sanorium koetjape*), banana (*Musa* spp.), longan (*Dimocarpus longan*), mango (*Mangifera indica*), coconuts (*Cocos nucifera*), champoo (*Syzygium samarangense*) and palm (*Arenga* spp.). These orchards are patchily distributed in a flat lowland area along natural streams adjacent to tropical forest. *Parkia* trees were grown semi-wild or cultivated in house yards, fruit orchards, roadside and forest edges. Wild banana patches were found along the forest ravine and in early stage rubber plantations. Indian trumpet flower (*Oroxylum indicum*), cotton trees (*Bombax ceiba*) and kapok trees (*Ceiba pentandra*) were sporadic food resources for nectarivorous bats in the study area.

Radio-tracking

Ten individuals of *E. spelaea* were radio-tracked, five of which foraged at *Parkia* patches and the other five at *Durio* patches. Bats were caught in mist nets as they emerged from the cave or at flowering patches of the trees at foraging areas. The tagged individuals comprised mature and immature males and females, none of which were pregnant or lactating. We used radio tags (PD-2C, Holohil Ltd., Canada, wt. 4.0 g with cable-tie collars; Biotrack, UK, wt. 2.26 g). The weight of the radio-tags comprises 4.3 to 6.9 % of bat's body weight. Bats were fed with sugar syrup and released after collar attachment. Portable receivers (TRX-1000S, and 3-element Yagi antenna; Wildlife Materials, Carbondale, Illionis, USA) were used to receive radio-signals. The bats were tracked mostly by walking through fruit orchards. We used the homing-in technique, i.e. the bats were followed to their feeding trees and night roosts (White & Garrott 1990; Amelon et al. 2009). The bats' positions were confirmed using the 'close approach' method i.e. the bats were directly observed either flying around the tree or landing on inflorescences to drink nectar (Law & Lean 1999).

Assessment of foraging movements and tree visitation

Eonycteris spelaea used one to three foraging areas, up to 8 km apart, in a night (Acharya et al. in press). The identified foraging areas were surveyed during the afternoon before each tracking night, potential food trees located and their flowering status noted. When the bats arrived in the area at night, the trees they visited were confirmed with the aid of radio-signals or through direct observation of bats visiting the trees. The foraging behaviour of a tagged bat was noted from its arrival to its departure from the foraging area. While foraging, *E. spelaea* established a night roost in each foraging area and foraged intermittently. The duration of each foraging movement is referred to as a foraging bout. The bats visited several flowers on one or more trees during each foraging bout.

We waited at flowering trees of *D. zibethinus* and *P. speciosa* and examined the bats visitation behaviour and their movement pattern along the patches. The foraging bout began when the bat flew away from the night roost to forage and ended when it returned to the night roost for resting. Whether the bat confined its feeding to a single tree or multiple trees was recorded for each foraging bout. While visiting a particular tree, whether the bat was alone or in a group was also recorded. The numbers of individuals involved in group visitations and the time of these visitations were also noted. Visits were defined as the landing of a bat on the flower, whereas hovering of a bat as it approached an inflorescence (which often shook the long pedicel of *P. speciosa*), was not recorded as a visit if the bat did not land on the capitulum. Light from head lamps and cameras was reduced to minimize disturbance of foraging bats.

Pollen load examination

The number of stigmas with pollen and the number of pollen grains deposited on the stigmatic surface by the bat during a single visit were examined for *D. zibethinus* and *P. speciosa* flowers. A durian inflorescence comprises up to a 100 tightly packed flowers, only some of which open in a night so that only some of them touch the bat's body surface during a single visit and this may result in an underestimate of pollen load. Hence opened flowers were thinned to adjust the cluster size to 5-10 flowers. All the flowers were carefully emasculated during the afternoon (before anthesis). The stigmas were first covered by drinking straws to prevent pollen mixing while removing anthers. The anthers were then excised using scissors. The emasculated inflorescences were enclosed in a plastic cage to exclude flower visitors. The plastic cage was removed at night until a single visit by a bat had occurred and it was then replaced. Capitula of *P. speciosa* were difficult to emasculate due to their complex structure and hence we caged the capitula in the late afternoon before anthesis, and again removed these at night. An infrared closed-circuit television (CCTV) camera was placed at least a metre away from the flower and the live video was monitored on a screen positioned at the base of the tree. Immediately after a bat had landed on a target flower and departed, we enclosed the flower again with the plastic cage and the stigmas were collected the next morning. We collected all stigmas of bat-visited inflorescences of durian, and randomly sampled fifty stigma tips of *P. speciosa*. Transverse sections of the stigma surface were mounted on slides using gelatinous fuchsine dye and the numbers of conspecific and heterospecific pollen grains were counted.

Bat survey at flowering patches

Mist nets of various lengths (6, 9, 12 or 18 m) were stretched between c. 4 m poles under or adjacent to the flowering patches of *D. zibethinus* and *P. speciosa* trees. Bats were captured for 3-4 hours during the night for a total of 36 mist net hours at six durian orchards and 40 mist net hours at five *P. speciosa* patches. The captured bats were identified to species using Francis (2008). Canopy foraging bats (*Pteropus spp.*) were counted directly against the sky background.

Data Analysis

The duration of foraging bouts was compared between durian and *Parkia* food patches using Wilcoxon Mann Whitney U tests. Individuals were identified first by capital letter for gender (M = male, F = Female), followed by a three digits to indicate the frequency of the transmitter (KHz) used to tag the bat. Mean ± SD are used throughout.

RESULTS

Foraging bouts and movement pattern

Soon after sunset *E. spelaea* arrived at a foraging area and flew around for 5 to 10 min. They then flew to a night roost and rested for 2-3 min before commencing regular bouts of foraging. In each foraging bout, the bats visited a number of

TABLE I.   Summary of feeding behaviour of radio-tagged *E. spelaea* on *D. zibethinus* and *P. speciosa*. Bat ID denotes first letter of the bat's gender (M = male, F = Female), followed by three digits to indicate the frequency of the transmitter (KHz) used to tag the bat. The values are given as mean ± SD throughout with number of samples in brackets.

Bat ID	Major food patch	Foraging bout duration [min]	Multiple tree visits per bout [%]	Number of conspecific trees visited per bout	Percentage of lone visits to a tree by the bat	Visitation rate: trees per hr (range, *N*)	night roost distance in m from the feeding trees (range, *N* feeding trees)
M240	*D. zibethinus*	8 ± 3 (46)	85 (45)	not available	59 (22)	3 ± 4 (0-13, 12)	107 ± 56 (50-232, 21)
F280a	*D. zibethinus*	12 ± 12 (46)	100 (45)	not available	52 (25)	6 ± 4 (0-14, 8)	336 ± 106 (536-184, 20)
F280b	*D. zibethinus*	11 ± 9 (20)	91 (20)	not available	31 (16)	9 ± 8 (0-28, 12)	464 ± 541 (92-1666, 14)
F288	*D. zibethinus*	10 ± 4 (36)	92 (25)	not available	44 (16)	3 ± 3 (0-6, 10)	797 ± 679 (91-1681, 14)
M480	*D. zibethinus*	10 ± 8 (44)	74 (42)	not available	77 (22)	12 ± 13 (0-59, 29)	146 ± 59 (48-282, 20)
F419	*P. speciosa*	13 ± 9 (35)	79 (29)	4 ± 2 (4)	81 (21)	7 ± 6 (0-16, 19)	247 ± 103 (50-382, 14)
F599	*P. speciosa*	10 ± 10 (40)	90 (20)	3 ± 1 (6)	70 (23)	8 ± 3 (4-11, 10)	157 ± 45 (97-235, 14)
M518	*P. speciosa*	7 ± 7 (51)	93 (28)	3 ± 1 (6)	71 (17)	16 ± 13 (0-42, 10)	876 ± 736 (87-1985, 17)
M619a	*P. speciosa*	10 ± 10 (25)	81 (21)	3 ± 1 (8)	74 (19)	7 ± 3 (3-14, 16)	109 ± 77 (31-245, 16)
M619b	*P. speciosa*	12 ± 10 (25)	88 (25)	3 ± 1 (5)	72 (25)	9 ± 4 (4-19, 17)	160 ± 140 (45-452, 16)

feeding trees and returned to the night roost. These roosts were established in each foraging area on a non-food tree with a dense canopy. The average distance between the feeding trees and primary night roosts was 372 ± 439 m (range 31-1896 m, $N = 166$ from ten night roosts of ten bats). Additional night roosts were established when the feeding trees were more isolated and where the bat spent a few minutes before moving off to forage in another patch, finally returning to the primary night roost.

The duration of 367 foraging bouts was recorded for ten individuals, 192 bouts from five bats at *D. zibethinus* patches and 176 bouts from five bats at *P. speciosa* patches. The average foraging bout duration in durian patches was 10 ± 8 min (range 1-48 min) and 10 ± 9 min (range 1-59 min) in *P. speciosa* patches. The time allocated for foraging bouts by individual bats was not significantly different between *D. zibethinus* and *P. speciosa* patches (Wilcoxon Mann Whitney U test, W = 18029, P = 0.22). Foraging bouts tended to be longer as the night progressed.

Eonycteris spelaea visited multiple conspecific flowering trees for 86% of 123 recorded foraging bouts in *P. speciosa* patches and 88% of 179 recorded foraging bouts in *D. zibethinus* patches (Tab. I). The percentage of single tree visitation during foraging bouts was low (14% for *P. speciosa* and 12% for *D. zibethinus*). On average, *E. spelaea* visited 3 ± 1 *P. speciosa* trees (range 1-7) during each

foraging bout. Durian trees were planted a few metres apart in orchards and hence it was difficult to assess the number of trees actually visited by the bat in each bout. However, we noticed that the tagged individuals confined themselves to a few durian trees during a particular foraging bout. The trees selected for such visits changed from night to night apparently based on flower availability. When foraging on *P. speciosa*, the bats did not concentrate on particular trees during foraging bouts and most of the trees were visited sequentially. Heterospecific tree visitations were not seen when the bats were feeding on *Parkia* or *D. zibethinus* during their major flowering periods, but were observed when a few flowers remained, after peak flowering. In addition to *D. zibethinus* and *P. speciosa*, several tagged bats were observed feeding on *O. indicum* and *Musa* spp.

Tree visitation

When a bat approached a flowering tree, it circled or flew beneath the canopy, passing close to the inflorescence, often hovering at exposed inflorescences of *P. speciosa*. However, *E. spelaea* was never seen extracting nectar from a flower while hovering. When feeding, it landed on an inflorescence transiently, for 1 to 2 seconds, and flew away. The visits were often repeated at the same tree and at the same or at a different inflorescence or, alternatively, the bats switched to another food tree with no rest. The inter-tree movement while foraging was not predictable but the bats repeatedly used identical flight paths when approaching feeding trees.

While visiting trees, tagged individuals arrived alone at *D. zibethinus* for 64% of 101 observations and for 82% of 105 observations at *P. speciosa* trees (Tab. 1). Groups of up to seven individuals visited *D. zibethinus* trees and up to three individuals visited *P. speciosa* trees. Audible vocalizations of *E. spelaea* were common in durian patches but rare in *P. speciosa*. However, we did not see any agonistic behaviour, like chasing between the bats when they visited flowers.

Bats visited *D. zibethinus* tree 7 ± 10 times per hour (ranging from no visits to 59 visits, 5 bats observed for 71 hrs) and *P. speciosa* trees for 7 ± 6 times per hour (ranging between no visits to 42 repeated visits per hour, 5 bats observed for 72 hrs ,Tab. 1).

Pollen load experiment

A total of 118 durian flowers in 30 inflorescences from 18 different trees were collected after single visits by a bat. The *Durio* inflorescences comprised of 4 ± 1 flowers. Ninety three percent of bat- visited stigmas were loaded with pollen grains. One hundred and ten of 118 (93%) stigmas received conspecific pollen while only eight of 118 (7 %) received heterospecific pollen. Individual stigmas received 11 ± 18 conspecific pollen grains (range 1-27). Heterospecific pollen grains were observed in five of 30 inflorescences, mostly *Musa* spp. and a single case of *P. speciosa*. Each stigma with heterospecific pollen had 4 ± 4 (range between 1 and 12) grains of *Musa* pollen.

Twelve capitula of *P. speciosa* were collected after a visit by a bat. Fifty stigmas were randomly examined from each capitulum. In total, 600 stigmas were observed under a microscope to determine the number with a pollen load. Three hundred and thirteen stigma tips had received pollen grains, of which 301(~96% of pollen load) were conspecific and 12 (~4%) were heterospecific (identified as *Musa* spp.). Of the 50 stigmas tips examined from each capitulum, 25 ± 10 (range 12-40) were loaded with conspecific pollen comprising 50 ± 20% (range 24-80) of observed stigmas. Only two percent of capitula were found with heterospecific pollen grains.

Bat survey

A total of 149 bats were captured at durian orchards, and the majority were *E. spelaea* with 123 bats (83%), followed by *Rousettus* spp. (15 bats, 10%), *Macroglossus sobrinus* (9 bats, 6%) and the remaining 2% included one each of *Cynopterus* spp. and *Megaerops* spp. Two individual *Pteropus* were observed at one of the capture sites, foraging at the top of tall durian trees. In *P. speciosa* patches, 66 bats were captured, *E. spelaea* comprised 63 (95%), *M. sobrinus* 2 (3%) and *Cynopterus* spp. 1 (2%).

DISCUSSION

It is clear that *E. spelaea* typically visits multiple conspecific trees during most feeding bouts and feeds actively throughout the night. Additionally, each single visit

of nectarivorous bats, most of which are presumed to be *E. spelaea* based on the capture records, resulted in the deposition of conspecific pollen onto most of the stigmas of their food plants. Repeated visits to the feeding trees for several hours during the night can thus result in extensive pollen transfer between the trees. As a result, this foraging behaviour promotes cross pollination of these chiropterophilous plants. Since *E. spelaea* deposits quantities of conspecific pollen onto the stigma during every visit, we can thus claim that it is a legitimate and effective pollinator of these plants. This confirms and extends previous studies which reported that *E. spelaea* is the principal pollinator of the investigated chiropterophilous plants in mainland Southeast Asia (Bumrungsri et al. 2008, 2009; Srithongchuay et al. 2008). Durian fruit set in open pollination was as low as 0-1.4% in an orchard where bats were not seen foraging (Honsho et al. 2004). Acharya et al. (in press) found that a tagged bat moved between three isolated durian patches, eight kilometres apart when the durian was in flower, and when flowering ceased, confined itself to a single foraging area with *Parkia* and banana. The evidence of patch to patch movement by *E. spelaea* is thus crucial to the pollination of spatially separated chiropterophilus plants that are usually patchy in space and time in the human-modified landscape.

Eonycteris spelaea shows generally similar foraging behaviour in *Durio* and *Parkia* patches. The average number of foraging bouts, the inter-tree movement patterns and the tree visitation rates for *Durio* and *Parkia* feeders were almost identical. The size of foraging areas for *E. spelaea* at *Durio* and *Parkia* orchards was not significantly different (Acharya et al. in press). This identical foraging behaviour may be a compromise between energy expenditure of the bats and the nectar parcelling strategy of different plants. *Durio* trees produce abundant flowers but less nectar per flower than *Parkia* which produces fewer flowers but a higher nectar volume (Bumrungsri et al. 2008, 2009). *Parkia* inflorescences received more visits as the tree density and floral abundance is generally lower as compared to mass flowering and gregarious planting of durian trees in orchards.

The heterospecific pollen deposition on stigmas suggests that *E. spelaea* is a generalist nectarivore that visit different tree species in a night. Bats require a mixed diet to acquire the variety of nutrients they require, since these vary between different plants and plant parts (Courts 1998). Nectar is rich in sugar and water but poor in calcium, protein and lipids (Barclay 2002; Nelson 2003). In this study, *Musa* spp. and *O. indicum* were recorded as alternative food resources for *E. spelaea* in *Durio* orchards and *Parkia* patches. *Parkia* spp. and *Musa* spp are the major food resources for *E. spelaea* in the study area (Bumrungsri et al. 2013). Flowering times and positioning of anthers and stigma of these chiropterophilous plants limit outcrossing between heterospecific plants. For *O. indicum*, pollen is deposited on the dorsal surface of visiting bats (Srithongchuay et al. 2008), thus avoiding contamination with heterospecific plants. With such a strategy, those uncommon chiropterophilous plants can still benefit from cross pollination by *E. spelaea*.

Eonycteris spelaea shows strong fidelity to its foraging area and tagged bats regularly visit the same foraging areas for at least four months (Acharya et al. in press). The availability of other chiropterophilous plants such as *Musa* and *O. indicum* with aseasonal and steady-state flowering benefits nectarivorous bats by providing continuity of food supply throughout the year. The provision of alternative food resources for nectarivorous bats in fruit orchards is thus recommended, as it may encourage regular visits by bats, increasing the reliability of the pollination service for chiropterophilous fruit crops. Since the effective pollination period (i.e. the duration that flowers are still able to set fruit) of these plants is typically short, lasting only one night (Honsho et al. 2007), the reliability of pollinators is thus crucial for fruit set.

In conclusion, *Eonycteris spelaea* is a legitimate and effective pollinator of both *Durio* and *Parkia* in an agricultural landscape. Its high mobility means that it is capable of pollinating crops in a patchy agricultural landscape as well as those native plants in fragmented forests distant from the roost cave. Hence local populations of *E. spelaea* are crucial for natural pollination of chiropterophilous plants across the landscape. Protection of local colonies of *E. spelaea* and maintaining local chiropterophilous plant populations in the ecosystem is vital for pollination reliability and contributes to sustained production of fruit crops. Cave tourism, quarrying and bushmeat consumption are common threats to the cave bats in Southeast Asia and elsewhere (Mickleburgh et al. 2009) and should be controlled. Public education about the ecosystem service provided by bats should be undertaken especially in areas adjacent to bat colonies.

REFERENCES

Acharya PR, Racey PA, Sotthibandhu S, Bumrungsri S (in press) Home-range and foraging areas of the dawn bat *Eonycteris spelaea* in agricultural areas of Thailand. Acta Chiropterologica.

Acharya PR, Racey PA, McNeil D, Sotthibandhu S, Bumrungsri S (2015) Timing of cave emergence and return in the dawn bat (*Eonycteris spelaea*, Chiroptera: Pteropodidae) in Southern Thailand. Mammal Study. The Mammal Society of Japan 40:47-52.

Amelon SK, Dalton DC, Millspaugh JJ, Wolf SA (2009) Radiotelemetry: Techniques and Analysis. In: Kunz, T.H., S. Parsons. Editors. Ecological and Behavioural methods for the study of bats. Baltimore: John Hopkins University Press.

Barclay RM (2002) Do plants pollinated by flying fox bats (Megachiroptera, Pteropodidae) provide an extra calcium reward in their nectar? Biotropica 34:168-171.

Bates PJJ, Harrison DL (1997) Bats of the Indian subcontinent. Harrison Zoological Museum, Sevenoaks, Kent, England 258 pp.

Brown MJ (1997) *Durio* – a bibiliographic review. International Plant Genetic Resource Institute, New Delhi.

Bumrungsri S, Harbit A, Benzie C, Sridith K, Racey PA (2008) The pollination ecology of two species of *Parkia* (Mimosaceae) in southern Thailand. Journal of Tropical Ecology 24:467-475

Bumrungsri S, Sripaoraya E, Chongsiri T, Siridith K, Racey PA (2009) The pollination ecology of durian (*Durio zibethinus*, Bombacaceae) in southern Thailand. Journal of Tropical Ecology 25:85-92.

Bumrungsri S, Duncan L, Colin H, Sripaoraya E, Kitpipat K, Racey PA (2013) The dawn bat, *Eonycteris spelaea* Dobson (Chiroptera: Pteropodidae) feeds mainly on pollen of economically important food plant in Thailand. Acta Chiropterologica 15:95-104.

Corlett RT (2004) Flower visitors and pollination in the Oriental (Indo-Malayan) Region. Biological Reviews 79:497-532.

Courts SE (1998) Dietary strategies of Old World fruit bats (Megachiroptera, Pteropodidae): how do they obtain sufficient protein? Mammal Review 28:185-194.

Dobat K, Peikrt-Holle T (1985) Blüten und Fledermäuse. Bestäubung durch Fledermäuse und Flughunde (Chiropterophilie). Senckenbergische Naturforschende Gesellschaft, Frankfurt am Main, Verlag Waldemar Kramer.

Faegri K, Van Der Pijl L (1979) The principles of pollination ecology. Pergamon Press, Oxford. 242 pp.

Fenster CB (1991) Gene flow in *Chamaecrista fasciculata* (Leguminosae). I. Gene dispersal Evolution 45:398-409.

Fleming TH, Muchhala N (2008) Nectar-feeding bird and bat niches in two worlds: pantropical comparison of vertebrate pollination systems. Journal of Biogeography 35:764-780.

Fleming TH, Sosa VJ (1994) Effects of nectarivorous and frugivorous mammals on reproductive success of plants. Journal of Mammalogy 75:845-851.

Fleming TH, Geiselman C, Kress WJ (2009) The evolution of bat pollination: a phylogenetic perspective. Annals of Botany 104:1017-1043.

Francis CM (2008) A field guide to the mammals of South-East Asia. New Holland Publishers (UK) Ltd. and Asia Books Co., Ltd, 392 pp.

Fujita MS, Tuttle MD (1991) Flying foxes (Chiroptera: Pteropodidae): threatened animals of key ecological and economic importance. Conservation Biology 5:455-463.

Honsho C, Yonemori K, Somsri S, Subhadrabandhu S, Sugiura A (2004) Marked improvement of fruit set in Thai durian by artificial cross-pollination. Scientia Horticulturae 101:399-406.

Honsho C, Somsri S, Tetsumura T, Yamashita K, Yonemori K (2007) Effective pollination period in durian (*Durio zibethinus* Murr.) and the factors regulating it. Scientia Horticulture 111:193-196.

Hopkins HCF (1984) Floral biology and pollination ecology of the Neotropical species of *Parkia*. Journal of Ecology 72:1-23.

Kingston T (2010) Research priorities for bat conservation in Southeast Asia: a consensus approach. Biodiversity and Conservation 19:471-484.

Kunz TH, Barun De Torrez E, Bauer D, Lobova T, Fleming TH (2011) Ecosystem services provided by bats. Annals of the New York Academy of Sciences 1223:1-38.

Law BS, Lean M (1999) Common blossom bats (*Syconycteris australis*) as pollinators in fragmented Australian tropical rainforest. Biological Conservation 91:201-212.

Lim TK, Luders L (1997) Boosting durian productivity report for RIRDC Project DNT – 13A. Canbera, Australia, Rural Industries Research and Development Corporation.

Mayfield MM, Waser NM, Price MV (2001) Exploring the 'most effective pollinator principle' with complex flowers: bumblebees and *Ipomopsis aggregata*. Annals of Botany 88:59-596.

Mickleburgh SP, Hutson AM, Racey PA (1992) Old World fruit bats. An action plan for their conservation. Gland, Switzerland, IUCN.

Mickleburgh S, Waylen K, Racey PA (2009) Bats as bushmeat: a global review. Oryx 43:217-234.

Molina-Freaner F, Eguiarte LE (2003) The pollination biology of two paniculate agaves (Agavaceae) from northwestern Mexico: contrasting roles of bats as pollinators. American Journal of Botany 90:1016-1024.

Nelson SL (2003) Nutritional ecology of old-world fruit bats: a test of the calcium constraint hypothesis. Doctoral dissertation, University of Florida. [online] URL: http://www.botany. hawaii.edu/basch/uhnpscesu/pdfs/sam/Nelson2003AS.pdf (accessed Feb 2015)

Petchmunee K (2008) Economic valuation and learning process construction: a case study of the cave nectarivorous bat (*Eonycteris spelaea* Dobson). M.Sc. Thesis (Environmental Science), Prince of Songkla University, Thailand (In Thai with English Abstract). [online] URL: http://doc2.clib.psu.ac.th/public13/thises/312072.pdf (accessed Feb 2015)

Primack RB, Silander J (1975) Measuring the relative importance of different pollinators to plants. Nature 255:143-144.

Schemske DW, Horvitz CC (1984) Variation among floral visitors pollination ability: a precondition for mutualism specialization. Science 225:519-521.

Soepadmo E, Eow BK (1976) The reproductive biology of *Durio zibethinus* Murr. Gardens Bulletin, Singapore 29:25-33.

Spears EE (1983) A direct measure of pollinator effectiveness. Oecologia 57:196-199.

Sripaoraya E (2005) The relationship between nectar secretion rates and visits by the cave nectarivorous bats (*Eonycteris spelaea* Dobson). MSc. Thesis, Prince of Songkla University, Thailand, 48 pp (In Thai with English Abstract).

Srithongchuay T, Bumrungsri S, Sripao-raya S (2008). The pollination ecology of the late-successional tree, *Oroxylum indicum* (Bignoniaceae) in Thailand. Journal of Tropical Ecology 24:477-484.

Start AN (1974) The feeding biology in relation to food sources of nectarivorous bats (Chiroptera: Macroglossinae) in Malaysia, PhD Thesis. University of Aberdeen, Scotland.

Stebbins GL (1970) Adaptive radiation of reproductive characteristics in angiosperms: I. Pollination mechanisms. Annual Reviews of Ecology and Systematics I:307-326.

Subhadrabandhu JMP, Verheij EWM (1991) *Durio zibethinus* Murray. In Coronel EWMRE (ed), PROSEA: Plant Resources of South-East Asia. No. 2. Edible Fruits and Nuts. Wageningen, 157-161.

Valmayor RV, Coronel RE, Ramirez DA (1965) Studies on the floral biology fruit set and fruit development in durian. Philippine Agric 48:355-366.

White GC, Garrott RA (1990) Analysis of wildlife tracking data. Academic Press, San Diego, California, USA. 373 pp.

Yetman D (2007) The great Cacti Ethnobotany and Biogeography. (Tuscon, AZ: The University of Arizona Press) pp. 218-9.

5

POLLINATION ECOLOGY AND FLORAL VISITOR SPECTRUM OF TURTLEHEAD (*CHELONE GLABRA* L.; PLANTAGINACEAE)

Leif L. Richardson[1,2*], and Rebecca E. Irwin[1,3]

[1]*Department of Biological Sciences, Dartmouth College, Hanover NH, 03755 USA*
[2]*Present address: Gund Institute for Ecological Economics, University of Vermont, Burlington, VT 05405 USA*
[3]*Present address: Department of Applied Ecology, North Carolina State University, Raleigh, NC 27695 USA*

Abstract—Many flowering plants engage in mutualistic interactions with animals in order to sexually reproduce, exchanging food rewards such as nectar and pollen for the service of pollen transfer between flowers. Floral reward variation strongly influences visitation patterns of both pollinating mutualists and non-mutualist consumers, with consequences for both male and female components of plant reproductive success. Despite the importance of pollination to ecological systems, the pollination ecology of many plants is poorly known. At seven sites over three years, we studied the mating system, floral visitors and pollen limitation of turtlehead (*Chelone glabra* L.), an eastern North America wetland herb. We found that the plant is autogamous, but requires pollinator visitation to set seed. *C. glabra* flowers are protandrous, with floral rewards that vary between male and female sex phases. We found diurnal variation in reward presentation that was a function of both floral phenology and consumer behaviour. *Bombus vagans* Smith, the most common visitor to *C. glabra* flowers, removed a large fraction of available pollen (> 36%) in single visits to newly opened flowers, and compared to other flower visitors, passively transported more pollen on flights between flowers and deposited more to conspecific stigmas, suggesting it was the most effective pollinator. The solitary bee *Hylaeus annulatus* L. made frequent visits to flowers, but contributed little to pollination due to morphological mismatch and because it avoided male-phase flowers. Despite high bee visitation rates, flowers were pollen limited for seed production, possibly indicating a negative effect of non-pollinating flower visitors on plant reproductive success.

Keywords: Nectar, Pollen, Pollen thievery, Pollen limitation, Protandry, Buzz pollination

INTRODUCTION

Evolutionary ecologists are fundamentally interested in factors that govern the abundance, distribution, and evolution of species. Species interactions such as mutualism, predation and competition are ubiquitous in nature, and partner traits often contribute to outcomes of these interactions, with consequences for species distribution, evolution (Thompson 1999), reproduction and ecosystem services provisioning (Garibaldi et al. 2015). For angiosperms, interactions with pollinators are critically important to fitness, with the majority of species benefiting from transfer of pollen within and between flowers by animal pollinators (Ollerton et al. 2011). Deficits in pollinator visitation can limit host plant reproduction (Ashman et al. 2004; Knight et al. 2005) and structure plant population dynamics (Biesmeijer et al. 2006; Lundgren et al. 2015), and interactions between plants and pollinators provide some of the best known examples of evolution and co-evolution by natural selection (Fenster et al. 2004). Nectar and pollen rewards and spatiotemporal variation in their presentation can have strong effects on the community of pollinators visiting flowers (Pleasants 1983; Thomson et al. 2000), and that pollinator community can host species that range from effective pollinators to those that transfer little pollen among flowers and plants (Hargreaves et al. 2012). Despite the wealth of knowledge on the ecology and evolution of plant-pollinator interactions, there are many flowering plant species for which the floral ecology and floral visitor spectrum remain understudied. Nonetheless, natural history studies of a plant's pollination biology can provide key insights into the importance of pollination mutualisms for a species' ecology and conservation.

The purpose of this study was to investigate the mating system and floral visitor spectrum of a protandrous plant reported to be pollinator dependent, *Chelone glabra* L. (Plantaginaceae; hereafter *Chelone*). We chose to study the pollination ecology of *Chelone* because it is a common flowering plant in wetlands in eastern North America and its floral rewards may be important in maintaining bee health (Richardson, Adler, et al. 2015). One prior study reported that *Chelone* is self-compatible and dependent on pollination by bumble bees, but presented only qualitative observational data (Cooperrider 1967). Our goal was to provide quantitative insight into the mating system and floral ecology of *Chelone* to put questions about the plant's floral rewards and floral visitors into a relevant natural history context. We addressed four questions: 1) How do nectar and pollen reward presentation change from male to female sex

*Corresponding author: leif.richardson@uvm.edu

phases, and what is the diurnal impact of insect foraging on nectar and pollen availability? 2) Who are the primary floral visitors to *Chelone* flowers, and what is the relative effectiveness of each flower visitor at transferring pollen from anthers to stigmas? 3) Can nectar and pollen foragers distinguish between male- and female-phase flowers, and does distinguishing among these flower types affect their likelihood of transferring pollen? 4) Are plants pollen limited for fruit and seed production? We predicted that, similar to other protandrous plants, pollen would function as the principal attractant for pollinators during male phase (Bertin & Newman 1993), allowing for pollen collecting visitors that specialize on male-phase flowers and thus may not transfer pollen to female-phase stigmas. We predicted that nectar and pollen harvesting bumble bees would act as pollinators because of an apparent morphological fit between worker caste bees and floral architecture (Cooperrider 1967), but that smaller non-bumble bee visitors would act as pollen thieves (Inouye 1980), consuming pollen but potentially missing contact with stigmas while doing so. Finally, given the potential for non-mutualist pollen removal by bees, we predicted that *Chelone* seed production would be pollen limited, with pollen-augmented flowers producing more seeds than those experiencing natural floral visitation rates (Ashman et al. 2004). Taken together, this research provides a comprehensive analysis of the floral ecology and floral visitor spectrum of a pollinator-dependent plant.

MATERIALS AND METHODS

Study system

Chelone glabra is a perennial herb native to eastern North America, occurring in circumneutral seepage swamps, marshes and anthropogenic wetlands (Gleason & Cronquist 1991; Nelson 2012). It commonly forms dense, near monotypic stands in open marshes (range: approx. 1-50 stems per m²). Despite clonal vegetative reproduction, *Chelone* genets can often be distinguished because of spatial separation between plants. Individual plants produce numerous stems (range: 1-20+ stems per plant), most terminating in a single racemose inflorescence. Inflorescences initiate 14.4 ± 0.9 SE (range: 6-28) flower buds, of which only 10.6 ± 1.0 SE open in a given flowering season (the others failing to complete development). The sympetalous, white- to rose-colored flowers average 3.0 ± 0.2 SE cm in length (range: 2.6.5-3.5 cm), with 47% of the length comprised of a prominent upper hood and lower standard petal. Flowers are zygomorphic, and the distal margins of upper and lower petal segments typically touch, creating a constricted flower entrance. Each day, 2.1 ± 0.1 SE flowers per stem are open simultaneously (range: 1-6 flowers). Flowers each have four anthers recessed within the hooded upper corolla lip in pairs that are nearly confluent until forced apart by large bees. Pollen sacs dehisce via longitudinal slits and are covered with a dense layer of woolly hairs (Straw 1966). A short staminode is also present. *Chelone* flowers are strongly protandrous. While in male phase (approx. 1 d), flowers have a short, recessed style and anthers that dehisce pollen, often onto the dorsum of foraging bees. While in female phase (approx. 2 d), the style

is elongated and recurved approx. 180°, placing the stigmatic surface near the corolla entrance and distal to anthers (L. L. Richardson, pers. obs.). Nectar is secreted by a hypogynous disk of nectary tissue, and production begins at anthesis (Straw 1966). *Chelone* is reported to be self-compatible, but because flowers are protandrous, pollinators are required to carry self-pollen between anthers and stigmas of flowers on the same plant (Cooperrider 1967). Bumble bees are reported to be the primary visitors to *Chelone* flowers (Pennell 1935; Cooperrider 1967; Heinrich 1975; Williams et al. 2014), but the degree to which they are pollinators and whether other bees contribute to *Chelone* pollination or act as pollen thieves is unknown. Bumble bees forage for both nectar and pollen, often sonicating or 'buzzing' *Chelone* flowers to shake pollen from the anthers (Heinrich 2004).

Chelone leaves and other vegetative tissues contain two secondary metabolites that deter generalist herbivores, the iridoid glycosides aucubin and catalpol (Bowers et al. 1993). However, numerous specialized herbivores use the plant as a host, including foliar herbivores (e.g., *Euphydryas phaeton* Drury (Nymphalidae: Lepidoptera) and *Tenthredo grandis* Norton (Tenthredinidae: Hymenoptera) (Bowers et al. 1993)) and predispersal seed predators (e.g., *Phytomyza cheloniae* Spencer (Agromyzidae: Diptera) and *Endothenia hebesana* Walker (Tortricidae: Lepidoptera) (Stamp 1987)). Seed predators are reported to attack nearly a quarter of *Chelone* fruits (Stamp 1987) and thus may influence the degree to which pollination translate into successful seed-bearing fruits. Depending on timing of attack, fates of fruits hosting seed predators include partial or complete consumption of matured seeds, or abortion before seeds mature (L. L. Richardson, pers. obs.).

Field methods

From 2011-2013, we studied the reproductive biology of *Chelone* at seven plant populations spread over 75 km in northern Vermont, USA (Appendix 1).

Phenology

To study individual flower phenology, in 2011 at two populations we followed flowers from anthesis to when corollas dropped, recording floral sex phase three times daily. At each site, we followed 2-5 flowers on each of 8 plants for a total of 33 flowers. To study population flowering phenology, in 2011 and 2012 we randomly selected 20 focal inflorescences ≥ 10 m apart along a linear transect through populations (two populations in 2011; one population in 2012). We censused inflorescences every 2-3 days, recording total number of flower buds, open flowers of each sex phase and developing fruits. We also noted the presence of herbivores and seed predators.

Mating system

We studied the mating system of *Chelone* in 2011. We randomly chose 20 plants at each of two populations, and on each plant assigned inflorescences to one of three treatments: 'outcrossing', 'selfing', and 'unpollinated'. We covered all inflorescences with bags made of wedding veil before flowering began. We visited populations every 2-3 days during the flowering season to apply treatments. For the outcrossing treatment, we collected fresh pollen from ≥ 5 plants in the population by sonication with an electric toothbrush and mixed it in a small Petri dish. We removed bags and applied this mixture with a clean pinhead to the stigmas of all open flowers before replacing the bag. We verified by microscopy that this method resulted in pollen grains sticking to stigmas (data not shown). For the selfing treatment, we removed bags, collected pollen by sonication from all open flowers on the same inflorescence, then re-applied this pollen to stigmas of the same flowers with a pinhead cleaned in ethanol. Because we visited the inflorescences every 2-3 days, we did not apply treatments to every flower on the inflorescence. Thus, each time we applied the outcrossing and selfing treatments, we marked the sepals of treated flowers with a black marker (Sharpie, Illinois, USA) so that fruits resulting from our treatments could be identified; we manipulated, on average, 52% of the flowers on inflorescences. Flowers of the unpollinated treatment did not receive pollen, but each time we applied the other treatments, we handled these inflorescences similarly, removing and replacing the bags and marking sepals of open flowers. We removed bags after flowering had ceased and collected infructescences when they were nearly mature. We later dissected infructescences, counting total number of fruits matured and number of seeds in each fruit. We also noted presence of predispersal seed predators in these fruits, but did not analyse whether pollination treatments affected attack rates because placing pollinator exclusion bags over inflorescences may have biased seed predator oviposition.

We conducted all statistical analyses (here and below) using JMP (version 11.2; SAS Institute, Inc. 2014) and R statistical software (R Core Team 2015) and when appropriate compared AIC scores among candidate models to select statistical models that best fit the data. We used linear mixed models ('lme4' library for R statistical software; Bates et al. 2014; R Core Team 2015) to analyse seed number per fruit, comparing Akaike Information Criterion (AIC) scores sequentially to select random and fixed effects for a best-fit model (Bolker et al. 2009). We used analysis of variance to compare the best-fit model with reduced models, then calculated chi-square statistics and significance values for the influence of fixed effects. The full model included log-transformed seed number per fruit as the response variable, pollination treatment and presence of pre-dispersal seed predators as fixed effects, and plant individual and population and individual nested within population as random effects. We used Tukey HSD post-hoc tests to identify statistically significant differences among pollination treatments. We excluded from this analysis fruits where pre-dispersal seed predators made seed counts impossible, but included attacked fruits where accurate counts were possible. Due to high rates of pre-dispersal seed predator attack (36% of fruits collected) and fruit abortion, we were unable to accurately calculate per cent of fruits maturing seed as a function of pollination treatments.

Flower visitors

From 2011-2013, we made collections of *Chelone* floral visitors at seven populations throughout the flowering season. In timed collections, we randomly moved through patches of flowering *Chelone*, collecting by net as many foraging insects as we could in a 30-minute period. Collection efforts occurred throughout daylight hours on days when bees were foraging, and we also made 3 hrs of collections between 2100-2400 hrs to look for nocturnal visitors. We made additional haphazard collections of flower-visiting insects not observed during standardized collecting events, and made observations of bee visits to flowers, noting nectar and pollen collecting behaviour and making sound recordings with a digital voice recorder (Sony, USA) to document any potential bee sonication of flowers. We pinned and identified all collected insects, except that the genus *Lasioglossum* was identified only to morpho-species and flies, sawflies and wasps were identified only to Order (Mitchell 1960, 1962; Michener et al. 1994; Droege et al. 2014; Williams et al. 2014). We used basic summary statistics to compare flower visit frequencies among insects we collected at flowers. We calculated the proportion of collecting events during which we collected each species and the relative abundance of each in the overall collection.

To assess flower visit frequency, at two populations in 2011 we made nine timed observations of bee visits to inflorescences when flower visitors were active. We watched 24-122 flowers at a time, recording every bee visit to flowers during a 60-90 minute period of time. In this work we recorded visits by *Bombus vagans* (Apidae), *Hylaeus annulatus* (Colletidae), *Lasioglossum* species (Halictidae), and several Lepidoptera and Diptera. We lump the *Lasioglossum* into a single taxon for analysis (here and below) because they could not be identified on the wing. To compare flower visit duration among visitor species, we watched individual bees as they foraged on *Chelone*, recording transitions between flowers and flower visit duration with a FileMaker Go database on an Ipad tablet computer (Apple, Inc; Filemaker Pro 12.0).

To assess whether bees preferentially visited male- or female-phase flowers, on three dates (August 5, 6, and 7, 2011) we examined each flower in a randomly selected group ($N = 77$, 80 and 122 flowers) and marked the sex phase on the corolla with black ink. The proportion of flowers in male phase was 0.13, 0.59 and 0.37, respectively. For one hour we observed bee visits to these patches of flowers, recording bee species and sex phase of each visited flower. We calculated proportion of male- and female-phase flowers visited by each bee and compared this to expected proportions if bees foraged randomly in the patch. We recorded visits by *B. vagans*, *H. annulatus*, and *Lasioglossum* species.

Nectar

We studied nectar volume and sugar concentration in two *Chelone* populations in 2011-2013. To assess nectar available to freely foraging floral visitors (nectar standing crop), we randomly selected inflorescences each from different plants and collected nectar from all open flowers (1-6 flowers on each of 105 plants, $N = 201$ total flowers) with capillary micropipettes (5 μL size; Drummond Scientific, Broomall, Pennsylvania, USA). Because nectar accumulates in a constricted area of the corolla base, we had to sample flowers destructively, but we were careful not to introduce phloem sap into samples. We used a refractometer (National Industrial Supply, Temecula, CA, USA) to measure sucrose-equivalent sugars, expressed as % Brix, and converted volume and concentration to calories (net energy expressed as kilocalories) present in each flower (Bolten et al. 1979). We recorded time of day, individual plant identity and flower sex phase. We log-transformed nectar volume and energy to meet assumptions of parametric statistics, and used ANCOVA to analyse nectar volume, sugar concentration and energy, considering in full models as fixed effects time of day (simplified as two categories: 'morning', 0900-1100hrs, and 'afternoon', 1300-1600hrs), date of collection, and flower sex and as random effects plant individual, population and individual nested within population.

We also studied correlation of floral morphology with nectar traits for a portion of those flowers (1-5 flowers on each of 75 plants, $N = 159$ total flowers) in 2013. These measurements were taken in the morning and afternoon across the two *Chelone* populations. In combination with the nectar traits, we measured corolla length (from the base of the calyx to corolla opening), lower petal length (from the corolla opening to distal tip) and maximum corolla width (i.e., horizontal distance across the corolla opening) with digital callipers to the nearest 0.01 mm. We then used a multivariate analysis to test for correlations between floral morphology and nectar traits, splitting the dataset by morning and afternoon collections and excluding seven multivariate outliers identified by Mahalonobis distances. We combined the data across the two *Chelone* populations for correlation analysis; analyses within populations showed similar qualitative patterns (data not shown).

Pollen

We studied pollen production and removal patterns by different floral visitors in 2011-2012 at three populations. We placed wedding veil bags over expanded flower buds to exclude visitors. We returned to plants 24 hr later to remove bags from newly opened, unvisited flowers. We collected anther sacs from these flowers after application of three types of treatments: single visits from pollen- and nectar-foraging bees ($N = 138$ flowers); multiple visits from bees over time periods of varying lengths (1-8 hours, $N = 123$ flowers); and unvisited controls ($N = 135$ flowers). We did not record plant identity, but most samples came from different individuals. After single visits we made field identifications of bees to the lowest taxonomic level possible and recorded behavioural observations, including whether bees had audibly sonicated the flower. We also assessed pollen available to foraging bees by collecting open, unmanipulated flowers ($N = 60$ flowers; hereafter "open" flowers).

We excised anthers from flowers with forceps and allowed them to air dry for two weeks in open Eppendorf tubes. We added 1,500 µL of 70% EtOH to dry samples and sonicated them in a water bath for 60 minutes. We then homogenized samples by vortexing and removed 225 µL to a clean container, which we diluted to 1,500 µL with additional EtOH. After vortexing, we removed 3 µL aliquots to a haemocytometer slide and counted all pollen grains at 10× magnification under a dissecting microscope. We made four counts of each sample, computed an average and

multiplied to obtain an estimate of total pollen grains present per flower.

We used linear models to compare log-transformed pollen counts from flowers at three sites between unvisited flowers, flowers visited once by foraging insects, and flowers open to insect visitation. We made pollen collections after single visits from 5 bee species; we present a statistical comparison of the two most common visitors (*H. annulatus* and *B. vagans*), and qualitatively summarize results from a smaller number of replicates from the other three species. We tested a full model that included treatment (unvisited, open and single visit flowers) as a fixed effect, and date and plant population as random effects. We used Tukey HSD post-hoc tests to compare means among treatments and among plant populations. To analyse depletion of pollen over time from male-phase flowers, we regressed anther pollen counts against time since anthesis, and compared linear and non-linear models to describe the data.

To investigate pollen transport by floral visitors, at three populations in 2011 we collected free-foraging bees into clean, cyanide kill jars as they left *Chelone* flowers, making note of whether they sonicated the last flower they visited. We immediately pulled bees from kill jars with forceps, removed and discarded hind legs and associated pollen loads (Michener 2000), and rubbed their dorsal sides on a microscope slide with approx. 1 cm² of fuchsin gel (Kearns & Inouye 1993) for 10 seconds. We added a cover slip and heated the slide until the gel melted, staining and fixing the pollen. We used a compound microscope at 40× magnification to make 5 pollen counts of randomly selected fields on each slide, distinguishing conspecific vs. heterospecific pollen using a pollen reference library. Pollen was dispersed approx. uniformly across the slides, so we then calculated mean number of *Chelone* and heterospecific pollen grains per microscope view as an index of the full sample present on the slide. We used ANOVA to compare pollen transport by bee species.

We also studied a component of female plant reproduction, pollen deposition to stigmas by different floral visitors, at two populations in 2011. We allowed bees to make single visits to virgin flowers (previously bagged in bud stage), noted whether bees sonicated the flowers, and then collected stigmas with forceps. We also collected stigmas from unvisited flowers as controls for pollen deposited by wind or experimental error. We mounted stigmas in fuchsin gel on microscope slides (Kearns & Inouye 1993), and counted *Chelone* and heterospecific pollen grains with a compound microscope. We present data here for unvisited controls ($N = 19$ flowers) and the most common visitor species, *B. vagans* ($N = 26$ flowers). We used ANOVA to assess whether *B. vagans* was an effective floral visitor, comparing pollen receipt by stigmas of flowers visited by *B. vagans* to pollen grains present on stigmas of unvisited control flowers.

Pollen limitation

We studied pollen limitation of plant reproduction at each of two and five populations in 2011 and 2012, respectively. Because *Chelone* may grow as a densely aggregated clonal plant with many stems, we could not apply treatments at the whole-plant level (Ashman et al. 2004); we instead identified pairs of inflorescences ($N = 20$ pairs per population in each year) we could confirm were the same randomly selected genet, and applied one of two treatments (pollen addition or open control) to each. Every 2-3 days, we collected pollen from ≥ 5 donor plants (as in the *Mating system* methods), then applied it to stigmas of all open flowers of inflorescences in the pollen addition treatment, marking the sepals of each flower we treated with black ink. We paired this with an open control treatment in which we handled and marked open flowers, but did not add any pollen. By visiting inflorescences every 2-3 days, we treated on average 52% of the flowers on inflorescences. Both treatments were open to natural floral visitation. We later collected infructescences, dissecting those fruits we had treated and marked to count mature seeds and assess seed predator damage.

We used a linear mixed model to analyse seed number per fruit. The full model accounted for our paired sampling design by including as random effects individual plant genet (i.e., from which a pair of inflorescences was included in the experiment) and individual nested within plant population. Fixed effects in the model included pollen addition treatment (pollen addition vs. control), plant population and their interaction. We included population as a fixed effect in this analysis because we were interested in asking how these particular populations responded to pollen supplementation and whether the magnitude of pollen limitation varied among them. Due to high rates of predispersal seed predator attack (57-75%; see Results), we were unable to calculate per cent of fruits maturing seed. However, we used a generalized linear mixed model to study whether the frequency of predispersal seed predator attack was dependent on whether we added supplementary pollen to stigmas. We investigated a full model that included as response variable whether a fruit contained evidence of predispersal seed predator attack (presence/absence of frass, larvae or pupae), as a fixed effect pollen addition treatment, and as random effects plant individual and population. To compare pollen limitation of *Chelone* to that reported for other plants, we calculated the Hedges' *g* effect size of the difference between pollen addition and control groups (Gurevitch et al. 2001; Knight, Steets, et al. 2005). An effect size of 0.2 would be considered a small effect of pollen supplementation, 0.5 medium, and >0.8 large (Cohen 1988).

RESULTS

Phenology

The *Chelone* flowering period was 7 July-20 September and averaged approx. 66 days, with a peak of flowering around 5 August across two populations in 2011 and 2012. Individual flowers were open (mean + 1 SD) 3.00 ± 0.70 days, with flowers functionally male during the first day, and in female phase thereafter. We found that anthesis could take place at any time of day, but that most flowers opened at night when pollinators were not active.

Mating system

We found a significant effect of pollination treatment on seed set per fruit ($\chi^2 = 15.57$, $P = 0.0004$; Fig. 1). Post-hoc tests revealed that relative to controls, seed set was increased by 1.5 times by addition of self pollen ($Z = 4.03$, $P = 0.0002$) and 1.4 times by addition of outcross pollen ($Z = 2.90$, $P = 0.01$), but selfing and outcrossing treatments were not significantly different from each other ($Z = 1.27$, $P = 0.41$; Fig. 1).

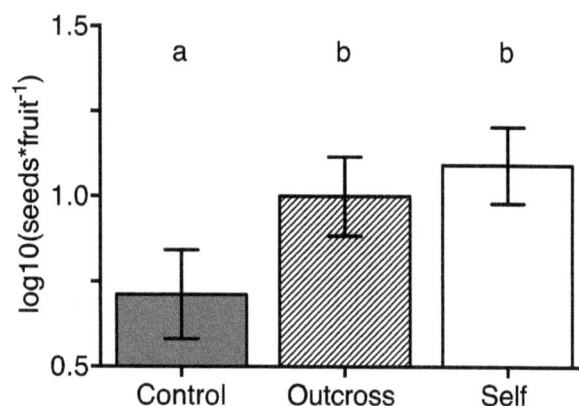

FIGURE 1. Log-transformed numbers of seeds produced per fruit in unpollinated control flowers, flowers that were outcrossed with pollen from other individuals, and flowers that received only self pollen. Data presented are means ± SE, and means that are significantly different are marked with different lower case letters.

Flower visitors

During 14 hours of daytime netting, we collected 18 species of solitary and social bees (Hymenoptera: Apoidea), and small numbers of flies (Diptera), sawflies (Hymenoptera) and wasps (Hymenoptera) foraging for nectar and/ or pollen at *Chelone* flowers (Tab. 1). We made a total of 3 hours of observations of flowers at night (2100-2400 hrs), but did not observe any nectar or pollen collecting visitors. Worker caste bumble bees were the most common visitors (75.4% of visits), and two species, *Bombus impatiens* and *B. vagans*, accounted for 71.6% of all collections. Bumble bees were observed to collect nectar, pollen or both resources from flowers, often audibly sonicating anthers while inside flowers. *B. vagans* workers quickly entered flowers after landing and crawled to the nectaries, making contact with both anthers and stigmas, but *B. impatiens* and other bumble bee species had difficulty forcing their way through the constricted floral entrance, typically inserting only their heads and forelegs, and making less contact with sexual parts of flowers. A variety of solitary bee species, most commonly *Hylaeus annulatus* females (6.8% of all visits), also collected nectar and pollen from flowers, but due to their smaller size, they frequently did not contact stigmas while in flowers. Sound recordings revealed that *H. annulatus* commonly sonicated anthers to release pollen. The three most common foragers varied significantly in length of their flower visits (*B. vagans* < *B. impatiens* < *H. annulatus*; Welch's test: $F_{2,59.8} = 3.68$, $P = 0.03$). *H. annulatus* flower visits were 1.7 times longer (mean = 7.1 seconds; Tukey test: $P = 0.05$) than those of *B. vagans*

Visitor	No. Collections	Fraction of collections present	Relative abundance
Apis mellifera	I	3.6	0.3
Augochlorella aurata	2	7.I	0.5
Bombus bimaculatus	I	3.6	0.3
B. borealis	2	7.I	0.5
B. fervidus	⁎		
B. griseocollis	⁎		
B. impatiens	9	28.6	2.4
B. ternarius	2	7.I	0.5
B. terricola	⁎		
B. vagans	265	100.0	71.6
Halictus rubicundus	I	3.6	0.3
Hylaeus annulatus	25	53.6	6.8
Lasioglossum (4 morphospecies)	4I	46.4	II.I
Megachile gemula	2	7.I	0.5
M. inermis	I	3.6	0.3
Sawfly sp.	I	3.6	0.3
Wasp sp.	2	7.I	0.5
Diptera spp.	I5	25.0	4.I

TABLE I.   A total of 370 insect specimens, including bees, other hymenoptera and flies, were collected during 28 30-minute observations at *Chelone glabra* flowers from 2011-2012 in seven sites. Asterisks indicate insect species collected outside of standardized collecting events.

(mean = 4.2 seconds), and other comparisons were not significantly different. We recorded pollen collection via sonication by both bumble bees and *H. annulatus*, and noted that for the latter, buzzing was often too quiet to hear without amplification. *B. vagans* workers demonstrated behavioural flexibility, often switching between sonication and passive pollen collection behaviours as they moved among plants. We also observed that *B. vagans* workers readily consumed nectar through holes chewed in corolla tissue by a florivore (*Tenthredo grandis*), and individuals switched between this behavior and 'legitimate' nectar foraging (*sensu* Inouye 1980).

When we investigated the frequency of flower visits to patches, one bee species, *B. vagans*, made 91.0% of visits during observations. The majority of other visits were made by *Lasioglossum* spp. (4.7%) and *H. annulatus* (3.1%). Each flower received 2.85 ± 0.5I SE bee visits per hour averaged across floral sex phases. However, individual bee foraging patterns with respect to flower sex phase did vary, and some bees visited one sex phase more often than expected by chance (range of proportional divergence from expected proportion of male flower visits of 0.5: -0.87 to 0.59). Overall, *H. annulatus* visited male-phase flowers significantly less often than expected ($t_{16} = -2.76$, $P = 0.01$), whereas neither *B. vagans* ($t_{50} = -0.55$, $P = 0.59$) nor *Lasioglossum* ($t_{11} = 0.49$, $P = 0.63$) preferentially visited flowers based on sex phase.

Nectar

Standing crop nectar volume, sugar concentration and energy content were each best described by models including as fixed effects time of day, flower sex phase and a time*sex interaction. Flowers open to insect visitation contained I.69 ± 0.17 SE µL nectar. We found that nectar standing crop volume was 2.4 times higher in the morning than the afternoon ($F_{1,142} = 17.63$, $P < 0.0001$) and 1.7 times higher in female- than in male-phase flowers ($F_{1,142} = 6.71$, $P = 0.01$), and there was a significant interaction between time of day and flower sex phase ($F_{1,142} = 5.28$, $P = 0.02$; Fig. 2a). Concentration of *Chelone* nectar sucrose-equivalent sugars was 34.38 ± I.17 SE % (range: 9.5-64.5%). Nectar sugars were I.46 times more concentrated in flowers sampled in the afternoon ($F_{1,84} = 35.19$, $P < 0.0001$; Fig. 2b), but other effects were not statistically significant ($F_{1,84} < 2.87$, $P > 0.09$). Standing crop nectar contained $3.76 \times 10^{-3} \pm 3.48 \times 10^{-4}$ kcal energy. Caloric reward of female-phase flowers was 168% higher than that of male-phase flowers ($F_{1,82} = 7.97$, $P = 0.006$; Fig. 2c), but other effects were not statistically significant ($F_{1,29} < 2.39$, $P > 0.13$).

We found that nectar standing crop was positively correlated with some measures of flower length (in the morning, petal length, and in afternoon, both corolla and petal length; Tab. 2). Nectar sugar concentration was not associated with other floral traits in the morning, but was negatively correlated with afternoon volume and petal length. Caloric reward of nectar was positively correlated with volume in both morning and afternoon samples. Caloric reward in morning samples was also positively correlated with corolla width and petal length; in afternoon samples nectar energy content was correlated positively with corolla length and negatively with sugar concentration. There was no correlation between corolla length and width either in morning or afternoon flowers. However, there was positive allometry between corolla and petal lengths in morning but not in afternoon (Tab. 2).

Pollen

Across populations, anthers of unvisited flowers contained I.30 ± 0.08 × I05 pollen grains. The best-fit model of pollen present in flowers included treatment

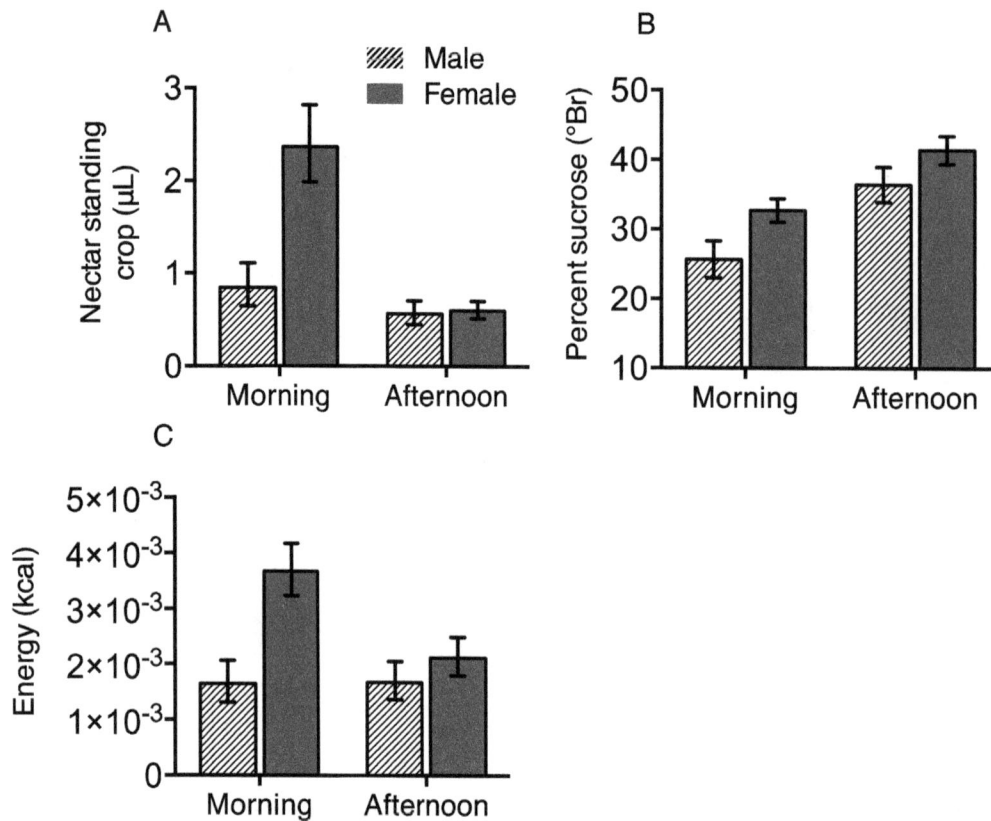

FIGURE 2. Least square means ± SE of A) nectar standing crop (μL), B) sugar concentration (Brix) and C) energy content (kilocalories) of male- and female-phase flowers sampled in the morning (at daily outset of flower visitation) and afternoon (≥ 5 hrs after floral visitation had begun).

TABLE 2. Correlations among flower morphology and nectar traits. Values are Pearson product-moment correlation coefficients. Bold values are statistically significant at $P \leq 0.05$ (*), $P \leq 0.001$ (**) and $P \leq 0.0001$ (***).

		Energy	Nectar volume	Nectar sugar	Corolla length	Corolla width
Morning	Energy					
	Nectar volume	**0.963***				
	Nectar sugar	0.171	-0.100			
	Corolla length	0.289	0.279	0.038		
	Corolla width	**0.308***	**0.291***	-0.282	0.015	
	Petal length	**0.357***	**0.357***	-0.263	**0.341***	**0.571***
Afternoon	Energy					
	Nectar volume	**0.938***				
	Nectar sugar	**-0.419**	**-0.704***			
	Corolla length	**0.367**	**0.344**	0.165		
	Corolla width	-0.155	0.060	-0.072	-0.070	
	Petal length	0.190	**0.309**	**-0.302***	-0.129	**0.558***

(unvisited vs. open vs. single visit flowers) as a fixed effect, and plant population as a random effect. Plant population explained 15.0% of the variance in the overall model, and a Tukey test revealed significant differences in pollen production among populations (Fig. 3a). There were significant differences in pollen grain number among unvisited and open flowers and those that had received single visits from two bee species, H. annulatus and B. vagans ($F_{3,308.2} = 21.77$, $P < 0.0001$, Tab. 3; Fig. 3b). Comparing

pollen remaining in anthers after bee visits to pollen in anthers of unvisited controls, H. annulatus ($N = 5$) removed 0.9% of pollen and B. vagans ($N = 123$) removed 36.6% of pollen in single visits. Despite this large mean difference in pollen removal, the difference between the two species was not statistically significant in a Tukey HSD post hoc test ($P = 0.62$), possibly due to small sample size for H. annulatus. Tukey tests further showed that mean pollen counts of flowers visited by H. annulatus were distinguishable from

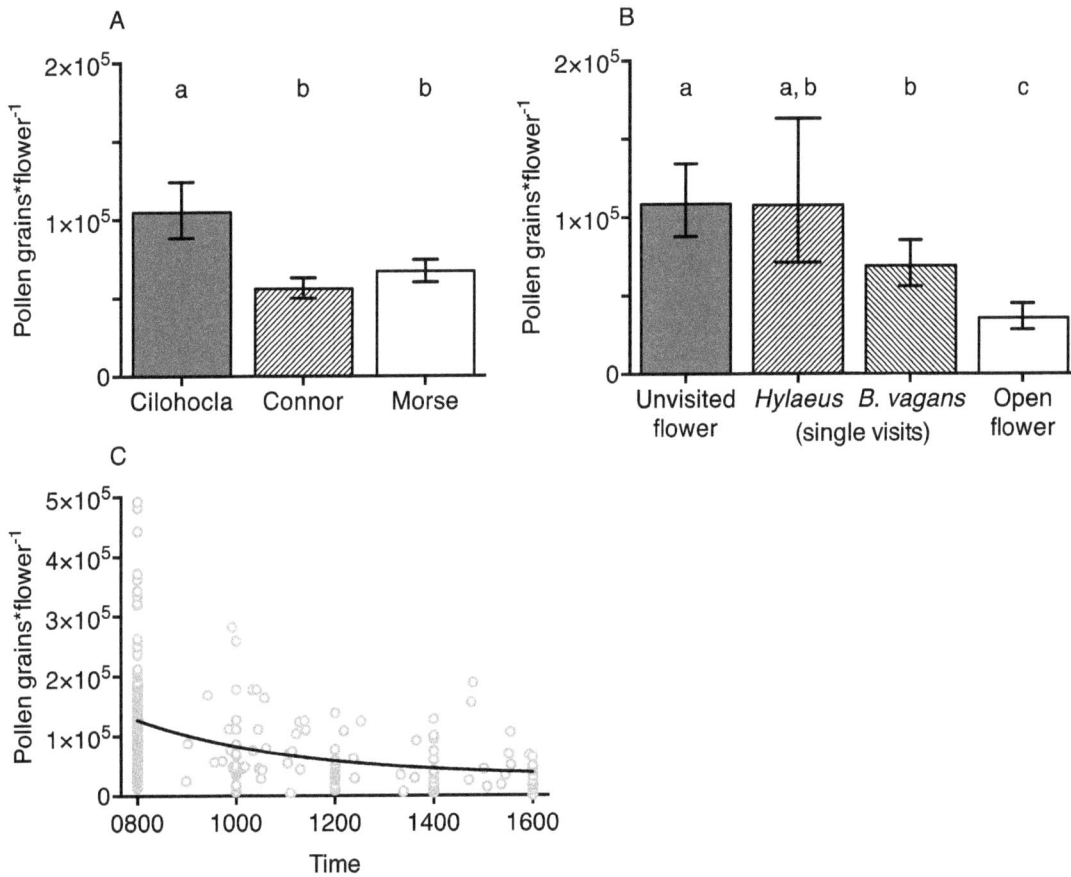

FIGURE 3.   A) Back-transformed least square means ± SE pollen grains present in anthers of unvisited, male-phase flowers from three *Chelone glabra* populations; B) Back-transformed least square means ± SE pollen grains present in anthers of unvisited flowers, those that received single visits from *Bombus vagans* or *Hylaeus annulatus*, and unmanipulated flowers open to multiple visits by pollen and nectar foragers; and C) raw numbers of pollen grains remaining in flowers exposed to insect visitation for varying lengths of time. For A) and B), means with different lower case letters are statistically different based on a Tukey HSD post-hoc test (*P* < 0.05).

TABLE 3.   Pollen remaining in *Chelone glabra* anthers (mean ± standard error) following single visits by bees compared with pollen in virgin flowers ('unvisited control') and those open to natural visitation by multiple visitors ('open control'). Due to small sample size, visits by *Bombus bimaculatus, B. borealis* and *Lasioglossum* sp. could not be statistically compared to those by *B. vagans* and *Hylaeus annulatus*.

Visitor	Pollen grains per flower		N
	Mean	SE	
Bombus bimaculatus	125,648	45,202	3
B. borealis	61,296	2,573	3
B. vagans	83,943	5,670	125
Hylaeus annulatus	137,333	65,575	5
Lasioglossum sp.	23,929	4,405	2
Open control	40,847	4,515	60
Unvisited control	129,543	8,138	135

those of flowers open to bee visitation (*P* = 0.02) but not unvisited flowers (*P* = 1.00). Pollen counts for *B. vagans* single visits were different from those for open and unvisited flowers (both comparisons: *P* < 0.0001; Fig. 3b).

The amount of pollen remaining in open, male-phase flowers was best modelled by a two-phase exponential decay function, and declined sharply in the first two hours after anthesis (R^2 = 0.186; Fig. 3c). Making the assumption that flowers open at night, the model demonstrates that individual flowers contribute little to plant male fitness after the first day they are open, when 85-90% of pollen is predicted to have been removed.

We found that *B. vagans* carried 7.6 and 26.3 times more *Chelone* pollen grains on their thoracic dorsum than *B. impatiens* or *H. annulatus*, respectively (Welch's test: $F_{2,5.18}$ = 44.55, *P* = 0.0005). *Chelone* pollen as a fraction of total pollen carried was significantly greater for *B. vagans* (82.9%) than *B. impatiens* (28.9%) or *H. annulatus* (17.3%; $F_{2,18}$ = 7.68, *P* < 0.004). There was no difference between *B. vagans* individuals that sonicated flowers and those that did not in numbers or per cent of *Chelone* pollen grains carried (F < 0.50, *P* > 0.49).

The best-fit model of pollen deposition to stigmas included treatment (single bee visit vs. unvisited control) as a fixed effect and plant population as a random effect. Stigmas of flowers visited by *B. vagans* had significantly more *Chelone* pollen deposited on them than those that had not received any visits ($F_{1,425}$ = 17.91, *P* < 0.0001; Fig. 4).

Heterospecific pollen accounted for 2.0% of pollen found on unvisited flower stigmas, and 2.5% of all pollen deposited by *B. vagans* workers.

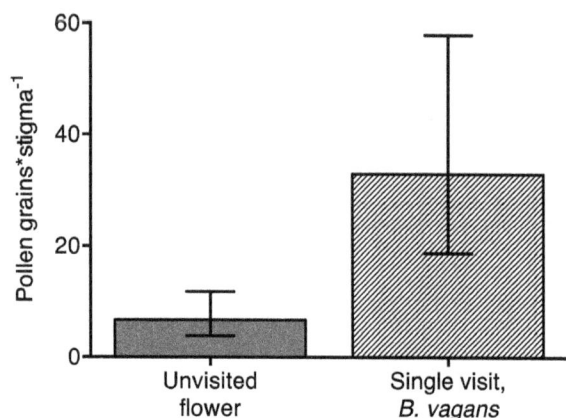

FIGURE 4.   Least square means ± SE pollen grains present on stigmas of unvisited flowers and those that had received single visits from *Bombus vagans*.

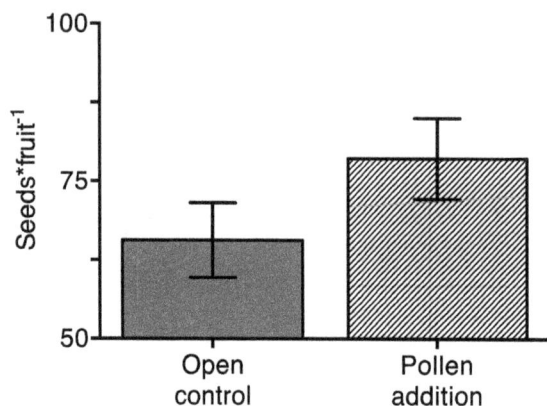

FIGURE 5.   Pollen addition to flowers increased seed set per fruit compared to open-pollinated control flowers. Bars are mean ± SE seeds per fruit.

Pollen limitation

In 2011, we lost >75% of all fruits to predispersal seed predators and herbivores; we therefore analyse only 2012 data here. The best-fit model for seed number in this experiment included as fixed effects pollen limitation treatment, plant population and their interaction, and as a random effect, individual plant nested within population. We found that overall, plants were pollen limited for seed set per fruit ($X_5^2 = 16.58$, $P = 0.01$; Fig. 5), with flowers in the pollen supplementation treatment producing 1.20 times more seeds than those in the open pollination treatment. Seed set per fruit varied among populations ($X_8^2 = 16.73$, $P = 0.03$), but the interaction between pollen supplementation and plant population was not statistically significant ($X_4^2 = 9.07$, $P = 0.059$). A Tukey HSD test showed significant pollen limitation in one population (Morse; t = 3.38, $P = 0.03$) but not the other four. The effect size, *d*, of the difference between pollination treatments across the five populations was 0.04 ± 0.09 SE (range: -0.10 ± 0.63 SE in the Valerie population to 0.26 ± 0.31 SE in the Morse population). Of the fruits collected during the 2012 pollen

limitation experiment, 59.7% were damaged by pre-dispersal seed predators, but there was no effect of pollen treatment on proportion of fruits attacked ($\chi^2 = 2.34$, $P = 0.13$).

DISCUSSION

Our results support previous qualitative reports that *Chelone* requires insect visitation to set seed and that it is self-compatible (Cooperrider 1967). We recorded visits from approx. 20 insect species to *Chelone* flowers, and show that the most common visitors vary in their ability to vector pollen between flowers. We conclude that *B. vagans* workers, the most common flower visitors, pollinate *Chelone* because they: 1) contacted both anthers and stigmas when they foraged; 2) visited both male- and female-phase flowers; 3) removed by sonication more than a third of all pollen from newly opened flowers; 4) passively carried greater numbers of *Chelone* and fewer numbers of heterospecific pollen grains than other bees when traveling between *Chelone* flowers, and 5) deposited pollen on stigmas while foraging. This result adds to a large body of research showing that some bumble bees can be highly effective pollinators while foraging (Winfree et al. 2007), especially when collecting pollen (Free 1993). By contrast, two other common flower visitors, *B. impatiens* and *H. annulatus*, were less effective pollinators of *Chelone*. Our observations suggest that relatively little pollen is transferred during visits by *B. impatiens* (in which bees do not fully enter flowers) and small, relatively hairless *H. annulatus* females, who often cling to anthers without contacting stigmas. Previous reports show that bumble bee species commonly differ in their value to plants as pollinators (Asada & Ono 1996; Thøstesen & Olesen 1996; King et al. 2013; Strange 2015), as demonstrated here for *B. impatiens* and *B. vagans*. Bumble bees often deposit more pollen in single visits than other flower visitors (e.g., Thomson et al. 2000; Javorek et al. 2002), but there are notable exceptions (e.g., King et al. 2013; Benjamin & Winfree 2014).

We show that *Chelone* is protandrous and that pollinator rewards vary according to sex phase and time since anthesis. When flowers open in male phase, bees may forage for pollen, but little nectar is available. The large initial removal of pollen by *B. vagans* workers we observed (mean: > 36%) falls within the range reported for bumble bees in other pollination systems (Wilson 1995; Thomson & Goodell 2001; Castellanos et al. 2003). In a survey of studies that reported single visit pollen removal measures, bumble bees removed 35.7 ± 19.0 SD % of pollen in virgin flowers (range: 6.7-80.0%; $N = 25$ bee-plant combinations reported in 10 studies; L. L. Richardson and R. E. Irwin, unpublished). This sample includes a range of bumble bee and plant species and collection by both sonication and other methods. Despite this large removal and initial exponential decrease in pollen availability, after an estimated > 20 visits (i.e., at the end of a full day in male phase), anthers still retain approx. 10% of their pollen. Similar exponential decay of pollen availability has been reported in other bumble bee-pollinated plants (Wilson & Thomson 1991). We speculate that the densely hairy anther sacs of *Chelone* flowers slow pollen removal by bees, allowing plants to export male gametes across a larger number of interactions with pollinators (Harder & Thomson 1989). Our data are consistent with pollen presentation theory, which holds that bee-pollinated plants with high rates of pollen forager visitation should mete out pollen in small doses rather than presenting all pollen at once, as many bird-pollinated plants do (Thomson et al. 2000). Additionally, the occluded anther sacs of *Chelone* may prevent pollen over-exploitation by sonicating bumble bees, who otherwise might remove the majority of pollen in one visit; by corollary, we expect that such an effect increases plant male function by allowing flowers to export pollen following larger numbers of bee visits. Buzz pollinated plants typically feature poricidal anthers (Buchmann 1983) rather than the longitudinally dehiscent type found in *Chelone* (Straw 1966), and buzz pollination is not commonly reported for the Plantaginaceae. Densely woolly anther sacs are also found in one subgenus of *Penstemon* (Dasanthera), to which *Chelone* is closely related. While two *Penstemon* species have been reported to be sonicated by pollinating bees (Cane 2014), they are not in the Dasanthera subgenus. Given our report of *Chelone's* pollination system, investigating buzz pollination in basal lineages of *Penstemon* which share these anther traits would provide additional evolutionary insight. Moreover, *Chelone* and its floral visitor community may provide a good system in which to test predictions of pollen presentation theory (Thomson et al. 2000).

We report the paradoxical observation that while pollen reward is greatest in newly opened, male-phase flowers, *H. annulatus* females that sonicate anthers to collect pollen preferentially visited older female flowers. Previous research has shown that in plants with temporally separate sex phases such as *Chelone*, pollen harvesters frequently avoid female flowers and so do not pollinate. For example, the dichogamous herb *Impatiens capensis* Meerb. is pollinated by nectar foraging bumble bees that visit flowers in both sex phases, yet pollen foraging honey bees and solitary bees avoid female flowers (Wilson & Thomson 1991). One hypothesis for why *H. annulatus* avoids male flowers, where the pollen reward is greatest, is that this small solitary species may not be large enough to loosen pollen from male phase flowers, and consequently visits older (female) flowers for pollen left after larger bees have repeatedly sonicated anthers. Additionally, these bees could be attracted to the higher nectar rewards we identified in female-phase flowers. Because individuals can restrict their visits to flowers of one sex phase, *H. annulatus* could potentially forage as a pollen or nectar thief (Baker et al. 1971; Inouye 1980), but the extent to which this takes place was beyond the scope of this study. Further research should investigate the cues by which this bee avoids male flowers, and should investigate the effect of *H. annulatus* on *Chelone* pollination success. However, because bee preferences may be affected by ratios of male- and female-phase flowers (Aizen 2001), such work should test *H. annulatus* foraging behavior over a range of flower sex phase ratios.

Similar to other studies of dichogamous plants we found that nectar reward, a function of both secretion and removal, varied according to flower sex phase, and was most abundant in female-phase flowers (Varga et al. 2013). There was strong diurnal variation in nectar volume, concentration and energy reward, and male- and female-phase flowers differed in these traits. Nectar volume declined sharply over the course of the day as foragers removed it, but nectar sugar concentration also dramatically increased, resulting in net energy rewards not predicted by volume alone. There are at least two mechanisms that might account for diurnal changes in *Chelone* sugar concentration. It is possible that flowers manipulate nectar sugar by selective resorption of the liquid in nectar or changes in secretion dynamics (Castellanos et al. 2002; Nepi & Stpiczyńska 2008). Alternately, nectar could passively become more concentrated by evaporation (Corbet et al. 1979). We find limited evidence suggesting evaporation affects nectar volume and concentration in *Chelone*: nectar volume was higher in flowers with greater distance between nectaries and flower opening, regardless of time of day, and nectar sugar concentration was negatively correlated with lip length, but only in afternoon samples, which had been exposed to approx. 5 hours of drying. However, while these associations suggest an influence of floral morphology on evaporation rates, morphology could also influence nectar volume by limiting foragers' access to nectar (Heinrich 2004) or filtering the visitor community, which might alter the nectar microbial community with consequences for the concentration of nectar sugars (Vannette et al. 2013). The interaction between floral morphology and reward quality warrants greater study in this system.

We found that 75% and 57% of fruits were attacked by predispersal seed predators in 2011 and 2012, respectively, 2-3 times that reported for *Chelone* in a previous study (Stamp 1987). Many fruits with evidence of seed predator oviposition (i.e., oviposition scars or larval feeding inside fruits) failed to mature any seeds or associated placenta tissue, suggesting that *Chelone* may abort pollinated fruits after attack. *Chelone* interactions with pollinators must be considered in light of this damage, as well as that caused by other herbivores, including Baltimore checkerspot butterfly larvae (*Euphydryas phaeton*), white-tailed deer (*Odocoileus virginianus* Zimmerman) and white-footed mice (*Peromyscus leucopus* Rafinesque). Like other plant parts, *Chelone* fruits contain the iridoid glycosides aucubin and catalpol (Richardson, Adler, et al. 2015), chemicals known to deter generalist herbivores from consuming the plant's leaves (Bowers et al. 1993). Interestingly, however, many of *Chelone's* herbivores, including its seed predators, are specialist feeders on plants containing these compounds. Iridoid glycosides have defensive and attractive functions in *Chelone* leaves (Bowers et al. 1993) and floral nectar (Richardson, Bowers, et al. 2015), respectively, but additional work is needed to clarify their role in developing fruits.

In a test of pollen limitation across five plant populations, we found that overall, *Chelone* was pollen limited, but this effect varied among populations and was only statistically significant in one of them. This result is broadly consistent with other research on angiosperm pollen limitation. For example, pollen limitation is commonly reported for plants with spatial or temporal separation of sexes (Ramsey & Vaughton 2000), small, fragmented populations with restricted gene flow (Knight, Steets, et al. 2005), and situations in which pollinators are harassed by antagonists (Knight, McCoy, et al. 2005), such as the parasitoid flies that hunt bumble bees at *Chelone* (Richardson, Bowers, et al. 2015). While the difference in seed set between means for pollen addition and control treatments was statistically significant, the effect size was small even in the most pollen limited population (Morse) and when compared to those reported in other work (Knight et al. 2006). It was beyond the scope of this study to determine why *Chelone*'s degree of pollen limitation was relatively low. One possibility is that our methods, in which we treated a subset of flowers on single inflorescences rather than all flowers on an individual plant, affected our estimate of pollen limitation. Some plant species can reallocate resources away from flowers that receive insufficient pollen, which may exaggerate measurements of pollen limitation of seed set (Knight et al. 2006). However, we expect this would have led to relatively high, not low estimates of pollen limitation. Additional pollen limitation experiments where all flowers on a plant are treated will be necessary to clarify the extent to which *Chelone* is pollen limited at the whole-plant level.

In conclusion, we document that *Chelone glabra* is self-compatible but requires insect visitation to set seed. We observed visits by a suite of nectar and pollen foraging insects that vary in their effectiveness as pollinators. We report evidence that the phenology of *Chelone* sex phase transition and reward presentation influence pollen transfer between male and female flowers; yet, one common floral visitor takes advantage of protandry to specialize on female-phase flowers. We show that despite high rates of floral visitation, *Chelone* is pollen limited for seed production. However, the outcomes of plant-insect interactions at flowers must be evaluated in light of the high rates of predispersal seed predation we observed. Our work demonstrates how interactions with mutualist and antagonist flower visitors combine to influence plant reproduction, and we project that outcomes of these processes have consequences for population dynamics of this wetland-dominant herb.

ACKNOWLEDGEMENTS

We thank R. Burten, A. Carper, A. Hogeboom and N. Jensen for field and laboratory help, and M. D. Bowers for comments on the manuscript. This research was funded by grants from the National Science Foundation (DEB-0841862) and the New England Botanical Club. Any opinions, findings and conclusions or recommendations expressed in this material are those of the authors and do not necessarily reflect the views of the National Science Foundation.

APPENDICES

Additional supporting information may be found in the online version of this article:

APPENDIX I.   Field sites where *Chelone glabra* research was conducted, 2011-2013.

REFERENCES

Aizen MA (2001) Flower sex ratio, pollinator abundance, and the seasonal pollination dynamics of a protandrous plant. Ecology 82:127–144.

Asada S'ichi, Ono M (1996) Crop pollination by Japanese bumblebees, *Bombus* spp. (Hymenoptera: Apidae): tomato foraging behavior and pollination efficiency. Applied Entomology and Zoology 31:581–586.

Ashman T-L, Knight TM, Steets JA, Amarasekare P, Burd M, Campbell DR, Dudash MR, Johnston MO, Mazer SJ, Mitchell RJ, Morgan MT, Wilson WG (2004) Pollen limitation of plant reproduction: ecological and evolutionary causes and consequences. Ecology 85:2408–2421.

Baker HG, Cruden RW, Baker I (1971) Minor parasitism in pollination biology and its community function: the case of *Ceiba acuminata*. BioScience 21:1127–1129.

Bates D, Maechler M, Bolker B, Walker S, Christensen RHB, Singmann H (2014) lme4: linear mixed-effects models using Eigen and S4. R package version 1.1-6. www.CRAN.R-project.org/package=lme4. [online] URL: http://cran.r-project.org/web/packages/lme4/index.html (accessed 4 June 2014).

Benjamin FE, Winfree R (2014) Lack of pollinators limits fruit production in commercial blueberry (*Vaccinium corymbosum*). Environmental Entomology 43:1574–1583.

Bertin RI, Newman CM (1993) Dichogamy in angiosperms. The Botanical Review 59:112–152.

Biesmeijer JC, Roberts SPM, Reemer M, Ohlemueller R, Edwards M, Peeters T, Schaffers AP, Potts SG, Kleukers R, Thomas CD, Settele J, Kunin WE (2006) Parallel declines in pollinators and insect-pollinated plants in Britain and the Netherlands. Science 313:351–354.

Bolker BM, Brooks ME, Clark CJ, Geange SW, Poulsen JR, Stevens MHH, White JSS (2009) Generalized linear mixed models: a practical guide for ecology and evolution. Trends in Ecology & Evolution 24:127–135.

Bolten AB, Feinsinger P, Baker HG, Baker I (1979) On the calculation of sugar concentration in flower nectar. Oecologia 41:301–304.

Bowers MD, Boockvar K, Collinge SK (1993) Iridoid glycosides of *Chelone glabra* (Scrophulariaceae) and their sequestration by larvae of a sawfly, *Tenthredo grandis* (Tenthredinidae). Journal of Chemical Ecology 19:815–823.

Buchmann SL (1983) Buzz pollination in angiosperms. In: Jones CE, Little RJ (eds) Handbook of Experimental Pollination Biology. Van Nostrand Reinhold, New York, NY, pp 73–113.

Cane JH (2014) The oligolectic bee *Osmia brevis* sonicates *Penstemon* flowers for pollen: a newly documented behavior for the Megachilidae. Apidologie 45:678–684.

Castellanos MC, Wilson P, Thomson JD (2002) Dynamic nectar replenishment in flowers of *Penstemon* (Scrophulariaceae). American Journal of Botany 89:111–118.

Castellanos MC, Wilson P, Thomson JD (2003) Pollen transfer by hummingbirds and bumblebees, and the divergence of pollination modes in *Penstemon*. Evolution 57:2742–2752.

Cohen J (1988) Statistical Power Analysis for the Behavioral Sciences. Academic Press.

Cooperrider T (1967) Reproductive systems in *Chelone glabra* var. *glabra*. The Proceedings of the Iowa Academy of Science 74:32–35.

Corbet SA, Willmer PG, Beament JWL, Unwin DM, Prŷs-Jones OE (1979) Post-secretory determinants of sugar concentration in nectar. Plant, Cell & Environment 2:293–308.

Droege S, Kolski S, Ascher JS, Pickering J (2014) Apoidea: identification keys to the bees of North America. Discover Life [online] URL: http://www.discoverlife.org/mp/20q?search=Apoidea#Identification (accessed 22 January 2015).

Fenster CB, Armbruster WS, Wilson P, Dudash MR, Thomson JD (2004) Pollination syndromes and floral specialization. Annual Review of Ecology, Evolution, and Systematics 35:375–403.

Free JB (1993) Insect pollination of crops. Academic Press, London, UK.

Garibaldi LA, Bartomeus I, Bommarco R, Klein AM, Cunningham SA, Aizen MA, Boreux V, Garratt MPD, Carvalheiro LG, Kremen C, Morales CL, Schüepp C, Chacoff NP, Freitas BM, Gagic V, Holzschuh A, Klatt BK, Krewenka KM, Krishnan S, Mayfield MM, Motzke I, Otieno M, Petersen J, Potts SG, Ricketts TH, Rundlöf M, Sciligo A, Sinu PA, Steffan-Dewenter I, Taki H, Tscharntke T, Vergara CH, Viana BF, Woyciechowski M (2015) Trait matching of flower visitors and crops predicts fruit set better than trait diversity. Journal of Applied Ecology, Online first. DOI: 10.1111/1365-2664.12530.

Gleason HA, Cronquist A (1991) Manual of vascular plants of northeastern United States and adjacent Canada. New York Botanical Garden, New York, New York, USA.

Gurevitch J, Curtis PS, Jones MH (2001) Meta-analysis in ecology. Advances in Ecological Research 32:199–247.

Harder LD, Thomson JD (1989) Evolutionary options for maximizing pollen dispersal of animal-pollinated plants. The American Naturalist 133:323–344.

Hargreaves AL, Harder LD, Johnson SD (2012) Floral traits mediate the vulnerability of aloes to pollen theft and inefficient pollination by bees. Annals of Botany 109:761–772.

Heinrich B (1975) Bee flowers: a hypothesis on flower variety and blooming times. Evolution 29:325–334.

Heinrich B (2004) Bumblebee economics, Revised edition. Harvard University Press, Cambridge, MA.

Inouye DW (1980) The terminology of floral larceny. Ecology 61:1251–1253.

Javorek SK, Mackenzie KE, Vander Kloet SP (2002) Comparative pollination effectiveness among bees (Hymenoptera: Apoidea) on lowbush blueberry (Ericaceae: Vaccinium angustifolium). Annals of the Entomological Society of America 95:345–351.

Kearns CA, Inouye DW (1993) Techniques for pollination biologists. University of Texas Press, Austin, TX.

King C, Ballantyne G, Willmer PG (2013) Why flower visitation is a poor proxy for pollination: measuring single-visit pollen deposition, with implications for pollination networks and conservation. Methods in Ecology and Evolution 4:811–818.

Knight TM, McCoy MW, Chase JM, McCoy KA, Holt RD (2005) Trophic cascades across ecosystems. Nature 437:880–883.

Knight TM, Steets JA, Ashman T-L (2006) A quantitative synthesis of pollen supplementation experiments highlights the contribution of resource reallocation to estimates of pollen limitation. American Journal of Botany 93:271–277.

Knight TM, Steets JA, Vamosi JC, Mazer SJ, Burd M, Campbell DR, Dudash MR, Johnston MO, Mitchell RJ, Ashman T-L (2005) Pollen limitation of plant reproduction: pattern and process. Annual Review of Ecology, Evolution, and Systematics 36:467–497.

Lundgren R, Lázaro A, Totland Ø (2015) Effects of experimentally simulated pollinator decline on recruitment in two European herbs. Journal of Ecology 103:328–337.

Michener CD (2000) The bees of the world. Johns Hopkins University Press, Baltimore, MD, USA.

Michener CD, McGinley RJ, Danforth BN (1994) The bee genera of North and Central America (Hymenoptera: Apoidea). Smithsonian Institution Press, Washington, D.C.

Mitchell TB (1960) Bees of the eastern United States. I. North Carolina Agricultural Experiment Station.

Mitchell TB (1962) Bees of the eastern United States. II. North Carolina Agricultural Experiment Station.

Nelson AD (2012) Chelone. In: Flora of North America Editorial Committee (ed) Flora of North America North of Mexico, provisional publication. New York and Oxford [online] URL: http://floranorthamerica.org/files/Chelone03f.CH%20for%20Prov%20Pub.pdf (accessed 9 October 2015).

Nepi M, Stpiczyńska M (2008) The complexity of nectar: secretion and resorption dynamically regulate nectar features. Naturwissenschaften 95:177–184.

Ollerton J, Winfree R, Tarrant S (2011) How many flowering plants are pollinated by animals? Oikos 120:321–326.

Pennell FW (1935) The Scrophulariaceae of eastern temperate North America. Academy of Natural Sciences of Philadelphia, Lancaster, PA.

Pleasants JM (1983) Nectar production patterns in Ipomopsis aggregata (Polemoniaceae). American Journal of Botany:1468–1475.

Ramsey M, Vaughton G (2000) Pollen quality limits seed set in Burchardia umbellata (Colchicaceae). American Journal of Botany 87:845–852.

R Core Team (2015) R: A language and environment for statistical computing. R Foundation for Statistical Computing, Vienna, Austria. ISBN 3-900051-07-0, www.R-project.org.

Richardson LL, Adler LS, Leonard AS, Andicoechea J, Regan KH, Anthony WE, Manson JS, Irwin RE (2015) Secondary metabolites in floral nectar reduce parasite infections in bumblebees. Proceedings of the Royal Society of London B: Biological Sciences 282:20142471.

Richardson LL, Bowers MD, Irwin RE (2015) Nectar chemistry mediates the behavior of parasitized bees: consequences for plant fitness. Ecology, Online first. DOI: 10.1890/15-0263.1.

SAS Institute, Inc. (2014) JMP®. SAS Institute, Inc., Cary, NC.

Stamp NE (1987) Availability of resources for predators of Chelone seeds and their parasitoids. American Midland Naturalist 117:265–279.

Strange JP (2015) Bombus huntii, Bombus impatiens, and Bombus vosnesenskii (Hymenoptera: Apidae) pollinate greenhouse-grown tomatoes in western North America. Journal of Economic Entomology 10.1093/jee/tov078:1–7.

Straw RM (1966) A redefinition of Penstemon (Scrophulariaceae). Brittonia 18:80–95.

Thompson JN (1999) The evolution of species interactions. Science 284:2116–2118.

Thomson JD, Goodell K (2001) Pollen removal and deposition by honeybee and bumblebee visitors to apple and almond flowers. Journal of Applied Ecology 38:1032–1044.

Thomson JD, Wilson P, Valenzuela M, Malzone M (2000) Pollen presentation and pollination syndromes, with special reference to Penstemon. Plant Species Biology 15:11–29.

Thøstesen AM, Olesen JM (1996) Pollen removal and deposition by specialist and generalist bumblebees in *Aconitum septentrionale*. Oikos 77:77–84.

Vannette RL, Gauthier M-PL, Fukami T (2013) Nectar bacteria, but not yeast, weaken a plant–pollinator mutualism. Proceedings of the Royal Society B: Biological Sciences 280:20122601.

Varga S, Nuortila C, Kytöviita M-M (2013) Nectar sugar production across floral phases in the gynodioecious protandrous plant *Geranium sylvaticum*. PLoS ONE 8:e62575.

Williams PH, Thorp RW, Richardson LL, Colla SR (2014) Bumble bees of North America: an identification guide. Princeton University Press, Princeton, NJ.

Wilson P (1995) Selection for pollination success and the mechanical fit of *Impatiens* flowers around bumblebee bodies. Biological Journal of the Linnean Society 55:355–383.

Wilson P, Thomson JD (1991) Heterogeneity among floral visitors leads to discordance between removal and deposition of pollen. Ecology 72:1503–1507.

Winfree R, Williams NM, Dushoff J, Kremen C (2007) Native bees provide insurance against ongoing honey bee losses. Ecology Letters 10:1105–1113.

6

Impact of Capitulum Structure on Reproductive Success in the Declining Species *Centaurea cyanus* (Asteraceae): small to self and big to flirt?

Laurent Penet*, Benoit Marion & Anne Bonis

UMR CNRS 6553, ECOBIO, Université de Rennes 1, Campus de Beaulieu, 35042 Rennes Cedex, France

Abstract—Attracting pollinators and achieving successful reproduction is essential to flowering plant species, which evolved different strategies to cope with unpredictable pollination service. The ability of selfing is most widespread and represents a reproductive insurance under varying conditions. In this study, we investigated reproductive success in *Centaurea cyanus*, a self-incompatible declining Asteraceae species. We measured seed set under outcrossing and autonomous selfing and assessed the impact of capitulum structure (i.e., the number of disc florets) on reproductive success. We report that the incompatibility system is either flexible or evolving a breakdown in this species, since autonomous selfing often resulted in production of few seeds. We also show that capitulum structure has a strong impact on reproduction, with smaller inflorescences presenting a better ability to self than larger ones, while larger inflorescences performed better than smaller ones when cross-pollinated. Variable capitulum structure in this Asteraceae species may therefore represent a reproductive strategy to achieve efficient reproduction under diverse pollination environments. Our results also suggest that this declining species might be disrupting its auto-incompatibility system in response to reduced habitats and declining population sizes.

Keywords: Asteraceae, capitulum structure, Centaurea cyanus, *ray and disc florets, reproductive success, self-incompatibility*

Introduction

Plant reproduction has fascinated scientists since ages (Darwin 1876). The phenotypic diversity evolved to achieve sexual reproduction in Angiosperms is simply dramatic and resulted, for instance, in extreme specialization of flowers into pollination syndromes (Hermann and Kuhlemeier 2011) and sophisticated mechanisms of selfing (Fenster and Martén-Rodríguez 2007). At the same time, many species evolved various strategies to avoid selfing: self-incompatibility (Levin 1996), dichogamy (Freeman et al. 1997) and separation of sexes on or between individual plants (Narbona et al. 2011), because selfing influences the probability of unmasking mutation loads in progeny and usually reduces individual plant fitness (Goodwillie et al. 2005). Self-pollen deposition may, however, be unavoidable and for self-incompatible species this often results into a decrease in seed production (e.g., Kameyama and Kudo 2009). The effect of self-pollen deposition is two-fold: first it represents a loss of pollen available to outcrossing (i.e., pollen discounting; Busch and Delph 2011), and second it interferes with outcross-pollen on stigmas (Dai and Galloway 2011). These effects may be enhanced in case of pollen limitation and in small populations (Busch and Schoen 2008). Plant mating system is mainly a pollinator driven process, though plant characteristics (e.g., asexual reproduction; Navascués et al. 2010) or other ecological interactions also play a role (e.g., nutriment availability; Helenurm and Schaal 1996), interactions with herbivores (Ivey and Carr 2005) or florivores (Ashman and Penet 2007; Penet et al. 2009). All these interactions are potentially threatened by the effects of climate change (Schweiger et al. 2010), such as loss of diversity (Sander and Wardell-Johnson 2011) and changes in pollinator specialization (Fontaine et al. 2008), and we may face a dramatic change in pollination service in the near future (Potts et al. 2011). It is therefore crucial to increase our knowledge about pollination and reproduction of plant species (Mayer et al. 2011), and particularly for rare or invasive species (Powell et al. 2011) to allow for a sensible management.

In rare or declining plant species, an understanding of mating system and pollination biology in addition to causes of decline is crucial to help in implementing conservation guidelines (Biesmeijer et al. 2011). Indeed, assessing selfing rates and plants tolerance to selfing is a first step toward managing viable reintroductions or reinforcement (Leducq et al. 2010), because it will bring light to density and relatedness effects in small populations (i.e., impacts of inbreeding depression and Allee effects; Ågren 1996; Leducq et al. 2010). In self-incompatible species, these effects are altered, because direct selfing is reduced (Busch et al. 2010), but consanguinity may still impact fitness via bi-parental inbreeding (Uyenoyama 1986; Elam et al. 2007). Many self-incompatible species have actually somehow a plastic expression of incompatibility and incompatibility breakdown has been reported in several families (Igic et al. 2008; Busch and Urban 2011). Ability to self in incompatible species is

---

*Corresponding author; email: laurent.penet@gmail.com

indeed associated with repeated extinction and colonization demographic events and it is often greater in small populations (Reinartz and Les 1994). Here we chose to investigate reproductive success in *Centaurea cyanus*, a self-incompatible species experiencing a dramatic decline in its native range since a few decades (e.g., Sutcliffe and Kay 2000; Pausic et al. 2010; Ulber et al. 2010), and considered as an invasive outside its native range (Muth and Pigliucci 2007; Jursik et al. 2009).

Like other species of Asteraceae, *Centaurea cyanus* expresses variation in capitulum size and structure (Jursik et al. 2009). Its capitula are structured with deep blue sterile ray florets advertising for pollinators and less showy fertile disc florets (Boršic et al. 2011). Variation in capitulum structure would reflect alternative investments into pollinator attraction vs. seed production, and we were interested in investigating the effect of reproductive efforts on reproductive success. In this study, we investigated reproductive success under controlled conditions and we asked the following specific questions: 1) how variable is capitulum composition in ray and disc florets, hereafter capitulum structure? 2) Is *C. cyanus* able of autonomous self-pollination and seed production? 3) Does capitulum structure variation impact seed set differently under outcross- and self-pollination? We were particularly interested in determining if variation in capitulum structure may represent a reproductive strategy in this species.

MATERIAL AND METHODS

Plant species

Centaurea cyanus (cornflower) is an annual plant from the Asteraceae family (= Compositae, Angiosperm Phylogeny Group 2003) with a wide distribution. The species originated in Caucasus (Boršic et al. 2011) and dispersed as a weed species associated with cereal crops since prehistorical times (Rösch 1998). Though initially not native to Europe, the species is now well naturalized and part of the flora. It may thus not be considered an alien anymore, given its ancient colonization in Europe.

Centaurea cyanus is generally plastic with regard to many traits (Muth and Pigliucci 2007) and grows either in crop fields (especially associated to wheat or canola) or along field margins, where it is expressing a usually high competitive ability (Wassmuth et al. 2009). Germination occurs in fall (Stilma et al. 2009). Flowering occurs from June to mid-summer and each plant produces lose inflorescences bearing many capitula typical of Asteraceae. Sterile peripheral ray florets are deep blue and surround several whorls of fertile tubular disc florets, each with an ovary of a single ovule. *Centaurea cyanus* is pollinated by diverse species of insects, though mostly bees (Carreck and Williams 2002) and it is reported as self-incompatible as a vast majority of species in Asteraceae (Charlesworth 1985).

Pollination treatments

Seeds from many (30 - 40) individuals of wild *C. cyanus* were sampled from discontinuous patches from three adjacent fields in Macon (Upper Normandy) during summer

2009 and bulked for storage at +4 C. We sowed seeds in autumn at the greenhouse facility of ECOBIO (University of Rennes I) and transplanted young plantlets into pots. We started the experiment in spring 2010 when experimental individuals were in bloom (n = 17 unrelated plants). We recorded the number of ray and disc florets of each experimental capitulum at opening. Capitula were individually marked and assigned to either manual cross-pollination or autonomous self-pollination. Each experimental plant received both treatments.

For cross-pollination, we collected pollen from several unrelated pollen donors (6 - 10 non-experimental plants). We used a gentle brush to cover receptive stigmas emerging from disc florets with the mix of pollen, thus outnumbering self-pollen grains already present on stigmas due to protandry (L. Penet, personal observation). In the selfing treatment, capitula were left without further manipulation until harvest. We collected achenes before their release from the capitulum. Each achene contains a single seed; we therefore estimated and counted the number of plump fertile seeds. Unfortunately, part of the capitula dried before maturation due to unexpected high temperatures in the greenhouse. For these, we therefore differentiated successfully fertilized ovules based on size and we considered those as seeds in further analyses. This maturation issue is unlikely to have biased our results since a similar number of capitula were damaged in both treatments and no difference in seed production was found between mature and damaged capitula within pollination treatments (data not shown). We collected 5 - 11 capitula per plant resulting in 64 cross- and 68 self-pollinated capitula.

Statistical analyses

We calculated reproductive success as the proportion of seeds per capitulum (i.e., seed set) and tested for a difference between cross- and self-pollination treatments. We conducted a two-way full-factorial analysis of covariance of seed set with pollination treatment as fixed factor and the number of disc florets (i.e., capitulum structure) as covariate. We added individual plant to account for maternal effects; the Ancova assumptions of homoscedasticity and normality of residuals were met. We included an interaction between pollination treatment and capitulum structure because variation in number of disc florets is expected to influence local pollen availability for autonomous selfing and probability of self-pollen transfer. We found that the interaction was significant. To correlate the number of disc florets and seed set while taking into account maternal effects, we conducted separate analyses of variance of seed set for outcross- and self-pollination treatments, with individual plants as sole factor, and extracted the residuals. We then calculated the correlation between the number of disc florets and the residuals for seed set for both pollination treatments. All analyses were conducted with R statistical package (R Development Core Team 2011).

RESULTS

Capitulum structure was highly variable both within and among *Centaurea cyanus* plants. The total number of florets

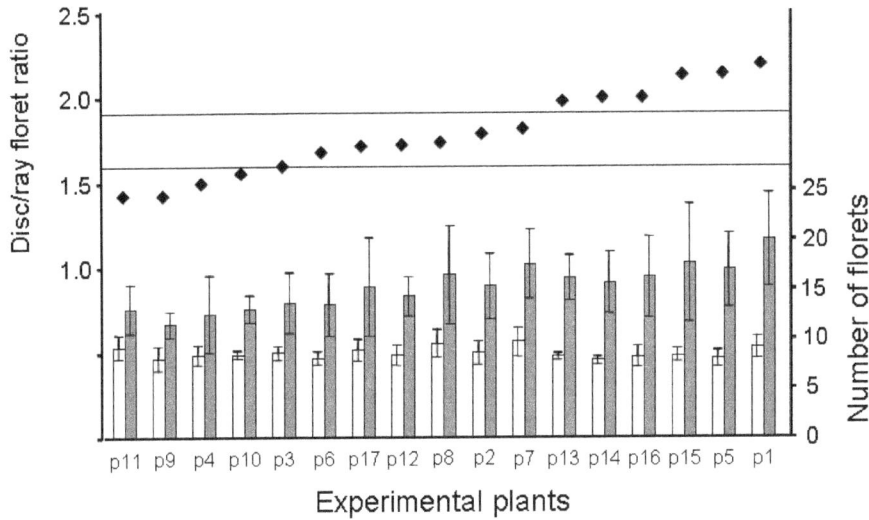

FIGURE 1. Variation in disc to ray florets ratio and number of floret types per inflorescence in the experimental *Centaurea cyanus* plants. Plants are ranked from lowest to largest individual mean disc/ray florets ratio (left y-axis); horizontal lines delineate lower and upper values segregating population in thirds. Mean ($\pm$ SE) number of ray florets, open bars, and disc florets, filled bars, per plant are presented (right y-axis).

per capitulum varied from 13 to 40 (mean $\pm$ se: 23.92 $\pm$ 0.42, n = 132) and the average number of florets at the plant level varied from 19 to 29 with a strong maternal effect ($F_{16,115}$ = 2.72, P = 0.001, n = 132; Fig. 1). Overall, we found that capitula bore a significantly greater number of fertile disc florets than sterile ray florets (grand-means $\pm$ se: 15.19 $\pm$ 0.54 and 8.49 $\pm$ 0.13, respectively, n = 17; paired t-test, t = 13.11, P < 0.0001) and that variance was also greater for the number of disc florets than for the number ray ones (Levene test, $F_{1,32}$ = 18.23, P = 0.0002). The ratio of disc to ray florets at the plant level varied continuously (Fig. 1).

The proportion of disc florets that set a seed was significantly affected by pollination treatment (Table 1). Despite a reported self-incompatibility system, most autonomously self-pollinated capitula (93%) set a few seeds

though significantly fewer than outcross-pollinated capitula (mean $\pm$ se seed set: 0.21 $\pm$ 0.02, and 0.66 $\pm$ 0.02, respectively). Interestingly, we found that the actual number of disc florets on the capitulum influenced seed set differently between pollination treatments (significant interaction in Table 1). Seed set increased with number of disc florets following cross pollination whereas it decreased with increasing number of disc florets in autonomously self-pollinated capitula (Fig. 2). Both correlations were statistically significant taking into account the maternal plant effect (r = 0.36, P = 0.004, and r = -0.30, P = 0.013, for outcrossing and selfing treatments, respectively). We found a significant maternal effect on seed set (Table 1), and high variation in pollination treatment effect among plants (after controlling for variance in number of disc florets, Fig. 3).

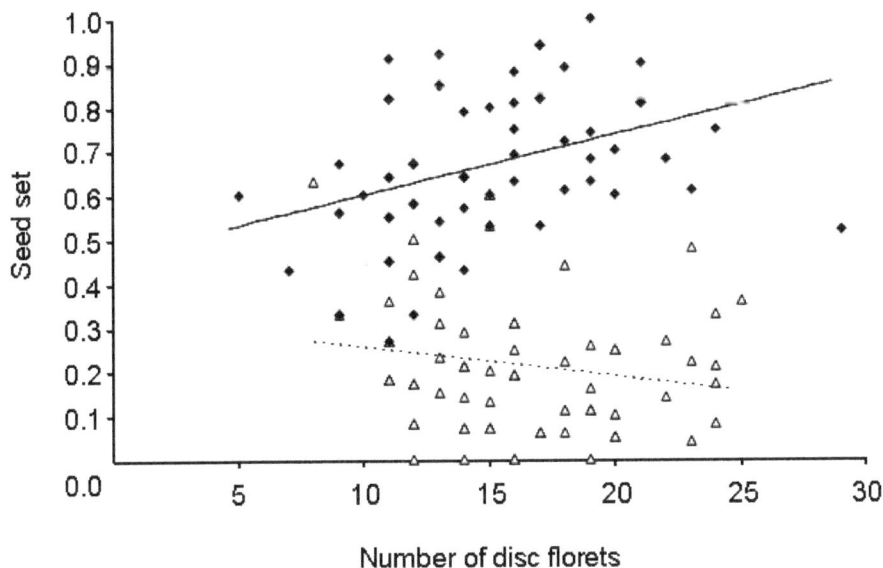

FIGURE 2. Correlation between seed set and number of disc florets per inflorescence following outcrossing and selfing in *Centaurea cyanus*. Pollination treatments: outcrossing (dark diamonds and continuous line), and autonomous selfing (white triangles and dotted line).

Source of variation	df	Mean Square	F-value	P-value
Pollination treatment	I	6.21	265.8	< 0.0001
Number disc florets	I	0.04	1.82	0.180
Treatment * Nb discs	I	0.33	14.22	0.0002
Experimental plant	16	0.05	2.01	0.018
Residuals	112	0.02		

TABLE I. Analysis of covariance of seed set per capitulum. Pollination treatments were manual outcrossing and autonomous selfing and the number of disc florets per capitulum was used as covariate.

DISCUSSION

Centaurea cyanus is classified as a self-incompatible species, but we document here that most capitula produced seeds following autonomous selfing under controlled conditions. Seed set was nevertheless low, and significantly smaller than in the outcross pollination treatment with high variation in ability to self among plants (Fig. 3). Most surprisingly, seed set was strongly influenced by capitulum structure though differentially between pollination treatments: small capitula with few disc florets performed better than large ones under selfing, and the reverse was found for large capitula with many disc florets (Fig. 2). This suggests that variation in capitulum structure might reflect a reproductive strategy to ensure reproduction. We discuss these findings in the light of conservation biology for this declining species.

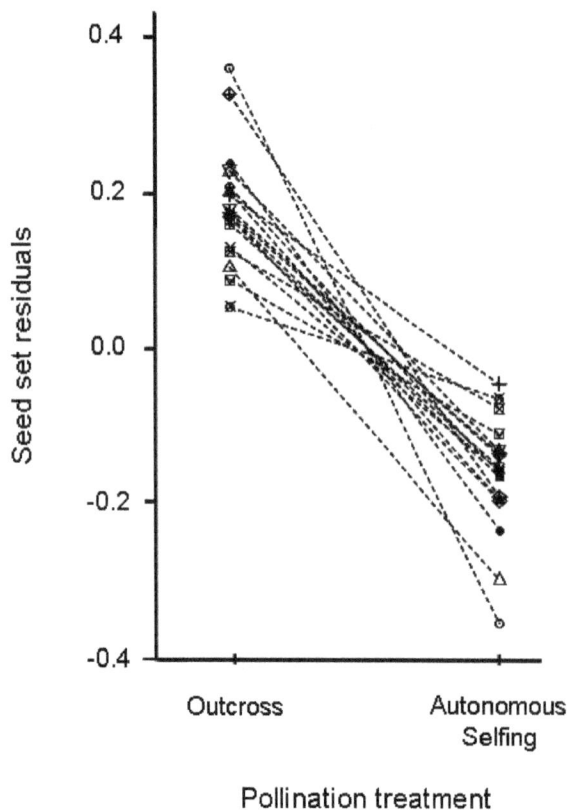

FIGURE 3. Individual plants' response to cross-pollination and autonomous selfing in *Centaurea cyanus*.

Seed set following autonomous selfing was low but non negligible since on average 21% of available ovules were fertilized and up to 50% of seed may be achieved. Thus, even if the plant would receive no visit from pollinators during its reproductive life, it would certainly ensure the production of enough seeds to allow for persistence in the field (at plant level, the average of 21% seed-set under selfing translate into several hundred seeds). Self-incompatible pollination systems are already known to be flexible and to vary depending on environmental factors (Reinartz and Les 1994), and many allegedly self-incompatible species are often reported as able to self (e.g., Mena-Ali and Stephenson 2007). Most interestingly, stigmas bend outward after they emerge from the stamen crown in *C. cyanus*, thus loading themselves with self pollen. This mode of stigma growth might have been selected for because it improves pollen receipt when pollinators visit the capitulum, but it may also be considered a striking pre-adaptation to selfing because it dramatically increases the odds of receiving self-pollen if anthers are still loaded (e.g., Penet et al. 2009). Moreover, a species experiencing population local decline – like cornflower – may be expected to evolve toward increased selfing ability, especially if pollination service becomes unpredictable (Goodwillie et al. 2005). The extent of inbreeding depression (i.e., the lower fitness of selfed progeny compared to outcrossed ones) is an important prospect of study to explore the consequences of evolving self-compatibility in this species.

Capitulum structure had a strong impact on seed set and capitula with fewer disc florets performed better than larger ones under autonomous selfing, while the opposite was found under outcrossing (Fig. 2, larger capitula performed better than smaller ones). In the outcross pollination treatment, we would have expected the reverse, since pollen loads were probably higher in smaller inflorescences, and it is easier to leave a flower un-visited when brushing inflorescences with pollen experimentally in a larger capitulum. This may reflect negative consequences of high competition for fertilization on stigmas if pollen loads were indeed greater in smaller capitula, or that stigmas in smaller capitula were less efficient in pollen receipt. Indeed, the number of disc florets, or inflorescence size, may also reflect individual variation in flower allometry, which is known to translate into reproductive differences between flowers (i.e., herkogamy, differential stigma size; Nishihiro and Washitani 2011), and into decreased pollen receipt and fertilization efficiency. To our knowledge, this is the first time that differences in capitulum structure are directly related to mating efficiency in Asteraceae, usually, an increase in inflorescence or flower size results in increased geitonogamy (i.e., pollinator-mediated selfing; e.g., Klinkhamer et al. 1989;) or an increase in seed predation (Fenner et al. 2002). This variation might thus reflect different reproductive strategies: small capitula, which are also probably less attractive to pollinators (Fenner et al. 2002), would still ensure reproduction via selfing, when pollination service is insufficient or uncertain. On the other hand, the larger and more attractive capitula would result in efficient outcross reproduction under good pollination conditions.

At the individual plant level, capitulum structure as a reproductive strategy appears more strongly than at population level: investment in reproduction (disc florets) compared to advertising (ray florets) showed a continuum in experimental plants (Fig. 1), and was essentially due to variation in disc flower number. Plants with lower values of disc:ray floret ratio (ratio < 1.6) may thus be considered as better "selfers" and those with higher values (ratio > 1.9) as mostly outcrossing. Nevertheless, individual plants showed high inter-capitulum variation in number of disc flowers (Fig. 1), even when they tended to have larger capitula, so that plants do not entirely rely on a single reproductive strategy (selfing or allogamy). When accounting for the influence of capitulum structure, plants showed differences in their response to selfing or outcrossing pollination treatments (Fig. 3), with some individuals being more self-incompatible than others. This gives further evidence that self-incompatibility is probably evolving a disruption and it opens the door to selection on other reproductive modes, like reproductive insurance via autonomous selfing, in this species (Reinartz and Les 1994).

This study documents a link between capitulum structure and reproductive strategies in an Asteraceae species probably experiencing a breakdown in its self-incompatibility system. Stigma bending during female phase is possibly pre-adaptive and facilitates evolution towards increased selfing ability. Autonomous selfing may be selected for in small populations, especially if individuals are related and consanguinity interferes with incompatibility (Busch and Schoen 2008). Since *Centaurea cyanus* is declining throughout its range due to massive use of herbicides, and even considered as locally endangered (Sutcliffe and Kay 2000; Pausic et al. 2010), these results open several pathways for further investigation. It would be of interest to evaluate levels of inbreeding depression experienced by selfed progenies, and whether it varies among lineages with differences in self-compatibility (e.g., Collin et al. 2009). Then, the extent and generality of an incompatibility breakdown within the species' range may drive local persistence of populations, especially if selfing ability is correlated to population size.

Finally, self-incompatibility breakdown and selfing-ability could be influenced by:

(1) Visits of different pollinator species and their relative efficiency to cross-pollinate (e.g., Ostevik et al. 2010); (2) inflorescence size and competition for pollination; (3) how visits translate into seed-set in natural conditions (e.g., as in Nienhuis and Stout 2009) and under which conditions a newly evolving selfing ability would perform better than outcross. Indeed, natural *Centaurea cyanus* seed set may be

lower *in natura* than seed sets for our cross pollination treatment. Investigating these interactions between plants and pollinators would inform us as to how the ecology of pollination could drive adaptations to new modes of reproduction.

ACKNOWLEDGEMENTS

The authors thank Carine L. Collin and two anonymous referees for comments on early versions of the manuscript; and Thierry Fontaine and Fouad Nassur for help in the greenhouse facility. This study is dedicated to Jean-Marie Penet (1950-2010), who made early suggestions about investigating mating consequences of variation in capitulum structure.

REFERENCES

Ågren J (1996) Population size, pollinator limitation, and seed set in the self- incompatible herb *Lythrum salicaria*. Ecology 77:1779-1790.

Angiosperm Phylogeny Group (2003) An update of the Angiosperm Phylogeny Group classification for the orders and families of flowering plants: APG II. Botanical Journal of the Linnean Society 141:399-436.

Ashman T-L, Penet L (2007) Direct and indirect effects of a sex-biased antagonist on male and female fertility: Consequences for reproductive trait evolution in a gender-dimorphic plant. The American Naturalist 169:595-608.

Biesmeijer JC, Sorensen PB, Carvalheiro LG (2011) How pollination ecology research can help answer important questions. Journal of Pollination Ecology 4:68-73.

Boršic I, Susanna A, Bancheva S, Garcia-Jacas N (2011) *Centaurea* sect. *Cyanus* : nuclear phylogeny, biogeography, and life-form evolution. International Journal of Plant Sciences 172:238-249.

Busch JW, Delph LF (2011) The relative importance of reproductive assurance and automatic selection as hypotheses for the evolution of self-fertilization. Annals of Botany doi:10.1093/aob/mcr219.

Busch JW, Joly S, Schoen DJ (2010) Does mate limitation in self-incompatible species promote the evolution of selfing? The case of *Leavenworthia alabamica*. Evolution 64:1657-1670.

Busch JW, Schoen DJ (2008) The evolution of self-incompatibility when mates are limiting. Trends in Plant Science 13:128-136.

Busch JW, Urban L (2011) Insights gained from 50 years of studying the evolution of self-compatibility in *Leavenworthia* (Brassicaceae). Evolutionary Biology 38:15-27.

Carreck NL, Williams IH (2002) Food for insect pollinators on farmland: insect visits to flowers of annual seed mixtures. Journal of Insect Conservation 6:13-23.

Charlesworth, D (1985) Distribution of dioecy and self-incompatibility in angiosperms. Pp 237-268 in JJ Greenwood and M Slatkin, eds. Evolution —essays in honor of John Maynard Smith. Harvard University Press, Cambridge, Massachusetts.

Collin CL, Penet L, Shykoff JA (2009) Early inbreeding depression in the sexually polymorphic plant *Dianthus sylvestris* (Caryophyllaceae): effects of selfing and biparental inbreeding among sex morphs. American Journal of Botany 96:2279-2287.

Dai C, Galloway LF (2011) Do dichogamy and herkogamy reduce sexual interference in a self-incompatible species? Functional Ecology 25:271-278.

Darwin CR (1876) The effects of cross- and self-fertilisation in the vegetable kingdom. Murray, London.

Elam DR, Ridley CE, Goodell K, Ellstrand NC (2007) Population size and relatedness affect fitness of a self-incompatible invasive plant. Proceedings of the National Academy of Sciences of the United States of America 104:549-552.

Fenner M, Cresswell JE, Hurley RA, Baldwin T (2002) Relationship between capitulum size and pre-dispersal seed predation by insect larvae in common Asteraceae. Oecologia 130:72-77.

Fenster CB, Marten-Rodriguez S (2007) Reproductive assurance and the evolution of pollination specialization. International Journal of Plant Sciences 168:215-228.

Fontaine C, Collin CL, Dajoz I (2008) Generalist foraging of pollinators: diet expansion at high density. Journal of Ecology 96:1002-1010.

Freeman DC, Doust JL, ElKeblawy A, Miglia KJ, McArthur ED (1997) Sexual specialization and inbreeding avoidance in the evolution of dioecy. Botanical Review 63:65-92.

Goodwillie C, Kalisz S, Eckert CG (2005) The evolutionary enigma of mixed mating systems in plants: Occurrence, theoretical explanations, and empirical evidence. Annual Review of Ecology Evolution and Systematics 36:47-79.

Helenurm K, Schaal BA (1996) Genetic load, nutrient limitation, and seed production in Lupinus texensis (Fabaceae). American Journal of Botany 83:1585-1595.

Hermann K, Kuhlemeier C (2011) The genetic architecture of natural variation in flower morphology. Current Opinion in Plant Biology 14:60-65.

Igic B, Lande R, Kohn JR (2008) Loss of self-incompatibility and its evolutionary consequences. International Journal of Plant Sciences 169:93-104.

Ivey CT, Carr DE (2005) Effects of herbivory and inbreeding on the pollinators and mating system of Mimulus guttatus (Phrymaceae). American Journal of Botany 92:1641-1649.

Jursik M, Holec J, Andr J (2009) Biology and control of another important weeds of the Czech Republic: Cornflower (Centaurea cyanus L.). Listy Cukrovarnicke A Reparske 125:90-93.

Kameyama Y, Kudo G (2009) Flowering phenology influences seed production and outcrossing rate in populations of an alpine snowbed shrub, Phyllodoce aleutica : effects of pollinators and self-incompatibility. Annals of Botany 103:1385-1394.

Klinkhamer PGL, de Jong TJ, de Bruyn G-J (1989) Plant size and pollinator visitation in Cynoglossum officinale. Oikos 54:201-204.

Leducq JB, Gosset CC, Poiret M, Hendoux F, Vekemans X, Billiard S (2010) An experimental study of the S-Allee effect in the self-incompatible plant Biscutella neustriaca. Conservation Genetics 11:497-508.

Levin DA (1996) The evolutionary significance of pseudo-self-fertility. American Naturalist 148:321-332.

Mayer C et al. (2011) Pollination ecology in the 21st Century: Key questions for future research. Journal of Pollination Ecology 3:8-23.

Mena-Ali JI, Stephenson AG (2007) Segregation analyses of partial self-incompatibility in self and cross progeny of Solanum carolinense reveal a leaky S-allele. Genetics 177:501-510.

Muth NZ, Pigliucci M (2007) Implementation of a novel framework for assessing species plasticity in biological invasions: responses of Centaurea and Crepis to phosphorus and water availability. Journal of Ecology 95:1001-1013.

Narbona E, Ortiz PL, Montserrat A (2011) Linking self-incompatibility, dichogamy, and flowering synchrony in two Euphorbia species: alternative mechanisms for avoiding self-fertilization? PLoS ONE 6:e20668 EP.

Navascués M, Stoeckel S, Mariette S (2010) Genetic diversity and fitness in small populations of partially asexual, self-incompatible plants. Heredity 104:482-492.

Nienhuis CM, Stout JC (2009) Effectiveness of native bumblebees as pollinators of the alien invasive plant Impatiens glandulifera (Balsaminaceae) in Ireland. Journal of Pollination Ecology 1:1-11.

Nishihiro J, Washitani I (2011) Post-pollination process in a partially self-compatible distylous plant, Primula sieboldii (Primulaceae). Plant Species Biology 26:213-220.

Ostevik KL, Manson JS, Thomson JD (2010) Pollination potential of male bumble bees (Bombus impatiens): Movement patterns and pollen-transfer efficiency. Journal of Pollination Ecology 2:21-26.

Pausic I, Skornik S, Culiberg M, Kaligaric M (2010) Weed diversity in cottage building material used in 19th century: Past and present of the plant occurence. Polish Journal of Ecology 58:577-583.

Penet L, Collin CL, Ashman T-L (2009) Florivory increases selfing: an experimental study in the wild strawberry, Fragaria virginiana. Plant Biology 11:38-45.

Potts SG, Biesmeijer JC, Bommarco R, Felicioli A, Fischer M, Jokinen P, Kleijn D, Klein A-M, Kunin WE, Neumann P, Penev LD, Petanidou T, Rasmont P, Roberts SPM, Smith HG, Sørensen PB, Steffan-Dewenter I, Vaissière BE, Vilà M, Vujić A, Woyciechowski M, Zobel M, Settele J, Schweiger O (2011) Developing European conservation and mitigation tools for pollination services: approaches of the STEP (Status and Trends of European Pollinators) project. Journal of Apicultural Research 50:152-164.

Powell KI, Krakos KN, Knight TM (2011) Comparing the reproductive success and pollination biology of an invasive plant to its rare and common native congeners: a case study in the genus Cirsium (Asteraceae). Biological Invasions 13:905-917.

Reinartz JA, Les DH (1994) Bottleneck-induced dissolution of self-incompatibility and breeding system consequences in Aster furcatus (Asteraceae). American Journal of Botany 81:446-455.

Sander J, Wardell-Johnson G (2011) Fine-scale patterns of species and phylogenetic turnover in a global biodiversity hotspot: Implications for climate change vulnerability. Journal of Vegetation Science 22:766-780.

Schweiger O, Biesmeijer JC, Bommarco R, Hickler T, Hulme PE, Klotz S, Kühn I, Moora M, Nielsen A, Ohlemüller R, Petanidou T, Potts SG, Pyšek P, Stout JC, Sykes MT, Tscheulin T, Vilà M, Walther G-R, Westphal C, Winter M, Zobel M, Settele J (2010) Multiple stressors on biotic interactions: how climate change and alien species interact to affect pollination. Biological Reviews 85:777-795.

Stilma ESC, Keesman KJ, van derWerf W (2009) Recruitment and attrition of associated plants under a shading crop canopy: Model selection and calibration. Ecological Modelling 220:1113-1125.

Sutcliffe OL, Kay QON (2000) Changes in the arable flora of central southern England since the 1960s. Biological Conservation 93:1-8.

R Development Core Team (2011) R: a language and environment for statistical computing. In. R Foundation for Statistical Computing, Vienna, Austria.

Rösch M (1998) The history of crops and crop weeds in southwestern Germany from the Neolithic period to modern times, as shown by archaeobotanical evidence. Vegetation History and Archaeobotany 7:109-125.

Ulber L, Steinmann HH, Klimek S (2010) Using selective herbicides to manage beneficial and rare weed species in winter wheat. Journal of Plant Diseases and Protection 117:233-239.

Uyenoyama MK (1986) Inbreeding and the cost of meiosis: The evolution of selfing in populations practicing biparental inbreeding. Evolution 40:388-404.

Wassmuth BE, Stoll P, Tscharntke T, Thies C (2009) Spatial aggregation facilitates coexistence and diversity of wild plant species in field margins. Perspectives in Plant Ecology, Evolution and Systematics 11:127–135

7

POLLINATION ECOLOGY OF *OREOCALLIS GRANDIFLORA* (PROTEACEAE) AT THE NORTHERN AND SOUTHERN ENDS OF ITS GEOGRAPHIC RANGE

Jenny A. Hazlehurst[1], Boris Tinoco[2], Santiago Cárdenas[2], and Jordan Karubian[1]

[1]*Department of Ecology and Evolutionary Biology, Tulane University, 400 Lindy Boggs Center, New Orleans LA 70118*
[2]*Escuela de Biología, Ecología y Gestión, Universidad del Azuay, Av. 24 de Mayo 7-77 y Hernán Malo, Cuenca, Ecuador*

Abstract—Geographic variation in pollination ecology is poorly documented, if at all, in many plant-pollinator systems. Great insights could be gained into the abiotic and biotic factors which impact the evolution of floral properties and their potential to lead to speciation by doing so, as both can vary naturally over the geographic range of a plant species. We characterized the pollination ecology of the Andean tree *Oreocallis grandiflora* (Family: Proteaceae) at the northern and southern ends of its range in Ecuador and Peru in terms of flower morphology, nectar properties, pollinators and plant reproduction. We found significant divergence in the two populations in terms of style length and flower openness, nectar standing crop and secretion rate, and pollinator community. We did not find a significant difference in the length of the pollen presenter or in nectar sucrose concentration by weight (% Brix). The observed divergence in floral traits between the two study populations may be related to a combination of factors, including genetic drift and isolation by distance, distinctive suites of pollinators, or heterospecific pollen competition, which future studies should further investigate. This study demonstrates that pollination ecology can vary substantially across the geographic range of a species, with implications for delimiting species and subspecific taxa.

Keywords: Andes, biogeography, floral traits, hummingbirds, mammals, Proteaceae

INTRODUCTION

The study of pollination ecology has played an important role in our current understanding of co-evolution (Cook & Rasplus 2003) and speciation (Kay & Sargent 2009), and also provides important baseline information to inform practical ecosystem-level conservation efforts (Pauw 2007) in a time of pollinator declines (Biesmeijer et al. 2006). However, more basic information is needed to improve our understanding in all of these areas. For example, the lack of data on plant-pollinator interactions has been identified as one of the main obstacles to understanding how zoophilic pollination may act as a mechanism of speciation (Kay & Sargent 2009). Studies that document pollination ecology at different points along a single species' geographic range are rare, and could provide insights for future research into the role of pollination in driving floral isolation and even speciation, as well as how pollination mutualisms adapt to changing conditions.

Intraspecific variation in floral morphology and pollination ecology may arise through several mechanisms, primarily genetic drift, abiotic selection, and biotic selection driven by pollinators, herbivores, and competing plant species. Environmental factors such as temperature and

precipitation are known to influence floral traits such as flower size (Sapir et al. 2002) and colour (Strauss & Whittall 2006). According to Stebbins' (1974) Most Effective Pollinator Principal (MEPP), selection should also favour floral traits that promote visitation by the most frequent and effective pollinator. The precise role of selection by pollinators in the speciation of Angiosperms is under debate, but many switches in pollinator syndrome have been documented in the literature. For example, the Neotropical genus *Costus* (Costaceae) has shifted from bee to hummingbird pollination multiple times (Kay & Schemske 2003), presumably because changes in the available pollinator community caused directional selection away from the original pollinator guild. Herbivores (Gómez et al. 2009a) and heterospecific pollen competition from other plant species in the community (Ashman & Arceo-Gómez 2013) may also exert selective pressure on specific floral traits such as flower number and pistil length. Conversely, in cases where the available pollinator community and environmental context are consistent and adequate levels of gene flow exist across a species range, pollination ecology and corresponding floral traits might be conserved due to stabilizing selection. At present, the lack of comparative studies between species and subpopulations limits our ability to distinguish between the frequency and likelihood of these alternative scenarios. Studies that document the pollination ecology of a species at different points in its geographic range will help to highlight possible

*Corresponding author: jhazlehu@tulane.edu

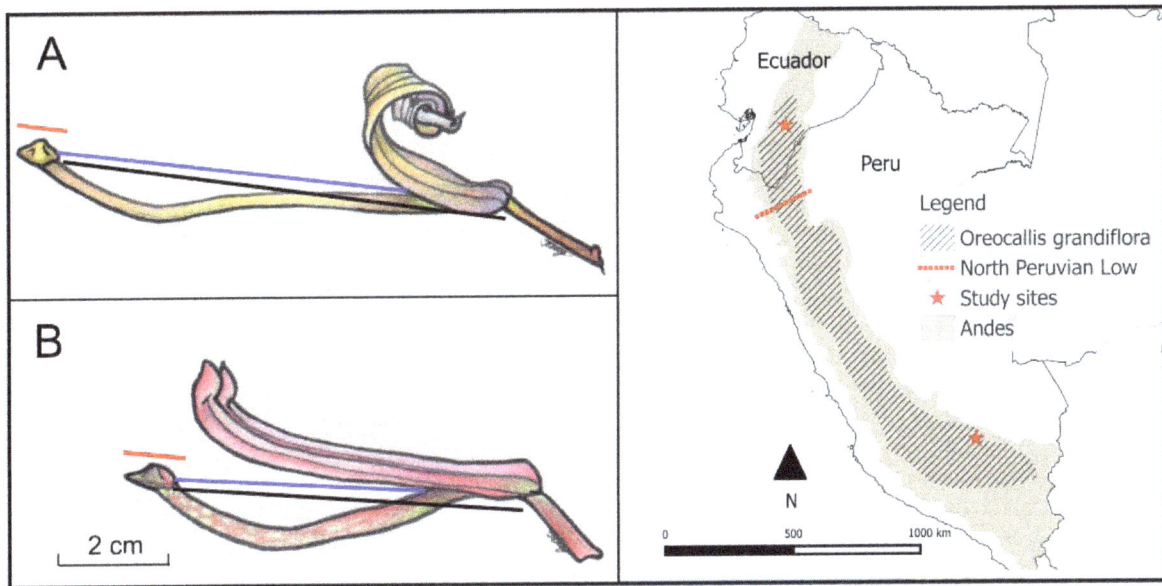

FIGURE I. There is marked variation in floral morphology between Peruvian (A) populations at the southern range limit and Ecuadorian (B) populations at the northern range limits of *Oreocallis grandiflora*. Blue lines indicate the straight line distance between the tip of the pollen presenter and the point at which nectar accumulates (PED), back lines indicate the style length (SL), and red lines indicate stigma height (SH). On the map the red drops represent the study site locations in Ecuador and Peru, the dotted line represents an estimate of the range of *O. grandiflora*, and the double black line represents the location of the Northern Peruvian Low.

questions and study systems for future research into these mechanisms of divergence. In this study, we describe relevant floral traits and pollination ecology of the Andean tree *Oreocallis*.

The Andes are an important and under-studied centre of plant-pollinator system diversity (Kay & Sargent 2009) which is increasingly threatened by habitat conversion and climate change (Ocampo Peñuela & Pimm 2015). *Oreocallis grandiflora* is a widespread and abundant plant species whose range spans a distance of over 1,500 km along the Andes mountain range (Fig. 1). Based upon intraspecific variation in the colour and pubescence of *O. grandiflora* inflorescences, Sleumer (1954) defined two distinct species; *O. mucronata*, with white, glabrous inflorescences, and *O. grandiflora* with pubescent, pink-red inflorescences. This two-species conclusion was also reached by Weston & Crisp (1999). However, subsequent herbarium analysis suggested that the variation in pubescence was continuous and not associated with colour, and the two species were condensed into *O. grandiflora* (Prance 2008). Yet, neither of these more recent, herbarium-based studies appears to have considered the geographic distribution of this variation, suggesting that characterizing both the degree of variation that may exist between spatially distinct populations as well as any differences in their pollinator community may provide insight into the taxonomic status of this species.

The overarching goal of this study was to describe the pollination ecology of *O. grandiflora* from two research projects at either extreme of its geographic range and to discuss potential explanations for geographic variation in floral traits. To do so, we assess variation in floral morphology, nectar properties, and pollinator community between two populations in qualitatively similar evergreen montane forest habitat (Tovar et al. 2013) that are situated

at the extremes of the species range along its north to south axis in the Andes, one in southern Ecuador and one in southern Peru.

MATERIALS AND METHODS

Study sites

Field work was conducted by two separate research teams over three years every July–November from 2012–2014 at the Wayqecha Biological Station (13°11'S, 71°35'W) in Manu National Park, Peru and from November–February from 2014–2015 at El Gullán Biological Station of the Universidad del Azuay in Ecuador (3°20'S, 79°10'W) at 3,000–3,400 m asl by the other team. At the time that most of the work was conducted, the respective Peruvian and Ecuadorian teams were unaware of each other. For this reason, some of the methods vary between the two sites, but the data are still compatible. The respective time frames correspond to the end of the dry season and the start of the rainy season in both habitats. The two sites are 1,378 km apart straight-line distance and located at the transition between the "evergreen montane forest" and "high elevation grasslands" biomes (Tovar et al. 2013). There exist several potential geographic barriers to gene flow for high-elevation plant species along the North-South axis of the Andes, including the North Peruvian Low (NPL), the lowest point in the Andes between Chile and Colombia (Fig. 1), which could disrupt gene flow and impact the degree of divergence in pollination ecology and floral traits.

Study species

The Andean firebush, *Oreocallis grandiflora* Lam. (Proteaceae), is a small tree up to 7 m in height that produces terminal flowered raceme inflorescences of 10–50 long, paired flowers that open sequentially in groups of 2–20 at a time from the base of the inflorescences towards the top. Flowers have a tubular to cylindrical perianth that opens into 4 segments. Flowers are also bisexual and have a relatively large pollen presenter 0.35–0.45 mm long, which refers to any structure other than the anthers that distributes pollen, and in this case is a modification of the style and stigma (Prance et al. 2008). Fruits are woody follicles that dehisce to reveal winged, wind-dispersed seeds. Flowering and fruiting occur simultaneously year-round in both Ecuador and Peru. *O. grandiflora* is especially common in disturbed soils along its range in the Andes from southern Peru to Central Ecuador and has been reported from 1,200–3,800 m asl (Prance et al. 2008). Data on pollinators is scarce, with no information on geographic variation and only three published hummingbird species as visitors (Prance et al. 2008). The pollination ecology of Neotropical Proteaceae in general is poorly documented, but there are more reported cases of entomophily than ornithophily (Prance et al. 2008), and one possible case of chiropterophily (Fleming et al. 2009). There are also a few examples of mixed pollination systems in Neotropical Proteaceae (Devoto et al. 2006, Chalcoff et al. 2008). Proteaceae globally exhibit a wide range of pollinator communities with several reported cases of pollination by non-flying rodents (Rourke & Wiens 1977), bats (Daniel 1976), and birds and insects (Mast et al. 2012).

Flower colour and morphology

We visually assessed petal colour by photographing flowers against grid paper as belonging to either the "pink" or "white" morph. We studied flower morphology at both sites by randomly sampling two flowers ($N_{Ecuador}$ = 73, N_{Peru} = 94) from individual *O. grandiflora* trees. Flowers were photographed on a 1 cm × 1 cm grid background and the following measurements were extracted from photos using the program tpsDig version 2.16 (Rohlf 2010): style length (SL; the straight-line distance from the base of the corolla along the longest axis to the base of the stigma), stigma height (SH; the longest distance across the stigma (also the pollen presenter), and the minimum straight-line distance between the pollen presenter and the intersection of the petals and the style, where nectar accumulates (PED), and the angle of flower openness (AO; the smallest angle between the petals and the style) (Fig. 1).

Nectar properties

To quantify nectar properties, standing crop and sucrose concentrations were measured in 2 to 5 flowers that were sampled from randomly selected *O. grandiflora* trees within the study sites ($N_{Ecuador}$ = 90, N_{Peru} = 107) and nectar volume and sucrose concentration were measured using 50 µL microcapillary tubes (Sigma-Aldrich Co., St.Louis, Missouri, USA) and a handheld sucrose refractometer (Bellingham and Stanley Ltd, Basingstoke, UK). To measure daily patterns in nectar secretion, we randomly selected four trees and placed mesh bags on four flowers on each tree, two per inflorescence, to exclude pollinators. Nectar secretion was then measured every two hours (Peru) and three hours (Ecuador) from 6 AM until 6 PM from opening until flower

dehiscence, 3 to 5 days. Nocturnal nectar secretion patterns were not quantified, because this data was originally recorded as part of research on diurnal hummingbird activity and nocturnal visitation was previously undocumented. Nectar was also extracted at the appropriate time interval before the first measurement to get an accurate reading at 6 AM. To measure nectar accumulation rates over 24 h, individual trees were randomly selected and one inflorescence on each tree bagged off from visitors. At 6 PM the evening prior to sampling all nectar was emptied from the flowers and after 24 h nectar volume was measured from the same flower.

Pollinator community

We documented the pollinator community and pollinator visitation rate of *O. grandiflora* in Peru and Ecuador by randomly selecting individuals for observation and then setting up digital camcorders (Sony Inc., New York, USA). At both sites, each plant was recorded for 2–6 h (depending on the available camera and the weather) in the morning and in the afternoon for a period of maximum period of five days. Videos were then reviewed manually and the time and identity of any floral visitors was recorded. Nocturnal pollination was opportunistically sampled at both sites using infrared-enabled trap cameras (EBSCO Inc., Birmingham, USA).

Reproduction of Oreocallis grandiflora

We used hand-pollination experiments to study how pollen source impacted fruit set, seed set, and mass. The Peruvian and Ecuadorian sites had slightly different protocols for the experiment. In Ecuador only fruit set was quantified, while in Peru fruit set was not quantified, but seed set and mass were. In Ecuador, 49 individual trees of *O. grandiflora* were randomly selected, and each flower on the same randomly selected inflorescence received one of the following hand-pollination treatments: Self-pollen (from the same inflorescence), natural self-pollination (a freshly opened flower isolated from visitation with a mesh bag), nearest-neighbour pollen, far pollen (pollen from plants > 1 km away), and a control treatment. Flowers were monitored monthly for three months and total fruit production was measured. In Peru, ten individuals of *O. grandiflora* of similar size and at least 20 m apart were selected. One inflorescence per tree was randomly selected to receive one of each of the following hand-pollination treatments in 5 freshly opened flowers: self-pollen, nearest-neighbour pollen, next-patch pollen, and far pollen. Nearest-neighbour pollen was collected from the nearest individual of *O. grandiflora*, next-patch pollen was collected from individuals 50–100 m away from the focal plant, and far pollen was collected from individuals 1 km away. The quantity of pollen to be applied was standardized as lying flat against a 1 cm × 1 cm square on grid paper. An applicator made out of hummingbird feathers was applied to the square and brushed against the stigma daily for four days after anthesis to simulate a hummingbird visit. After treatment, flowers were bagged. For both sets of experiments, fruit development was monitored monthly and collected once ripe. Fruit were dried in the sun until dehiscence and seeds were extracted and counted. Each seed was then measured along the longest axis and weighed to 0.000 g.

Statistical analyses

All statistics were conducted in R version 3.2.3 (R Core Team 2015). To assess variation between the populations in these parameters, we conducted a principal components analysis (PCA) and a linear discriminant analysis (LDA) on log-transformed morphological values using the package MASS (Venables & Ripley 2002). In order to determine which principal components to use in further comparative analysis, we used a broken-stick null model (Jackson 1993) with the package 'BiodiversityR' (Kindt & Coe 2005). We then conducted a two-tailed t-test using the package 'stats' (R Core Team 2015) to test whether significant principal component scores varied significantly between the Peruvian and Ecuadorian populations.

To determine the effects of site (Ecuador or Peru) on nectar standing crop we used two separate models to first analyse all the data for the presence or absence of nectar using a generalized linear mixed model (GLMM) with a binomial distribution using the package 'lme4' in R (Bates et al. 2015), then analysing only the log-transformed non-zero data using a general linear mixed model (GLMM) with a Gaussian error distribution using the package 'nlme' (Pinheiro et al. 2015). In both steps flowers were nested within plants if multiple flowers were sampled from the same individual, and time of day was included as a fixed effect. To determine the effects of site on nectar sucrose by weight concentration (% Brix) we used the square-root of Brix values as the dependent variable and site as fixed effect, with flower nested within plant as the random effect using a GLMM with a Gaussian error distribution with the package 'nlme' (Pinheiro et al. 2015). To determine the effects of site on 24 h nectar accumulation rates, we used the package 'stats' (R Core Team 2015) to conduct a nested ANOVA with log-transformed nectar volume as the dependent variable and site, flower, and date as the fixed effects and individual tree as the random effect. To determine the effect of time of day on nectar secretion rate we independently analysed the data from Ecuador and Peru, since they were collected at different sampling intervals. We used a GLMM with a Gaussian error distribution with square-root transformed nectar volume as the dependent variable, time of day as the fixed effect and flower nested within inflorescence within individual tree as the random effect using the package 'nlme' (Pinheiro et al. 2015). To compare daily nectar secretion patterns between the two sites while correcting for the differences in sampling intervals we summed nectar secretion between 6 AM and 12 PM and took the mean of sucrose concentrations recorded during that time. We then used a Gaussian GLMM with square-root transformed cumulative nectar secretion as the response variable and site as the fixed effect, with individual tree as the random effect, using the package 'nlme' (Pinheiro et al. 2015). To analyse the mean sucrose concentration (% Brix) over this 6 h period, we used the same analysis method but with % Brix as the response variable, country as the fixed effect, and tree as the random effect.

To compare the diurnal pollinator community between the sites we used a nonmetric multidimensional scaling (NMDS), with Bray-Curtis distance matrix to ordinate Ecuadorian and Peruvian plants in relation to the community of pollinators. One hr visitation rates per inflorescence were used as the quantitative link between plants and pollinators. Differences in the community of pollinators between Ecuadorian and Peruvian plants were tested with a non-parametric Manova (Anderson 2001) using the same distance matrix employed in the MNDS with the "vegan" package (Oksanen et al. 2016). We also calculated hourly visitation rates, Shannon's Diversity Index (H') and Pielou's Evenness (J') on a per-plant basis by pooling the observations for each plant, then taking the averages for all plants for each site and for the two sites pooled together using Excel. We compared visitation rates using a one-way ANOVA with the package 'stats' with square-root transformed visitation rate as the dependent variable, country as the fixed effect, and hours of observation per plant as the random effect. We analysed the effect of the "closed" treatment on fruit set in Peru using a Gaussian LMM with logit-transformed proportions of fruit set as the dependent variable, and treatment as the fixed effect and individual tree as the random effect using the package 'nlme' (Pinheiro et al. 2015).

We analysed the effect of the different hand-pollination treatments on seed set using a Gaussian LMM with seed set as the dependent variable and pollen treatment as the fixed effect. To test the effect of treatment on seed mass, we used a Gaussian LMM with square-root transformed seed mass as the dependent variable and pollen treatment as the predictor variable with the package 'nlme' (Pinheiro et al. 2015). In both tests inflorescence was nested within individual tree as the random effect. To analyse the Ecuadorian data, we used a binomial general linear mixed model (GLMM) in the package 'lme4' (Bates et al. 2015) with the presence or absence of fruit per hand-pollinated flower as the dependent variable, treatment as the fixed effect, and treatment nested within plant as the random effect.

Results

Flower colour and morphology

Flowers from the Peruvian and Ecuadorian populations exhibited striking differences in colour and morphology (Fig. 1, Tab. 1). Peruvian individuals of *O. grandiflora* all presented obviously magenta flowers while Ecuadorian individuals presented white-green flowers. In general, Ecuadorian flower morphology was characterized by longer style length (SL) and the minimum straight-line distance between the pollen presenter and the intersection of the petals and the style (PED), and a wider angle of openness (AO), while Peruvian flowers were characterized by shorter SL and PED with a smaller AO (Tab. 1). This finding is corroborated by Principal Components Analysis (PCA), which showed distinct grouping of Ecuadorian and Peruvian flowers in the morphospace (Fig. 2). The first component (PC1) accounted for 62% of the variation, the second (PC2) for 23%, the third (PC3) for 13% and the fourth (PC4) for 1% (See Tab. 2). Only the first principal component had an eigenvalue greater than would be expected at random. The coefficients for contribution to PC1 were as follows: pollination efficiency distance was 0.60, stigma length was

TABLE I. Summary of floral traits and pollination ecology of O. *grandiflora* in southern Ecuador and northern Peru. Mean values with standard errors for Ecuadorian and Peruvian populations of O. *grandiflora* and the combination of both. Asterix ($*$) indicates significant differences between the Ecuadorian and Peruvian populations < 0.05, $**$ indicates P-values < 0.01, $***$ indicates P-values less than 0.001. Numbers within parentheses represent sample sizes.

	Ecuador	Peru	Combined
Morphology			
Stigma length (mm)$*$	5.3 ± 0.4 (73)	3.3 ± 0.5 (94)	4.2 ± 0.1 (167)
Pollination efficiency distance (mm)$*$	4.6 ± 0.0 (73)	2.1 ± 0.0 (94)	3.2 ± 0.1 (167)
Angle of openness (°)$*$	25.0 ± 0.6 (73)	34.1 ± 0.6 (94)	29.0 ± 0.6 (167)
Stigma height (mm)	0.4 ± 0.6 (73)	0.4 ± 0.6 (94)	0.4 ± 0.0 (167)
Colour$*$	white-green	magenta	N/A
Nectar Properties			
Percent containing nectar$***$	94% (90)	37% (107)	63% (197)
Standing crop (µL)$*$	15.1 ± 1.5 (84)	10.9 ± 1.4 (39)	13.8 ± 1.2 (123)
Sucrose (%Brix)	27.8 ± 1.6 (84)	30.0 ± 3.0 (39)	28.5 ± 1.6 (123)
24-hr secretion (µL)$**$	31.7 ± 3.5 (36)	12.6 ± 1.0 (41)	21.5 ± 2.0 (77)
Pollination			
Visits per hour$*$	0.80 ± 0.19 (151)	0.64 ± 0.11 (295)	0.72 ± 0.11 (446)
Shannon's diversity index	0.31 ± 0.06 (17)	0.53 ± 0.09 (27)	0.44 ± 0.06 (44)
Pielou's evenness index	0.16 ± 0.03 (17)	0.30 ± 0.05 (27)	0.24 ± 0.04 (44)

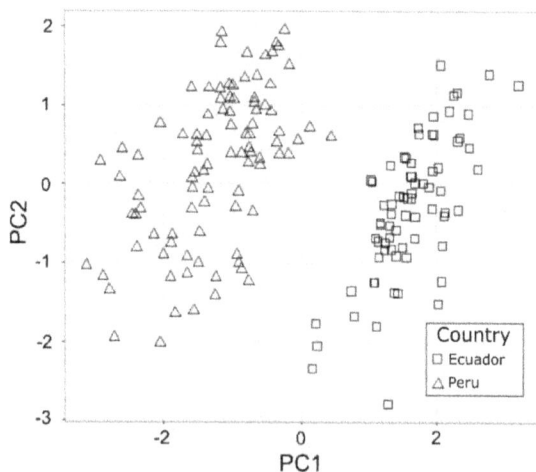

Figure 2. PCA results for a comparison of floral morphology in Peruvian ($N = 94$) and Ecuadorian ($N = 73$) populations of *Oreocallis grandiflora*. Squares represent Ecuadorian samples and triangles represent Peruvian samples.

0.60, angle of openness was 0.48, and stigma height was 0.21. A two-tailed student's t-test of PC1 values by site was significant ($T = 28.05$, $DF = 165$, $P < 0.01$). This grouping was also confirmed by a linear discriminant analysis (LDA), which had a 100% success rate at identifying specimens by country of collection.

Nectar properties

Flowers in Peru were found to be significantly more likely to be empty when randomly sampled than flowers in Ecuador, even when time of day was accounted for ($Z = -4.5$, $DF = 193$, $P < 0.001$). When nectar was present, Ecuadorian flowers had significantly more nectar than did Peruvian flowers ($T = -2.25$, $DF = 41$, $P < 0.05$, $R^2_{marginal} = 0.06$, $R^2_{conditional} = 0.49$). There was no significant difference in sucrose concentration by weight between Ecuadorian and Peruvian flowers ($T = -0.60$, $DF = 41$, $P = 0.55$, $R^2_{marginal} < 0.01$, $R^2_{conditional} = 0.21$) (Tab. 1). Ecuadorian flowers also had significantly higher 24-hr nectar accumulation rates than did Peruvian flowers (Tab. 1; $F_{1, 17} = 12.40$, $P < 0.001$). Nectar secretion rates in Peru varied significantly by time of day between 6:00 am and 6:00 pm ($T = -3.59$, $DF = 33$, $P < 0.00$, $R^2_{marginal} = 0.41$, $R^2_{conditional} = 0.61$) with highest secretion rates at 6:00 am and then dropping off during the day (Fig. 3). Nectar secretion rates in Ecuador also varied significantly by time of day, with the secretion rate significantly higher in the afternoon ($T = 4.94$, $DF = 141$, $P < 0.00$, $R^2_{marginal} = 0.04$, $R^2_{conditional} = 0.45$) (Fig. 3). The mean cumulative nectar volume secreted between 6 am and noon in Ecuador was significantly greater in Ecuador ($T = -3.25$, $DF = 32$, $P < 0.01$, $R^2_{marginal} = 0.18$, $R^2_{conditional} = 0.51$).

Pollinator community

Pollinator visitation rates were significantly higher in Ecuador than in Peru (Tab. 3; $F_{1,41} = 4.52$, $DF = 41$, $P < 0.05$). There were significant differences in community of diurnal pollinators between Peru and Ecuador plants (Fig. 4; $F_{1,39} = 4.86$, $P < 0.001$). *Aglaeactis cupripennis* was the most common visitor in both Peruvian and Ecuadorian plants (63% of visits in Peru, and 44% of visits in Ecuador), but other important hummingbird visitors were exclusive for each location, specifically *Boissoneaua matthewsii*, *Coeligena violifer*, *Heliangelus amethysticolis* in Peru and *Coeligena iris*, *Heliagelus viola*, *Lesbia nuna*, and *Lesbia victoria* and *Ramphomicron microhynchus* in Ecuador (Tab. 4). Both the Shannon's diversity and Pielou's evenness indices were greater in Peru (Tab. 3). Nocturnal trap cameras revealed the

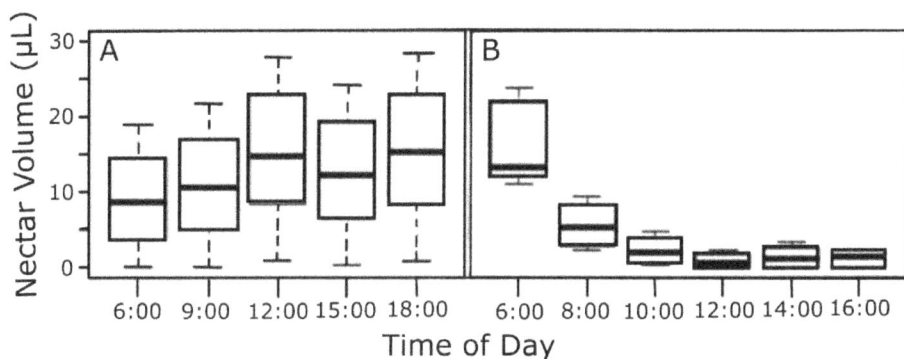

FIGURE 3. A comparison of nectar secretion rates of *Oreocallis grandiflora* in (A) Ecuador over 3-hr intervals ($N = 16$) and (B) Peru over 2-hr intervals ($N = 16$).

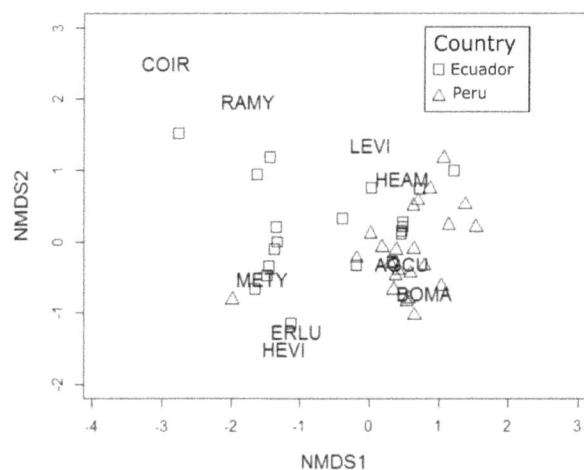

FIGURE 4. A comparison of hummingbird pollinator community of *Oreocallis grandiflora* in Ecuador ($N = 175$ hrs) versus Peru ($N = 294$ hrs). The analysis is based on a nonmetric multidimensional scaling ordination using visitation rates of the hummingbird species.

TABLE 2. Principal components analysis of floral morphology loading scores, standard deviation, and proportion of variance explained for each measured floral trait.

Measure	PC1*	PC2	PC3	PC4
Angle of openness	0.48	-0.20	0.85	0.72
Pollination efficiency distance	0.60	-0.13	-0.33	0.72
Stigma length	0.60	-0.05	-0.39	-0.70
Stigma height	0.21	0.97	0.11	0.04
Standard deviation	1.58	0.97	0.73	0.19
Proportion of variance	0.62	0.23	0.13	0.01

presence of bat and rodent visitors in Ecuador (Tab. 3) but no evidence of nocturnal visitation in Peru ($N = 50$ hours).

Reproduction of Oreocallis grandilfora

Results of the hand-pollination experiments in Ecuador showed no impact of pollen treatment on fruit set ($\chi^2 = 0.20$, $DF = 4$, $P > 0.5$). Results of the hand-pollination experiments in Peru suggest that self-pollen results in lower seed set compared to the far treatment ($T = -2.31$, $DF = 34$, $P < 0.05$, $R^2_{marginal} = 0.08$, $R^2_{conditional} = 0.18$) but not

compared to the nearest-neighbour or next-patch treatments. The self-pollen treatment also had significantly lower seed mass than the far treatment ($T = -4.40$, $DF = 34$, $P < 0.01$, $R^2_{marginal} = 0.18$, $R^2_{conditional} = 0.62$) as did the nearest-neighbour treatment ($T = -2.30$, $DF = 34$, $P < 0.05$) (Fig. 5).

DISCUSSION

Ecuadorian and Peruvian populations of *O. grandiflora* exhibited significant differences in a suite of characteristics relevant to pollination. The Peruvian flowers were all magenta in colour while those in Ecuador were white with a green tinge (Fig. 1). We also observed significant separation in morphospace between the two populations, and style length (SL) and the minimum straight-line distance between the pollen presenter and the intersection of the petals and the style (PED) were most responsible for the variation between the populations. There was significantly higher nectar

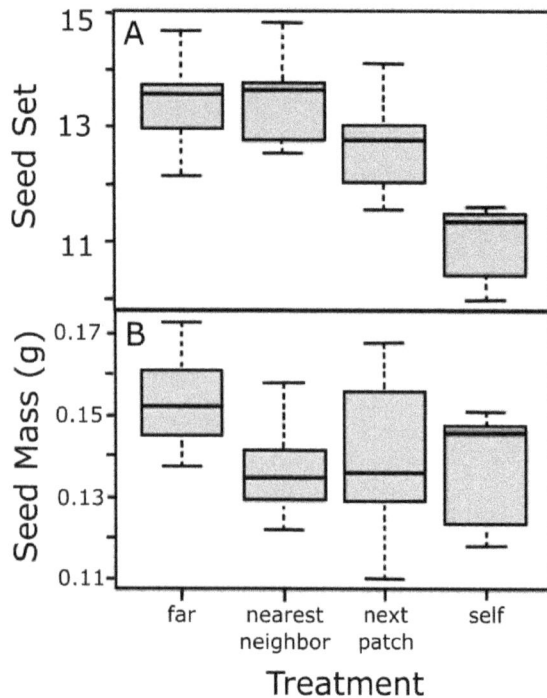

FIGURE 5. The effects of pollen source on (A) seed set and (B) seed mass in *Oreocallis grandiflora* from Peru. Sample sizes for the treatments were as follows: far (19), nearest neighbour (15), next patch (17), self-pollen (19). The Ecuadorian data was not included because no significant effects were found, though self-pollen and autogamous self-pollen treatments were successful at producing fruit as in Peru.

volume in Ecuador and greater 24 h and morning nectar secretion, and in Peru there were significantly more flowers with no nectar present. There was some variation in daily nectar secretion patterns (Fig. 3) though they are not directly comparable due to differences in sampling. Pollinator visitation rates were significantly higher in Ecuador than in Peru. The Peruvian diurnal pollinator community was both more diverse and more even than the Ecuadorian. Our hand-pollination experiments revealed that selfing was possible in both populations, and that in Peru pollen source may impact plant reproduction. In Ecuador however, hand-pollination experiments showed no effect of pollen source on fruit-set.

The two study populations are over 1,500 km apart following the species' geographic range in the Andes, and therefore variation in floral traits is expected. However, the extent of the variation that we observed, especially in terms of floral morphology, suggests that the *O. grandiflora* system may provide a good study system for future research into the evolutionary processes that shape floral morphology, nectar properties, and pollinator community. Our data is limited by

the fact that we do not have replicate populations at each latitude, but the authors have opportunistically observed similar suites of floral characteristics, especially in terms of colour and style length, across populations at both latitudes (J. Hazlehurst and B. Tinoco, *personal communication*). Nonetheless, it cannot be known to what extent the observed differences in floral traits segregate geographically without standardized, repeated study populations at both latitudes. It is also possible, as has been found in other species of montane Angiosperms, that variation in floral characteristics and pollinator community follow a mosaic geographic pattern (Gómez et al. 2009b).

Our results suggest that this system is ripe for future research into the evolutionary mechanisms driving the observed variation in floral traits and nectar properties. Here, we discuss these relevant selective forces as areas of future research for which the *O. grandiflora* system would be suitable. Genetic drift is a possible source of variation in floral traits between the populations, however it is unlikely that genetic drift is the sole explanation for the divergence we observed because of the critical importance of floral morphology and pollination ecology on plant fitness (e.g. Murcia 1990, Armbruster 2014), but we recommend that future research on this system might include a genetic component that could better assess a potential role for drift. Both colour and flower size (correlated with style length) were divergent in our study populations and both traits are known to be subject to abiotic selection pressures in other systems. In terms of flower colour, increased anthocyanin pigmentation imbues flowers with increased heat and drought tolerance (Strauss & Whittall 2006), as does decreased flower size (Sapir et al. 2005).

Biotic selection pressures driven by differences in pollinator community have been demonstrated to significantly impact a suite of floral traits such as flower colour, style length, and nectar volume in other systems (e.g., Temeles et al. 2009, Whittall & Hodges 2007), including in *Embothrium coccineum,* an Andean Proteaceae which also shows distinct pollinator communities in different populations (Chalcoff et al. 2008). Many of the observed differences in diurnal pollinator community between our study populations represented substitution of one species by another morphologically similar member of the same genus outside the range of the original species and may therefore not represent significant shifts in selection pressure on floral traits. For example, *Coeligena violifer* in Peru is replaced by *C. iris* in Ecuador (Tab. 4). Additionally, a single species, *A. cupripennis,* was the dominant visitor in both populations, and it is possible that individuals move along the entire geographic range of *O. grandiflora*, thereby promoting pollen flow between the two populations. While it is known that the North Peruvian Low acts as a dispersal barrier to other high-elevation species of hummingbirds (Chaves et al. 2011), it is unknown if it also acts as a barrier to the hummingbird pollinators of *O. grandiflora*. Future research should determine if these differences in diurnal pollinator community are sufficient to produce differential selection pressures on relevant floral traits in *O. grandiflora* and the degree to which the NPL may or may not impact gene flow.

Several of the floral traits that varied between our study populations are known to be subject to pollinator-driven selection pressure. Flower colour is frequently selected for by pollinators (Fenster et al. 2004; Cronk & Ojeda 2008). For example, whitish flowers such as those found in the Ecuadorian population, are typically associated with mammal- or insect- pollination (Fenster et al. 2004), while

TABLE 3.   Observed and recorded visitors of *O. grandiflora* in Ecuador and Peru. Numbers represent the percentage of visits observed during this study, a line (-) represents circumstantial observation, "yes" indicates that the species is limited to either Ecuador or Peru, while "no" indicates it may occur at both.

Species	Ecuador	Peru	Range restricted?
Diurnal Visitors (Trochilidae)			
Aglaeactis cupripennis	45	63	no
Boissoneaua matthewsii	0	15	yes
Chalcostigma mulsant	-	0	yes
Chalcostigma ruficeps	0	-	yes
Coeligena violifer	0	3	yes
Coeligena iris	4	0	yes
Colibri coruscans	-	13	no
Heliangelus amethysticollis	0	1	yes
Heliangelus viola	9	0	yes
Lesbia nuna	3	-	no
Lesbia victoriae	4	0	yes
Metallura tyrianthina	34	4	no
Patagona gigas	-	0	no
Ramphomicron microhynchus	1	0	yes
Diurnal Visitors (Thraupidae)			
Diglossa brunneiventris	0	-	yes
Diglossa cyanea	-	-	no
Diglossa humeralis	-	0	yes
Diglossa mystacalis	0	-	yes
Nocturnal Visitors			
Anoura geoffroyi (Bat)	-	0	yes
Cricetidae spp. (Rodent)	-	0	yes

pink-magenta flowers, such as those found in Peru, suggest ornithophily. Hummingbird pollinator bill length and shape has been shown to exert selection pressure on floral traits such as style length (Temeles et al. 2009). Style length (Stroo 2000) and nectar volume (Opler 1983), which also varied significantly between our study populations, both scale with pollinator body size and are therefore larger in mammal-pollinated plants; in the current study, both were larger in Ecuador. Overall, the Ecuadorian study population exhibited more floral traits associated with mammal pollination. Indeed, visitation of *O. grandiflora* by nocturnal rodents and bats has been observed in other studies in Ecuadorian populations (Cardenas 2016). In comparison, no nocturnal pollination was observed despite some effort in the Peruvian population.

Future studies should systematically document nocturnal pollination in *O. grandiflora* across its range. In addition to pollinator selection, heterospecific pollen competition has also been shown to select for style length in other systems (Ashman & Arceo-Gómez 2013). A visual inspection of hummingbirds with similar or identical body sizes visiting *O. grandiflora* at the two sites revealed that in Peru *O.* *grandiflora* pollen was deposited on the gorget feathers, whereas in Ecuador it was deposited on the belly or chest, presumably due to the differences in style length between the plant populations (J. Hazlehurst, *personal observation*). Future research into the role of pollinator community, heterospecific pollen competition, and abiotic factors on selection for floral traits should consider using *O. grandiflora* as a study system given the potentially interesting variations we observed between our study populations.

In terms of plant reproduction, both the Ecuadorian and the Peruvian populations were capable of selfing, and autogamous selfing in particular was observed at both sites. There was no significant effect of pollen treatment on fruit set in the Ecuadorian population, however there was a significant positive effect of the far treatment on seed set and mass in Peru as compared to the self-pollen treatment. Selfing in other Proteaceae can result in poor pollen tube growth while outcrossing results in positive effects on fruit set (Fuss & Sedgley 1991). It is probable that *O. grandiflora* is protandrous, like many Proteaceae, and in natural conditions selfing is avoided when pollen is removed from the presenter by visiting pollinators before the stigma becomes receptive. When pollen is not removed, the pollen presenter may act as a kind of "bet-hedging" strategy, wherein selfing is a last-ditch effort at reproduction should no outcrossed pollen be available.

Our findings suggest that, as recommended by Prance (2008), the two-species question raised by Sleumer (1954) for the *Oreocallis* genus should be considered using an analysis of living plants in the field. An informal inventory by the authors of online records of *O. grandiflora* specimens collected across the range of the species at the Missouri Botanical Garden suggests a change in flower colour from magenta to white as one moves from south to north (Appendix I), though there are scattered reports of southern white populations and northern magenta populations. It is important to note that these reports may not accurately represent geographic variation in flower colour, as colour can be subjective and the reports were collected opportunistically. In reality the transition may be clinal, abrupt, or a mosaic based on local abiotic and biotic variables. Indeed, other montane Angiosperms show a geographic mosaic pattern in floral traits as well as pollinator community (Gómez et al. 2009*b*). An analysis of pollinator community, nectar properties, style morphology, pubescence, colour, and genetics should be undertaken along the geographic range of *O. grandiflora* to aid in resolving the taxonomic status of this species.

Conclusion

We found that the pollination ecology of *Oreocallis grandiflora* fit generally within what has been reported in other members of the Proteaceae, though it is of note that autogamous selfing was possible. We found significant divergence between the Ecuadorian and Peruvian populations in terms of floral morphology, nectar volume, nectar secretion rates, and daily patterns of nectar production. Based on our observations, we suggest that the *O. grandiflora* system could be an ideal study system for further study on the abiotic and biotic factors that shape floral traits and pollination ecology, for example pollinator-driven selection and heterospecific pollen competition. While replication of study populations at both the northern and southern latitudes of *O. grandiflora*'s range are necessary to quantify if and how the variation in floral traits and pollinator community that we observed correlates with geography, the authors nonetheless recommend further testing of the *O. mucronata* species concept. This study is a clear example of how ecology can reveal important differences between plant populations not evident from herbarium specimens alone.

ACKNOWLEDGEMENTS

Field work was facilitated by Omar Landázuri and students from the Universidad del Azuay, many wonderful volunteers, Percy Chambi Porroa and Lucas Pavan. Work was funded by a grant from Decanato de Investigaciones at Universidad del Azuay, the National Geographic Society, a NSF Doctoral Dissertation Improvement Grant (#1501862), a Louisiana Board of Regents Fellowship, and a writing fellowship from the Dept of Ecology & Evolutionary Biology at Tulane University. We would also like to thank the Ministries of the Environment of Peru and Ecuador.

APPENDICES

Additional supporting information may be found in the online version of this article:

APPENDIX I. Map of *O. grandiflora* specimen colour

REFERENCES

Anderson MJ (2001) A new method for non-parametric multivariate analysis of variance. Australian Ecology 26:32–46.

Armbruster WS (2014) Floral specialization and angiosperm diversity: phenotypic divergence, fitness trade- offs and realized pollination. Annals of Botany, Plants 6:1–24.

Ashman TL and Arceo-Gómez G (2013) Toward a predictive understanding of the fitness costs of heterospecific pollen receipt and its importance in co-flowering communities. American Journal of Botany 100:1061–1070.

Bates D, Maechler M, Bolker B, Walker S (2015) Fitting linear mixed-effects models using lme4. Journal of Statistical Software 67:1–48.

Biesmeijer JC, Roberts SPM, Reemer M, Ohlemüller R, Edwards M, Peeters T, Schaffers AP, Potts SG, Kleukers R, Thomas CD, Settele J, Kunin WE (2006) Parallel declines in pollinators and insect-pollinated plants in Britain and the Netherlands. Science 313:351–354.

Cárdenas Calle S (2016) Ecología de polinización de *Oreocallis grandiflora* (Lam.) R. Br. (Proteaceae) en un matorral montano del sur del Ecuador. Tesis de licenciatura. Escuela de Biologia Universidad del Azuay.

Chalcoff VR, Ezcurra C, Aizen MA (2008) Uncoupled geographical variation between leaves and flowers in a South-Andean Proteaceae. Annals of Botany 102:79–91.

Chaves, JA, Weir JT, Smith TB (2011) Diversification in Adelomyia hummingbirds follows Andean uplift. Molecular Ecology 20:4564–4576.

Cook JM, Rasplus J (2003) Mutualists with attitude: coevolving fig wasps and figs. Trends in Ecology and Evolution 18:241–248.

Cronk Q, Ojeda I (2008) Bird-pollinated flowers in an evolutionary and molecular context. Journal of Experimental Botany 59:715–727.

Daniel MJ (1976) Feeding by the short-tailed bat (Mystacina tuberculata) on fruit and possibly nectar. New Zealand Journal of Zoology 3:391–398.

Devoto M, Montaldo NH, Medan D (2006) Mixed hummingbird–long-proboscid-fly pollination in "ornithophilous" Embothrium coccineum (Proteaceae) along a rainfall gradient in Patagonia, Argentina. Australian Ecology 31:512–519.

Fenster CB, Armbruster WS, Wilson P, Dudash MR, Thomson JD (2004) Pollination syndromes and floral specialization. Annual Review of Ecology and Systematics 35:375–403.

Fleming TH, Geiselman C, Kress WJ (2009) The evolution of bat pollination: a phylogenetic perspective. Annals of Botany 104:1017–1043.

Fuss AM, Sedgley M (1991) Pollen tube growth and seed set of *Banksia coccinea* R.Br. (Proteaceae). Annals of Botany 68:377-384.

Gómez JM, Perfectti F, Bosch J, Camacho JP (2009a) A geographic selection mosaic in a generalized plant–pollinator–herbivore system. Ecological Monographs. 79:245–263.

Gómez JM, Abdelaziz M, Camacho JP, Muñoz-Pajares AJ, Perfectti F (2009b) Local adaptation and maladaptation to pollinators in a generalist geographic mosaic. Ecology Letters. 12:672–82.

Jackson DA (1993) Stopping rules in principal components analysis: a comparison of heuristical and statistical approaches. Ecology 74:2204–2214.

Kay KM, Sargent RD (2009) The role of animal pollination in plant speciation: Integrating ecology, geography, and genetics. Annual Review of Ecology, Evolution, and Systematics 40:637–656.

Kay KM, Schemske DW (2003) Pollinator assemblages and visitation rates for 11 species of Neotropical *Costus* (Costaceae). Biotropica 35:198–207.

Kindt R, Coe R (2005) Tree diversity analysis. A manual and software for common statistical methods for ecological and biodiversity studies. World Agroforestry Centre (ICRAF), Nairobi, Kenya. Website http://www.worldagroforestry.org/treesandmarkets/tree_diversity_analysis.asp/ (accessed 02 October 2015).

Mast AR, Milton EF, Jones EH, Barker RM, Barker WR, Weston PH (2012) Time-calibrated phylogeny of the woody Australian genus Hakea (Proteaceae) supports multiple origins of insect-pollination among bird-pollinated ancestors. American Journal of Botany 99:472–487.

Murcia C (1990) Effect of floral morphology and temperature on pollen receipt and removal in Ipomoea trichocarpa. Ecology 78:1098–1109.

Ocampo-Peñuela N, Pimm S (2015) Elevational ranges of montane birds and deforestation in the Western Andes of Colombia. Plos ONE 10:1–14.

Oksanen J, Guillaume-Blanchet F, Kindt R, Legendre P, Minchin PR, O'Hara RB, Simpson GL (2016) vegan: Community ecology package. R package version 2.3-3. R Foundation for Statistical Computing, Vienna, Austria. Available at https://CRAN.R-project.org/package=vegan/ (accessed 02 October 2015).

Opler PA (1983) Nectar production in a tropical ecosystem. In: Bentley B, Elias T (eds) The Biology of Nectaries. Columbia University Press, New York, pp 30–79.

Pauw A (2007) Collapse of a pollination web in small conservation areas. Ecology 88:1759–1769.

Pinheiro J, Bates D, DebRoy S, Sarkar D, R Core Team (2015) nlme: Linear and nonlinear mixed effects models. R package version 3.1-122. R Foundation for Statistical Computing, Vienna, Austria. Available at http://CRAN.R-project.org/package=nlme (accessed 02 October 2015).

Prance GT, Plana V, Edwards KS, Pennington RT (2008) Proteaceae. Flora Neotropica, monograph 100. Hafner, New York, New York.

R Core Team (2015) R: A language and environment for statistical computing, version 3.2-3. R Foundation for Statistical Computing, Vienna, Austria. Website https://www.R-project.org/ (accessed 02 October 2015).

Rohlf FJ (2010) TpsDig, version 2.16. Department of Ecology and Evolution, State University of New York at Stony Brook, New York, USA. Website http://life.bio.sunysb.edu/morph/ (accessed 02 October 2015).

Rourke J, Wiens D (1977) Convergent floral evolution in South African and Australian Proteaceae and its possible bearing on pollination by non-flying mammals. Annals of the Missouri Botanical Garden 64:1–17.

Sapir Y, Shmida A, Fragman O, Comes HP (2002) Morphological variation of the Oncocyclus irises (Iris: Iridaceae) in the southern Levant. 139:369–382.

Sleumer HO (1954) Proteaceae americanae. Botanische Jahrbuecher fuer Systematik 76:139–211.

Stebbins GL (1974) Flowering plants. Harvard University Press, Cambridge, Massachusetts, USA.

Strauss SY, Whittall JB (2006) Non-pollinator agents of selection on floral traits. Ecology and evolution of flowers, Oxford University Press, Oxford, pp120–138.

Stroo A (2000) Pollen morphological evolution in bat pollinated plants. Plant Systematics and Evolution 222: 225–242.

Temeles EJ, Koulouris CR, Sander SE, Kress WJ (2009) Effect of flower shape and size on foraging performance and trade-offs in a tropical hummingbird. Ecology 90:1147–1161.

Tovar C, Arnillas CA, Cuesta F, Buytaert W (2013) Diverging responses of tropical Andean biomes under future climate conditions. PLoS one 8.5:e63634.

Venables WN, Ripley BD (2002) Modern Applied Statistics with S, 4th ed. Springer, New York, USA.

Weston PH, Crisp MD (1999) Cladistic biogeography of Waratahs (Proteaceae, Embothrieae) and their allies across the Pacific. Australian Systematic Botany 7:225–249.

Whittall JB, Hodges SA (2007) Pollinator shifts drive increasingly long nectar spurs in columbine flowers. Nature 447:706–710.

8

THE FORGOTTEN POLLINATORS – FIRST FIELD EVIDENCE FOR NECTAR-FEEDING BY PRIMARILY INSECTIVOROUS ELEPHANT-SHREWS

Petra Wester*

Institute of Sensory Ecology, Heinrich-Heine-University, Universitätsstr. 1, 40225 Düsseldorf, Germany

Abstract—Pollination of plants by non-flying mammals, such as mice (Rodentia), is a rarely observed phenomenon. Previously, elephant-shrews (Macroscelidea), small African mammals looking similar to mice, but not being related to them, were believed to be purely insectivorous and occasional flower visits of elephant-shrews in captivity were interpreted as a by-product of the search for insects. Only recently it was demonstrated that under lab conditions elephant-shrews regularly lick nectar from flowers. However, field observations of flower-visiting elephant-shrews and their role as pollinators were completely missing. Here I present the first evidence for flower visits and nectar consumption for elephant-shrews in the field. With video camcorders and infrared lights I recorded Cape rock elephant-shrews (*Elephantulus edwardii*) beside Namaqua rock mice (*Micaelamys namaquensis*) visiting flowers of the Pagoda lily (*Whiteheadia bifolia*, Asparagaceae) under natural conditions in the Namaqualand of South Africa. With their long tongues, the elephant-shrews visited the flowers non-destructively, definitely licking nectar, but not eating insects. The footage clearly shows that the elephant-shrews' fur around their long noses touches the pollen-sacs and the stigmas of the flowers and that the animals' fur is being dusted with pollen. As the elephant-shrews visited several flowers of different plants, it is obvious that they transfer pollen between the plants. This observation contributes to the knowledge about the behaviour of these representatives of a unique clade of small African mammals – especially in their natural habitat. With their behavioural and anatomical uniqueness, it is not unlikely that elephant-shrews even play a role as selective force driving floral evolution.

Keywords: Elephant-shrews, Macroscelidea, field evidence, nectar consumption, pollination, Whiteheadia bifolia, Asparagaceae

INTRODUCTION

Pollination of flowers by non-flying mammals is one of the most recently discovered interactions between animals and plants. This unusual and understudied phenomenon mainly includes marsupials, primates as well as rodents (Buchmann & Nabhan 1996; Carthew & Goldingay 1997; Wester et al. 2009). Especially in South Africa, in the recent years, several studies accumulated evidence that mice (order Rodentia) regularly pollinate flowers and that specific plants are adapted to pollination by these animals (Wiens & Rourke 1978; Johnson et al. 2001; Wester et al. 2009; Johnson & Pauw 2014). Until recently, elephant-shrews, small African mammals looking similar to mice, but belonging to a separate order (Macroscelidea, within the superorder Afrotheria), were often believed to be purely insectivorous (Perrin 1997). From occasional flower visits of elephant-shrews in captivity, in which the animals never lapped nectar from the nectar reservoir of the *Protea* (Proteaceae) flowers presented to them (Wiens et al. 1983), it was presumed that the animals fed on insects when visiting flowers (Fleming & Nicolson 2002, 2003). However, knowledge built up that also plant material, such as leaves, fruits and seeds, is eaten by elephant-shrews (van Deventer &

Nel 2006). Only recently, nectar-feeding by elephant-shrews was shown through laboratory experiments for *Whiteheadia bifolia* (syn. *Massonia bifolia*, Asparagaceae, previously in Hyacinthaceae, *Elephantulus edwardii*, Wester 2010), a plant that was previously described to be pollinated by mice (Wester et al. 2009), *Cytinus visseri* (Cytinaceae; *E. brachyrhynchus*, Johnson et al. 2011) and *Hyobanche atropurpurea* (Orobanchaceae; *E. edwardii*, Wester 2011). From these studies, it was only inferred that flower visits take place in the field, because the scats and the fur around the snouts of elephant-shrews captured near the plants carried pollen of the corresponding flowers (Wester 2010, 2011; Johnson et al. 2011; see also Wiens et al. 1983). However, field observations of flower-visiting elephant-shrews and their role as pollinators were completely missing.

As it was known that elephant-shrews visited *W. bifolia* flowers in the lab (Wester 2010), this plant was chosen for observations in the field. In order to test under natural conditions whether elephant-shrews visit flowers for nectar or for preying on insects, *W. bifolia* plants were monitored with the help of video camcorders and infrared light sources in the Namaqualand of South Africa.

MATERIALS AND METHODS

Observations were carried out on the farm Pendoornhoek (S 30°11.672' E 18°00.385', elevation 1085

*Corresponding author: Westerpetra3@gmail.com

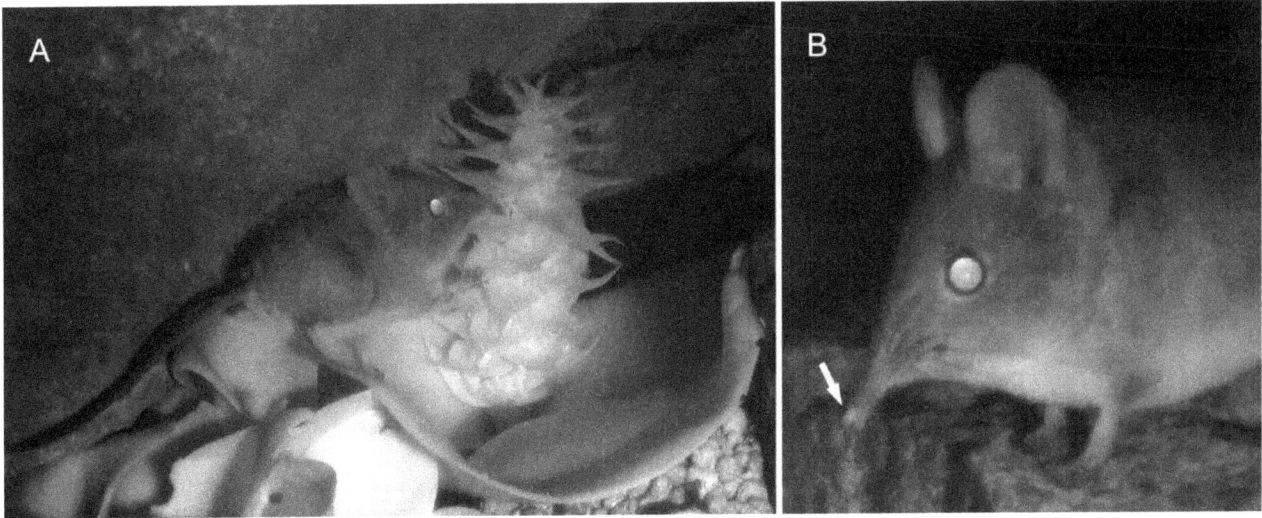

FIGURE I.   First field observation of elephant-shrews as pollinators. (A) A Cape rock elephant-shrew (*Elephantulus edwardii*, Macroscelididae) licks nectar from flowers of the Pagoda Lily (*Whiteheadia bifolia*, Asparagaceae) growing in rock crevices in the Namaqualand of South Africa. (B) The elephant-shrew has pollen (see arrow) on its nose after a flower visit. Both still images are from infrared video footage.

m), 7 km east of Kamieskroon in the Kamiesberg mountain range (western Northern Cape of South Africa), where *Whiteheadia bifolia* grows scattered in shady rock crevices. Six *W. bifolia* plants (one to two at a time per camera) with about 5 to 10 open, nectar-containing flowers per plant, were observed for potential visitors at different places. The observations were carried out with four video camcorders (Sony HDR-XR550) with additional self-made infrared light sources (using one to three 1 Ampère SMD LEDs emitting 940 nm light) using 12V/18Ah lead-acid batteries as power source. The camcorders and light sources were positioned about 70-100 cm away from the plants and running non-stop (5.5 to 13 hours). No motion or heat sensor for automatic triggering was used to avoid data loss due to mis-triggering or shutter lag. The plants were observed from 22nd to 31st August 2014 for 72 hours in total (about 22 hours during the day and 50 hours at night) over 5 days and nights between 01:00 pm and 07:30 am. The licking frequency (in-and-out flicking of tongue) could be determined in detail only for the elephant-shrews, but not for rodents, as the movement of the elephant-shrews' long tongue was clearly visible.

RESULTS

Cape rock elephant-shrews (*Elephantulus edwardii*, Macroscelididae) keenly visited the flowers of the six observed individuals of the Pagoda lily (*Whiteheadia bifolia*; Fig. 1A, Appendix I). The visits took place during all of the five days/nights between 6:45 pm and 3:45 am, mainly at late sunset and during early evening. Altogether 30 flower visits during 7 foraging bouts (sequence of flower visits) were observed. A foraging bout lasted 1.3 to 27.0 seconds (15.6 seconds on average) and included 1 to 8 flower visits (4 visits on average). With two exceptions, the flowers were visited only once. A flower visit lasted 0.5 to 7.5 seconds (3 seconds on average). With their long tongues the elephant-shrews licked the viscous nectar that is located between the ovary and the six stamens (Fig. 1A, Appendix I). Licking by

the elephant-shrews led the inflorescences to slightly wobble. The elephant-shrews licked 2 to 28 times (9.4 times on average) per flower visit with a licking frequency mostly about 5 Hz (up to about 8 Hz). When visiting a flower and licking nectar, the animals' long and flexible nose was between the stamens and the style, and touched the pollen-sacs and the stigmas of the flowers (Fig. 1A, Appendix I). Thereby, they were dusted with pollen on their nose (mostly the distal half; Fig. 1B). As far as it was noticeable in at least three of the foraging bouts, the elephant-shrews already had pollen on their nose before they visited the flowers (Appendix I). During almost all of the foraging bouts it was clearly visible that the animals accumulated more and more pollen on the fur around their noses in the process of visiting the flowers. Sometimes the elephant-shrews only briefly sniffed at a flower, but did not visit it (probably due to lacking of nectar). The elephant-shrews visited the flowers non-destructively, not consuming pollen or insects directly and they did not eat floral parts. Depending on the size of the inflorescence and the position of the flowers, the animals sometimes stood upright on their hind legs, sometimes additionally leaning on the long bracts with one or two of their forepaws to reach the upper flowers (Fig. 1A, Appendix I).

Namaqua rock mice (*Micaelamys namaquensis*, formerly *Aethomys namaquensis*, Muridae) visited several *W. bifolia* flowers of 4 different individual plants exclusively in the dark (between 7:00 pm and 06:40 am during two nights). Altogether 50 flower visits (length: 0.3 to 17.0 seconds, 4.2 seconds on average) during 10 foraging bouts (length: 7.2 to 50.0 seconds, 27.2 seconds on average) could be observed. A foraging bout included one to nine flower visits (5.5 on average). With six exceptions, the flowers were visited only once. The mice mostly licked nectar like the elephant-shrews, accumulating pollen on the fur around their snout, touching the pollen-sacs (Appendix II). However, at least in two foraging bouts, the mice nibbled at the flowers. At least one time it was visible on the footage that the mice ate the pollen-sacs of a relatively young flower (just opening)

(Appendix III). It is very likely that during these two foraging bouts the mice were feeding on the pollen-sacs of the young flowers visited as the animals were chewing after the flower visits. One time, a large bract was eaten in between the flower visits. Licking frequency of the mice was similar to that of the elephant-shrews, however, not clearly measurable as the tongue was mostly hidden by the head or snout. During the flower visits the mice mostly stood upright on their hind legs and leaned on the long bracts (one time on the stamens) with their forepaws (one time additionally with one hindpaw), causing the inflorescences to slightly wobble.

No other visitors were observed except one ant that was crawling on a bract during one video sequence. An elephant-shrew, that was lapping nectar at the same plant, did not prey on the insect.

DISCUSSION

The present study clearly shows at the example of *Whiteheadia bifolia* that elephant-shrews visit flowers non-destructively for nectar in their natural habitat. While licking nectar, the elephant-shrews touched the pollen-sacs and stigmas and were dusted with pollen on the fur of their long noses. As the elephant-shrews visited several flowers of different plants, they certainly play a role in transferring pollen between the plants. Although elephant-shrews are primarily insectivorous (Perrin 1997; Skinner & Chimimba 2005), one elephant-shrew preferred nectar over an ant crawling on the same plant.

Beside elephant-shrews, mice were also observed to visit the flowers of *Whiteheadia bifolia*. Whereas nectar drinking mice already were directly observed and photographed in the field (*Micaelamys namaquensis* at *W. bifolia*, Wester et al. 2009), video-graphed with motion activated camera traps (Striped field mouse, *Rhabdomys pumilio* at *Protea foliosa*; Melidonis & Peter 2015) or video-graphed with camcorders in combination with surveillance systems based on body heat and motion-sensing (unidentified rodent at *Liparia parva*, Fabaceae; Letten & Midgley 2009), the present study provides the first field evidence for flower visits and nectar consumption by elephant-shrews. Furthermore, this study gives meaning to former laboratory experiments showing that elephant-shrews visit flowers for nectar, and pollen found on animals captured near flowering plants (Wester 2010, 2011). Since pollen evidence provides no information about the specific behaviour of the animals at the plants, for instance, whether they eat pollen on purpose, remove pollen of their fur via grooming, touch the stigmata and cause pollination, or eat or destroy flowers. The importance of elephant-shrews for pollination of specific plants becomes only apparent via or in combination with direct field observations.

In most plants with direct or indirect evidence for pollination by elephant-shrews, mice have also been found to play a role as pollinators (*W. bifolia*: Wester et al. 2009; Wester 2010; this study; *Protea* spp.: Wiens et al. 1983; *Cytinus visseri*: Johnson et al. 2011). Both animal groups are keen on nectar, and their facial and cranial morphology fits to the floral structure of these plants. As they touch the reproductive organs of the flowers, they are capable of pollen transport in the fur around their snout, enabling pollen transfer between the plants (Fleming & Nicolson 2002; Wester et al. 2009; Wester 2010; this study). The behaviour in both small mammal groups is very similar except that mice can act destructively on flowers. Whereas *M. namaquensis* never ate or damaged flowers of *W. bifolia* in the Cederberg study (Wester et al. 2009, for other plant species see Wiens et al. 1983; Kleizen et al. 2008; Biccard & Midgley 2009), in the present study the mice sometimes fed on floral parts. Similar behaviour was found for instance in *R. pumilio*, that did not act destructively at flowers as observed by Johnson et al. (2011), but sometimes or often did so in other studies (Wiens et al. 1983; Biccard & Midgley 2009; Melidonis & Peter 2015). As destructive behaviour of *M. namaquensis* occurred only rarely in the present study and mostly legitimate flower visits took place, the species is interpreted as a successful pollinator of *W. bifolia*. Elephant-shrews have never been observed performing destructive behaviour at flowers (*W. bifolia* & other species; Wiens et al. 1983; Wester 2010, 2011; Johnson et al. 2011; this study).

Elephant-shrews and mice that are known as pollinators are omnivorous, primarily insectivorous or feed also on plant material other than nectar and pollen (Skinner & Chimimba 2005), thus they are not dependent on flowers that are temporarily restricted. In contrast, small mammal-pollinated plants depend on their pollinators and show characters that have likely evolved as adaptations to these pollinating animals (e.g. geoflory, visual inconspicuousness, bowl-shaped, robust flowers with easily accessible nectar and specific smell; see also Wiens et al. 1983; Wester et al. 2009; Wester 2010; Johnson & Pauw 2014).

With the first field evidence for flower visits and nectar consumption by elephant-shrews, the present study contributes to the knowledge about the behaviour of these remarkable representatives of a unique clade of small African mammals – notably in their natural environment. Given their behavioural and anatomical peculiarities, it is not unlikely that these (almost) forgotten pollinators even play a role as a unique selective force driving floral evolution. Future studies have to show how effective elephant-shrews are as pollinators.

ACKNOWLEDGEMENTS

I thank Lita Cole (Kamieskroon) for locality information, Ria and Koos Beukes (Pendoornhoek) for locality information and the permission to work on their property, Hanneline Smit-Robinson (BirdLife South Africa) for elephant-shrew identification advice, Klaus Lunau (Heinrich-Heine-University Düsseldorf) for providing camcorders and the Northern Cape Department of Tourism, Environment and Conservation for the necessary permits (FAUNA 1210/2014, FLORA 084/2013).

APPENDICES

Additional supporting information may be found in the online version of this article:

APPENDIX I.   With its long tongue *Elephantulus edwardii* licks nectar from *Whiteheadia bifolia* flowers, getting dusted with pollen on its nose. Infrared video.

APPENDIX II. *Micaelamys namaquensis* licking nectar from *Whiteheadia bifolia* flowers, getting dusted with pollen on its nose. Infrared video.

APPENDIX III. *Micaelamys namaquensis* feeding on pollen-sacs of *Whiteheadia bifolia* flowers. Infrared video.

REFERENCES

Biccard A, Midgley JJ (2009) Rodent pollination in *Protea nana*. South African Journal of Botany 75:720-725.

Buchmann SL, Nabhan GP (1997) The forgotten pollinators. Island Press, Washington.

Carthew SM, Goldingay RL (1997) Non-flying mammals as pollinators. Trends in Ecology and Evolution 12:104-108.

Fleming PA, Nicolson SW (2002) How important is the relationship between *Protea humiflora* (Proteaceae) and its non-flying mammal pollinators? Oecologia 132:361-368.

Fleming PA, Nicolson SW (2003) Arthropod fauna of mammal-pollinated *Protea humiflora*: ants as an attractant for insectivore pollinators? African Entomology 11:9-14.

Johnson CM, Pauw A (2014) Adaptation for rodent pollination in *Leucospermum arenarium* (Proteaceae) despite rapid pollen loss during grooming. Annals of Botany 113:931-938.

Johnson SD, Burgoyne PM, Harder LD, Dötterl S (2011) Mammal pollinators lured by the scent of a parasitic plant. Proceedings of the Royal Society B-Biological Sciences 278:2303-2310.

Johnson SD, Pauw A, Midgley J (2001) Rodent pollination in the African lily *Massonia depressa* (Hyacinthaceae). American Journal of Botany 88:1768-1773.

Kleizen C, Midgley JJ, Johnson SD (2008) Pollination systems of *Colchicum* (Colchicaceae) in southern Africa: evidence for rodent-pollination. Annals of Botany 102:747-755.

Letten AD, Midgley JJ (2009) Rodent pollination in the Cape legume *Liparia parva*. Austral Ecology 34:233-236.

Melidonis CA, Peter CI (2015) Diurnal pollination, primarily by a single species of rodent, documented in *Protea foliosa* using modified camera traps. South African Journal of Botany 97:9-15.

Perrin M (1997) Cape rock elephant shrew, *Elephantulus edwardii*. In: Mills G, Hes L (eds) The Complete book of Southern African mammals. Struik Winchester, Cape Town, p. 66.

Skinner JD, Chimimba CT (2005) The mammals of the Southern African subregion. 3rd ed. Cambridge University Press, Cambridge.

van Deventer M, Nel JAJ (2006) Habitat, food, and small mammal community structure in Namaqualand. Koedoe 49:99-109.

Wester P (2010) Sticky snack for sengis: the Cape rock elephant-shrew, *Elephantulus edwardii* (Macroscelidea) as a pollinator of the Pagoda lily, *Whiteheadia bifolia* (Hyacinthaceae). Naturwissenschaften 97:1107-1112.

Wester P (2011) Nectar feeding by the Cape rock elephant-shrew *Elephantulus edwardii* (Macroscelidea) - a primarily insectivore pollinates the parasite *Hyobanche atropurpurea* (Orobanchaceae). Flora 206:997-1001.

Wester P, Stanway R, Pauw A (2009) Mice pollinate the Pagoda Lily, *Whiteheadia bifolia* (Hyacinthaceae) - first field observations with photographic documentation of rodent pollination in South Africa. South African Journal of Botany 75:713-719.

Wiens D, Rourke JP (1978) Rodent pollination in southern African *Protea* species. Nature 276:71-73.

Wiens D, Rourke J, Casper B, Rickart E, Lapine T, Peterson C, Channing A (1983) Nonflying mammal pollination of southern African Proteas: a non-coevolved system. Annals of the Missouri Botanical Garden 70:1-31.

9

COMPARATIVE FLORAL ECOLOGY OF BICOLOUR AND CONCOLOUR MORPHS OF *VIOLA PEDATA* L. (VIOLACEAE) FOLLOWING CONTROLLED BURNS

Peter Bernhardt[1], Retha Edens-Meier[2], Dowen Jocson[1], Justin Zweck[1], Zong-Xin Ren[3], Gerardo R. Camilo[1], Michael Arduser[4]

[1]*Department of Biology, Saint Louis University, St. Louis, MO, USA 63103*
[2]*School of Education, Saint Louis University, St. Louis, MO, USA 63103*
[3]*Key Laboratory for Plant Diversity and Biogeography of East Asia, Kunming Institute of Botany, Chinese Academy of Sciences, 132 Lanhei Road, Kunming, Yunnan 650201, P. R. China*
[4]*325 Atalanta Ave., Webster Groves, MO, USA 63119*

Abstract—We compared pollinators, pollination rates and seed set of bicolour and concolour morphs in self-incompatible, *Viola pedata* over two seasons. The two populations grew on a wooded slope (CR) vs. an exposed glade (SNR) and were of unequal sizes. Both were burned in 2014. The number of flowers produced by concolour plants at SNR was higher in 2014 while the number of flowering bicolour plants increased significantly at CR in 2015. Petal temperatures, regardless of site, showed that the dark purple, posterior petals of bicolours were consistently warmer than their own mauve-lilac, anterior (lip) petals and the all mauve petals of concolours. Major pollen vectors were polylectic/polyphagic bees (Andrenidae, Apidae and Halictidae) but females of *Andrena carlinii* dominated at both sites. Bees foraged on flowers upside down or right side up but neither mode correlated with either morph. Bees foraged preferentially on concolour at both sites. Pollen tube counts were higher in concolours at both sites with a marginally greater number of pollen tubes penetrating concolour ovules regardless of site or year. While both populations produced more seeds in 2014 SNR plants always produced more seeds than CR plants. The increasing numbers of bicolour plants at CR in 2015 suggested that bicolours may equal or outnumber concolours as dark petals offer additional warmth to ecto-thermic pollinators foraging in a cooler, shady forest vs. an open, sunny glade. Subtle environmental factors may give a floral trait a selective advantage influencing fitness when an unbalanced polymorphism persists in discrete and localized populations.

Keywords: Bees, bicolour, concolour, morphs, ovules, pistils, pollen tubes, posterior petals

INTRODUCTION

Colour polymorphisms have been well documented in flowers of unrelated species. Unlike plants with heterostylous flowers (e.g. *Linum*, see Armbruster et al. 2006) most colour morphs are interpreted as unbalanced polymorphisms (sensu Futuyma 2013) as the frequencies of 2-4 colour morphs vary broadly with natural distribution over time (Irwin & Strauss 2005; Pelligrino et al. 2008). A population's shift in colour morph frequencies may have more than one explanation (Rausher 2008). While the population's response to the selective foraging of its dominant pollinators is anticipated (Epperson & Clegg 1987; Irwin & Strauss 2005; Malberla & Nattero 2011; Russell et al. 2016) there are other factors. These may include florivory/herbivory (Carlson & Holsinger 2013; de Jager & Ellis 2014; Sobral et al. 2016), differential rates of self-pollination in discrete morphs (Fehr & Rausher 2004), variation in inflorescence display (Gomez 2000), genetic trends that cause unidirectional changes in

pigments (Rausher 2008) and/or an indirect response to selection related to pleiotropic, non-floral traits (Armbruster 2002).

Curiously, *Viola* species have not been used to study variation in frequencies of floral colours although they are recorded throughout the genus in North America and Europe (McKinney 1992; Hildebrandt et al 2006; Mereda et al. 2008; Pellegrino et al 2008; Marcussen & Borgen 2011). Instead flowers of *Viola* species were more likely to be used as model systems in genetic variation (Clausen 1926; Culley 2002), systematic variation (Nieuwland & Kaczmarek 1914; McKinney 1992), developmental morphology (Johri et al 1992; Weberling, 1989), molecular development (Wang 2008) and reproductive ecology (Gurevitch et al. 2006; Winn & Moriuchi 2009). In some *Viola* species endemic to Europe petal colour frequency is predictable according to whether populations grow on old soils polluted by zinc or lead (Hildebrandt et al. 2006). Gradations in petal colour in the "zinc" violets are also results of a past history of interspecific introgression (Migdalek et al. 2013).

There should have been a continuous interest in uniting demographic studies of colour morphs in *Viola* species with

---

*Corresponding author: bernhap2@slu.edu

their pollination ecology as interpretations of floral adaptations in their chasmogamous flowers started in the 19th century (Darwin 1876; Müller 1883). However the much later work of Beattie (1969; 1971ab; 1972; 1974; 1976) remains the definitive introduction to pollination mechanisms and breeding systems in this genus. Specifically, Beattie's observations showed that *Viola* petals and pedicels changed positions and angles over their respective lifespans encouraging visits by many pollinators representing at least three insect Orders (Diptera, Hymenoptera, Lepidoptera). Some insects were more likely to forage for nectar in an inverted position acquiring ventral depositions of pollen (sternotribic) while others foraged after landing prone on the liplike anterior petal receiving dorsal depositions (nototribic). Beattie (1974) attributed these foraging behaviours to the evolution of two overlapping syndromes based on floral architecture, petal ornamentation and modifications of the terminal surfaces of anthers and pistils.

With more than 400 species in the genus *Viola* (Mabberley, 1997) it is not surprising that a few recent studies challenge the earlier descriptions of generalist entomophily reported by Davidse (1968) and Beattie (1969; 1971; 1974). Herrera (1990; 1993) concluded that floral traits of *V. cazorlensis* most probably evolved under disruptive selection as only one hawkmoth, *Macroglossum stellatarum* was the primary pollinator. Freitas & Sazima (2003) found that flowers of two, Neotropical, high elevation species produced little or no nectar and depended primarily on pollen harvesting bees (*Anthenoides*; Andrenidae). This conflicts with the generalization that insects forage on *Viola* flowers for nectar exclusively, and this ends in the passive deposition of pollen onto the vector's body (sensu Bernhardt 1996).

In fact, *Viola pedata* shows floral characteristics atypical for the genus that may make it ideal for studies on colour morph frequencies. As it produces no cleistogamous flowers and appears to be the only *Viola* species studied so far with a self-incompatible (late-acting) breeding system (Becker & Ewart 1990). Unlike the much used *Ipomoea purpurea* (Fehr & Rausher 2004) it is an obligate out-crosser. It is not known to hybridize with allied, acaulescent species (McKinney 1992). Floristic taxonomists recorded a white-flowered form (alba) in *V. pedata*, a second form in which all petals are lavender-mauve (concolour) and a third in which three petals are lavender-mauve while the two posterior petals are dark purple (bicolour). These last two forms are so distinct that children living in the Missouri Ozarks called the concolour forms hens and bicolour forms roosters (Steyermark 1963) but are these morph frequencies pollinator driven? Carroll & Goldmann (1994) found that dusky winged skippers (*Erynnis* species; Lepidoptera) spent the same amount of time foraging on bicolour and concolour morphs of *V. pedata*. In fact, these insects foraged on each morph in proportion to morph frequencies found in the same population.

We require more information on the pollination ecology of *V. pedata* as insects observed foraging on its flowers in past studies were not always identified to species and collectors failed to note whether euthanized foragers carried the pollen of the host flower. Gibson & Davie (1901) proposed that the species was pollinated by a combination of Lepidoptera and long tongue bees. Beattie (1974) was far more specific recording 63% of all visits to the flower by bees in the family, Andrenidae (species unidentified) and 21% to Lepidoptera including day-flying hawkmoths (*Hemaris*; Sphingidae). In contrast, Carroll & Goldman (1994) described a more specialized system in Missouri based on small Lepidoptera. For Beattie (1974), floral presentation in *V. pedata* expressed intermediate floral characters based on its floral architecture and observations of how insects foraged on the flower. About 36% of all flowers of *V. pedata* received insects landing on the anterior (lip-like) petal leading to dorsal (nototribic) depositions of pollen while 64% foraged in an inverted manner leading to ventral (sternotribic) depositions.

This paper attempts to address and clarify eight interrelated questions regarding the pollination dynamics and relative fitness of two, colour morphs of *Viola pedata* in two populations in Missouri following exposure to controlled burns. First, do colour morphs offer the same numbers of flowers/plant (floral presentation) over time? Second, as dark colours absorb heat more efficiently, are the exposed and reflexed, deep purple petals of bicolour flowers warmer than the nodding, mauve-lilac posterior (lip) petal on the same flower and in the posterior petals of concolour flowers? Third, does pollinator diversity vary according to colour morph and site over time? Fourth, do pollinators approach and then forage on bicolour and concolour flowers in the same way? Fifth, are pollinators more likely to be generalist or specialist foragers as this species produces both nectar and pollen as edible rewards? Sixth, do natural rates of pollination (pollen tubes penetrating pistils) vary in two colour morphs according to site and season? Seventh, do bicolour and concolour morphs produce the same number of ovules/pistil over time? Finally, do bicolour and concolour morphs produce the same numbers of seeds in different sites over time? Ultimately, these combined results will allow us to ask which morph is fittest according to site, season, pollinator activity and history of fire-regime.

MATERIALS AND METHODS

Study sites and field states

The first site used from 5/2/13 – 5/7/13, from 4/14/14 – 5/27/14 and from 4/2/15 – 6/3/ 15 was on Oak Ridge, adjacent to the 11 km Sugar Creek trail within Cuivre River State Park (CR), Lincoln County, Missouri. Less than 200 flowering plants of *V. pedata* were found on the slope under mixed hardwood (*Quercus* dominated) forest annually for all three seasons (Fig. 1). The site is usually burned in late winter in alternate years by park staff and was burned in 2014 but not in 2013 or 2015. Voucher specimens were deposited in the herbarium of the Missouri Botanical Garden (MO).

The second site used from 4/16/14 – 5/28/14 and from 4/8/15 – 6/4/15 was at Shaw Nature Reserve (SNR) at Gray Summit, Franklin County (Fig. 2). The population under study was confined to the relatively sterile,

FIGURE 1.   Broad habitat view of field site at Oak Ridge, Cuivre River State Park, April 2015 (R. Edens-Meier, photographer).

FIGURE 2.   Broad habitat view of field site at Shaw Nature Reserve, April 2015. Note the dead tillers of grass remaining from the previous autumn when the glade is not burned (Justin Zweck, photographer).

highly drained, sandy soil lower edges of the dolomite glade (Crescent Glade) dominated by grasses and forbs with some shrubs (*Rhus* species). In 2014 and 2015 this site produced > 1,000 flowering, basal rosettes of *Viola pedata*. The site was burned in 2014 by Shaw employees but not in 2015. Burns are conducted in the winter at 3-year intervals, coinciding with the burning of surrounding woodlands. However, some years do not burn well due to the sparseness of fuel in a true glade flora. Prior to 2014 this site was burned last in 2010 (James Trager, personal communication). Voucher specimens were deposited as above. Pooled field observations, bagging and collection

hours (see below) at both sites by co-authors from 2013 – 2015 totalled approximately 112 hours.

Morphs frequency and flowers per morph

We documented when the first flowers opened and the last flowers wilted at both sites in 2014 and 2015. Morphs of *V. pedata* at CR were restricted to the bicolour and concolour forms (McKinney 1992). Only two plants with white morphs were observed at the SNR in 2014 and 2015. Neither was measured and we did not take any time to observe insect visitation of these flowers. In 2014 and 2015 we kept counts of the number of rosettes of bicolour and concolour morphs (Figs. 3 & 4) at each site. We also counted the number of flowers produced within each rosette. However, as the SNR population was so large we also established a 13.1 × 8.0 m quadrat where the majority of bicolour morphs were found and counted the number of bicolours in 2014 and 2015.

Attractants, rewards and petal temperatures

Viola pedata is among the few species native to North America lacking the characteristic trichome tufts (beards) towards the bases of both lateral petals (McKinney 1992). Flowers of both morphs were dissected to see if they had nectar glands attached to the connective filaments of two stamens as in most *Viola* spp. (Beattie 1974). To determine if the flowers produced a discernible scent we smelled flowers of each morph on plants between 10 AM – Noon. We also placed 1 or 2 flowers of the same morph in clean glass vials, capped the vial and then smelled the contents 20 and 30 minutes later. In 2015 we measured the spur lengths of living flowers of concolour and bicolour morphs, remaining attached to their pedicels, at the Shaw Nature Reserve using electronic digital calipers (Fisher Scientific).

As *V. pedata* is a vernal flowering species the heat generated by solar energy once absorbed and retained by flower petals, may reward ectothermic, insect pollinators (see review in Willmer 2011). One presumes that dark colours (deep purple) absorb more solar energy as heat than light (mauve – lavender). Petal temperatures of both morphs were recorded using an Omega Type T Thermocouple Cu-CuNi HH-25TC Thermometer, Range -80°C to 400°C and an Omega Hypodermic Tissue Probe MP1-30 ½-T-G-60 SMPW-M were used to determine petal temperatures. In each flower of each morph the tissue probe was inserted into one of the posterior petals by carefully weaving the probe through tissue three times (Fig. 3). The temperature was recorded after one minute. The same procedure was also used to determine the petal temperature for the lower, nodding, mauve-lilac, anterior (lip) petal in the same flower. Before taking each petal temperature we recorded the time of day, the ambient temperature and whether the sky was sunny or cloudy.

Floral foragers

Insect visitors were observed *in situ* at both sites. Videos of floral foragers were made by R. Edens-Meier using a Sony Full HD 1080 Handycam, HDR-CX760V, 24.1 Megapixels at the CR site. Three of these videos are shown on Youtube; https://www.youtube.com/results?search_query=viola+pedata. Insect foragers were netted only when they were observed visiting one or more flowers of *V. pedata* and could be observed extending proboscides down the floral tube or manipulating anthers. From 4/6 – 4/7/2013 at CR and from 2014 -2015 all insects collected were euthanized using fumes of ethyl acetate and they were pinned, labelled and identified. From 2014-2015 insects were always euthanized in separate jars according to the morph on which they were captured.

Pollen load analyses

To identify and record pollen carried by foragers each specimen euthanized within 24 hours was first placed on a

FIGURE 3. Bicolour flower of *Viola pedata*, with thermocoupler probe inserted through the posterior petals (R. Edens-Meier, photographer).

FIGURE 4. Concolour flower of *Viola pedata* (R. Edens-Meier, photographer).

glass slide and bathed in 1-2 drops of ethyl acetate and/or the scopal load was removed with a probe and added to the slide surface. Grains left on the slide following ethyl acetate evaporation were stained in Calberla's fluid and mounted with glass cover slips for light microscopy after the stain dried. All techniques for washing, staining, mounting, observing grains and co-referencing the label on the slide to the label under the pinned insect followed Bernhardt et al. (2014). As more than one insect was euthanized in the same morph jar pollen of a known species was considered present on a slide when > 25 grains of that morphotype were counted. Therefore, pollen loads from each foraging insect were classified as one of the following; No pollen (< 25 grains of *V. pedata*), pure loads (> 25 gains of *V. pedata*), mixed load (> 25 grains of *V. pedata* +> 25 grains of at least one other co-blooming species) and alien load (> 25 grains of other species but no *V. pedata*).

Landing orientation and foraging bouts on morphs by bees

Bees were the most frequent visitors to these flowers (see below). At the CR site in 2014 and 2015 we observed how bees oriented themselves upon approach and landed on the flowers of each morph prior to feeding on nectar or collecting pollen. Two modes were observed. The bee could land directly on the liplike anterior petal or on a lateral petal of the same flower. This was recorded as right side up (Beattie, 1976 used the term, nototribe). Otherwise, the bee landed on one or both of the posterior petals continued to cling to the posterior petals with its third pair of legs and so reached the anther cone or spur by foraging upside down (Beattie 1976 used the term sternotribe). Insects that landed on the flowers but failed to forage for nectar or pollen were not recorded. While we observed bees foraging on flowers at the SNR site in 2014 and 2015 the population was so large in both years it was not possible to discern when a bee actually entered the site to start foraging and when it exited (see above).

In 2015 a new protocol was added. We followed bees of several species (most were females of *Andrena carlinii*) at CR to determine how many flowers of each morph they visited during a foraging bout. Bouts were recorded when the bee entered the field site and visited its first flower within the population. Counts stopped when the bee either left the site or was observed foraging on the flowers of another species (see Bernhardt & Montalvo 1979). This protocol could not be used at SNR due to its size as described above.

Natural rates of ovule number, pollination (tubes penetrating pistil tissue) and seed set

Pedicels of both morphs in both populations were selected at random and tagged with jeweller's tags while flowers were in bud each week. As the populations at each site remained in flower less than three weeks each year tagging occurred over a two-week period at both sites in 2014 and 2015. Only one pedicel/basal rosette was tagged and the perianth of the tagged bud was allowed to complete its floral lifespan. Each bicolour or concolour flower was collected one to three days after we observed the wilting of petals. The wilted flower was fixed in 3:1 95% ethanol:glacial acetic acid for 2-6 hours then preserved in 70% ethanol. Pistils were excised, softened in sodium sulfite solution under incubation and squashed in decolourized aniline blue prior to viewing under epifluorescence using the Zeiss Axioskop 40 as described in Edens-Meier et al. (2010) to view pollen tubes germinating on the stigmas and penetrating styles, ovaries and ovules. Due to the rigidity of the style each pistil had to be softened for 60 minutes at 42°C but, as the stigma and style are so small, it was not necessary to split them lengthwise to view pollen tube progress. The ovary was butterflied with a scalpel, prior to squashing to permit a count of the number of ovules and observe pollen tube penetrations of micropyles.

To record seed set basal rosettes were selected at random. An open flower in each rosette was then selected at random and tagged, as above. After the petals wilted we bagged the pedicel in a marked organza bag. After four weeks we recovered and collected as many bags as possible recording the number of mature, filled seeds in each bag. Seeds produced at each site were donated to the seed bank maintained by the Missouri Botanical Garden at the Shaw Nature Reserve.

Statistical analyses

The overall design of this study is a three-way factorial design with colour morph (concolour vs bicolour) of the flower as a fixed effect, whereas location (SNR vs CR) and year (2014 vs 2015) were considered random effects. We used the package lme4 (v. 1.1-7) in the R computational environment (v. 3.1.0, R Core Team 2014) in order to perform the mixed effects ANOVA's. Given that all the data collected from the flower squashes were counted data, e.g., number of pollen tubes, we used a squared root transformation to meet the assumptions of the test.

RESULTS

Morphs' frequency and the number of flowers per morph

At the CR (Cuivre River) site in 2014 (burn year) we counted a total of 28 bicolour morph plants and 32 plants with concolour flowers (approximately 1:1.02 morph ratio). The difference in morph ratios was not significant (binomial test, $P = 0.5654$). In 2015 we counted 161 plants producing bicolour flowers and 88 producing concolours, resulting in a ratio of 1.8: 1.0 respectively. This ratio was a significant deviation from one to one ($P < 0.0001$). At the SNR (Shaw Nature Reserve) in 2014 and 2015, we stopped counting flowering basal rosettes after 1000 and morph ratios remained self-consistent in both seasons (bicolour 40.0: concolour 1.0). In 2015, the main area where bicolour was most common, produced 25 bicolour plants and 71 concolour plants.

While the number of flowers per plant at the CR site was significantly lower in 2014 than in 2015 there was no significant difference between the number of flowers/plant in bicolour vs. concolour morphs (Fig. 5). In contrast, the Mean number of flowers/concolour plant at SNR was far higher in the burn year of 2014 (6.9 flowers/plant) compared to the bicolour plants (2.7). In 2015, when the site was not burned, there were no significant differences between the morphs (bicolour 3.7: concolour 4.0; Fig. 5).

Floral phenology and presentation (attractants, rewards and petal warmth)

Populations at both sites remained in bloom for 14 to 20 days each year, flowering from mid-April to early May. Once the corolla of either morph opened at each site it wilted within seven days or less. Melanism in the bicolour flower varied at the CR site where they were most common in 2015. In most plants only the two, top posterior flowers were dark purple but we found some other specimens in which melanism extended to the tips of the lateral petals as described and illustrated in the two species of Eurasian, melanium pansies by Clausen (1926). We also found one bicolour plant in which all five petals showed some degree of melanism. In the absence of lateral beards the anther cone

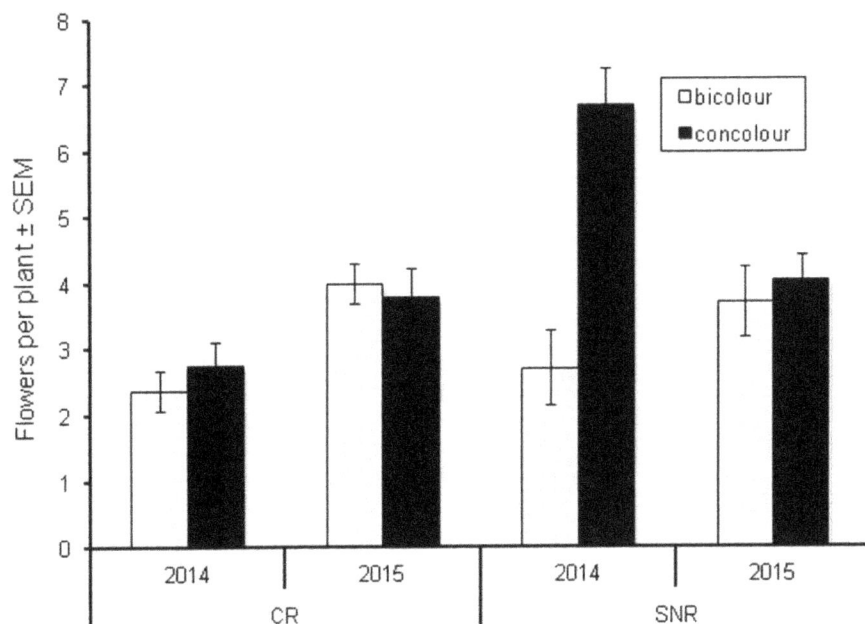

FIGURE 5. Mean number of flowers per plant of both morphs of *Viola pedata* at two sites and two years (2014 and 2015), Missouri, USA.

was fully visible at anthesis in both morphs. While the style is straight and lacks a rostellum (as noted by Beattie 1974) we also report that it is also very stiff and persists for several days following the withering of the corolla.

No discernible scent was detected in the flowers of either morph at either site and that includes sampling flowers in capped vials. Dissections of both morphs showed that each of the two, lower, anther connectives in each flower wore a large, elongated, green gland that protruded up to 66% the length of each narrowly, keeled spur. Spur lengths of bicolour and concolour morphs at SNR were greater than 5.0mm, but differences between morphs were not statistically significant ($t = 0.81$, $P = 0.427$); (concolour, 5.2 ± 0.70; $N = 18$) and (bicolour 5.4 ± 0.71; $N = 0.71$).

Petal temperature data at both sites was pooled. Regardless of colour morph the two, top (posterior) petals were anywhere between 1 – 3 degrees warmer than the lower, nodding, and often shaded, anterior (lip) petal ($F = 6.36$, $P = 0.0179$). For the concolour morph, though, there was no statistically significant difference between either of the two, petal temperatures. In contrast, the deep purple, posterior petals were always ~3 degrees warmer than the flower's anterior petal and 2 degrees warmer than the posterior petals of any concolour flower at either site ($t = 2.26$, $P = 0.0317$). Thus, it seems likely that the overall temperature differences we recorded were driven by the bicolour morph. Under shady conditions (e.g., under CR tree canopy or cloudy periods at SNR) the differential between the posterior and anterior petals decreased in magnitude but remained consistent.

Forager and foraging diversity, and pollen load analyses

Collections of insects at CR began on 5/2/13 when we observed bees on both morphs. We netted and euthanized 12 specimens that day and all were identified as females of *Andrena carlinii* but we did not record the morphs on which

each bee was caught. Two *A. carlinii* collected on 5/2/13 failed to carry the pollen of *V. pedata* but did carry the pollen of *Oxalis violacea*. The ten remaining *A. carlinii* collected the same day carried mixed loads of *V. pedata* pollen with the pollen of *O. violacea* and/or *Hypoxis hirsuta*. On 5/6/13 and 5/7/13 we caught one female *A. carlinii* and one female *Andrena nasonii* on concolour morphs respectively. The single *A. nasonii* carried the pollen of *Viola pedata* mixed with the pollen of *O. violacea*. In contrast, the female *A. carlinii* carried only grains of *H. hirsuta*.

An additional seven bee species were collected and identified on *V. pedata* when the contents of both sites were pooled from 2014-2015 but females of *Andrena carlinii* remained the most commonly recorded foragers on both morphs at both sites in both seasons (Tab. I; Fig. 6). This brought the total of bees captured from 2013-2015 to 56. While three more bee species were collected at SNR, compared to CR, two of those species represented single captures over two seasons. As concolour morphs represented the vast majority of the SNR population (see above) it is not surprising that we collected only four bees, including a male of *Anthophora ursina*, on the bicolour morph over two seasons (Tab. 2). Over a two-year period the number of bees collected at both sites had almost identical pollen loads. At CR, 68% of bees collected carried the host flower's pollen while 70% carried the host flower's pollen at SNR (Tab. I).

Females of *A. carlinii* rarely visited more than one or two open flowers produced in the same rosette at CR but they commonly visited more than one rosette in the same patch or clump. We observed that, if an *A. carlinii* landed on the same flower for a second time during the same foraging bout it did not stay long enough to drink nectar or collect pollen a second time. In 2015, at both sites, we noted that, sometimes, *A. carlinii* often appeared to lose its sense of direction if it landed on the two posterior petals, regardless of morph. When these bees were unable to find the common

TABLE I.   Pollen loads of bees collected on *Viola pedata* at Cuivre River and Shaw Nature Reserve, Missouri, USA, 2013-2015.

Location Species (sex)	Morphs visited Bicolor/Concolor	*Viola*	Pollen Loads *Viola* + other spp.	Other species only	No Pollen
CR 2013					
Andrena carlinii (f)	NA*	0	11	3	0
CR 2014 – 2015					
Andrena carlinii (f)	0/6	1	10	5	0
A. nasonii (f)	1/0	0	0	0	1
A. nasonii (m)	0/1	0	1	0	0
A, perplexa (f)	0/1	0	1	0	0
A. pruni (f)	0/1	0	1	0	0
Anthophora ursina (f)	0/1	0	0	1	0
Anthophora ursina (m)	1/1	0	2	0	0
Sub Totals	2/11	1	15	6	1
SNR 2014 – 2015					
Andrena carlinii (f)	2/8	2	6	2	0
A. cressoni (f)	0/2	0	2	0	0
A. perplexa (f)	0/1	0	1	0	0
Anthophora ursina (f)	1/2	0	1	2	0
Augochlorella aurata (f)	0/1	0	0	0	1
Lasioglossum bruneri (f)	0/1	0	1	0	0
L. cressoni (f)	1/0	0	0	0	1
Sub Totals	4/15	2	11	4	2
Grand Totals	6/24	3	37	13	3

*NA = Not assessed

entrance to the anther cone or spur they flew to another flower or left the site.

We observed a total of 39 foraging bouts at the CR site by females of *A. carlinii* from 4/10 – 5/2/2015. On 4/14/15 we observed that one, female, *A. carlinii* visited 31 flowers on 27 basal rosettes over a 20 minute period. It visited 4 bicolour flowers and 10 concolour flowers before resting on an inflorescence of an *Antennaria* species and probing the florets ending the first bout. It then began a second bout returning to *V. pedata*, less than 30 seconds later, to forage on 10 bicolour flowers and seven concolour flowers before leaving the site.

In 2014 the ratio of concolour to bicolour flowers was almost equal at the CR site (see above). In that season, we observed that bees of four species and varying sizes visited 17 flowers of bicolour and 25 flowers of concolour. At the same site in 2015, when bicolour morphs greatly outnumbered concolours (see above), we observed bees of different species and sizes visiting 51 flowers of bicolour and 59 concolour flowers. Over a two-year period of observation at SNR, where the ratio of bicolour to concolour was always 40:1, only six observations of bee foraging observed were to bicolour morphs.

Male bees of all species collected were observed probing the spur regardless of site or morph. They were not observed to forage on the anther cone, regardless of colour morph. In contrast, after females of each *Andrena* species foraged for nectar, they were observed to clasp the cone of anthers around the style repeatedly while they scraped out pollen with their forelegs. These grains were transferred to their scopae (Figs.6, 7; and see https://www.youtube.com/watch?v=190rBkdvVXE).

While gynes of *Bombus* species nested at both sites they only hovered near the flowers, or touched the petals momentarily without foraging, then flew away. Pollen loads

FIGURE 6.   Female of *Andrena carlini* foraging upside down on a bicolour flower. (Zong-Xin Ren, photographer).

TABLE 2.   Insect-mediated rates of pollination in *V. pedata* at Cuivre River and Shaw Na-ture Reserve, Missouri, USA, in 2014 and 2015.

Location	Year	Morph	Number of flowers	Grains /tubes in Pistils Absent
CR	2014	Bicolour	18	10 (55.6%)
		Concolour	31	8 (25.8%)
	2015	Bicolour	15	9 (60.0%)
		Concolour	15	5 (33.3%)
Grand total			79	32 (40.5%)
SNR	2014	Bicolour	32	4 (12.5%)
		Concolour	34	1 (3%)
	2015	Bicolour	15	5 (33.3%)
		Concolour	15	0 (0%)
Grand total			96	10 (10.4%)

TABLE 3.   Comparative averages of ovule, pollen tube penetration, and seed production of *V. pedata* at two sites over two years. Values in parentheses represent the standard error of the mean, except for seed production, which represents the range.

Location	Year	Morph	Ovules	Pollen tubes	Pollen tubes	Seed production
CR	2014	Bicolour	32.9 (2.3)	24.1 (4.2)	11.4 (2.5)	4.5 (0–27)
		Concolour	37.0 (2.2)	23.5 (3.8)	12.4 (2.5)	3.7 (0–38)
	2015	Bicolouru	34.1 (3.8)	18.8 (7.3)	3.7 (2.1)	6.5 (0–37)
		Concolor	29.0 (2.4)	17.7 (3.1)	7.1 (1.8)	6.0 (0–38)
SNR	2014	Bicolour	31.1 (2.4)	27.3 (3.6)	14.3 (2.1)	15.3 (0–38)
		Concolour	38.2 (2.7)	28.2 (2.6)	16.2 (1.9)	21.1 (0–40)
	2015	Bicolour	21.1 (2.7)	19.8 (3.2)	6.7 (1.4)	13.5 (0–35)
		Concolour	24.7 (3.5)	44.8 (9.1)	16.4 (4.3)	12.5 (0–40)

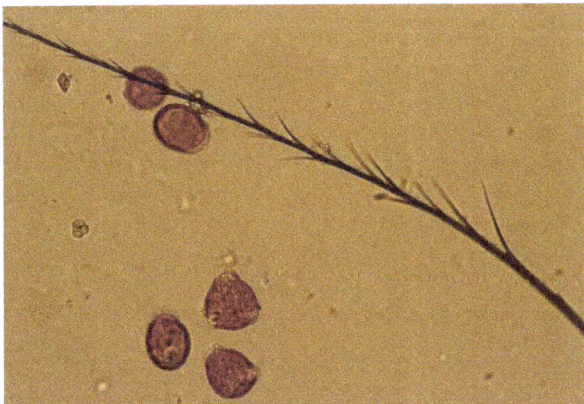

FIGURE 7.   Pollen of *Viola pedata* and scopal hair of *Andrena carlini*. (D. Jocson, photographer)

of male bees, regardless of species, indicated they were all polyphagic while pollen loads of females indicated that most were polylectic (see above). While a male of the oligolectic species *Andrena violae* was collected on coblooming *Viola palmata* in 2013 at CR we did not catch this species on *V. pedata* at either site over the next two seasons.

Small bombylid flies were observed at both sites in both years but they did not contact the anther cone or stigma while they foraged. They did not carry significant loads of the host flower's pollen (see above) in 2014 and we stopped collecting them.

As visits by Lepidoptera at either site were so infrequent; we observed but did not collect them. We did not observe *Erynnis* species at either site. In 2014 we observed one visit by *Papilio glaucus* in which the butterfly visited two flowers (one flower on each morph). An unidentified *Papilio* species with black forewings and greenish-black hind wings visited two flowers on two rosettes (morphs not recorded). *Papilio* species were observed most commonly foraging for nectar on co-blooming *Phlox divaricata* at CR. An unidentified, yellow-winged member of the Pieridae (resembling *Phoebis sennae*) visited three flowers of concolour. In all three cases butterflies did not land on the posterior petals and did not forage in an inverted position. They landed on the lateral and/or anterior petals extending their proboscides under the anther cone to reach the spur.

Orientation of bees on morphs

In 2014 at the CR site we recorded 59 visits of bees to flowers of *V. pedata*. That was the year we collected bees within less than 60 seconds of their entry into the site while they foraged on their first flowers. We did not wait for them to finish their bouts (see above). A total of 36 (0.64) visits were made using the right side up orientation. In contrast, when we allowed CR bees to finish their foraging bouts in

2015 we counted 104 visits to these flowers in which (0.27) of the landing orientations were right side up.

Therefore, when observations of the first visit of a bee to a *V. pedata* at the CR site are combined for 2014 and 2015 the bee was more likely to land on the anterior (lip) petal right side up, and insert its proboscis under the anther cone, than it was to land on the posterior petals first and then forage upside down when that foraging bout began. However, as the same bee visited additional flowers during the same foraging bout in 2015 at CR the more likely it would change its foraging pattern from right side up to upside down. As these bouts progressed and ended in 2015 a total of 76 (0.74) of these orientations were made upside down. When a bee foraged upside down we did observe and record infrequent cases in which its third pair of legs clung to one or both of the two lateral petals. This usually occurred when posterior petals were askew (see, https://www.youtube.com/watch?v=qNAcOkcI9Ak). In the majority of observations, a female of *Andrena carlinii* clutched the posterior petals with its third pair of legs and the bee's abdomen also appeared to contact these petals (Fig. 6; https://www.youtube.com/watch?v=I90rBkdvVXE) regardless of colour morph.

In 2014 and 2015 there were no statistically significant differences in bee orientation to bicolour vs, concolour morphs at the Cuivre River site. All but four bee orientations observed at SNR from 2014-2015 were always to concolour morphs and were always made upside down with one exception. The collection of the male *Anthophora ursina* in 2014 was made after the bee landed right side up on the bicolour flowers. As related above, we were not able to determine when bees began and ended foraging bouts on *V. pedata* at SNR. Regardless of morph, bees foraging upside down at CR and SNR continued to cling to the two posterior petals via their third pair of legs.

Ovule number and rates of pollination

The number of ovules in ovaries (Tab. 3) was consistent between sites ($F = 0.0158$, $P = 0.9972$), but not years ($F = 35.3$, $P < 0.0001$). In 2015 ovule production was consistently lower at SNR compared to SNR in 2014 and at CR in 2014 and 2015 (Fig. 8). There was no difference between the numbers of ovules inside an ovary between morphs at either site in either year ($F = 0.7863$, $P = 0.3831$).

Results of squashes and fluorescence analyses (Tab. 3, Figs. 10-13) at both sites indicated that rates of insect-mediated pollination were far higher at the SNR vs. CR, regardless of year ($F = 5.7$, $P < 0.0001$). However, at both sites, analyses of pollen tubes in pistils showed that bicolour morphs were visited less frequently compared to concolour morphs regardless of year (Tab. 2). Rates of pollination did not vary much, at either site, according to whether the area was burned or not (Tab. 3). At CR the number of pistils lacking pollen grains or pollen tubes increased slightly in both morphs the year after the burn (2015). At SNR, though, the number of bicolour pistils that were not pollinated in 2015 increased over 20% the year after the burn (2015). In contrast, the number of concolour pistils at

SNR lacking grains and tubes in 2015 actually declined slightly with100% of pistils analyzed containing pollen tubes (Tab. 2, 3).

However, the mean number of pollen tubes actually germinating and penetrating a pistil did not vary between sites ($F = 0.04$, $P = 0.8345$) or years ($F = 1.75$, $P = 0.1957$). We did detect weak interaction effects between morphs and years ($F = 3.87$, $P = 0.0519$). This effect seems to be driven by the increased number of tubes in the style in the concolour morph at SNR in 2015 (Fig. 9).

As this is a species with late-acting self-incompatibility the number of pollen tubes that actually entered the ovary and penetrated ovule micropyles within 7 days after the flowers opened, required comparison. At both sites, ovule penetration was higher in 2014 (burn year) compared to 2015 ($F = 35.3$, $P < 0.0001$; see also Tab. 3). The average number of tubes penetrating ovules in 2014 and 2015 was marginally higher in concolour pistils regardless of site ($F = 3.9143$, $P = 0.0581$; Fig. 10).

Seed set

Seed production varied greatly between sites, morphs and years (Tab. 3). There were no main effects in seed set among years ($F = 0.6592$, $P = 0.6877$), location ($F = 0.3571$, $P = 0.6910$), or flower morph ($F = 0.2404$, $P = 0.4818$). There was a significant interaction between year and location ($F = 5.9292$, $P = 0.0161$), with seed set at SNR in 2014 being significantly higher than in 2015 and it was significantly higher compared to either year at CR.

DISCUSSION

Variation in colour morph frequencies

Unlike studies on zinc violets (Hildebrandt et al 2006) we can't attribute variation in colour morph frequencies in our populations to either a history of interspecific hybridization or soil pollution. However morph ratios in *V. pedata* may vary, at least in part, on other environmental factors according to habitat. We note that irregular burning regimes had little negative effect on morph frequencies in a rocky glade (SNR). A glade burn may be very hot but it must also be brief as only dead, thin stems of forbs and grass culms provide fuel. In contrast a positive but short-term effect, caused by the brief release of micronutrients in ash after rain probably stimulated reproductive effort (flower production) in concolour morphs at SNR. This has been well studied in some geophytes native to Mediterranean biomes in Australia and South Africa. Cyclical burns stimulate flowering in many herbaceous species but they belong to floras that evolved with cyclical fires (Le Maitre & Brown 1992; Lamont & Downes 2011). In fact cyclical fire regimes also occur in xeric, North American glades. In the absence of fires the herbaceous vegetation is succeeded by woody species (Martin & Houf 1993).

Compare this to our mixed hardwood forest (CR) where burns were performed in alternate years due to a greater accumulation of biomass represented by fallen branches and leaf detritus as fuels. This residue may smolder for hours following late-winter - early spring burns (unpublished

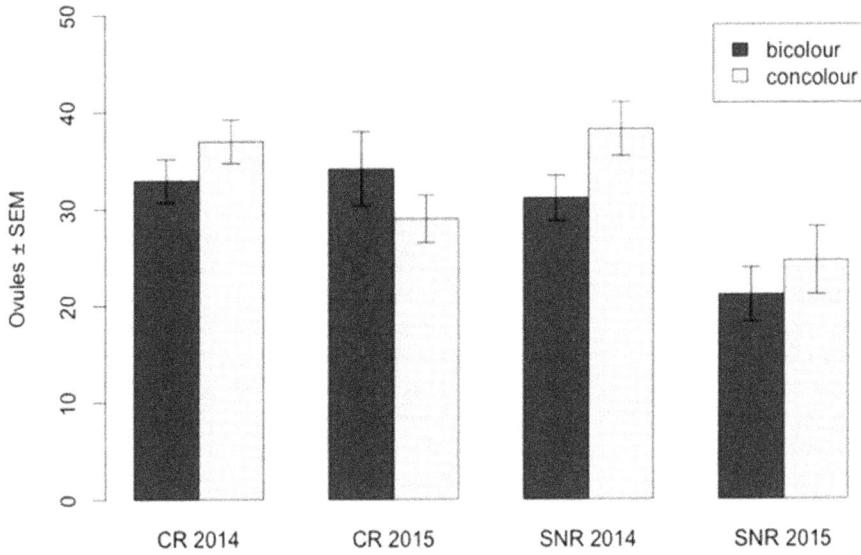

FIGURE 8. Mean number of ovules per flower in two morphs of *Viola pedata* at two sites and two years in Missouri, USA.

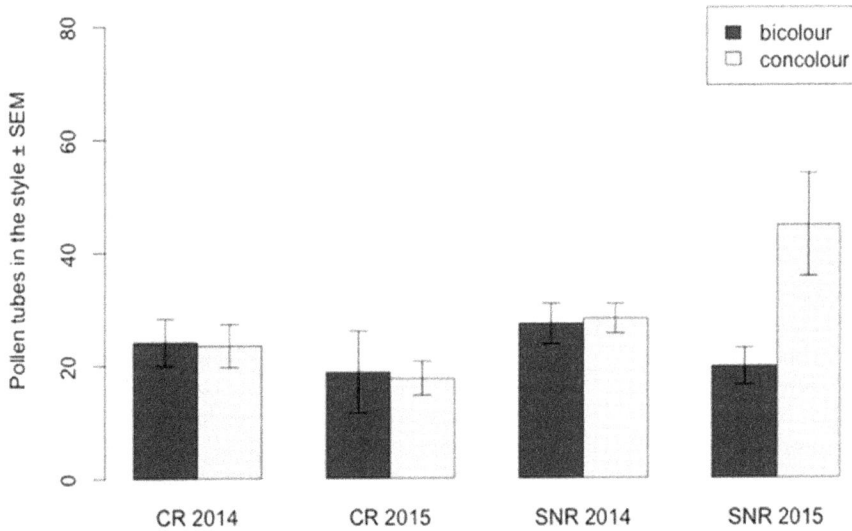

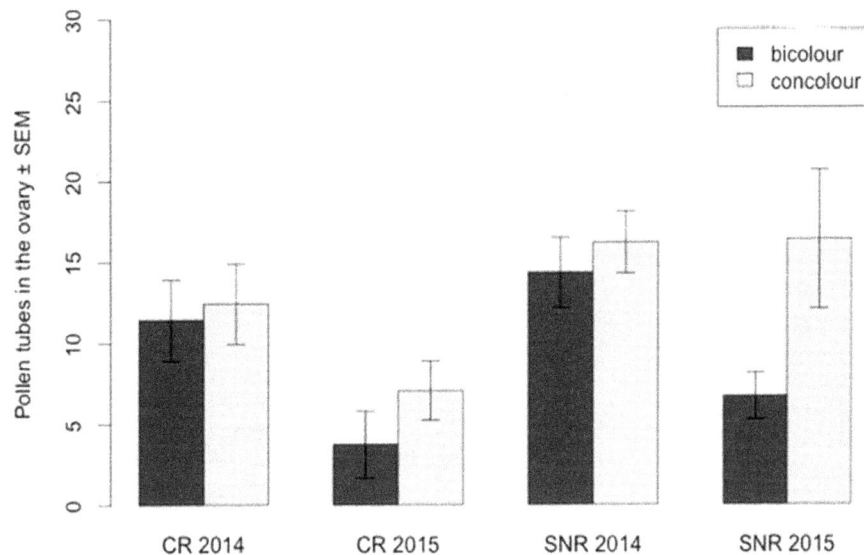

FIGURE 9. Mean number of pollen tubes germinating and penetrating styles in two morphs of *Viola pedata* at two sites and two years in Missouri, USA.

FIGURE 10. Mean number of pollen tubes penetrating ovules in two morphs of *Viola pedata* at two sites and two years in Missouri, USA.

observations). Perhaps it damaged more budding rosettes poised to flower in 2014 as bicolour rosettes appeared in twice the numbers (compared to concolours) at CR in 2015.

Comparative lack of pollinator diversity and foraging preferences between sites

We now have a third record of active (vector mediated) pollen collection (sensu Bernhardt, 1996) in a *Viola* species. The difference between this study and Freitas & Sazima (2003) was that their Neotropical species offered pollen as their primary (only?) reward. This did not appear to be the case in *V. pedata*. It maintained prominent nectar glands and spurs in both morphs. Our populations reflected the potential plasticity of generalist pollination systems found in *Viola* species in general. In other parts of its range Lepidoptera and male bees visit *V. pedata* exclusively for nectar (Beattie 1974; Carroll & Goldman 1994). While the apices of anthers of *V. pedata* wear the modified "snow shovels" (sensu Beattie 1974), associated with maximization of passive release of pollen onto nectar foraging insects, our females of *A. carlinii* remained active pollen collectors regardless of site. Bisexual flowers pollinated by a combination of nectar-drinkers and active pollen-collectors are not unique. Some Neotropical *Fuchsia* species are pollinated by a combination of hummingbirds and *Bombus* species. The birds consume only nectar (passive anther contact) while *Bombus* species drink nectar but also make active pollen collections (Bernhardt & Montalvo 1979; Breedlove 1969).

Within our two sites studied over three seasons the dominant pollinators were female, short-tongued bees in the family, Andrenidae as noted previously by Beattie (1974). We note that, at both sites over two years, females of *A. carlinii* dominated visits despite obvious differences in habitat landscapes and differences in prescribed burn cycles (see above). This is not surprising when the distribution of this bee is reviewed. Its North American range far exceeds the distribution of *V. pedata*. *Andrena carlinii* is polylectic also foraging on flowers of *Vaccinium* species and ephemeral woodland herbs. It nests in woodlands but is also native to open sites, not obscured by vegetation (Schrader & LaBerge 1978). This explains its presence in a mixed hardwood forest and in an exposed glade. In general, our insect collections failed to show that pollen dispersal of *V. pedata* at either site depended either on oligolectic bees (e.g. *Andrena violae*) or on insects foraging exclusively for nectar.

Beattie (1969; 1971b; 1972; 1974) provided ample evidence that many temperate zone, *Viola* species have generalist pollination systems. However, we should also consider the possibility that *V. pedata* may have a regionally narrow spectrum of pollinators when its populations are discontinuous, discrete and disjunctive. In our case we were disappointed repeatedly by the comparative lack of participation by native Lepidoptera at both of our sites compared to the observations of Carroll and Goldman (1994) in Missouri. Over three seasons at two sites we failed to observe the diurnal sphingid moths first described visiting *V. pedata* in West Virginia by Beattie (1974). Potential variation in guilds of anthophilous insects, based on the broad distribution of *V. pedata*, must be expected as this

species is recorded in almost half of the eastern, continental United States and south-eastern Canada. We wonder whether a combination of insecticide use and global warming over the past two decades accounted for the absence of *Erynnis* species at our sites?

Variation in bee orientation on morphs vs. morph preference

The orientation of bees on flowers of *V. pedata* appears to have little or nothing to do with foraging on either colour morph. Bee orientation mode appears to be based on when an insect begins and then completes its foraging bouts. We presume that when a bee first enters the site it first flies down to the flower and lands prone on the anterior petal. As the bout progresses the same bee appears to be more likely to fly above the flowers, lands directly on the posterior petals and then inverts while its hind legs continue to cling to the posterior petals. Consequently, our observations of orientation and landing differed from that of Beattie (1974). If one records only the visit of a bee to the first *V. pedata* flower it visits during a foraging bout then the right side up mode of visitation will dominate leading to some nototribic deposition of pollen as dorsal regions of its head and thorax contact the anther extensions. However, if one watches a full bout this right side up orientation declines as foraging and cross-pollination progresses. The majority of these later visits must be inverted leading to stenotribic depositions (sensu Beattie 1972).

As we did not measure the nutrients in the nectar and pollen grains of bicolour vs. concolour morphs we can't account for the bees' preferences for concolours. If both sites remain under their current maintenance programs it appears likely that frequencies in the bicolour morph should continue to decline at SNR as concolour is preferred by bees at both sites. Why then, should bicolour morphs persist at CR and increase and surpass numbers of concolour rosettes in 2015? One reason we suggest is that, while bicolour is less preferred by pollinators, field observations and pollen tube analyses showed that native bees continued to visit this morph at CR in the non-burn year (2015) with 40% of bicolour pistils containing pollen tubes. Residual warmth in the purple, posterior petals may mean that pollen-collecting bees will continue to forage on the anthers of some bicolour flowers as microhabitats become shadier and cooler at different times of the day due to changes in the angle of sunlight coupled with the irregular density of the forest canopy. Video and photos indicate that these dark purple petals may warm parts of the bee's third pair of legs and its abdomen. In contrast, at SNR the herbs on the floor of the glade do not stand under trees and are probably more exposed to more direct sunlight for far longer periods especially after a controlled burn (see below).

Floral warmth vs. foraging preferences

There is further precedence for this interpretation in Beattie (1971a) who studied *Viola glabella* in a much shadier, conifer forest. He noted that the pollinators were itinerant foragers on these yellow flowers visiting only during those brief periods when plants stood in direct sun. Beattie recorded ambient temperature, not floral temperature, but also used a light meter. In addition, Bernhardt et al (2014) did not take ambient temperatures but noted that small to medium size-bees didn't visit generalist, food mimic, *Cypripedium montanum*, when these flowers were in deep shade in the course of the day.

We suggest that the warmer, deep purple petals of bicolour may be selectively advantageous but only under very specific environmental conditions. They may encourage some pollinators to forage in situ for longer periods after ambient temperatures start to drop as light gaps shift over the day. During some extended foraging bouts (see above) bees appeared more likely to forage first on the dark, bicolour petals at the CR site. These flowers are not heliotropic or paraboloid in shape (see Kevan 1972; 1975). Therefore, a bee that visits one flower cannot engage in long-term flower basking (sensu Heinrich 1993) and visits several genotypes over relatively short periods effecting cross-pollination. As *V. pedata* has such a broad distribution a useful, future exercise may be to record morph frequencies according to habitat, light intensity and whether dominant woody species are deciduous or evergreen.

Otherwise, there remains only one more untested possibility. Some pollinators (sphingid moths?) may prefer bicolour across parts of the range of *V. pedata* but those pollinators were not observed or collected in our populations for two years. If this is the case than skewed frequencies of colour morphs of *V. pedata* may be driven, in part, by resident pollinator preferences and may be more common than anticipated. Currently, publications that follow the density and diversity of specific pollinators throughout the broad, natural distributions of one animal-pollinated species remain uncommon (but see, Espindola et al. 2011). It is also intriguing to note that Steyermark (1963) reported a pure white population in Barton Co., Missouri and "mostly white" specimens from Polk County but never reported when or which insects visited the flowers.

Variation in ovule production pollination rates and seed set

Ovule production does not vary much between morphs in this species but burning in 2014 appeared to stimulate ovule production in concolours at SNR. Once again we credit the potential release of micronutrients in a habitat in which there is little soil and these plants grow between cracks in rocks. By killing or depressing the growth of some taller plant species that shade *V. pedata* our plants at SNR may have had greater access to water and sunlight channelling vernally produced sugars into greater ovule production. Once again, these are also the standard explanations for increased flower production in fire-cycle habitats in temperate Australia (see above). In contrast, the CR population grows under trees, is subjected to daily shading throughout vernal growth periods and was unlikely to manufacture enough carbohydrate to compete with ovule production at SNR over a two-year period converting fewer ovules into seeds.

As we noted previously, bees are more likely to visit concolour flowers. It comes as no surprise, then, that more concolour pistils contained pollen tubes in 2014 at both sites. In 2015 the sheer number of pollinated pistils at SNR was > 0.50 higher in concolours than bicolours.

As the sheer number of flowering rosettes increases one presumes that pollinators will visit fewer plants as they become satiated more rapidly. This may also result in fewer compatible exchanges of pollen. When the number of flowering rosettes increased at CR in 2015 the number of ovules containing pollen tubes dropped dramatically. One wonders whether the 2014 burn also stimulated visitation by itinerant pollinators as fire should have removed debris or glade thatch at SNR making flowers blooming at about a centimetre above the ground more visible to foragers? Furthermore, concolour plants at SNR produced their largest visual display of flowers following the 2014 burns. Flowers of concolours were at a maximum visual presentation at SNR that year while, at CR, the ratio of the two colour morphs was almost identical.

However, once a flower was pollinated the sheer number of tubes per pistil was usually the same regardless of morph. We interpret this as evidence that a pollinator usually leaves the same number of viable grains on a receptive stigma, regardless of colour morph, restricting the number of pollen tubes that reach the ovary. Of course, when pistils express some form of late acting self-incompatibility the mere presence of tubes in entering ovules does not guarantee seed set. In some late acting SI expressed by some unrelated angiosperms (Kenrick & Williams 1986; Sage et al. 1999; Sage & Sampson 2003; Ramos et al 2005) recognition and rejection of shared alleles may continue to occur after tubes enter respective micropyles. Seed set at CR did not vary significantly over two years regardless of annual changes in morph frequencies. At SNR, while seed set in 2014 (burn year) was significantly higher than in 2015 we note that rates of pollen tube penetration were far higher in concolour flowers in 2015.

There are, of course, a number of alternative explanations for lower rates of seed set in any population from year to year. However, when we compare high pollen tube penetration of pistils at SNR in 2015 vs. lower seed production in the same season it should suggest an increase in insect mediated, self-incompatible pollinations based on bees visiting more than one flower on the same plant (geitonogamy) or crosses between genets sharing one or more of the same SI alleles (Kenrick et al. 1982). In 2014, at SNR, individual plants produced more flowers and this could have increased the frequency of geitonogamous crosses.

Once again, the 2014 burn at SNR appeared to benefit *V. pedata* producing more seed than in 2015. This also suggests that seed production is higher in an open glade compared to shady woodland. Fertility rates in *V. pedata* may be more dependent on the growth habits of surrounding vegetation (trees vs. grasses and forbs) according to burn cycle.

In conclusion, like most unbalanced polymorphisms the morphs expressed by *V. pedata* vary in frequency due to differences in regional modes of selection (see review in Futuyma 2013). As in other species with colour-based morphs (Rausher 2008) foraging preferences by the dominant pollinators may drive differential rates of reproductive success because acts of cross-pollination in this species appear to be assortative according to morph preference (sensu Richards 1986; Rymer et al. 2010). While bees prefer concolour flowers the warmer posterior petals of bicolours may provide a novel adaptation increasing their fitness but only when they grow in specific habitats. We also note that burn regimes may also play a role in fitness benefitting flower and ovule production in the concolour morph but, once again, this selective advantage may occur only within a specific habitat. As the number of flowers produced by a plant and the number of ovules in an ovary must be interpreted as floral traits we suggest they may be linked directly to the colour polymorphism instead of as a nonadaptive pleiotropic effect (Armbruster 2002).

ACKNOWLEDGEMENTS

We would like to thank the staff of the Shaw Nature Reserve and Cuivre River State Park for allowing us to tag and bag populations. We are especially grateful to James Trager (SNR) and Bruce Schuette (CR) for showing us the larger populations at respective sites and informing us as to burned sites and burn cycles. We are grateful to Larry Meier for maintaining equipment and helping with temperature measurements and videography and to Ms. Courntey Dvorsky and Ms. Kelli Frye for their lab work. Dr Zong-Xin Ren's work at St. Louis University was funded by the Chinese Academy of Sciences.

REFERENCES

Armbruster WS (2002) Can indirect selection and genetic context contribute to trait diversification? A transition-probability study of blossom-colour evolution in two genera. Journal of Evolutionary Biology 15: 468-486.

-----, Perez-Barrales RM, Arroyo J, Edwards ME, Vargas P (2006) Three-dimensional reciprocity of floral morphs in wild flax (*Linum suffuticosum*): a new twist on heterostyly. New Phytologist 171:581-590.

Beattie AJ (1969) Pollination ecology of *Viola*. I. Contents of stigmatic cavities. Wat-sonia 7:142-156.

----- (1971a) Itinerant pollinators in a forest. Madrono 21:120-124.

----- (1971b) Pollination mechanisms in *Viola*. New Phytologist 70:343-360.

----- (1972) The pollination of *Viola*. 2, Pollen loads of insect-visitors, Watsonia 9:13-25

----- (1974) Floral evolution in *Viola*. Annals of the Missouri Botanical Garden 61:781-793.

----- (1976) Plant dispersion, pollination and gene flow in *Viola*. Oecologia 25:291-300.

Becker RE, Ewart LC (1990) Pollination, seed set and pollen tube growth investigations in *Viola pedata* L. Acta Horticulturae 272: 33-36.

Bernhardt P (1996) Anther adaptations in animal-pollination. In: D'Arcy WG, Keating RC (eds) The Anther: Form, function and phylogeny. Cambridge University Press, Cambridge, pp 192–220.

-----, Edens-Meier R, Westhus EJ, Vance, N (2014) Bee-mediated pollen transfer in two popula-tions of *Cypripedium montanum* Douglas ex Lindley. Journal of Pollination Ecology 13:188-202.

-----, Montalvo EA (1979) The pollination of *Echeandia macrocarpa* (Liliaceae). Brittonia 31:64-71.

Breedlove DE (1969) The systematics of *Fuchsia* section *Encliandra* (Ona-graceae). University of California Publications in Botany 53:1-69.

Davidse G (1968) A biosystematic investigation of the intermountain yellow violets. MSc Thesis, Utah State University, Logan, Utah.

Carlson JE, Holsinger KE (2013) Direct and indirect selection on floral pigmentation by pollina-tors and seed predators in a colour polymorphic South African shrub. Oecologia 171: 905-919.

Carroll SB, Goldman P (1994) Analysis of a flower colour polymorphism in *Viola pedata* (birdfoot violet). Proceedings of the North American Conference on Savannas and Barrens. https://archive.epa.gov/ecopage/web/html/carroll.html.

Clausen J (1926) Genetical and cytological investigations on *Viola tricolor* L. and *V. arvensis* Murr. Hereditas 8:1-156.

Culley TM (2002) Reproductive Biology and delayed selfing in *Viola pubescens* (Violaceae), an understory herb with chasmogamous and cleistogamous flowers. Interna-tional Journal of Plant Sciences 163:113-122.

De Jager ML, Ellis AG (2014) Floral polymorphism and the fitness implications of attracting pollinating and florivorous insects. Annals of Botany 113: 213-222.

Darwin C (1876) The effects of cross and self fertilization in the vegetable kingdom John Murray, London.

Edens-Meier RM, Vance N, Luo YB, Li P, Bernhardt P (2010) Pollen-pistil interactions in North American and Chinese *Cypripedium* L. (Orchidaceae). International Journal of Plant Sciences 171:370-381.

Epperson BK, Clegg MT (1987) Frequency-dependent variation for outcrossing rate among flower-colour morphs of *Ipomoea purpurea*. Evolution 411: 1302-1311.

Espindola A, Pellissier L, Alvarez N (2011) Variation in the proportion of flower visitors of *Arum maculatum* across its distributional range in relation with community-based climatic niche analysis. Oikos 120:728-734.

Fehr C, Rausher (MD 2004) Effects of variation at the flower-colour A locus on mating system parameters in *Ipomoea purpurea*. Molecular Evology 13: 1839-1847.

Freitas L, Sazima M (2003) Floral Biology and pollination mechanisms in two *Viola* species – from nectar to pollen flowers? Annals of Botany 91:311-317.

Futuyma DJ (2013) Evolution. Third edition. Sinauer Associates, Inc. Sunderland, Massachusetts.

Gibson WH, Davie EE (1901) Blossom hosts and insect guests: How the heath family, the bluets, the figworts, the orchids and similar wild flowers welcome the bee, the fly, the wasp, the moth and other faithful insects. Newson and Company, New York.

Gomez JM (2000) Phenotypic selection and response to selection in *Lobularia maritima*: Importance of direct and correlational components of natural selection. Journal of Evolutionary Biology 13:689-699.

Heinrich B (1993) The hot-blooded insects: Strategies and mechanisms of thermoregulation. Harvard University Press. Cambridge, Massachusetts.

Herrera CM (1990) The adaptedness of the floral phenotype in a relict endemic, hawkmoth-pollinated violet. I. Reproductive correlates of floral variation. Biological Journal of the Linnean Society 40:263-274.

----- (1993) Selection on floral morphology and environmental determinants of fecundity in a hawk moth-pollinated violet. Ecological Monographs 63:251-275.

Gurevitch J, Scheiner S, Fox, G (2006) The Ecology of Plants. Second edition. Sinauer Associ-ates, Inc. Sunderland, Massachusetts.

Hildebrandt U, Hoef-Emden K, Backhausen S, Bothe H, Bozek M, Siuta A, Kuta E (2006) The rare, endemic zinc violets of Central Europe originate from *Viola lutea* Huds. Plant Systematics & Evolution 257:205-222.

Irwin RE, Strauss SY (2005) Flower colour microevolution in wild radish: Evolutionary response to pollinator-mediated selection. The American Naturalist 165:225-237.

Johri BM, Ambegaokar JR, Srivastava PS (1992) Comparative Embryology of Angiosperms. Volume I. Springer-Verlag, Berlin.

Kenrick J, Kaul V, Williams EG (1986) Self-incompatibility in *Acacia retinodes*. Site of pollen-tube arrest is the nucellus. Planta 169:245-50.

Kevan P (1972) Heliotropism in some Arctic flowers. The Canadian Field Naturalist 86:41- 44.

----- (1975) Sun-tracking solar furnaces in high arctic flowers: significance for pollination and insects. Science 189:723-726.

Lamont BB, Downes KS (2011) Fire-stimulated flowering among resprouters and genophytes in Australia and South Africa. Plant Ecology 212:2111-2125.

Le Maitre DC, Brown PJ (1992) Life cycles and fire-stimulated flowering in geophytes. In: van Wilgen BW, Richardson DM, Kruger FJ, van Hensbergen JH (eds) Fire in South African Mountain Fynbos, Springer0Verlar, New York, pp 145-160.

Mabberley DJ (1997) The plant-book. Second edition. Cambridge University Press, Cambridge.

Malberla R, Nattero J (2011) Pollinator response to flower colour polymorphism and floral dis-play in a plant with a single-locus flower colour polymorphism: Consequences for plant re-production. Ecological Research 27: 377-385.

Marcussen T, Borgen L (2011) Species delimitation in the Ponto-Caucasian *Viola sie-heana* complex, based on evidence from allozymes, morphology, ploidy levels and crossing experiments. Plant Systematics and Evolution 291:183-196.

Martin P, Houf GF (1993) Glade grasslands in southwest Missouri. Rangelands 15:70-73.

McKinney LE (1992) A taxonomic revision of the acaulescent blue violets (*Viola*) of North America. Botanical Research Institute of Texas, Inc., Fort Worth, Texas.

Migdalek G, Wozniak M, Slomka A, Godsik B, Jedrzejcyk-Korycinska M, Rostanski A, Bothe H, Kuta E (2013) Morphological differences between violets growing at heavy metal polluted and non-polluted sites. Flora 208:87-96.

Mereda P, Hodalova I, Martonfi P, Kucera J, Lihova J (2008) Intraspecific variation in *Viola suavis* in Europe: Parallel evolution of white-flowered morphotypes. Annals of Botany 102:443-462.

Müller H (1883) The fertilization of flowers. London.

Nieuwland JA, Kaczmarek RM (1914) Studies in *Viola*, I: Proposed segregates of *Viola*. The American Midland Naturalist 8:207-217.

Pellegrino G, Bellusci F, Musacchio A (2008) Double floral mimicry and the magnet species effect in dimorphic co-flowering species, the deceptive orchid *Dactylorhiza sam-bucina* and rewarding *Viola aethnensis*. Preslia 80:411-422.

Ramos RR, Venturieri GA, Cuco SM, Castro NM (2005) The site of self-incompatibility action in cupassu (*Theobroma grandiflorum*). Brazilian Journal of Botany 28:569-578.

Rausher MD (2008) Evolutionary transitions in floral colour. International Journal of Plant Sci-ences. 169:7-21.

Richards AJ (1986) Plant Breeding Systems. Allen & Unwin, Boston, Massachusetts.

Russell AL, Newman SR, Papaj DR (2016) White flowers finish last: Pollen foraging bumble bees show biased learning in a floral colour polymorphism. Evolutionary Ecology (2016). Doi:10.1007/s10682-016-9848-1.

Rymer PD, Johnson SD, Savolainen V (2010) Pollinator behavior and plant speciation: can as-sortative mating and disruptive selection maintain distinct morphs in sympatry. New Phytologist 188:426-436.

Sage T, Sampson B (2003) Evidence for Ovarian Self-incompatibiolity as a cause of self-sterility in the relictual woody angiosperm, *Pseudowintera axillaris* (Winteraceae). Annals of Botany 91:807-816.

Sage T, Strumas F, Cole WW, Barrett SCH (1999) Differential ovule development following self- and cross-pollination: the basis of self-sterility in *Narcissus triandrus* (Amayllidaceae). American Journal of Botany 86: 855-870.

Schrader MN, LaBerge WE (1978) The nest biology of the bees: *Andrena* (*Melandrena*) *regularis* Malloch and *Andrena* (Melandrena) *carlini* (Hymenoptera: Andrenidae). Biological notes; no. 108. State of Illinois, Dept. of Registration and Education, Natural History Survey Division.

Sobral M, Losada M, Veiga T, Gruitian J, Guitian J, Guitian P (2016) Flower colour preferences of insects and livestock: Effects on *Gentiana lutea* reproductive success. PeerJ 4:e1685; DOI 10.7717/peerj.1685.

Steyermark J (1963) Flora of Missouri. Iowa State Press, Iowa City.

Wang Y (2008) Molecular biology of flower development in *Viola pubescens*, a species with the chasmogamous-cleistogamous mixed breeding system. PhD Dissertation, Ohio Uni-versity. UMI 3302717.

Weberling F (1989) Morphology of Flowers and Inflorescences. Cambridge University Press. Cambridge.

Willmer P (2011) Pollination and Floral Ecology. Princeton University Press. Princeton, New Jersey.

Winn AA, Moriuchi KS (2009) The maintenance of mixed mating by cleistogamy in the perennial violet *Viola septemloba* (Violaceae). American Journal of Botany 96:2074-2079.

10

Preliminary Studies on Ornithophilous Floral Visitors in the Australian Endemic *Passiflora herbertiana* Ker Gawl. (Passifloraceae)

Shawn Krosnick[1,*], Tim Schroeder[2], Majesta Miles[3] and Samson King[4]

[1]*Department of Biology, Tennessee Technological University, P.O. Box 5063, Cookeville, TN, 38505, U.S.A.*
[2] *Department of Biochemistry and Chemistry, Southern Arkansas University, 100 East University Street, Magnolia, AR, 71753, U.S.A.*
[3]*Department of Biology, Southern Arkansas University, 100 East University Street, Magnolia, AR, 71753, U.S.A.*
[4]*Department of Neurobiology and Anatomy, Wake Forest School of Medicine, Winston-Salem, NC, 27517, U.S.A.*

Abstract—The pollination biology of the Australian endemic species *Passiflora herbertiana* (*Passiflora* subgenus *Decaloba*, supersection *Disemma*, section *Disemma*) was investigated in a single population growing in the Witches Falls section of Mount Tamborine National Park, Queensland. Three native honeyeaters were observed at the flowers, including Lewin's Honeyeater (*Meliphaga lewinii*), the Noisy Miner (*Manorina melanocephala*), and the Eastern Spinebill (*Acanthorhynchus tenuirostris*). Visitation began at 07:30 and ended by 15:30 each day. The most frequent visitor was Lewin's Honeyeater. Flowers typically began anthesis in the afternoon, with a small number of flowers opening in the early morning. Flowers remained open between four and five days, even after successful pollination. Both the age of the flower and the amount of sun exposure were determined to affect perianth colour change from pale yellow to salmon-pink. Andromonoecy was observed infrequently in the population; most plants exhibited bisexual flowers, but a small number of individuals exhibited both hermaphroditic and male flowers with short styles held permanently erect. Controlled hand pollinations indicated that *P. herbertiana* is self-compatible but is not autogamous. Pollen tubes required at least 48 hours to reach the most apical ovules within the ovary. These data provide new insights into the evolution of ornithophily in the Old World *Passiflora*.

Keywords: Australia, Disemma, honeyeater, ornithophily, Passiflora herbertiana, Passifloraceae

Introduction

Passiflora L. is a genus of ca. 525 vines, lianas and small trees distributed throughout Mexico, Central and South America, with an additional 24 species endemic to Asia, Southeast Asia, and the Austral Pacific. Insect pollination is thought to be ancestral for *Passiflora* (MacDougal 1994), but hummingbird and bat pollination syndromes have evolved independently in multiple clades (Kay 2001; Krosnick et al. 2013). For example, subgenus *Decaloba* (DC.) Rchb. includes at least four unrelated lineages of hummingbird-pollinated species. These species have tube-shaped flowers with bright red, yellow, or pink colouration, produce large quantities of dilute sucrose-rich nectar, have horizontal to erect floral orientation at anthesis, and lack floral scent. While most species of *Passiflora* are self-incompatible, a few species are self-compatible (MacDougal 1994; Koschnitzke & Sazima 1997; Amela Garcia & Hoc 1998; Kay 2003).

The Australian native species *Passiflora herbertiana*, *P.*

cinnabarina, and *P. aurantia* (supersection *Disemma* (Labill.) J. M. MacDougal & Feuillet, section *Disemma* (Labill.) J. M. MacDougal & Feuillet) exhibit many of the same floral features as New World hummingbird-pollinated species (e.g. *P. murucuja* L. and *P. tulae* Urb.). The Australian species are known to be self-compatible in cultivation, and two species (*P. herbertiana* and *P. aurantia*) exhibit floral colour change with age. No studies have been performed to document avian visitors to the Australian species in their native ranges. Thus, these species provide a unique opportunity to examine the evolution of ornithophilous pollination syndromes and self-compatibility in Australia. As a first step towards understanding these species in greater detail, an investigation into the reproductive biology of *P. herbertiana* was conducted. The objectives were to 1) document floral phenology, 2) determine types of floral visitors, 3) examine floral nectar constituents and 4) perform controlled hand pollinations to determine the nature of self-compatibility in this species.

Materials and Methods

Study site

Fieldwork was carried out between June 30 and July 9, 2011 at a large population of *P. herbertiana* growing in the

*Corresponding author: skrosnick@tntech.edu

Witches Falls section of Mount Tamborine National Park, Queensland, Australia (27°93'S, 153°18'E; elevation 559 meters). The population consisted of ca. 150 individual plants, most being non-reproductive juveniles growing in the dense shade of the rainforest canopy. Three mature plants of *P. herbertiana* were chosen along the forest margin. The selected individuals were mature lianas ca. 3-10 meters long, each climbing up into nearby trees and displaying numerous flowers in various stages of maturity, and located ca. 35 meters apart from one another (plant 1, *SK 735*; plant 2, *SK 734*; plant 3, *SK 736*; vouchers at MO).

Reproductive phenology of Passiflora herbertiana

Nineteen floral buds across all three plants were tagged and covered with mesh bags for observation and controlled pollinations. Flowers were examined three times per day (07:00, 12:00, 16:00) to determine patterns of anthesis.

Assessment of self-compatibility - controlled pollination experiment

Pollen tube germination and rates of ovule fertilization were used to assess self-compatibility in *P. herbertiana*. Hand-pollinations were carried out on 15 bagged flowers across all three plants. Seven flowers were self-pollinated and eight were outcrossed with pollen from the other two vines in the population. Outcrossed flowers were emasculated while in bud to prevent contamination of selfed pollen. As soon as floral buds opened enough to expose the stigmas, they were assumed receptive and pollination was performed. The stigmas of out-crossed flowers were evenly dusted with pollen using a single anther removed from the donor flower. In selfed individuals, pollinations were performed once the anthers had begun dehiscing, and pollen from a single anther was divided among all stigmas. The remaining anthers were left on the flower to serve as potential pollen donors in other pollinations. Flowers were collected 48 hours after pollination, fixed in FAA for 24 hours, and transferred to 70% ethanol for long term storage. Pollen tube growth was examined with aniline blue staining following Kay (2003) and Bernhardt (1982) on a ZEISS Primostar iLED microscope using a fluorescence LED module at 455 nm and filter set 67.

Analysis of nectar concentration and constituents

Nectar concentration was assessed for 16 flowers sampled across all three plants using a hand-held refractometer. Samples were collected at 16:00 each day using 50 µL microcapillary tubes, and concentrations were recorded at ambient temperature (18°C to 22°C). Twelve samples were collected for analysis of sugar constituents using High Performance Liquid Chromatography (HPLC) with 20 or 50 µL microcapillary tubes, quantified, blotted on filter paper, and stored with silica gel. Samples were dissolved by rinsing filter paper 3-5 times with 4 ml of Milli-Q Direct purified water (Millipore, Billerica, MA) heated to 95°C. Samples were then diluted to 50 ml and analyzed using a Shimadzu Prominence twin pump HPLC system with a Shodex Asahipak NH2P-50 4E column (Shimadzu, Columbia, MD). Each day of analysis, calibration curves of known sugar amounts were constructed using standard solutions from 400 ppm (0.4 g/L) to 2000 ppm (2.0 g/L) of glucose, fructose and sucrose (working concentration 100 µg/g). Data were integrated using SHIMADZU LCSolutions software and peak areas were regressed against sugar weight. For each sample, percentages of glucose, fructose and sucrose of the total sugar weight were calculated.

Floral visitors

Observations of floral visitors were conducted between 06:00 and 17:00 each day for the 10 days of the study. Animals were identified using field guides and confirmed with photographs from the site. Videos were taken to document visitor behaviour (see Appendix 1).

RESULTS

Reproductive phenology

Floral buds take ca. 21 days to develop to maturity from initiation along the shoot tip. Most flowers opened overnight ($N = 15$), but a small number opened over the course of the day ($N = 4$). In night-opening flowers, the afternoon before full anthesis (ca. 16:00) sepals split to 1-2 cm at the apex (Fig. 1A), at which point the styles and anthers became visible. This was defined as the start of

FIGURE 1. Anthesis pattern of *Passiflora herbertiana* over five days. (A) Onset of anthesis: Sepals open 1-2 cm in the afternoon, styles initially held at ca. 20° angle from androgynophore. (B) Next morning: perianth begins to open; anthers start to dehisce, styles will eventually become level with stamens. (C) 48-72 hours: perianth becomes fully expanded; the styles continue descent downward. (D) Last day of anthesis (stamens removed): perianth begins to close; styles fully extended downward, eventually curling around the ovary.

anthesis or "0 hours." The three styles are initially held at a 20° angle relative to the androgynophore. By 07:00 the next morning, the sepals are nearly expanded and at least one anther is dehiscing; the petals are still oriented towards the centre of the flower (Fig. 1B). Throughout the first full day of anthesis, the styles gradually bend downward at a 45° angle, eventually becoming aligned with the stamens. Day-opening flowers begin with a 1-2 cm split at 07:00; by 15:00 that same day, flowers are completely open and anthers are dehiscing. Day-opening flowers follow the same stylar movement patterns as night-opening flowers. All flowers remain open both day and night for 48-72 hours. Sepals and petals open wider each day, eventually becoming perpendicular to the androgynophore (Fig. 1C). Stigmas gradually bend down towards the ovary while the stamens remain at the same general position. All flowers remain open up to five days, with most flowers closing on day four or five. By the end of the last day, the stigmas bend down and curl around the ovary, the anthers shrivel and the perianth closes (Fig. 1D).

Three flowers were observed with styles 4-6 mm shorter than normal flowers (Fig. 2); in these flowers styles remained fully upright throughout anthesis. No pollinations were performed with short-styled individuals. Both sun exposure and flower age are important factors contributing to floral colour in *P. herbertiana*. Regardless of age, flowers in full shade remain pale yellow, while those in full sun are salmon-pink. In partial shade, colour and flower age are more closely correlated, where yellow (Fig. 2A) changes to salmon pink or reddish (Fig. 2B) during the final days of anthesis.

FIGURE 2. Short style morph in *Passiflora herbertiana*. (A) Young flower with short, erect styles that lack normal movement patterns. Note anther dehiscence. (B) Older flower with short styles that remain in original position and anthers that have released all available pollen (photo credit: R. Van Raders).

Nectar concentration and nectar constituents

The mean sugar concentration of nectar in *P. herbertiana* was 19.49% ± 1.85 (1 S.D., $N = 16$). HPLC analysis indicated that all samples consisted of 100% sucrose with no detectable peaks of glucose or fructose.

Self-compatibility

Evidence for compatible pollen transfer in *P. herbertiana* was considered to be pollen tube germination and growth through stylar tissue followed by fertilization of ovules. All selfed and outcrossed flowers exhibited pollen germination on the stigma and growth through stylar tissue (Fig. 3A-B).

Pollen tubes in *P. herbertiana* require longer than 48 hours to reach all ovules within the ovary. All 15 hand-pollinated flowers showed pollen tubes present the entire length of the style (Fig. 3B), but only six had fertilized ovules and these were present at the apex of the ovary. Fertilization was visible (Fig. 3C) in both hand-outcrossed and -selfed flowers. The mean number of ovules per ovary was 476.27 ± 107.23 (1 S.D., $N = 17$). Three outcrossed flowers had fertilized ovules visible (3-20 ovules; 0.71-4.75% fertilization), as did three self-pollinated flowers (2-55 ovules; 0.34-11.96%).

Floral visitors

Three avian species were observed at *P. herbertiana*: Noisy Miners (*Manorina melanocephala* Latham; Fig. 4A), Lewin's Honeyeaters (*Meliphaga lewinii* Swainson; Fig. 4B), and Eastern Spinebills (*Acanthorhynchus tenuirostris* Latham; Fig. 4C). Regardless of species, visits were 30-60 seconds long and consisted of the bird probing the nectar chamber at the base of the flower several times. Lewin's Honeyeaters spent 2-4 minutes total on an individual vine, visiting multiple flowers. Eastern Spinebills targeted a flower from a distance, probed that flower, and immediately left. Noisy Miners perched on a vine for several minutes before visiting a flower, quickly probed it, and departed. All species appeared to establish some contact with the androgynophore while probing the flower. Frequency of floral visitation (Fig. 5) varied among the three bird species. Lewin's Honeyeaters were most frequent, followed by Eastern Spinebills, and then Noisy Miners. Visitation began at approximately 07:30 and continued until 15:30; no preference for mornings or afternoons was observed.

DISCUSSION

This is the first study to document ornithophilous floral visitation in Australian native *Passiflora*. This study highlights the presence of floral features that may be associated with pollination in *P. herbertiana*, including the production of dilute, sucrose-rich nectar and changes in floral colour. These data also confirm early self-compatibility in *P. herbertiana*. New insights into the reproductive biology of *P. herbertiana* include the presence of day- and night-opening flowers, andromonoecy, and an extended 4-5 day anthesis.

Floral phenology

Passiflora herbertiana has two flowering events per year (Krosnick 2006); the largest occurs in December–February, with a smaller flowering event in May–July. In the current study, onset of anthesis took ca. 18 hours (Fig. 1A-B); it is possible temperature played a role in the length of onset. Staggered anthesis (night versus day-opening flowers) may limit gene flow among ramets (Borchsenius 2002), or may result from daily variations in temperature. Onset of anthesis appeared to be correlated with observed visits of pollinators (Fig. 5), as pollinators did not visit flowers before 07:30 or after 15:30.

Passiflora herbertiana exhibits perianth colour change throughout anthesis, which may signal decreased availability of nectar rewards and may be associated with fertilization of

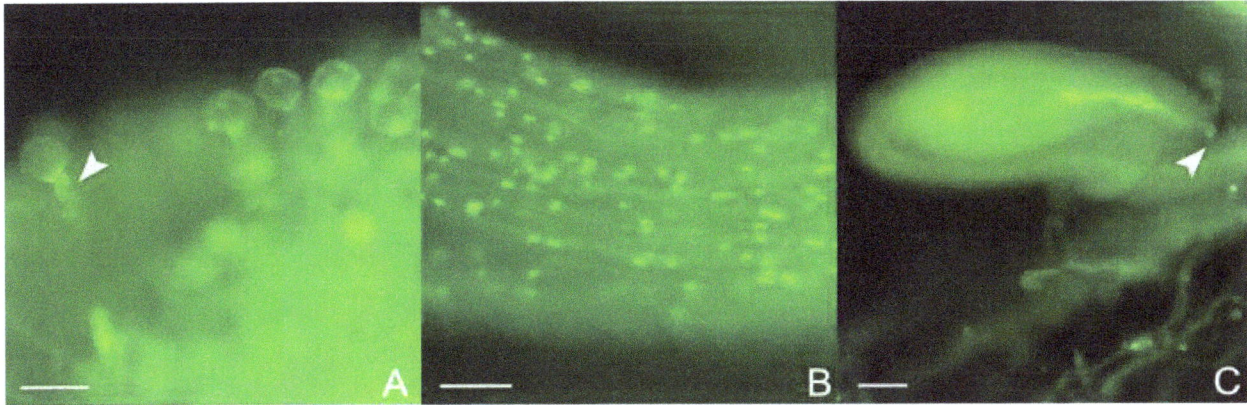

FIGURE 3.   Assessment of self-compatibility in *Passiflora herbertiana*. (A) Pollen grains germinating on the style. Arrow indicates pollen tube. (B) Example of pollen tubes germinating through the centre of the style in a self-pollinated individual. (C) View of a fertilized ovule in a self-pollinated individual. Arrow indicates pollen tube penetrating the micropyle and reaching the nucellus. Scale bars: a, c = 50 μm; b = 100 μm.

FIGURE 4.   Floral visitors to *Passiflora herbertiana*. (A) Noisy Miner. (B) Lewin's Honeyeater. (C) Eastern Spinebill.

ovules (Delph & Lively 1989; Weiss 1995). Anthocyanins are often responsible for floral colour change (Weiss 1995), and provide UV protection to plant tissues (Mazza et al. 2000). Flowers exposed to full sun were salmon-coloured while those in shade were pale yellow. In full sun, anthocyanins may accumulate more quickly as a protective response. Most flowers received indirect sunlight exposure and accumulated pigmentation gradually as the flower aged (Fig. 2). Thus, colour change may serve a dual purpose as both a UV sunscreen and a signal to pollinators in *P. herbertiana*.

Floral rewards and associated pollinators

Nectar produced by *P. herbertiana* is comprised exclusively of sucrose. This is consistent with hummingbird-pollinated *Passiflora mathewsii* (Mast.) Killip and *P. murucuja* (Krosnick unpublished data), both of which have sucrose-dominant nectar. Hummingbirds require sucrose-rich nectar (Baker & Baker 1982, 1983; Lotz & Schondube 2006) due to their high energetic needs. Passerines were thought to prefer hexose sugars (Baker & Baker 1982, 1983), but recent studies (Dupont et al. 2004; Johnson & Nicolson 2008) indicate they are varied in their preferences.

Honeyeaters have been shown to be nearly as efficient at assimilating sucrose as hummingbirds (Lotz & Schondube 2006), suggesting that the Australian honeyeaters observed in the present study should have little difficulty in utilizing nectar from *P. herbertiana*.

Honeyeaters are the dominant floral visitors in Australia, and they have been described as both generalists and opportunists: they visit many species of flowers, including non-natives (Ford et al. 1979; Recher 1981). Little information is known about nectar preferences and patterns of visitation in Eastern Spinebills, Lewin's Honeyeaters, or Noisy Miners. Mitchell and Paton (1990) suggest that even though honeyeaters are a diverse lineage, they place similar selective pressures on nectar traits including nectar volume and sugar content. The three species observed visiting *P. herbertiana* are not at all closely related based on the most recent Melaphagid phylogeny (Joseph et al. 2014), yet exhibit similar preferences for the dilute, sucrose-dominant nectar in *P. herbertiana*.

Self-compatibility

Floral morphology and phenology in *P. herbertiana* are consistent with outcrossing, yet this species is clearly self-

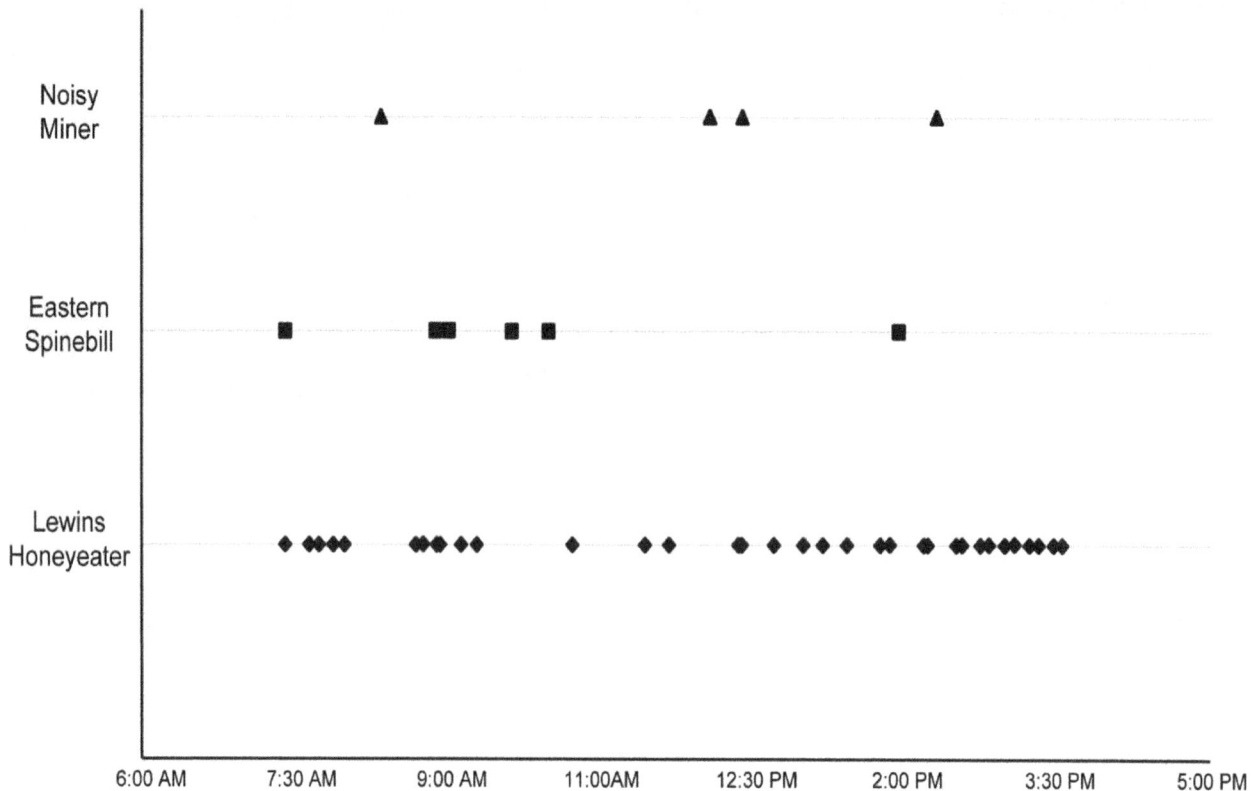

FIGURE 5. Visitation patterns of birds on *Passiflora herbertiana*. Floral visits generally began by 7:30 and stopped by 15:30 each day. Lewin's Honeyeaters were most frequent with visits documented fairly consistently throughout the day. Eastern Spinebills were the next most frequent, followed by Noisy Miners.

compatible. Pollinator-mediated selfing may provide an alternative when other pollen sources are lacking (Barrett 2002). Late-acting self-incompatibility was not examined in this study, and if present could have a significant effect by eliminating self-fertilized progeny.

Reproductive biology

Similar to *P. aurantia* and *P. cinnabarina*, flowers of *P. herbertiana* remain open for three to five days (Krosnick 2006). Most *Passiflora* typically open for just one day, but some hummingbird-pollinated species in the Andean supersection *Tacsonia* have flowers that remain open for up to three days (MacDougal 1994). This may be an adaptation to infrequent visits from pollinators, or may result from slow rates of pollen tube growth due to cooler temperatures at higher elevations. In June–July, *P. herbertiana* was just beginning to undergo fertilization of ovules at 48 hours post-pollen transfer (Fig. 3), but this might occur more quickly during December–February when ambient temperatures are higher. Stigmas in other *Passiflora* have been shown to be receptive quite early during anthesis (MacDougal 1994).

Styles in *P. herbertiana* are notable in their movement patterns. As the end of anthesis approaches, the styles in most *Passiflora* return to their original vertical orientation, regardless of whether or not pollination has occurred (MacDougal 1994). In *P. herbertiana*, the styles continue downward (Fig. 1C-D); this movement might promote self-

pollination as the flower ages. Moreover, short-styled floral morphs were observed (Fig. 2) that lacked any stylar movement. This phenomenon has been noted sporadically across *Passiflora* (MacDougal 1994). Andromonecious plants were documented in *P. incarnata* (Dai & Galloway 2012); in that study, male flowers appeared hermaphroditic but had short, erect styles that never bent downward, making self-pollination nearly impossible. More studies are needed in *P. herbertiana* to examine the effect short-style morphs may have on population dynamics.

Conclusion

The similarities observed in NW hummingbird-pollinated *Passiflora* and Australian *P. herbertiana* provide an example of convergence of both floral form and function in response to similar pollination syndromes. Many details remain to be examined for *P. herbertiana*, including the frequency and significance of andromonecy, seasonal variation in pollen tube growth rates, timing of stigma receptivity, and rates of floral nectar production during anthesis. Floral visitation is not the same as effective pollination, and field studies specifically designed to examine pollen transfer during floral visits are needed to address interactions between *P. herbertiana* and its visitors. To understand the evolution of reproductive syndromes in the Australian *Passiflora*, studies are also needed for *P. cinnabarina* and *P. aurantia*. All three species are relatively widespread, and studies across their entire range should reveal new plant-pollinator interactions in addition to those documented here. It will be especially interesting to see if the *Passiflora*-honeyeater association described for *P. herbertiana* is supported in the other Australian species.

APPENDICES

Additional supporting information may be found in the online version of this article:

APPENDIX I. This video shows footage of Lewin's Honeyeaters visiting various *P. herbertiana* flowers at the field study location (Witches Falls section of Mount Tamborine National Park, Queensland, Australia).

ACKNOWLEDGEMENTS

The authors would like to thank Elma Kay, Lucinda McDade, Sibyl Bucheli, and Erin Tripp for advice on field sampling, Mackenzie Taylor for advice on fluorescence techniques, and the SAU Natural Resources Research Center for facilities access. Andrew Ford, Allen Trumbull-Ward, Glen Leiper, Craig Robbins, Neville Walsh, Bob Makinson, and Brendan Lepschi all provided assistance with field logistics. We thank John MacDougal and Lucinda McDade for critical comments on the manuscript, and Ranier Van Raders for providing an image of a short-styled *P. herbertiana*. Field studies were conducted under the authority of the Queensland Parks and Wildlife Service Department of Environment and Resource Management, Permit WITK09408311. This research was supported by National Science Foundation Grant 0717151 and REU Supplement to S. Krosnick and L. McDade.

REFERENCES

Amela Garcia MT, Hoc PS (1998) Biología floral de *Passiflora foetida* (Passifloraceae). Revista de Biología Tropical 46 (2):191-202.

Baker HG, Baker I (1982) Chemical constituents of nectar in relation to pollination mechanisms and phylogeny. In: Niteki MH (ed) Biochemical aspects of evolutionary biology. University of Chicago Press, Chicago, pp 131-171.

Baker HG, Baker I (1983) Floral nectar sugar constituents in relation to pollinator type. In: Jones CE, Little RJ (eds) Handbook of experimental pollination biology. Van Nostrand Reinhold, New York, pp 117-141.

Barrett SCH (2002) The evolution of plant sexual diversity. Nature Reviews Genetics 3 (4):274-284

Bernhardt, P (1982) Interspecific incompatibility amongst victorian species of *Amyema* (Loranthaceae). Australian Journal of Botany 30 (2):175-184.

Borchsenius F (2002) Staggered flowering in four sympatric varieties of *Geonoma cuneata* (Palmae). Biotropica 34 (4):603-606.

Dai C, Galloway LF (2012) Male flowers are better fathers than hermaphroditic flowers in andromonoecious *Passiflora incarnata*. New Phytol. 193 (3):787-796.

Delph LF, Lively CM (1989) The evolution of floral color change: Pollinator attraction versus physiological constraints in *Fuchsia excorticata*. Evolution 43 (6):1252-1262.

Dupont YL, Hansen DM, Rasmussen JT, Olesen JM (2004) Evolutionary changes in nectar sugar composition associated with switches between bird and insect pollination: the Canarian bird-flower element revisited. Functional Ecology 18 (5):670-676.

Ford HA, Paton DC, Forde N (1979) Birds as pollinators of Australian plants. New Zealand Journal of Botany 17 (4):509-519.

Johnson SD, Nicolson SW (2008) Evolutionary associations between nectar properties and specificity in bird pollination systems. Biological Letters 4 (1):49-52.

Joseph L, Toon A, Nyari AS, Longmore NW, Rowe KMC, Haryoko T, Trueman J, Gardner JL (2014) A new synthesis of the molecular systematics and biogeography of honeyeaters (Passeriformes: Meliphagidae) highlights biogeographical and ecological complexity of a spectacular avian radiation. Zoologica Scripta 43 (3):235-248.

Kay E (2001) Observations on the pollination of *Passiflora penduliflora*. Biotropica 33 (4):709-713.

Kay E (2003) Floral evolutionary ecology of *Passiflora* (Passifloraceae): subgenera *Murucuia, Pseudomurucuja* and *Astephia*. Ph.D. Dissertation, Saint Louis University, St. Louis.

Koschnitzke C, Sazima M (1997) Floral biology of five species of *Passiflora* L. (Passifloraceae) in a semideciduous forest. Revista Brasileira de Botanica 20 (2):119-126.

Krosnick SE (2006) Phylogenetic relationships and patterns of morphological evolution in the Old World species of *Passiflora* (subgenus *Decaloba:* supersection *Disemma* and subgenus *Tetrapathea*). Ph.D. Dissertation, The Ohio State University, Columbus.

Krosnick SE, Porter-Utley KE, MacDougal JM, Jorgensen PM, McDade LM (2013) New insights into the evolution of *Passiflora* subgenus *Decaloba* (Passifloraceae): phylogenetic relationships and morphological synapomorphies. Systematic Botany 28 (3):1-22.

Lotz CN, Schondube JE (2006) Sugar preferences in nectar- and fruit-eating birds: Behavioral patterns and physiological causes. Biotropica 38 (1):3-15.

MacDougal JM (1994) Revision of *Passiflora* subgenus *Decaloba* section *Pseudodysosmia* (Passifloraceae). Systematic Botany Monographs 41:1-146.

Mazza CA, Boccalandro HE, Giordano CV, Battista D, Scopel AL, Ballare CL (2000) Functional significance and induction by solar radiation of ultraviolet-absorbing sunscreens in field-grown soybean crops. Plant Physiology 122 (1):117-125.

Mitchell RJ, Paton DC (1990) Effects of nectar volume and concentration on sugar intake rates of Australian Honeyeaters (Meliphagidae). Oecologia 83 (2):238-246.

Recher HF (1981) Nectar-feeding and its evolution among Australian vertebrates. In: Keast A (ed) Ecological Biogeography of Australia. Dr. W. Junk, The Hague, pp 1637-1648.

Weiss MR (1995) Floral color change: a widespread functional convergence. American Journal of Botany 82 (2):167-185.

11

A METHOD FOR UNDER-SAMPLED ECOLOGICAL NETWORK DATA ANALYSIS: PLANT-POLLINATION AS CASE STUDY

Peter B. Sørensen[1,*], Christian F. Damgaard[1], Beate Strandberg[1], Yoko L. Dupont[2], Marianne B. Pedersen[1], Luisa G. Carvalheiro[3,4], Jacobus C. Biesmeijer[4], Jens Mogens Olsen[2], Melanie Hagen[2], Simon G. Potts[5]

[1]*Aarhus University, Bioscience, Vejlsøvej 25, P.O.Box 314, 8600 Silkeborg, Denmark*
[2]*Aarhus University, Bioscience, NyMunkegade 114, 8000 Aarhus C, Denmark*
[3]*Institute of Integrative and Comparative Biology, University of Leeds, Leeds LS2 9JT, UK*
[4]*NCB-Naturalis, postbus 9517, 2300 RA, Leiden, The Netherlands*
[5]*University of Reading, School of Agriculture, Policy and Development, UK*

Abstract—In this paper, we develop a method, termed the Interaction Distribution (ID) method, for analysis of quantitative ecological network data. In many cases, quantitative network data sets are under-sampled, i.e. many interactions are poorly sampled or remain unobserved. Hence, the output of statistical analyses may fail to differentiate between patterns that are statistical artefacts and those which are real characteristics of ecological networks. The ID method can support assessment and inference of under-sampled ecological network data. In the current paper, we illustrate and discuss the ID method based on the properties of plant-animal pollination data sets of flower visitation frequencies. However, the ID method may be applied to other types of ecological networks. The method can supplement existing network analyses based on two definitions of the underlying probabilities for each combination of pollinator and plant species: (1), p_{ij}: the probability for a visit made by the ith pollinator species to take place on the jth plant species; (2), q_{ij}: the probability for a visit received by the jth plant species to be made by the ith pollinator. The method applies the Dirichlet distribution to estimate these two probabilities, based on a given empirical data set. The estimated mean values for p_{ij} and q_{ij} reflect the relative differences between recorded numbers of visits for different pollinator and plant species, and the estimated uncertainty of p_{ij} and q_{ij} decreases with higher numbers of recorded visits.

Keywords: Ecological network, Bayesian method, plant-animal pollination data analysis, under-sampled data sets

INTRODUCTION

Plant-pollinator interactions are important for maintenance of biological diversity, and pollination is a valuable ecosystem function for both wild plant communities and agricultural production (Potts et al. 2010). Hence, anthropogenic changes to the environment have negative effects on plants and pollinators, and hence pollination is seen as an ecological network (e.g. Biesmeijer et al. 2006; Hegland et al. 2009). To better understand mechanisms behind such consequences, modelling, interpretation and handling of pollination data as ecological networks are necessary steps (Potts et al., 2011).

During the past decade, ecologists have become increasingly interested in ecological networks, and network analysis is applied to complex patterns of interactions among species in food webs, mutualistic and host-parasite networks (reviewed by Ings et al. 2009). The application of methods of the network analysis has gained new insights into their topological patterns, e.g. degree distributions (Jordano et al. 2003; Vazquez 2005), nestedness (Bascompte et al. 2003;

Dupont et al. 2003), modularity (Olesen et al. 2007), small world properties (Olesen et al. 2006), patterns of generalization/specialization (i.e. level of degree) (Bascompte et al. 2006; Olesen and Jordano 2002; Vazquez and Aizen 2003), tolerance to species extinction (Fortuna and Bascompte 2006; Memmott et al. 2004), and phenological shifts (Kaiser-Bunbury et al. 2010; Memmott et al. 2007). Network data are often qualitative, i.e. include only presence/absence information about species and links; however, quantitative networks, which include link strength, i.e. visitor/visitation frequencies, are becoming increasingly available, and network descriptors based on quantitative data have been developed (e.g. Bersier et al. 2002). Such network descriptors as well as the outcome of studies on ecological networks (e.g. extinction simulations) are highly susceptible to the overall number of interactions detected (e.g. stability; see Banasek-Richer et al. 2009; Dormann et al. 2009).

The validity of an interpretation derived from a description using network theory depends on the properties of the underlying empirical data, and the different sampling methods used in pollination network field studies have different constraints that deviate from randomness (e.g. Gibson et al. 2011). Gathering pollination network data sets is resource and time consuming and they are nearly always

*Corresponding author, email: pbs@dmu.dk

under-sampled because species or interactions easily escape observation (Olesen et al. 2010; Vazquez et al. 2009). Moreover, pollination networks are temporally highly dynamic, i.e. species and interactions are continuously changing (Alarcón et al. 2008; ; Dupont et al. 2009; Olesen et al. 2008; Petanidou et al. 2008). In most empirical studies, data collection is plant focused (Olesen et al. 2010), i.e. a fixed number of plant species are observed for visiting pollinators. This may impede our understanding of network organization and function. In particular, interactions may remain undetected because most flower visits are rare, of short duration, and usually do not leave traces on the flower. Thus, the number of recorded visits is only a small subset of the actual visits made by a species (Blüthgen et al. 2010; Goldwasser and Roughgarden 1997). Obviously, the problem of under sampling is most severe for larger networks. As a rule of thumb, the sampling effort has to increase in proportion to the number of interactions, i.e. combinations of a pollinator species and a plant species.

The following question is addressed in this paper in order to facilitate further progress for application of network data:

How can we improve the applicability of data sets to support network analysis without misinterpretation due to sampling bias and inadequate number of data records?

The problem of under sampling has been statistically investigated and modelled (Dormann et al. 2009; Vazquez and Aizen 2003) and assuming random sampling. In this paper, we propose and discuss a Bayesian approach called the ID method and a general concept model to link experiment and data analysis to see how this can supplement existing methods and, thereby, contribute to better applicability. Thus, the hypothesis of this paper is:

A Bayesian approach and a conceptual model can improve the applicability of data sets by setting up a description of the probability for a record to involve a specific combination of a visitor (pollinator species) and a receptor (plant species)!

This paper will describe the suggested methods and discuss the outcome under on the following headlines:
- How can the conceptual model clarify the governing assumptions underpinning all application of under-sampled data sets in any type of network analysis?
- What are the governing assumptions underpinning the ID method compared to the alternatives?
- How can the ID method increase the understanding in network analysis?
- How easy is the application of the ID method?
- The method is described in the next paragraph followed by the discussion to address the questions above.

METHODS

Two definitions of underlying probabilities of visits are applied for each visit:

- $p_{i,j}$: *the pollinator focused probability*. Out of all visits by the ith pollinator species, $p_{i,j}$ is the probability of a visit

to take place in the jth plant species. This is a measure of a pollinator species preference for a plant species.

- $q_{i,j}$: *the plant focused probability*. Out of all visits done in the jth plant species, $q_{i,j}$ is the probability of a visit to be done by the ith pollinator. This is a measure of pollinator species i's preference for visiting plant species j, relative to the preference of other pollinator species to visit the same plant species.

Thus, the visits to each plant species is considered as a multinomial process, where the individual pollinator "decides" to visit a plant species with some unknown probability. The task of this paper is to estimate possible intervals for this probability based on empirical data.

Conceptual model

The conceptual model is described based on sets, where the set containing all single visits between a single pollinator and plant species that took place in the area and period of study is denoted A. Every single visit is an element in the set A. Set A is divided into subsets as $Apoll_i$ and Apl_j, where $Apoll_i$ is the subset of A, containing all elements in A where pollinator i performs the visits and Apl_j is a subset of A, containing all elements in A where the plant j is being visited. All visits will involve one and only one pollinator species (i) and plant species (j), respectively, and thus are single elements that belong to both the subsets $Apoll_i$ and Apl_j. A subset of A is defined as the set containing all *recorded* (collected) visits and denoted as set B, and for set B the subsets $Bpoll_i$ and Bpl_j are defined for respective pollinator and plant species. The conceptual model is illustrated in Fig. 1 for three pollinator species and four plant species. Hence, if a visit is recorded in the data set, then the visit is an element that belongs to both set A, $Apoll_i$ and Apl_j and set B, $Bpoll_i$ and Bpl_j, respectively.

If all visits in set B are random observations from A without bias for any pollinator or plant species, then B is claimed to be randomly collected. Thus, a random collection assumes that the data collector is not more likely to record visits by some species of pollinators, e.g. large conspicuous bumble bees, than others, e.g. small flies. A fully random collection also assumes that the plant species are randomly selected. Thus, if plant species Pl1 receives twice as many visits as plant species Pl2, then the probability of an observer, in a fully random collection, to observe a visit by pollinator of Pl1 is twice as high as the probability of observing a pollinator visiting Pl2.

Thus, for the "ideal" random observer, every visit in the study area is equally likely to be observed, and set B is a random selection of some of the elements in set A. However, for empirical data sets, the set B is rarely a random subset of A and its applicability for analysis is, thus, constrained or to some degree uncertain. Two typical cases of "non-randomness" or bias can be defined in the following way:

Pollinator focused sampling, has random sampling within $Bpoll_i$, but not between different pollinators and, thus, only allows estimating $p_{i,j}$. This can be illustrated in Fig. 1 as a situation where an element placed in $Apoll1$ is

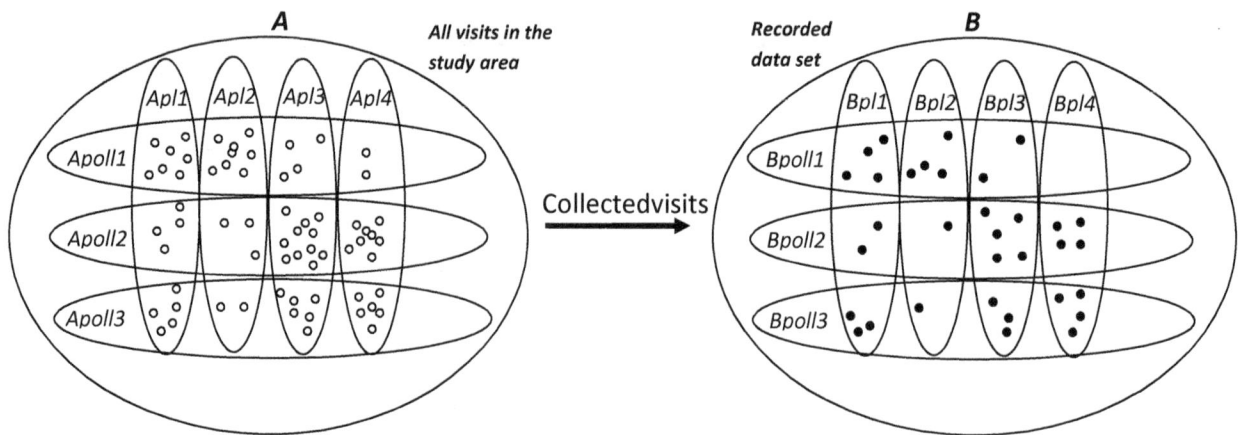

Figure 1. Illustration of the concept model, including three pollinator species and four plant species, with the set of all interactions between pollinator and plant species (A) and a subset of recorded interactions (B). The definition of sets in the model concept follows the definitions in the text, where every single dot (   ) represents is a visit.

more likely to be collected than an element placed in *Apoll2*, but within *Apoll1* the likelihood for an element to be collected is the same for all plant species (*Apl1-4*). This type of randomness will be denoted 'pollinator focused sampling' and can be used only to estimate p_{ij}. Data generated with a pollinator focus, e.g. by tracking pollinator individuals visiting flowers, could be considered as pollinator focused sampling. Another reason behind this type of non-randomness (bias) could be that the observer has more focus on the large conspicuous bumble bees than on small flies. If this type of sampling is applied for generating data to feed network models, then a high connectivity for one pollinator species compared to others can be an artefact due to an extra intensive sampling effort for that specific pollinator species.

Plant focused sampling has random sampling within *Bpl*$_j$, but not between different plant species and, thus only allows estimating q_{ij}. This can be shown in Fig. 1 as a situation where an element in *Apl1* is more likely to be collected than an element in *Apl2*, but within *Apl1* the likelihood of a collection from one of the pollinator species (*Apoll1-4*) is the same. A sampling method that will mimic this situation is an approach, where the plant species are recorded by an observer who is waiting for the pollinators to arrive to the focused upon plant individual. If this type of sampling is applied for generating data to feed network models, then a high connectivity for one plant species compared to other plants in the data set can be an artefact due to an extra intensive sampling effort for that specific plant species.

The data set *B* is used to make a data table (Table 1) by summing the number of elements for every combination of pollinator and plant species. This table is termed an interaction frequency matrix. Row *i* in Table 1 contains all visits by *Bpoll*$_i$ and column *j* contains all visits received by *Bpl*$_j$. The value v_{ij} is, thus, the number of visits, equivalent to the number of elements in *Bpoll*$_i \cap$ *Bpl*$_j$.

Model equations

Number of total recorded visits by the *i*th pollinator species on any plant species in the data set is

$$V_i = \sum_{j=1}^{M} v_{ij} \qquad \text{1a}$$

Number of total recorded visits to plant species *j* by any pollinator species is

$$W_j = \sum_{i=1}^{N} v_{ij} \qquad \text{1b}$$

Eqs. 1a and b are based on the definitions in Table 1.

A row of $v_{i,1}, v_{i,2}, ..., v_{i,M}$ values in Table 1 can be considered as a vector v_j. The number   shows that the pollinator *i* has visited the plant *j* a total of v_{ij} times. When pollinator *i* makes a visit, then the probability for this visit to take place in plant *j* is p_{ij} If it is *a priori* known that pollinator species *i* will never visit plant species *j*, then this combination of *i* and *j* is denoted "null" in Table 1, and v_{ij} must necessarily have been recorded as zero in the data set. This situation will occur if plant and pollinator species mismatch e.g. in phenology or morphology or in season (Olesen et al. 2010). It follows that the probability for a pollinator species to visit the plant species must be zero for all "null" combinations of *i* and *j*, thus, $v_{ij} \equiv 0$ for all combinations of *i* and *j* values having a "null" for v_{ij}. On the other hand, a value of $v_{i,j} = 0$ will not necessarily be a "null" value, as it could imply that the pollinator species *i* so seldom visiting plant *j* that such a visit is not recorded in the data set or it may be an unknown null value.

In conclusion, if the p_{ij} values are known, it will be possible to set up a statistical multinomial model to predict the distributions of possible numbers of visits made by a pollinator species to different plant species in a data set. The challenge is that the p_{ij} values are unknown and, hence, should be estimated from an empirical data set (Table 1). The Dirichlet distribution can estimate the distribution of possible p_{ij} values for the multinomial distribution based on empirical data (Frigyik et al. 2010). Thus, it follows that if we have the correct values for p_{ij}, then we can estimate the distribution of possible v_{ij} values, using a multinomial distribution as a statistical model based on the total number

Pollinator species	Plant species								Total number of non-null plant species	Total number of visits
	1	2	3	4	-	j	-	M		
1	$v_{1,1}$	$v_{1,2}$	$v_{1,3}$	$v_{1,4}$	-	$v_{1,j}$	-	$v_{1,M}$	m_1	V_1
2	$v_{2,1}$	null	$v_{2,3}$	$v_{2,4}$	-	null	-	$v_{2,M}$	m_2	V_2
3	$v_{3,1}$	$v_{3,2}$	$v_{3,3}$	$v_{3,4}$	-	$v_{3,j}$	-	$v_{3,M}$	m_3	V_3
4	null	$v_{4,2}$	$v_{4,3}$	$v_{4,4}$	-	$v_{4,j}$	-	$v_{4,M}$	m_4	V_4
-	-	-	-	-	-	-	-	-	-	-
i	$v_{i,1}$	$v_{i,2}$	$v_{i,3}$	$v_{i,4}$	-	$v_{i,j}$	-	$v_{i,M}$	m_i	V_i
-	-	-	-	-		-		-	-	-
N	$v_{N,1}$	$v_{N,2}$	$v_{N,3}$	$v_{N,4}$	-	$v_{N,5}$	-	$v_{N,M}$	m_N	V_N
Total number of non-null plant species	n_1	n_2	n_3	n_4		n_j		n_M		
Total number of visits	W_1	W_2	W_3	W_4	-	W_j	-	W_M		

Table I. The data matrix of visitation data (interaction frequency matrix), with the recorded number of visits by pollinator i onto plant j. N and M are numbers of pollinator and plant species in the data set, respectively. The null values are used for combinations of pollinator species and plant species where visits are known to be impossible.

of observations. However, because the p_{ij} values are unknown, we can use the data set to find possible values based on the assumption that a multinomial distribution is more likely to result in the observed data set for some p_{ij} values compared to others. In Bayesian terms, this means that the Dirichlet distribution can be used to find this distribution of p_{ij} values as the conjugate prior for the multinomial distribution (Frigyi et al. 2010). However, this paper will not go deeper into the background of Bayesian analysis and will, thus, take this statement for granted. For comprehensive data sets (high sampling effort), the Dirichlet distribution will be "narrow" and, thus, estimate the p_{ij} value as narrow (certain) intervals, while a sparse data set (low sampling effort) will result in a broad and more uncertain estimate of the p_{ij} values.

The Dirichlet distribution Dir (α) for p_i, where p_i is the vector of the probabilities $p_{i,1}, \ldots p_{i,m}$, and α is the parameter vector $\alpha_1, \ldots, \alpha_m$ is

$$f_{V_i}(p_i, \alpha) = \frac{\Gamma\left(\sum_{j=1}^{M} \alpha_j\right)}{\prod_{j=1}^{M} [\Gamma(\alpha_j)]} \cdot \prod_{j=1}^{M} (p_{i,j})^{\alpha_j - 1} \qquad 2$$

Where and $\Gamma(\ )$ is the gamma function (Evans et al. 2000). If there are no data (a priori) to consider, then the Dir (α) is assumed to have unified distributions for all $p_{i,1}, \ldots p_{i,m}$, which is equivalent to stating that "no data" is "no knowledge". The Dirichlet distribution yields a unified distribution for $p_{i,1} \ldots p_{i,m}$ when: $\alpha_1, \ldots, \alpha_m = 1$ and acts as conjugate prior for the multinomial distribution by

Dir($\alpha_i + v_i$)(Frigyik et al., 2010), where v_i is the vector of $v_{i,1}, v_{i,2}, \ldots, v_{i,M}$ (Countings in Table I). Thus, using $\alpha_1, \ldots, \alpha_m = 1$ for Dir($\alpha_i + v_i$), the distribution function becomes:

$$f_{V_i}(p_i, v_i) = \frac{\Gamma(V_i + m_i)}{\prod_{j=1}^{M} [\Gamma(v_{i,j} + 1)]} \cdot \prod_{j=1}^{M} (p_{i,j})^{v_{i,j}} \qquad 3$$

Both p_i and v_i are only defined for j values that are not "null" in the data set.

All the considerations above can be repeated for the probabilities $q_{i,1}, \ldots q_{i,M}$ and the vector v_i of $v_{1,j}, v_{2,j}, \ldots, v_{N,j}$ in order to investigate the probabilities for different pollinator species to visit plant species j. This yields a similar equation for $q_{i,j}$,

$$f_{W_j}(q_j, v_i) = \frac{\Gamma(W_j + n_j)}{\prod_{j=1}^{M} [\Gamma(v_{i,j} + 1)]} \cdot \prod_{i=1}^{N} (q_{i,j})^{v_{i,j}} \qquad 3b$$

Where q_i is the vector of the probabilities $q_{i,1} \ldots q_{i,m}$. Both q_i and v_i are only defined for i values that are not "null" in the data set.

The following necessary relations are true for the probabilities:

$$\sum_{j=1}^{M} p_{i,j} = 1 \qquad 4a$$

The probability for a pollinator to visit any possible plant when it makes a visit is 1

$$\sum_{i=1}^{N} q_{i,j} = 1 \qquad \qquad 4b$$

The probability of a plant receiving a visit from any possible pollinator when it gets a visit is 1

It can be shown (Frigyik et al., 2010) that the density distribution (marginal distributions) of $p_{i,j}$ and $q_{i,j}$, respectively, can be described by the beta function as:

$$f_{p_{ij}}(p_{i,j}) = beta(1 + v_{i,j}; m_i + V_i - v_{i,j} - 1) \qquad 5a$$

$$f_{p_{ij}}(q_{i,j}) = beta(1 + v_{i,j}; n_i + W_i - v_{i,j} - 1) \qquad 5b$$

The Beta distribution has some simple statistical properties (see e.g. Evans et al., 2000). Hence, using Eqs. 5a and b, we find mean (E) and variance (VAR) for $p_{i,j}$ and $q_{i,j}$:

$$E(p_{i,j}) = \frac{1 + v_{i,j}}{m_i + V_i} \qquad \qquad 6a$$

$$E(q_{i,j}) = \frac{1 + v_{i,j}}{n_i + W_i} \qquad \qquad 6b$$

$$VAR(p_{i,j}) = \frac{E(p_{i,j}) \cdot (1 - E(p_{i,j}))}{1 + m_i + V_i} \qquad 7a$$

$$VAR(q_{i,j}) = \frac{E(q_{i,j}) \cdot (1 - E(q_{i,j}))}{1 + n_j + W_j} \qquad 7b$$

Increasing values for V_i and W_j will lead to a greater increase in the nominator relative to the denominator in Eqs. 7a and 7b, and VAR(), therefore, will decrease when the number of records is increased. Hence, p_{ij} and q_{ij} become increasingly precisely estimated for an increasing number of records. This also applies to cases where additional records are not related to specific pollinator or plant species (different i or j value).

It is possible to make a simplified uncertainty assessment of the under-sampled data sets based on the binominal distribution and p_{ij} or q_{ij} respectively.

$$f_{vij} = Bin(v_{i,j}, V_i, p_{i,j}) \qquad \qquad 8a$$

$$f_{vij} = Bin(v_{i,j}, W_j, q_{i,j}) \qquad \qquad 8b$$

Where

$$E(v_{i,j}) = V_i \cdot p_{i,j} \qquad \qquad 9a$$

$$E(v_{i,j}) = W_j \cdot q_{i,j} \qquad \qquad 9b$$

If the values of p_{ij} and q_{ij} are assumed to be known or estimated using Eq. 6a or b, respectively, then it is possible to estimate the interval of "realistic" $v_{i,j}$ values that can be

recorded out of all V_i or W_j records for pollinator i. The variance of v_{ij} can be estimated in cases where the normal approximation is valid: $V_i \cdot p_{i,j} \gg 1$ or $W_j \cdot q_{i,j} \gg 1$ and $V_i \cdot p_{i,j} \cdot (1 - p_{i,j}) \gg 1$ or $W_j \cdot q_{i,j} \cdot (1 - q_{i,j}) \gg 1$ as

$$VAR(v_{i,j}) \approx V_i \cdot p_{i,j} \cdot (1 - p_{i,j}) \qquad 10a$$

$$VAR(v_{i,j}) \approx W_j \cdot q_{i,j} \cdot (1 - q_{i,j}) \qquad 10b$$

Combining Eqs. 6a and b with 10a or 10b yields a simple rough estimate for the variance of v_{ij}:

$$VAR(v_{i,j}) \approx \frac{V_i(1 + v_{i,j})}{m_i + V_i} \cdot \left(1 - \frac{1 + v_{i,j}}{m_i + V_i}\right) \qquad 11a$$

$$VAR(v_{i,j}) \approx \frac{W_j(1 + v_{i,j})}{n_i + W_j} \cdot \left(1 - \frac{1 + v_{i,j}}{n_j + W_j}\right) \qquad 11b$$

If the normal approximation is valid, then it will also be true that $v_{i,j} \gg 1$ and, in many cases, also $V_i \gg m_i$ or $W_j \gg n_j$, yielding the following simple but rough estimate for the variance:

$$VAR(v_{i,j}) \approx v_{i,j} \cdot \left(1 - \frac{v_{i,j}}{V_i}\right) \qquad 12a$$

$$VAR(v_{i,j}) \approx v_{i,j} \cdot \left(1 - \frac{v_{i,j}}{W_j}\right) \qquad 12b$$

NUMERICAL EXAMPLE FOR ILLUSTRATION

The principle of the method is best illustrated by a simple artificial numerical example. Real data sets will, typically, be much larger, so a smaller numerical example is chosen for the purpose of illustration. The data set includes three pollinator species (rows) and two plant species (columns) (Table 2).

The distribution of p and q is calculated using Eqs. 5a and b, respectively, and the results are shown in Fig 2a-f.

The Poll 1 and Pl 1 combination in Table 2 shows a situation where Poll 1 most frequently visits this plant and, thus, a density distribution (Fig. 2a) for $p_{1,1}$ that is located mainly above 0.5. On the other hand, the plant species receives more visits from Poll 2, so the value of $q_{1,1}$ is smaller

Table 2. Illustrative data set for three pollinator and two plant species

	Pl1	Pl2	Total
Poll1	10	5	15
Poll2	50	3	53
Poll3	0	6	6
Total	60	14	

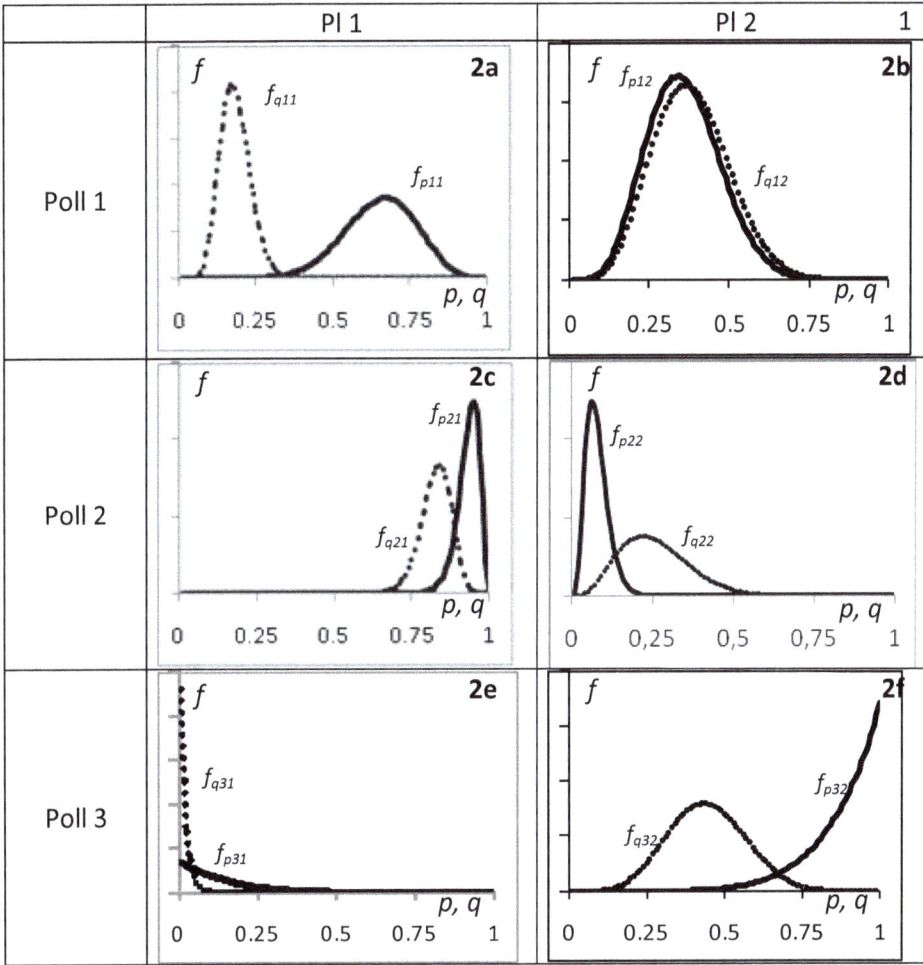

Figure 2a-f. Graphic display of eqs. 5 a and b for the data in Table 2, where the y-axis is the probability density for p and q, and the x-axis is the values of p and q (continuous line: function of p, dotted line: function of q).

than 0.5. The limited number of total recorded visits for Poll I results in a wider distribution (larger VAR()) of the $p_{1,1}$value compared to the stronger (smaller VAR()) determination of the $q_{1,1}$ value. A similar relation, where the strength of the estimation is highly different, is also shown for the Poll 2 and Pl 2 combination (Fig. 2d), but the roles of pollinator and plant are reversed. The Poll 2 and Pl I combination (Fig. 2c) shows a situation with a larger number of records, yielding a good determination of both probabilities. Poll 3 only visited Pl 2 and was only recorded six times in total. From the data, one may conclude that Poll 3 is not visiting Pl I. However, due to the low number of records, this may simply be a result of small sample size. The curve for $p_{3,1}$ in Fig. 2e shows that the probability for Poll 3 to visit Pl I, when Poll 3 is visiting either Pl I or 2, is less than 0.25, but markedly above zero. However, if the question is reversed, i.e. 'what is the probability of a visit to Pl I by Poll 3' ($q_{3,1}$), the result is dramatically different, i.e. a probability close to zero is highly probable due to a high number (60) of visits observed at Pl I, but none were by Poll3, so in this case the ID method may have identified an unknown "null value".

The Dirichlet distribution function (Eq. 3b) for Pl 2 is shown in Fig. 3.

The dynamics of the multivariate probability are shown in Fig 3. For instance, the distribution for $q_{1,2}$ (visits by Poll

I to Pl 2) depends on the value of $q_{3,2}$, (visits of Poll 3 to Pl 2). Hence, if the $q_{3,2}$ value is high, then it leaves smaller likelihood and variation for $q_{1,2}$, because $q_{3,2}+q_{1,2}+q_{2,2} = 1$, which forces the density distribution for $q_{1,2}$ to level out for larger values of $q_{3,2}$.

The distribution functions listed in Fig. 2a-f indicate the sampling uncertainty for the data in Table 2, where a wide

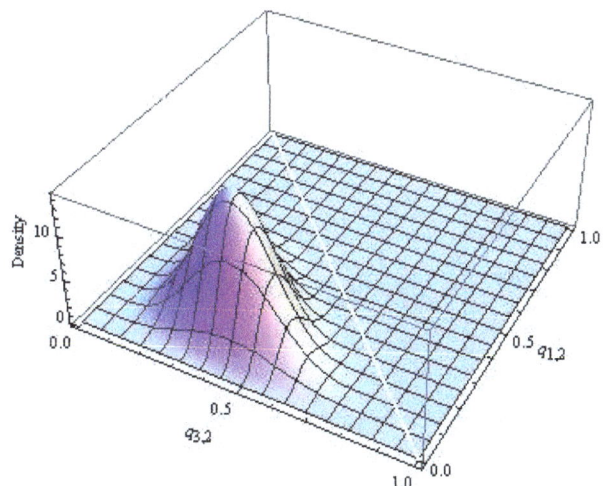

Figure 3. The Dirichlet distribution function for Pl 2 showing $q_{3,2}$ and $q_{1,2}$, where $q_{2,2}$ has been determined for all combinations of $q_{3,2}$ and $q_{1,2}$ as $1-q_{3,2}-q_{1,2}$.

distribution predicts a high uncertainty due to a limited number of records. However, it is also possible to make a simple assessment of the uncertainty based on Eqs. 11a or 12a if relatively many records ($V_i \cdot p_{i,j} \gg 1$ and $V_i \cdot p_{i,j} \cdot (1 - p_{i,j}) \gg 1$) are available. This can be considered as a valid approximation for e.g. Pl1 in Table 2.

For cases in which few records are found (e.g. v_{22} in Table 2 there are only three recorded visits), the normal approximation is invalid and the Eqs. 11a and 12a are useless. So, in this case, the uncertainty of the recorded number needs to be assessed based on the Eqs. 5a or b and 8a or b. For v_{22} in Table 2, two questions could be "what values can v_{22} take if 53 visits are recorded by poll2" or "what values can v_{22} take if 14 visits are received by Pl2". A nested Monte Carlo algorithm is used to find the answers. If the function F(x) is the accumulated density distribution function for the parameter x, then the value of F(x) is defined for the interval 0-1, and the principle in the Monte Carlo algorithm is to let the computer draw a number at random within the interval of 1-0- and then use the inverse F(x), F^{-1}(x) to find the corresponding value of x. This can be done in a simple spreadsheet without a comprehensive mathematical effort if the inverse functions exist in the software. The principle is firstly to compute a value of p_{22} or q_{22} using a random number (0-1) as input to the inverse accumulated Beta distribution and the parameters defined in Eqs. 5a or b. Secondly, the obtained values of p_{22} or q_{22} are used as input to Eq. 8a or 8b to draw a value of v_{22}. The sequential procedure is repeated e.g. 10 000 times to make a set of v_{22} values. The results are shown in Fig. 4 using both p_{22} (Eqs. 5a and 8a) and q_{22} (Eqs. 5b and 8b). Fig. 4 shows that the recorded value of v_{22} for any re-sampled data set of this size will be in the interval of 0-9 (or 10) visits, with 1-4 visits being the most likely values.

It is also possible to re-sample a whole data set and use these re-sampled data to test robustness of network descriptors calculated based on the data. The replication can be repeated thousands of times to find the percentile of the calculated descriptors, and the following example demonstrates how the ID method is easily applied for this purpose. The principle of the simulated re-sampling is to let the computer "sample" the data set: (1) Estimate the probability of "observing" a visit in the next sample for each combination of pollinator and plant species; (2) Use that probability to let the computer draw (decide) which combination to be sampled, as described in the text above Figure 4; (3) Repeat the item 1 and 2 until the number of data records is similar to the number in the original data set. The probability of "observing" a visit ($Ps_{i,j}$) is calculated as:

$$Ps_{i,j} = Q_i \cdot p_{i,j} \qquad\qquad 13$$

Where Q_i is the probability of the simulated "observer" observing the pollinator species i without distinguishing between the plant species involved. The reasoning behind Eq. 13 is that the probability of observing the ith pollinator species on the jth plant species is equal to the probability of the ith pollinator species to be observed as a visitor for any plant species and multiplied with the probability for the ith pollinator to visit the plant species j when the pollinator species is observed. The Dirichlet distribution can be used to estimate the Q_i value based on the $V_1, V_2, \cdots, V_i, \cdots, V_N$ values defined in Table 1. Instead of estimating the q_{ij} as the probability of pollinator species i to visit the plant species j, we are now estimating the probability of pollinator species i to visit any plant species. So the principle of using the Dirichlet distribution remains for the merged data:

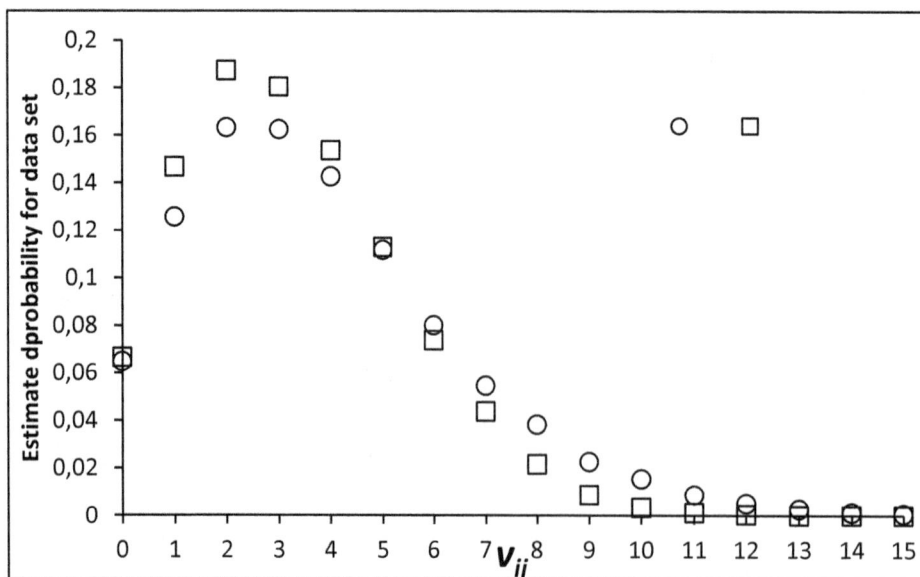

Figure 4. Nested Monte Carlo estimation of the probability for getting different v_{22} values in a re-sampled data set, where in total 53 recordings are made of pollinator species 2 (to estimate p_{22}) and 14 recordings are made for plant species 2 (to estimate q_{22}).

$$f_{V_i}(\boldsymbol{Q}, \boldsymbol{V}) = \frac{\Gamma(S+N)}{\prod_{i=1}^{N}[\Gamma(V_i+1)]} \cdot \prod_{i=1}^{N}(Q_{i,j})^{V_i} \qquad 14$$

where S is the total number of visits in the data set:

$$S = \sum_{i=1}^{N} V_i \qquad 15$$

Where Q and V are the vectors: $Q_1, Q_2 \cdots, Q_i, \cdots, Q_N$ and $V_1, V_2, \cdots, V_i, \cdots, V_N$ respectively.

Thus, the probability of obtaining a record of the ith pollinator and the jth plant species is a product of two probabilities, each being estimated by the data set using a Dirichlet distribution. It is possible to generate a random number that follows the Dirichlet distribution using an inverse Gamma distribution (see the algorithm in Frigyik et al. 2010). The principle of generating a single simulated data set, based on the Eqs. 3a, 13 and 14, is illustrated in Figure 5 for the data set in Table 2. The procedure in Figure 5 can be repeated to make a larger number of simulated data sets.

DISCUSSION

Clarification of governing assumptions for application of under-sampled data sets

The governing assumptions underpinning application of under-sampled data sets are evaluated using a conceptual model (Figure 1). This model can help to specify the type of probability that can be estimated based on the data depending on how the data are collected. An ideal data set is a random sample of visits without considering the pollinator or plant species, however, such a data set is difficult to obtain. If the governing assumption of sampling randomness is not fulfilled, then it conflicts with many descriptors that have been calculated based on the network analysis. Despite this, many cases of data collection are plant focused (Olesen et al. 2010) and this will only support the calculations of descriptors for each plant species separately. If the data are completely randomly sampled, then the ID method can estimate meaningful values for both q and p. However, in case of plant focused sampling, only q is meaningful, and in case of pollinator focused sampling, only p is meaningful. This problem of missing randomness should be consulted as a preliminary step before application of nearly any mutual network analysis method. This involves a careful description of the data collection protocols to display any form of potential bias. The conceptual model can help to clarify the usefulness of data set in network analysis by specifying the meaning of pollinator focused and plant focused data sets, respectively.

Assumptions underpinning the ID method

In plant-pollinator networks, some interactions never occur (termed forbidden links), for instance due to morphological and phenological mismatching (Jordano et al. 2003; Olesen et al. 2010). It may be well known that some pollinator species in the data set avoid visiting some plant species in the data set, and in this case it will improve the predictive power of the ID method to set the values in the

data set (Table 1) as "null". The a priori assumption of the remaining "allowed" combinations of pollinator and plant species is that the plant species are equally likely to be visited by any allowable pollinator species, and all pollinator species are equally likely to visit any allowable plant species. This is described in Eqs. 6a and b, where the expectation for p and q is $1/m_i$ and $1/n_j$, respectively, if there are no records in the data set ($V_i = W_j = 0$). The estimated probabilities can deviate more and more strongly from being equal distributed as the amount of data records increases.

A strength of the ID method is that, in contrast to other methods described in e.g. (Dormann et al. 2009), there is no need to assume any distribution of the data (log Normal or others) to be valid. Such additional assumptions open up for two types of uncertainties: (1) The structural uncertainty, where the form of the assumed distribution function may not be correct as description of the variability, e.g. it may allow nearly infinite high sampling values or more or less unknown truncations, (2) Parametric uncertainty, where the values of the distribution parameters (mean value, standard deviation etc.) may not be known for certain. This does not mean the ID method is always the best choice, as this depends on the condition of the data set and other sources of information in the particular case. If there are information available to parameterise and validate assumed distribution functions the statistical method as presented in (Dormann et al. 2009) could turn out to be as good or event better than the ID method. The ID method has a potential to be used especially when the validity of additional assumptions, others than given in the concept model, are insufficiently documented.

The ID method represents the simplest form for Bayesian approach, and in future activities it may be possible to develop more complex methods that better can take different type of a priori biological knowledge into account.

Methodological outcome as a contribution to better understanding

The ecological interpretation of p and q depends on the temporal and spatial scale of the data. If the data are collected over a few days in a local site where all pollinators have been foraging on the same plants and under more or less constant weather conditions, then p will reflect the real behaviour of the pollinators when they are choosing between different species of plants, and q will reflect a joint result of both pollinator abundance and behaviour. On the contrary, if the data are collected during a longer period, then some of the recorded plant species in the data set may not have been flowering synchronically during the investigation period (Olesen et al. 2010). In this type of data, a high p_{ij} value can either be due to the fact that plant species j is attractive compared to other plant species, or due to the fact that plant species j was the only one to blossom and, thus, to be visited during a critical period within the data collection. Similarly, a high q_{ij} value can be due to the fact that either plant species j is attractive to pollinator species i compared to other pollinator species, or because pollinator species i was the only pollinator to be active during the flowering period of plant species j. For larger areas, the recorded pollinators may

The recorded data
with 74 records
(Table 2)

	Pl1	Pl2	Total (**V**)
Poll1	10	5	15
Poll2	50	3	53
Poll3	0	6	6

74 sets of Q_i values are drawn from
the Dirichlet distribution (Eq. 14)
using inverse Gamma distribution

74 sets of $p_{i,j}$ values are drawn from
the Dirichlet distribution (Eq. 3a)
using inverse Gamma distribution

$p_{i,j}$	1		2		-	-	74	
	Pl1	Pl2	Pl1	Pl2	-	-	Pl1	Pl2
Poll1	0.595	0.405	0.670	0.330	-	-	0.516	0.483
Poll2	0.954	0.046	0.907	0.093	-	-	0.886	0.113
Poll3	0.001	0.999	0.298	0.702	-	-	0.338	0.661

Q_i	1	2	-	-	74
Poll1	0.218	0.176	-	-	0.206
Poll2	0.688	0.730	-	-	0.687
Poll3	0.094	0.094	-	-	0.107

Joint probability for
sampling the poll and
pl combination in each
of the records 1,2..74

$p_{i,j} \cdot Q_i$ (Eq. 13)	1		2		-	-	74	
	Pl1	Pl2	Pl1	Pl2	-	-	Pl1	Pl2
Poll1	0.130	0.089	0.118	0.058	-	-	0.106	0.100
Poll2	0.657	0.031	0.663	0.068	-	-	0.610	0.078
Poll3	0.000	0.093	0.028	0.065	-	-	0.036	0.070

Selection	1		2		-	-	74	
	Pl1	Pl2	Pl1	Pl2	-	-	Pl1	Pl2
Poll1	x				-	-		x
Poll2			x		-	-		
Poll3					-	-		

Counts the
number of
selections

Countings	Syntethic data set	
	Pl1	Pl2
Poll1	6	7
Poll2	51	4
Poll3	0	6

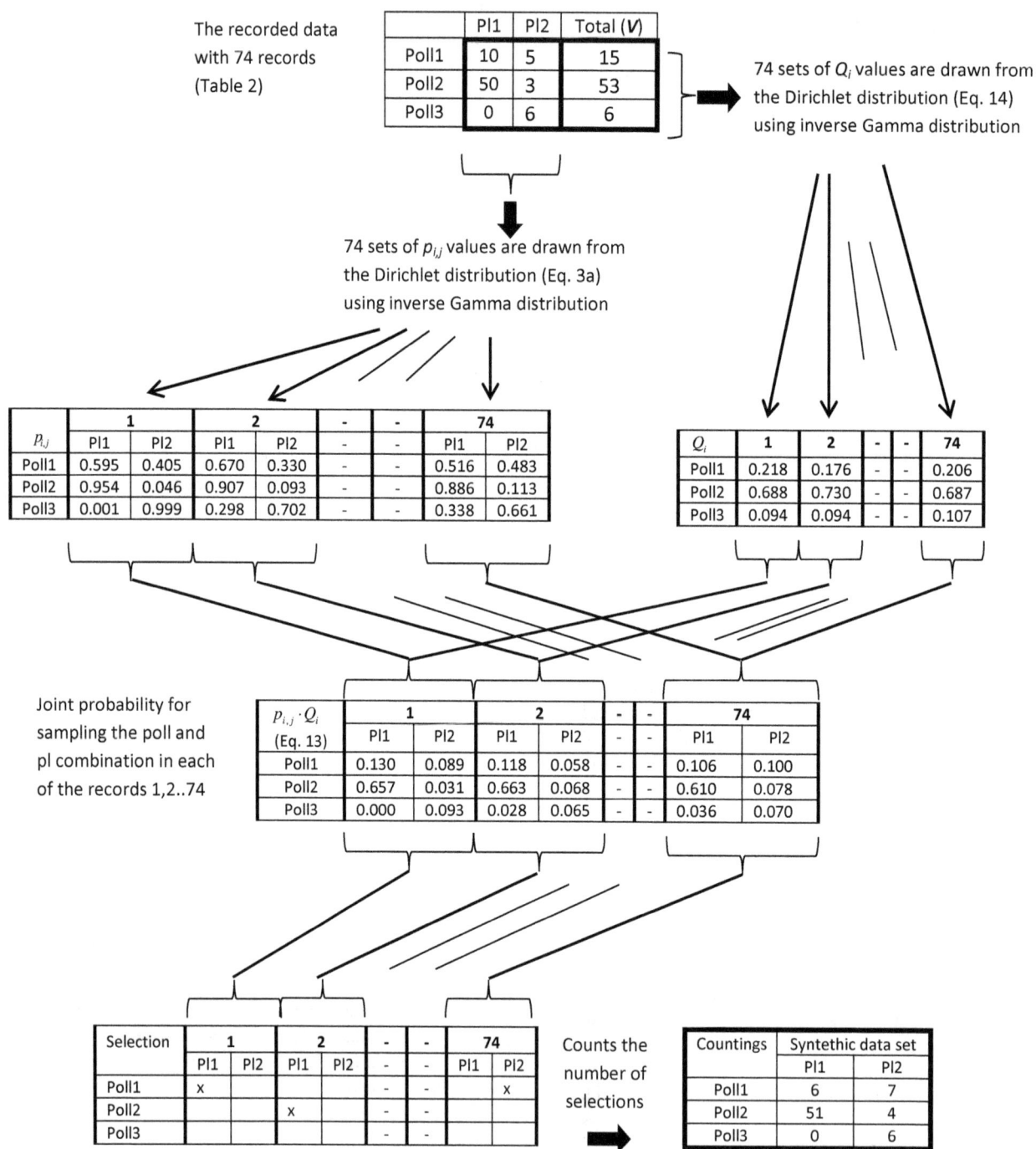

Figure 5. Principle of resampling of 74 records to generate a synthetic data set on basis of the data set shown in Table 2. The original data set is used in the inverse Gamma distribution to draw stocastgically 74 set of respectively, Q_i and $p_{i,j}$ values from their respective Dirichlet distributions. The values of Q_i and $p_{i,j}$ are multiplied (Eq. 13) and used to make a biased random selection of which pollinator to "sample". Thus, if $Q_1 \cdot p_{1,1} = 0.130$ then there is a 13.0 % change to "sample" the Poll1/Pl1 combination. Finally all the "synthetic" observations are counted to yield the synthetic data set for 74 observations.

not have been foraging in the same local area. In this case, the availability of plant species may not have been the same for different pollinator species, depending on their foraging radius and local abundance. In all cases, the values of p and q disclose important ecological information, and the certainty of the estimates will show the statistical usefulness of the under-sampled data for any quantitative interpretation. In all cases, the ID method can compile the data set to find statistical information about the interactions, but the interpretation of the results depends on the actual background of the data set.

The probabilistic property of respectively p_{ij} and q_{ij} makes them directly applicable for the entropy (Shannon) based indexes (Dormann et al., 2009). The ID method can, in contrast to existing approaches, generate synthetic data for construction of networks whiteout assuming any density function to govern the recorded number of visits (see Figure 5) and without assuming any fixed number of observations for pollinators and/or plants other than a fixed total number of records. The simulated data set can test any network calculation. e.g. the d' and H_2' indexes suggested by (Blüthgen et al. 2006), using the real data set and many (more than 1 000) of the simulated data sets, respectively.

Application

The ID method has a general relevance for many resource-consumer networks for which the conceptual model (Fig 1) and data sets as defined in Tab. 1 apply. An add in for Excel 2010 and a related short tutorial, is made as supplementary material to this paper that runs the algorithm in Fig. 5. An empirical but close approximation to the invers gamma distribution is used in this add in to speed up the calculations and the add in will be continuously extended in the future. For all interested parties, it is possible to attend a mailing list by sending an e-mail to the first author of this paper. Definitely, software exists that can handle the Dirichlet distribution directly, e.g. Mathematica (http://www.wolfram.com/mathematica/) or R (http://www.r-project.org/).

ACKNOWLEDGMENTS

This project was conceived and developed as part of STEP (Status and Trends of European Pollinators), which is funded by the European Commission as a Collaborative Project within Framework 7 under grant 244090 – STEP– CP – FP (Potts et al. 2011).

REFERENCES

Alarcón R, Waser N M, Ollerton J (2008) Year-to-year variation in the topology of a plant-pollinator interaction network. Oikos 117: 1796–1807.

Banašek-Richter C, Bersier LF, Cattin MF, Baltensperger R, Gabriel JP, Merz Y, Ulanowicz E, , Tavares AF, Williams DD, De Ruiter PC, Winemiller KO, Naisbit RE (2009) Complexity in quantitative food webs. Ecology 90: 1470-1477.

Bascompte J, Jordano P, Melián C.J, Olesen JM (2003) The nested assembly of plant-animal mutualistic networks. Proceedings of the National Academy of Sciences of the USA 100: 9383-9387.

Bascompte J, Jordano P Olesen JM (2006) Asymmetric coevolutionary networks facilitate biodiversity maintenance. Science 312: 431-433.

Biesmeijer JC, Roberts SPM, Reemer M, Ohlemüller R, Edwards M, Peeters T, Schaffers AP, Potts SG, Kleukers R, Thomas CD, Settele J, Kunin WE (206) Parallel Declines in Pollinators and Insect-Pollinated Plants in Britain and the Netherlands. Science 21:351-354.

Bersier LF, Banašek-Richter C, Cattin MF (2002). Quantitative descriptors of food web matrices. Ecology 83: 2394-2407.

Blüthgen N, Menzel F, Blüthgen N (2006) Measuring specialization in species interaction networks. BMC Ecology 6: 9.

Blüthgen N (2010) Why network analysis is often disconnected from community ecology: A critique and an ecologist's guide. Basic and Applied Ecology 11, 185-195.

Dormann CF, Blüthgen N, Fründ J, Gruber B (2009) Indices, graphs and null models: Analyzing bipartite ecological networks. The Open Ecology Journal 2:7-24.

Dupont YL, Hansen DM, Olesen JM (2003) Structure of a plant - flower-visitor network in the high-altitude sub-alpine desert of Tenerife, Canary Islands. Ecography 26:301-310.

Dupont YL, Padrón B, Olesen JM Petanidou T (2009) Spatio-temporal variation in the structure ofpollination networks. Oikos 118:1261-1269.

Evans M, Hastings N, Peacock B (2000) Statistical Distributions (3rd ed.). Wiley Series in Probability and Statistics, John Wiley Sons, Inc.

Fortuna, M.A. &Bascompte, J. (2006) Habitat loss and the structure of plant–animal mutualistic networks. Ecology Letters 9: 281-286.

Frigyik BA, Kapila A, Gupta MR (2010) Introduction to the Dirichelt Distribution and Related Processes, UWEE Technical Report Number UWEETR-2010-0006.Department of Electrical Engineering, University of Washington.

Gibson RH, Knott B, Eberlein T, Memmott J (2011) Sampling method influences the structure of plant – pollinator networks. Oikos 120: 822-831.

Goldwasser L, Roughgarden J (1997) Construction and analysis of a large Caribbean food web. Ecology 74: 1216-1233.

Hegland SJ, Nielsen A, Lázaro A, Bjerknes AL, Totland Ø (2099) How does climate warming affect plant–pollinator interactions? Ecology Letters 12:184-195.

Ings TC, Montoya JM, Bascompte J, Bluthgen N, Brown L, Dormann CF, Edwards F, Figueroa D, Jacob UI, Jones JI, Lauridsen RB, Ledger ME, Lewis HM, Olesen JM, van Veen FJF, Warren PH Woodward G (2009) Ecological networks - beyond food webs. Journal of Animal Ecology 78: 253-269.

Jordano P, Bascompte J Olesen JM (2003) Invariant properties in coevolutionary networks of plant-animal interactions.Ecology Letters 6: 69-81.

Kaiser-Bunbury CN, Valentin T, Mougal J, Matatiken D Ghazoul J (2010) The tolerance of island plant–pollinator networks to alien plants. Journal of Ecology 99: 202-213.

Memmott J, Waser N, Price MV (2004) Tolerance of pollination networks to species extinctions. Proceedings of the Royal Society of London Series B 271: 2605-2611.

Memmott J, Craze PG, Waser NM, Price MV (2007) Global warming and the disruption of plant–pollinator interactions. Ecology Letters 10: 710-717.

Olesen JM Jordano P (2002) Geographic patterns in plant-pollinator mutualistic networks. Ecology 83: 2416-2424..

Olesen JM, Bascompte J, Dupont YL Jordano P (2006) The smallest of all worlds: Pollination networks. Journal of Theoretical Biology 240: 270–276.

Olesen JM, Bascompte J, Dupont YL Jordano P (2007) The modularity of pollination networks. Proceedings of the National Academy of Sciences of the USA 104: 19891–19896.

Olesen JM, Bascompte J, Elberling H Jordano P (2008) Temporal dynamics in a pollination network. Ecology 89, 1573-1582.

Olesen JM, Bascompte J, Dupont YL, Elberling H, Jordano P (2010) Missing and forbidden links in mutualistic networks. Proceedings of the Royal Society of London Series B-Biological Sciences 278: 725-732.

Petanidou T, Kallimanis AS, Tzanopoulos J, Sgardelis SP Pantis JD (2008) Long-term observation of a pollination network: fluctuation in species and interactions, relative invariance of network structure and implications for estimates of specialization. Ecology Letters 11, 564 575.

Potts SG, Biesmeijer JC, Kremen C, Neumann P, Schweiger O, Kunin WE (2010) Global pollinator declines: trends, impacts and drivers. Trends in Ecology & Evolution 24:345-353.

Potts SG, Biesmeijer JC, Bommarco R, Felicioli A, Fischer M, Jokinen P, Kleijn D, Klein AM, Kunin WE, Neumann P, Penev LD, Petanidou T, Rasmont P, Roberts SPM, Smith HG, Sørensen PB, Steffan-Dewenter I, Vaissière BE, Vilà M, Vuji□ A, Woyciechowski M, Zobel M, Settele J and Schweiger O (2011)

Developing European conservation and mitigation tools for pollination services: approaches of the STEP (Status and Trends of European Pollinators) project. Journal of Apicultural Research 50:152-164.

Vazquez DP (2005) Degree distribution in plant-animal mutualistic networks: forbidden links or random interactions? Oikos 108: 421-426.

Vazquez DP, Aizen MA (2003) Null model analyses of specialization in plant-pollinator interactions. Ecology, 84:2493-2501.

Vázquez DP, Blüthgen N, Cagnolo L, Chacoff N P (2009) Uniting pattern and process in plant–animal mutualistic networks: a review. Annals of Botany 103: 1445–1457.

12

Potential pollinators of understory populations of *Symphonia globulifera* in the Neotropics

Andre Sanfiorenzo*[1], Manuel Sanfiorenzo, Ronald Vargas[3], Lisette Waits[2] and Bryan Finegan[4]

[1]*Department of Agriculture Technology University of Puerto Rico-Utuado, Puerto Rico*
[2]*Department of Fish and Wildlife Sciences, University of Idaho, 83844-1136, United States*
[3]*Scientific department, Organization for Tropical Studies (OTS) La Selva, Puerto Viejo de Sarapiquí, Costa Rica*
[4]*Forests, Biodiversity and Climate Change Program,, Tropical Agricultural Research and Higher Education Center (CATIE), Turrialba 30501, Costa Rica*

Abstract—One difference between the forest canopy and the understory is that animals pollinate the majority of understory species in the tropical wet forest. Pollinators active in the understory are also different from those in the forest canopy and are adapted to the mesic conditions underneath the canopy. We used video cameras to observe flowers of understory *Symphonia globulifera* (Clusiaceae) in tropical wet forests of Costa Rica. We quantified the timing, frequency and behaviour of flower visitors to explore their potential contribution to pollination. A total of 82 flower visits were observed during 105 h of observations. Flowers were visited by ten insect species and one hummingbird species; the most active time period was between 1200-1700 h followed by the time between 0500-1000 h. The time period with fewer visitors was 1700-2200 h, during this period, we observed flowers being visited several times by a bushcricket (Tettigoniidae). The most frequent flower visitors were the stingless bee *Tetragonisca angustula* and the hummingbird *Phaethornis longirostris*; both came in contact with anther and stigma during visits. We observed different flower visitors from those reported for canopy populations of *S. globulifera*. Insects predominated, in contrast to observations in canopy populations of *S. globulifera,* where perching birds predominated. We also documented the consumption of pollen by visiting insects. These findings highlight differences in flower visitors between the forest canopy and the understory for the same tree species and contribute to better understanding of the pollination ecology of understory tropical wet forest species.

Keywords: Video observations, hummingbird, Neotropics, Orthoptera, Tetragonisca angustula, Phaethornis longirostris

Introduction

Tropical wet forests (TWF) are characterized by the immense diversity of taxa and complex vertical and horizontal structure. One characteristic that differentiates TWF from other forest biomes is that animals pollinate the great majority of tree species (Bawa et al. 1985; Dick et al. 2008). Differences among pollinators regarding behaviour and homes-range size create variation in the distance pollen is transported. In addition, the pollinator communities can differ between the several forest strata (Dick et al. 2008). Insects are the most important pollinator groups; vertebrates, such as birds and bats, also serve as pollinators, but for a smaller fraction of TWF species (3-11%) (Dick et al. 2008; Fleming et al. 2009). Among insects, bees constitute the most important group in number and diversity of plant species pollinated (Bawa 1990).

Pollinator-community surveys have traditionally been performed by direct observation and, more recently, by photography and continuous video recording (e.g. Bawa 1990; Quesada et al. 2003; Tschapka 2003; Lortie et al. 2012;

Padyšáková et al. 2013). Identification of flower visitors and estimation of the frequency of visits are critical for evaluating animal pollination and obtaining an understanding of the plant-animal interactions that facilitate plant reproductive success (Bawa 1990; Vazquez et al. 2005). There are two key components of pollinator activity that determine pollinator performance: frequency and effectiveness of flower visits (Ne'eman et al. 2010). Visit frequency can be simply defined as the number of visits to a flower per unit of time. Effectiveness, also called efficiency, is open to various interpretations and it relates to the pollinator's behaviour during flower visits (visit duration, contact with reproductive structures), and the amount of pollen carried away and deposited on receptive stigmas (Sahli & Conner 2006; Ne'eman et al. 2009). Meta-analyses of plant-pollinator datasets indicate that the most frequent flower visitors generally account for >50% of the total pollination service (Vazquez et al. 2005; Sahli & Conner 2006). Visitation frequency has been suggested as an accurate surrogate of pollinators contribution to overall reproductive success (Vazquez et al. 2005; Sahli & Conner 2006; Ne'eman et al. 2009). However, existing data on plant-pollinator interactions have been derived from studies mostly of herbs and shrubs (Vazquez et al. 2005; Sahli & Conner 2006). Additional data are needed from animal-pollinated trees to achieve a more comprehensive understanding of the

*Corresponding author: asanfiorenzo@gmail.com

relationship between visitation frequency and pollinator importance.

In this study we identified the animals that visited flowers of *Symphonia globulifera* (Clusiaceae) in the Caribbean slope of Costa Rica. *Symphonia globulifera* has a broad distribution, being found throughout the Neotropics and in Africa. Perching birds and hummingbirds have been suggested to be the most important pollinators, at least in observations of populations in which adults reach the canopy (Degen et al. 2004; da Silva Carneiro et al. 2007; Dick & Heuertz 2008). We surveyed understory *S. globulifera* flowering trees in mature lowland TWF forest sites of Costa Rica, using video cameras to identify flower visitors. *Symphonia globulifera* occurs only as an understory tree in our study area, while in other regions it is a canopy tree (Degen et al. 2004; da Silva Carneiro et al. 2007; Dick & Heuertz 2008). Thus, we hypothesized that flower visitors of understory populations would be different from those of the canopy populations. We quantified the timing, frequency and behaviour of flower visitors. Visitation frequency and foraging behaviour are examined to explore the potential contribution to pollination

by the observed flower visitors. We also discuss differences between the results of our study and those of studies of canopy populations of *S. globulifera*.

MATERIALS AND METHODS

Study area

We conducted this study in three mature forest patches in Sarapiquí County, Heredia Province, in the Caribbean lowlands of northern Costa Rica, centred at 10.440588 N, -84.115308 W. The study area is a 100 km² polygon that contains all three research sites (Fig. 1). This area is characterized by elevation that ranges from sea level to 300 m a.s.l.; terrain is a mixture of alluvial terraces, swamplands, and steep hills (Sesnie et al. 2008). Mean annual temperatures average 24°C and mean annual precipitation is 4000 mm per year (Sesnie et al. 2008). Land use is dominated by pasture, and recently pineapple cultivation has increased greatly. Other crops are also present intermixed with mature and secondary forest patches (Shaver et al. 2015).

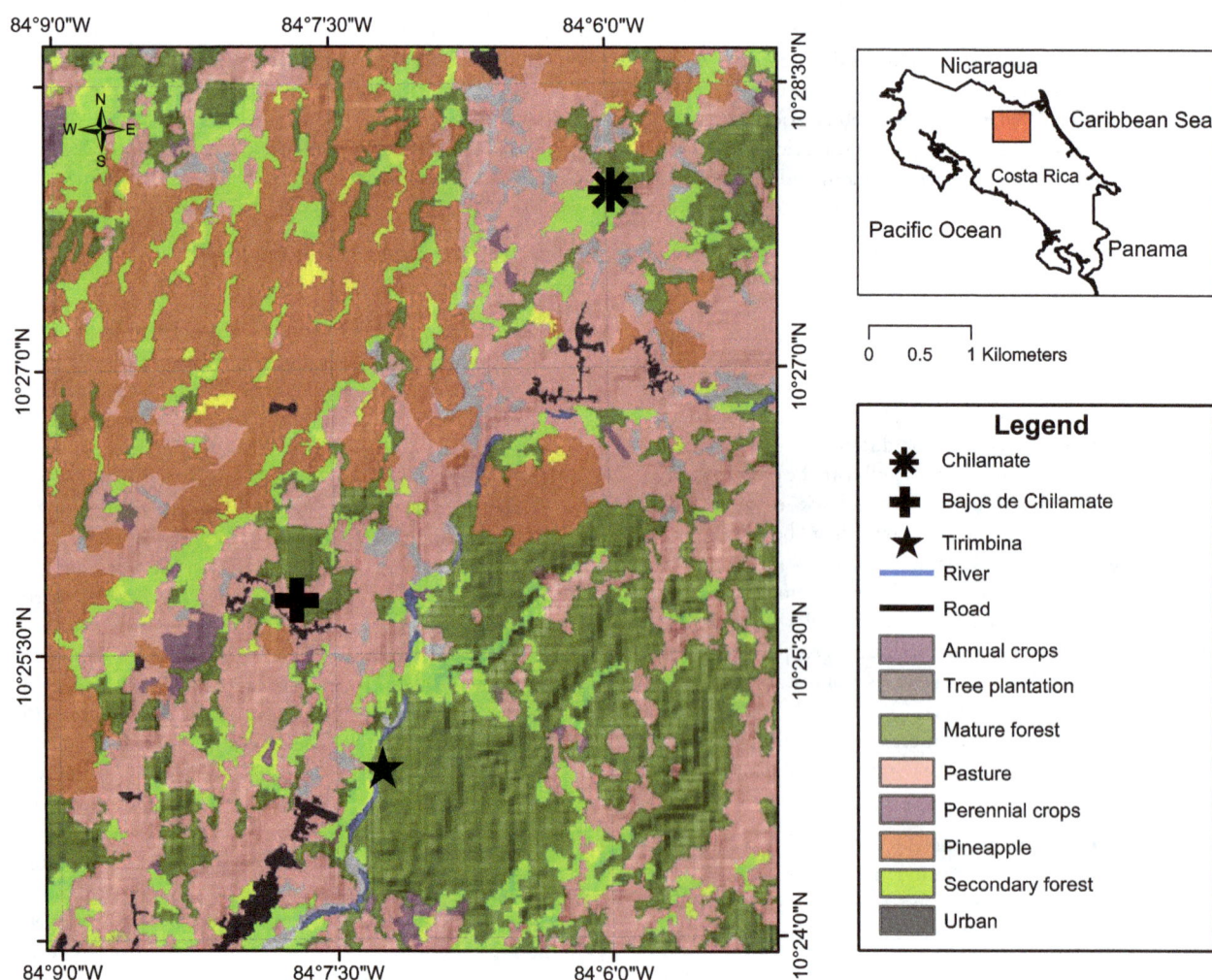

FIGURE 1. Study area, land cover and three sites where flower observations were performed. Land cover data source from Shaver et al. 2015.

Study species

Symphonia globulifera (Clusiaceae) is a shade-tolerant tree species distributed in rain forests across the Neotropics and equatorial Africa (Dick & Heuertz 2008). It is the only recognized species in its genus found outside of Madagascar, where 16 *Symphonia* species are present (Abdul-Salim 2002). Although *S. globulifera* are typically large canopy trees (Degen et al. 2004; Woodward 2005; Dick & Heuertz 2008), populations in the Sarapiquí region in Costa Rica occur only as understory trees, with a minimum reproductive size of 1 cm diameter at breast height (dbh, 1.3 m; personal observation). In French Guiana *S. globulifera* are large canopy trees that exist in two distinct sympatric forms, one with big leaves and the other with small leaves; they are treated as separate species by local forestry managers (Degen et al. 2004; Dick & Heuertz 2008). None of this morphological variation has yet been considered sufficient to merit splitting of *S. globulifera* into more than one Neotropical species (Dick et al. 2003; Dick & Heuertz 2008).

Inflorescences of *S. globulifera* consist of 1-15 axillary, bisexual flowers (Aldrich et al. 1998; Woodward 2005). Flowers are scarlet red, odourless, globose in shape, and more or less vertically oriented. At anthesis, petals contort and form a chamber in which nectar accumulates. Access to the interior chamber for flower visitors is only possible at the apex between the incurved petals and the staminal tube. The staminal tube surrounds the pistil; the anthers are inserted at the lobes of the staminal tube and open abaxially to display pollen immersed in a sticky, oily substance (Bittrich & Amaral 1996; Gill et al. 1998). A previous study found an unsaturated fatty acid methyl ester (methyl nervonate) to be the only component of the oil in which pollen is immersed. This secretion was thought to protect the pollen against foragers since no pollen foraging was observed (Bittrich et al. 2013). A well-developed nectary surrounds the base of the staminal tube. The stigma is shaped like a five-lobed star, with small pores at the apices of each lobe (Bittrich & Amaral 1996). Partial self-compatibility has been reported by Bittrich & Amaral (1996); however, the development of seeds through maturity was not followed, and the total number of viable seeds was not provided.

Pollination of *S. globulifera* flowers was described as mediated by sunbirds, wasps, bees, and butterflies in Africa (Oyen 2005). In French Guyana, perching birds have been described as potential pollinators (Gill et al. 1998). In Colombia, Brazil and Costa Rica, hummingbirds have been suggested as potential pollinators (Bittrich & Amaral 1996; Lasprilla & Sazima 2004). All the previous studies were carried out on canopy populations. Visits by euglossine (Apidae: Euglossini) and meliponine (Apidae: Meliponini) bees have been documented for the understory population in our study area (Rincón et al. 1999). Pascerralla (1992) conducted flower observations in the same area as the present study and reported hummingbirds as frequent flower visitors and nectar thieves, because no contact with fertile parts of flowers was observed. This author instead suggested Lepidoptera were probable pollinators, based on flower shape and plant distribution. *Symphonia globulifera* seeds are contained in large 4–5 cm drupes and are dispersed by bats

and monkeys (Aldrich et al. 1998). The species is usually > 90% outcrossed (Degen et al. 2004; da Silva Carneiro et al. 2007), although some degree of self-fertilization (> 10%) has been documented in canopy populations in disturbed habitats in Costa Rica (Aldrich et al 1998).

Fieldwork

Trees were chosen based on the availability of flowers and accessibility, in three mature forest sites that offered security for the video recording equipment. In total, 25 flowers were observed, six flowers from one tree in Tirimbina, nine flowers from one tree in Chilamate and ten flowers from two trees (five each) in Bajos de Chilamate. Video recordings were performed during May and June 2013, using a Sony Digital Handycam HDR-SR10 with supplemented infrared light at night. The cameras were placed inside waterproof cases, sufficiently close (less than 3 m) to the flower to allow clear vision of the anthers and stigma. Video recordings were made during flower anthesis in three time periods: 0500-1000 h, 1200-1700 h and 1700-2200 h. In total, 105 hours of video recordings were analysed to assess flower visits, 35 hours from each time period.

Data Analysis

Video observations and analyses were performed using Adobe Premier software, through visual identification of the arrival of flower visitors. Animals observed were only considered visitors if they touched the stigmas or anthers or consumed nectar. Visits in which no fertile-part contacts were made or no nectar was consumed were not considered further. Most of these latter cases were by ants roaming around the flower petals. For each pollinator visit, the following data were recorded: duration of visit, whether stigma or anthers were contacted, and whether pollen or nectar was obtained. We considered pollen or nectar consumption if the buccal apparatus of the visitor touched the anther or accessed the nectar chamber and feeding behaviour was displayed (Sakamoto et al. 2012). Still images from the video were selected and used for identification. We calculated the visitation rate for each species, defining it as visits per flower per hour for each single recording period, then averaged across all observation periods.

RESULTS

A total of 82 visits to *S. globulifera* flowers were observed during the 105 hours of evaluated video recordings. The flowers were visited by ten insect species and one hummingbird species (Tab. 1, Fig. 2). We were unable to confidently identify two species, one small flying insect, probably a small hymenopteran or dipteran, and a nocturnal lepidopteran probably of the family Geometridae. These two visitors accounted for one observation each and were not considered in further analysis.

We observed four species of bees (Hymenoptera: Apidae): *Tetragonisca angustula* and three species of *Trigona*. Various ants (Formicidae) were observed: *Pseudomyrmex*, *Crematogaster* and *Solenopsis*. One wasp in the genus *Polybia* (Vespidae) was also present. Additionally, we recorded one species of hermit hummingbird *Phaethornis longirostris*

TABLE I. Species visiting flowers of *S. globulifera*, including number of visits and frequency of visitation (*N* of visits per species/ total *N* of visits).

Class, Order	Family	Species	*N*	Frequency
Insecta, Hymenoptera	Apidae	*Trigona* sp. 1	2	0.03
		Trigona sp. 2	1	0.01
		Trigona sp. 3	5	0.03
		Tetragonisca angustula	25	0.29
	Formicidae	*Pseudomyrmex* sp. 1	1	0.01
		Crematogaster sp. 1	7	0.10
		Solenopsis sp. 1	17	0.25
	Vespidae	*Polybia* sp. 1	3	0.04
Orthoptera	Tettigoniidae	Tettigoniidae sp. 1	9	0.09
Aves, Apodiformes	Trochilidae	*Phaethornis longirostris*	12	0.13

(Trochilidae: Phaethornithinae) and one bushcricket (Tettigoniidae), Tettigoniidae sp. 1.

Considering all observations together, the bee *Tetragonisca angustula* was the most frequent flower visitor, followed by the ant *Solenopsis* sp. 1, which was present on many occasions during diurnal observation periods. The hummingbird *Phaethornis longirostris* ranked third in visitation frequency with 12 observed visits. Other flower visitors were observed with lower visitation frequencies (Tab. I). Visitation activities varied among time periods (Tab. 2). Visitors were more abundant during the 1200-1700 h time period with seven species recorded during this period. We observed *Pseudomyrmex* sp. 1, *Crematogaster* sp. 1, *Polybia* sp. 1 only during this period. In contrast, *Trigona* sp. 3, *Tetragonisca angustula*, *Solenopsis* sp. 1 and *Phaethornis longirostris*, were observed during two time periods (0500-1000 h; 1200-1700 h). The least active time period was between 1700-2200 h; the only visitor observed more than once during 1700-2200 h was the bushcricket (Tettigoniidae sp. 1), which was observed only during this observation period.

Visitation rate calculated as the average number of visits per hour reveals the number of interactions per unit of time. The highest visitation rate was for the bee *Tetragonisca angustula* with 0.28 visits flower^{-1} h^{-1}, followed by the ant *Solenopsis* sp. 1 with 0.21 visits flower^{-1} h^{-1} and the hummingbird *Phaethornis longirostris* with 0.13 visits flower^{-1} h^{-1} (Tab. 3). Other diurnal visitors showed lower visitation rates, some of which represent a single visit (Tab. 1, 3). During the 1700-2200 h time period Tettigoniidae accounted for 0.9 visits flower^{-1} h^{-1}.

Foraging behaviour during flower visits varied between species (Tab. 3). Eight species came in contact with the anthers during flower visits; only two ant species (*Pseudomyrmex* sp. 1 and *Crematogaster* sp. 1) did not touch the anthers while visiting flowers. Considering the species that touched the anthers, six were observed consuming pollen, that is, their buccal apparatus touched the anther area. We observed that seven species came in contact with the stigma while consuming pollen or nectar during flower visits. Seven species consumed nectar from flowers; the ants

Pseudomyrmex and *Crematogaster* visited flowers to consume nectar and did not touch the anther or stigma. The hummingbird *P. longirostris* was the only species capable of accessing the internal chambers formed by the flower petals where nectar is accumulated; it used its long beak and tongue to consume the available nectar. During the short visits by *P. longirostris*, we witnessed direct contact between the upper beak and anthers and stigmas. We observed ants consuming nectar residues in the locations where *P. longirostris* had inserted its beak, immediately after the latter had visited. In general, visits by the ant *Solenopsis* sp. 1 consisted of constant roaming around the flower, and we observed pollen and nectar consumption during flower visits. They moved over the anther multiple times during a visit; in some cases, individuals remained near or on the petals of the flower for the entire filming period. The other ant species, *Pseudomyrmex* sp. 1, and *Crematogaster* sp. 1, were less frequent visitors, but showed a similar behaviour of roaming around the flower and consuming nectar.

The stingless bee *T. angustula* was the most frequent flower visitor, with a mean visit duration of 110 seconds. This bee spent most of the time eating and collecting pollen; most of its body touched the anther, and on many occasions the abdomen and legs contacted the stigma. Three species of *Trigona* bees also visited *S. globulifera* flowers; these visits were less frequent and their duration was shorter. One species of wasp, *Polybia* sp. 1, was also observed three times; it consumed pollen and roamed around the flower coming in contact with the stigma. Tettigoniidae sp. 1 was the only visitor during the 1700-2200 h observation period. It was observed after sunset touching the anther and stigma, this species accounts for the longest duration of visits with a mean value of 515 seconds. During its visits, Tettigoniidae sp. 1 spent most of the time consuming pollen and many parts of the upper body came in contact with anthers and on some occasions touched the stigma.

DISCUSSION

Nine insect species and one hummingbird were the most common and abundant flower visitors for *S. globulifera* in the understory populations of the Sarapiquí region in Heredia, Costa Rica. In the forest understory, insects and

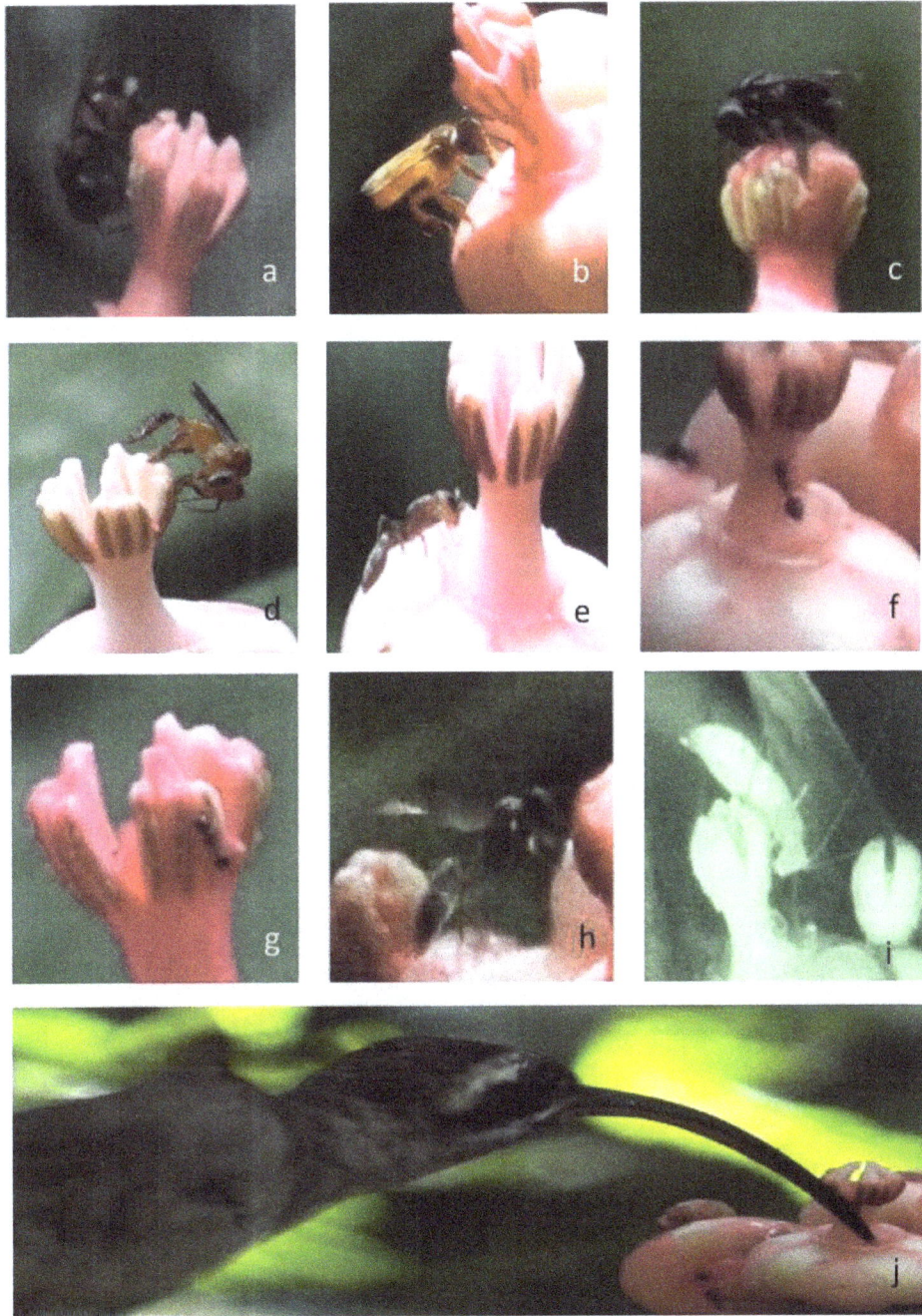

FIGURE 2. *Symphonia globulifera* flower visitors (a) *Trigona* sp. 1, (b) *Trigona* sp. 2, (c) *Trigona* sp. 3, (d) *Tetragonisca angustula*, (e) *Pseudomyrmex* sp. 1, (f) *Crematogaster* sp. 1, (g) *Solenopsis* sp. 1, (h) *Polybia* sp. 1, (i) Tettigoniidae sp. 1, (j) *Phaethornis longirostris*.

hummingbirds were the most frequent flower visitors of *S. globulifera*, in contrast to canopy populations of *S. globulifera* in French Guyana, where perching birds are reported as the main pollinators (Gill et al. 1998). Similarly, hummingbirds were suggested as the main potential pollinator in Brazil (Bittrich & Amaral 1996).

Flower visitors observed in this study are known to play important roles in the pollination of many plants in the TWF understory. Bees (Apidae) are often the most frequent visitors of flowers and the predominant pollinators for most plants and ecosystems (Neff & Simpson 1993; Winfree et al. 2011). Hummingbirds (Trochilidae) are found only in the Americas,

and include 328 flower-visiting species (Winfree et al. 2011). In TWF hummingbirds are responsible for the pollination of herbaceous monocots in the genus *Heliconia* and also regularly visit flowers from a wide range of other species (Lasprilla & Sazima 2004). In some cases, hummingbirds have also been reported as nectar thieves and not true pollinators (Pascarella 1992; Muchhala et al. 2008; Hadley et al. 2014). Ants visiting flowers are usually considered non-pollinating insects (Hull & Beattie, 1988; Dutton & Frederickson 2012; Chacoff & Aschero 2014). However, there is evidence that ants can sometimes be pollinators since they are common flower visitors and are able to carry pollen that results in seed

TABLE 2. Number and frequency (*N* of visits per species in time period/ total *N* of visits in time period) of observed *S globulifera* flower visits by time period.

Species	0500-1000 h		1200-1700 h		1700-2200 h	
	N	Frequency	N	Frequency	N	Frequency
Trigona sp. 1	2	0.07	0	0.00	0	0
Trigona sp. 2	I	0.03	0	0.00	0	0
Trigona sp. 3	I	0.03	4	0.09	0	0
Tetragonisca angustula	II	0.37	14	0.33	0	0
Pseudomyrmex sp. 1	0	0.00	I	0.02	0	0
Crematogaster sp. 1	0	0.00	7	0.16	0	0
Solenopsis sp. 1	I0	0.33	7	0.16	0	0
Polybia sp. 1	0	0.00	3	0.07	0	0
Tettigoniidae sp. 1	0	0.00	0	0.00	9	I
Phaethornis longirostris	5	0.17	7	0.16	0	0
TOTAL	30		43		9	

TABLE 3. Foraging-behaviour data for visitors on *S. globulifera* flowers.

Species	Visitation rate (Number of visits/hour)	% of visits that				Visit duration (seconds)		
		contacted stigma	contacted anther	fed on nectar	fed on pollen	Mean	SD	Min-Max
Trigona sp. 1	0.02	I00	I00	0	50	II	I.4	I0-12
Trigona sp. 2	0.01	0	I00	0	0	5	0	5-5
Trigona sp. 3	0.06	80	80	20	80	11.5	13.4	2-21
Tetragonisca angustula	0.28	80	84	I6	76	118.2	142.6	4-562
Pseudomyrmex sp. 1	0.01	0	0	100	0	181	0	181-181
Crematogaster sp. 1	0.06	0	0	100	0	74.57	95.5	16-289
Solenopsis sp. 1	0.21	71	88	29	76	143.65	105.1	18-453
Polybia sp. 1	0.03	100	100	0	67	18	11.5	7-30
Tettigoniidae sp. 1	0.09	78	100	33	100	515	545.3	115-1445
Phaethornis superciliosus	0.13	83	90	100	0	8.44	II	I-36

set (de Vega et al. 2009; Ashman & King 2005; Kawakita & Kato 2002). Neotropical tettigoniine bushcrickets are well known nocturnal florivores (Armbruster et al. 1997; Wardhaugh 2015) and are usually not considered to be pollinators (Schuster 1974; Proctor et al. 1996); their consumption of *S. globulifera* pollen in our observations suggests such a relationship here. However, Micheneau et al. (2010) reported that in wet lowlands forests the orchid *Angraecum cadetii* may be pollinated by leaf-rolling crickets (Orthoptera: Gryllacrididae). Furthermore, pollination by nocturnal visitors has been documented previously in the Clusiaceae; the cockroach *Amazonina platystylata* (Blattoidea: Blattidae) has been identified as the pollinator of *Clusia sellowiana* and *Clusia blattophila* in wet tropical forests of French Guyana (Vlasáková et al. 2008; Vlasáková 2015),

although in these cases the insects feed primarily on special secretions instead of eating pollen. This raises the question as to whether the staminal secretions of *S. globulifera* might also play a role as reward for pollination services.

We observed that most insects came in contact with the anther and displayed pollen consumption behaviour during flower visits. For these flower visitors, it appears that the reward for visits was the pollen and oil solution present at the anther (Bittrich & Amaral 1996). This provides evidence of consumption for the unsaturated fatty acid methyl ester (methyl nervonate) in which pollen is immersed. This evidence was not available before, and absence of such observations led researchers to conclude that this substance provides protection against pollen foraging (Bittrich et al. 2013).

Many flower visitors made contact with the stigma and may therefore be potential pollinators. It is during this stigmatic contact that transfer of pollen, resulting in ovule fertilization, could occur. Visits from *Pseudomyrmex* and *Crematogaster* ants did not involve contact with the anther or stigma and probably reflect nectar foraging without any potential contribution to pollination. *Solenopsis* sp. I ants, one of the most frequent flower visitors, displayed similar behaviour, although they moved all over the flower, and we observed pollen consumption and brief contact with the stigmas. However, we observed that individuals from this species tended to stay in a single group of flowers for many hours, exhibiting opportunistic behaviour wherein ants seemed to be consuming nectar residues left on flower petals after hummingbird feeding. For these reasons we conclude that the potential of ants as pollinators for *S. globulifera* is minimal. The presence of ants did not seem to discourage other flower visitors, since in many occasions flower visits occurred with ants roaming on the petals.

Our results contrast markedly with studies of canopy populations of *S. globulifera*. In undisturbed lowland TWF of French Guyana, the most frequent and persistent flower visitors were five perching bird species of the family Thraupidae (Gill et al. 1998). Hummingbirds were also reported as regular flower visitors, but no insects were observed, and all flower visits were diurnal (Gill et al. 1998). In contrast, in Sarapiquí, a bushcricket (Tettigoniidae sp. I) visited flowers during the 1700-2200 h period. In disturbed lowland TWF of Brazil, also for canopy populations, two species of trochiline hummingbirds (Trochilidae: Trochilinae) were the most frequent flower visitors (Bittrich & Amaral 1996). Insect visitors were also observed including *Trigona* bees. *Trigona* bees displayed destructive behaviour by chewing petals to access nectar, damaging or completely destroying the flowers; therefore, they acted as nectar thieves not pollinators for these populations (Bittrich & Amaral 1996).

We quantified interactions using visits per flower per hour; this metric allowed us to identify species with the most frequent interactions and therefore with greater potential for the pollination of *S. globulifera.*, assuming nearly equal efficiencies across pollinator species. Our results suggest that, considering foraging behaviour and visitation rates, the bee *T. angustula* and the hummingbird *P. longirostris* had the greatest potential contribution to the pollination of *S. globulifera*. Most flower visitors exhibited foraging behaviour that involved at least occasional contact with anther and stigma, also suggesting possible contributions to pollination. According to some research, the most frequent visitors usually contribute the most to the plant's reproductive success, even when their effectiveness is relatively low (Vazquez et al. 2005; Sahli & Conner 2006). However, flower visits do not necessarily indicate pollination; flower visitors are not always effective at both picking up and depositing pollen (Armbruster et al. 1989; Waser et al. 1996; Fenster et al. 2004; Ne'eman et al. 2010). Parameters such as visitation frequency, behaviour, morphology and effective pollen movement determine the pollination potential of flower visitors (Armbruster et al. 1989; Ne'eman et al. 2010).

Tetragonisca. angustula was the most frequent flower visitor. Behaviour during flower visits involved the consumption of pollen; in many instances their body parts came in contact with the stigmas of the flowers. This species had the highest visitation rate (0.28 visits flower^{-1} h^{-1}), more than twice that of the hummingbird (0.13 visits flower^{-1} h^{-1}). In this sense this is the flower visitor with the strongest interaction with *S. globulifera* flowers in this landscape. Not only is it a more frequent flower visitor than the hummingbird, the duration of visits is also longer, allowing for lengthier flower interaction time and contact with the flower stigmas. This stingless bee is distributed from Mexico to Argentina, one of the most widespread bee species in the Neotropics (Freitas et al. 2009; Camargo & Pedro 2013). They are generalists in their habits and have been identified as pollinators of many Neotropical plant species (Braga et al. 2012).

Hummingbirds were observed as frequent flower visitors of this understory tree population, which is consistent with observations in populations of canopy *S. globulifera* across the Neotropics (Bittrich & Amaral 1996; Gill et al. 1998; Lasprilla & Sazima 2004). The behaviour of *P. longirostris* during flower visits suggests they are potential pollinators because they contact anthers and stigmas while consuming nectar from flowers. The visitation rate for this species was 0.13 visits flower^{-1} h^{-1}. Visits were short (mean 8 seconds); however, we observed contact between the upper beak and the anther and stigma on more than 80% of visits. *Phaethornis longirostris* is a known *Heliconia* specialist (Snow & Texeira 2005). Evidence suggests this hummingbird species is tolerant of some degree of forest fragmentation (Hadley & Betts 2009; Volpe et al. 2014). Interestingly, *Phaethornis longirostris* is generally associated with understory habitats, not canopies, of mature and old secondary forests (Skutch & Dunning 1979; Johnsgard 1997).

The use of video cameras in this study allowed us to identify flower visitors and meticulously observe their behaviour during flower visits. The use of 16 frames-per-second and the high-definition video permit us to document flower visitors as ant's and bees. Our video recordings showed that hummingbirds do come in contact with flower reproductive parts and should not be considered as nectar thieves for this species. This differs from the direct visual observations of Pascarrela (1992), where no contacts with flower reproductive parts could be seen during hummingbird flower visits. This result may simply have been caused by the difficulty of visual observation of short hummingbird flower visits in the forest understory. Video records of flower visitors also provided evidence of pollen consumption not previously reported. Disadvantages are also associated with camera observation, such as the inability to observe visitor activities before and after the focal-flower visit, including movements among flowers and among trees. Future studies could address this by including human field observations or the use of more cameras, filming wider areas that cover the whole tree canopy. However, this would also increase the cost and logistic complexity associated with installing equipment in the field.

Conclusion

Previous studies of canopy populations of *Symphonia globulifera* described perching birds and hummingbirds as the most frequent flower visitors, with no insects reported as potential pollinators. We documented a different community of visitors to flowers of an understory population of *S. globulifera*. Twelve species, belonging to 5 families, were observed visiting flowers: 11 species of insects and one hummingbird. This suggests a shift in flower visitors between canopy and understory populations, emphasizing the difference between canopy and understory dynamics even for the same species. This difference in pollinators could generate a partial barrier to gene flow between canopy and understory populations, given that the most frequent pollinators observed in this study are associated with the forest understory rather than the forest canopy. Species observed visiting flowers in both canopy and understory populations are *Trigona* bees, suggested to be primarily nectar thieves in some instances. Nonetheless, these bees could be responsible for some pollen exchange between canopy and understory population. Further research in this area is needed to better understand gene flow between canopy and understory population of *S. globulifera*.

Based on visit frequency and rates of contact with fertile structures, the most important potential pollinators of *S. globulifera* understory populations were *T. angustula* and *P. longirostris*. Hummingbirds and bees, even if tolerant to forest fragmentation, require forest habitat to persist in the landscape (Brosi et al. 2008; Volpe et al. 2014). Thus, it is likely that forest fragmentation and subsequent land uses in the matrix can influence patterns of movement for these species and consequently the exchange of pollen for *S. globulifera* and other understory species throughout the landscape.

The present study increases our understanding of flower visitors and pollination in the tropical forest understory, specifically for *S. globulifera*. Further research on the deposition of pollen by each species can contribute to a more in-depth evaluation of individual pollinators' contributions to overall reproductive success. Effects of forest fragmentation should also be assessed in terms of loss of pollinators and reduction of *S. globulifera* populations in order to achieve a better understanding of the biological consequences of fragmentation in tropical wet forests.

ACKNOWLEDGEMENTS

Thanks to the Tirimbina Biological Reserve, the Rodríguez family and Segura family. Thanks also to the interdisciplinary SJLS-Team Bala for support on the design and execution of this project. This research was supported by the National Science Foundation under IGERT grant Award No. 0903479 and National Science Foundation under CNH grant Award No. 1313824.

REFERENCES

Aldrich PR, Hamrick JL, Chavarriaga P, Kochert G (1998) Microsatellite analysis of demographic genetic structure in fragmented populations of the tropical tree *Symphonia globulifera*. Molecular ecology 7:933–44.

Armbruster WS, Howard JJ, Clausen TP, Debevec EM, Loquvam JC, Matsuki M, CerendoloB, Andel F (1997) Do Biochemical Exaptations Link Evolution of Plant Defense and Pollination Systems? Historical Hypotheses and Experimental Tests with *Dalechampia* Vines. The American Naturalist 149:461-484.

Armbruster WS, Keller S, Matsuki M, Clausen T (1989) Pollination of *Dalechampia magnoliifolia* (Euphorbiaceae) by Male Euglossine Bees. American Journal of Botany 76:1279-1285.

Arriaga-Weiss SL, Calmé S, Kampichler C (2008) Bird communities in rainforest fragments: guild responses to habitat variables in Tabasco, Mexico. Biodiversity and Conservation 17:173–190.

Ashman T, King E (2005) Are flower-visiting mutualists or antagonist? A study in a Gynodioecious wild strawberry. American Journal of Botany 92:891–895.

Bawa KS (1990) Plant-Pollinator Interactions in Tropical Rain Forests. Annual Review of Ecology and Systematics 21:399–422.

Bawa KS, Bullock SH, Perry DR, Coville RE, Grayum MH (1985) Reproductive Biology of Tropical Lowland Rain Forest Trees. II. Pollination Systems, American Journal of Botany 72:346–356.

Bittrich V, Amaral MCE (1996) Pollination biology of *Symphonia globulifera* (Clusiaceae). Plant Systematics and Evolution 200:101–110.

Bittrich V, Nascimento-Junior JE, Amaral MCE, de Lima Nogueira PC (2013) The anther oil of *Symphonia globulifera* L.f. (Clusiaceae) Biochemical Systematics and Ecology 49:131–134.

Braga JA, Conde MM, Barth OM, Lorenzon MC (2012) Floral sources to *Tetragonisca angustula* (Hymenoptera: Apidae) and their pollen morphology in a Southeastern Brazilian Atlantic Forest. International Journal Tropical Biology 60:1491–1501.

Brosi BJ, Daily GC, Shih TM, Oviedo F, Durán G (2008) The effects of forest fragmentation on bee communities in tropical countryside. Journal of Applied Ecology 45:773–783.

Pedro SRM, de Camargo JMF (2013) Stingless Bees from Venezuela. In: Vit P, Pedro MSR, Roubik D (eds) Pot-Honey: A legacy of stingless bees. Springer, New York, pp 73–86.

Chacoff NP, Aschero V (2014) Frequency of visits by ants and their effectiveness as pollinators of *Condalia microphylla* Cav. Journal of Arid Environments 105:91–94.

Degen B, Bandou E, Caron H (2004) Limited pollen dispersal and biparental inbreeding in *Symphonia globulifera* in French Guiana. Heredity 93:585–91.

Dick CW, Abdul-salim K, Bermingham E (2003) Molecular Systematic Analysis Reveals Cryptic Tertiary Diversification of a Widespread Tropical Rain Forest Tree. The American naturalist 162:691-703.

Dick CW, Hardy OJ, Jones FA, Petit RJ (2008) Spatial Scales of Pollen and Seed-Mediated Gene Flow in Tropical Rain Forest Trees. Tropical Plant Biology 1:20–33.

Dick CW, Heuertz M (2008) The complex biogeographic history of a widespread tropical tree species. Evolution; international journal of organic evolution 62:2760–74.

Didham RK, Ghazoul J, Stork NE, Davis J (1996) Insects in fragmented forests: a functional approach. Trends in ecology & evolution 11:255–60.

Dutton EM, & Frederickson ME (2012). Why ant pollination is rare: new evidence and implications of the antibiotic hypothesis. Arthropod-Plant Interactions 6:561-569.

Eckert CG, Kalisz S, Geber M, Sargent R, Elle E, Cheptou P-O, Goodwillie C, Johnston MO, Kelly JK, Moeller D, Porcher E, Ree RH, Vallejo-Marín M, Winn A (2010) Plant mating systems in a changing world. Trends in Ecology & Evolution 25:35–43.

Fenster CB, Armbruster WS, Wilson P, Dudash MR, Thomson JD (2004) Pollination syndromes and floral specialization. Annual Review of Ecology, Evolution, and Systematics 35:375–403.

Fleming TH, Geiselman C, Kress WJ (2009) The evolution of bat pollination: a phylogenetic perspective. Annals of Botany 104:1017– 1043.

Freitas BM, Imperatriz-Fonseca VL, Medina LM, Kleinert AD, Galetto L, Nates-Parra G, Quezada-Euán JJG (2009) Diversity, threats and conservation of native bees in the Neotropics. Apidologie 40:332-346.

Gill GE, Fowler RT, Mori SA (1998) Pollination Biology of Symphonia globulifera (Clusiaceae) in Central French Guiana. Biotropica 30:139–144.

Hadley AS, Betts MG (2009) Tropical deforestation alters hummingbird movement patterns. Biology Letters 5:207–10.

Hadley AS, Frey SJK, Robinson WD, Kress WJ, Betts MG (2014) Tropical forest fragmentation limits pollination of a keystone understory herb. Ecology 95:2202–2212.

Hull DA, Beattie AJ (1988). Adverse effects on pollen exposed to Atta texana and other North American ants: implications for ant pollination. Oecologia 75:153-155.

Johnsgard PA (1997) The hummingbirds of North America. Smithsonian Institution Press. Washington DC, pp 278.

Kawakita A, Kato M (2002) Floral biology and unique pollination system of root holoparasites, Balanophora kuroiwai and B. tobiracola (Balanophoraceae). American Journal of Botany 89:1164-1170.

Kearns CA, Inouye DW, Waser NM (1998) Endangered Mutualims: The Conservation of Plant-Pollinator Interactions. Annual Review of Ecology and Systematics 29:83–112.

Lasprilla LR, Sazima M (2004) Interacciones planta-colibri en tres comunidades vegetales de la parte suroriental del parque nacional natural Chibiquete, Colombia. Ornitologia Neotropical 15:183-190.

Lortie CJ, Budden AE, Reid AM (2012) From Birds to Bees: Applying video observation techniques to invertebrate pollinators. Journal of Pollination Ecology 6:125–128.

Micheneau C, Fournel J, Warren BH, Hugel S, Gauvin-bialecki A, Pailler T, Strasberg D, Chase MW (2010) Orthoptera, a new order of pollinator. Annals of Botany105: 355–364.

Morse WC, Schedlbauer JL, Sesnie SE, Finegan B, Harvey CA, Hollenhorst SJ, Kavanagh KL, Stoian D, Wulfhorst JD (2009) Consequences of Environmental Service Payments for Forest Retention and Recruitment in a Costa Rican Biological Corridor. Ecology And Society 14:1-23.

Muchhala N, Caiza A, Vizuete JC, Thomson JD (2008) A generalized pollination system in the tropics: bats, birds and Aphelandra acanthus. Annals of botany 103:1481–1487.

Ne'eman G, Jürgens A, Newstrom-Lloyd L, Potts SG, Dafni A (2009) A framework for comparing pollinator performance: effectiveness and efficiency. Biological reviews of the Cambridge Philosophical Society 85:435–451.

Neff JL, Simpson BB (1993) Bees, pollination systems and plant diversity. In: LaSalle J, Gauld ID (eds) Hymenoptera and Biodiversity. C.A.B. International, Wallingford, pp 143-167.

Oyen LPA (2005). Symphonia globulifera L.f. Record from Protabase. In: Louppe D, Oteng- Amoako AA, Brink M, (eds) PROTA (Plant Resources of Tropical Africa/Ressources végétales de l'Afrique tropicale). Wageningen: Netherlands <http://database.prota.org/search.htm> accessed on Feb. 15 2017.

Padyšáková E, Bartoš M, Tropek R, Janeček Š (2013) Generalization versus Specialization in Pollination Systems: Visitors, Thieves, and Pollinators of Hypoestes aristata (Acanthaceae). PLoS ONE 8:1–8.

Pascarella JB (1992) Notes on flowering phenology, nectar robbing and pollination of Symphonia globulifera L. f. (Clusiaceae) in a lowland rain forest in Costa Rica. Brenesia 38: 83-86.

Proctor M, Yeo P, Lack A (1996) The natural history of pollination. Timber Press, Portland, Oregon.

Quesada M, Stoner KE, Rosas-Guerrero V, Palacios-Guevara C, Lobo J a (2003) Effects of habitat disruption on the activity of nectarivorous bats (Chiroptera: Phyllostomidae) in a dry tropical forest: implications for the reproductive success of the Neotropical tree Ceiba grandiflora. Oecologia 135:400–406.

Rincón M, Roubik DW, Finegan B, Delgado D, Zamora N (1999) Understory bees and floral resources in logged and silvicultural treated Costa Rican rainforest plots. Journal of the Kansas Entomological Society 72:379–393.

Sahli HF, Conner JK (2006) Characterizing ecological generalization in plant-pollination systems. Oecologia 148:365–372

Sakamoto RL, Morinaga S, Ito M, Kawakubo N (2012) Fine-scale flower-visiting behavior revealed by using a high-speed camera. Behavioral Ecology and Sociobiology 66:669– 674.

Schuster JC (1974) Saltatorial Orthoptera as Common Visitors to Tropical Flowers Biotropica 6:138-140

Sesnie S, Gessler P, Finegan B, Thessler S (2008) Integrating Landsat TM and SRTM-DEM derived variables with decision trees for habitat classification and change detection in complex Neotropical environments. Remote Sensing of Environment 112:2145–2159.

Shaver I, Chain-Guadarrama A, Cleary K, Sanfiorenzo A, Santiago-García RJ, Finegan B, Hormel L, Sibelet N, Vierling L, Bosque-Pérez N, DeClerck F, Fagan ME, Waits LP (2015) Coupled social and ecological outcomes of agricultural intensification in Costa Rica and the future of biodiversity conservation in tropical agricultural regions. Global Environmental Change 32:74–86.

da Silva Carneiro F, Magno Sebbenn A, Kanashiro M, Degen B (2007) Low Interannual Variation of Mating System and Gene Flow of Symphonia globulifera in the Brazilian Amazon. Biotropica 39:628–636.

Skutch, AF, Dunning, JS (1979). Aves de Costa Rica. Editorial Costa Rica, San José.

Snow DW, Texeira DL (2005) Hummingbirds and their flowers in the coastal mountains of southeastern Brazil. Journal of Ornithology 123:446–450

Stiles FG (1980) The annual cycle in a topical wet forest hummingbird community. Ibis 122:322-43.

Steffan-Dewenter I, Münzenberg U, Bürger C, Thies C, Tscharntke T (2002) Scale- Dependent Effects of Landscape Context on Three Pollinator Guilds. Ecology 83:1421–1432.

Tschapka M (2003) Pollination of the understorey palm Calyptrogyne ghiesbreghtiana by hovering and perching bats. Biological Journal of the Linnean Society 80:281–288.

Vazquez DP, Morris WF, Jordano P (2005) Interaction frequency as a surrogate for the total effect of animal mutualists on plants. Ecology Letters 8:1088–109.

de Vega C, Arista M, Ortiz PL, Herrera CM, Talavera S (2009) The ant-pollination system of Cytinus hypocistis (Cytinaceae), a Mediterranean root holoparasite. Annals of Botany 103:1065–1075.

Vlasáková B, Kalinová B, Mats HG, Gustafsson, HT (2008) Cockroaches as Pollinators of Clusia aff. sellowiana (Clusiaceae) on Inselbergs in French Guiana. Annals Botany 102: 295-304.

Vlasáková B (2015) Density dependence in flower visitation rates of cockroach-pollinated Clusia blattophila on the Nouragues inselberg, French Guiana. Journal of Tropical Ecology: 31:95-98.

Volpe NL, Hadley AS, Robinson WD, Betts MG (2014) Functional connectivity experiments reflect routine movement behavior of a tropical hummingbird species. Ecological Applications 24:2122–2131.

Wardhaugh CW (2015) How many species of arthropods visit flowers? Arthropod-Plant Interactions 9: 547-565.

Waser NM, Chittka L, Price MV, Williams NM, Ollerton J (1996) Generalization in pollination systems, and why it matters. Ecology 77:1043–1060.

Winfree R, Bartomeus I, Cariveau DP (2011) Native Pollinators in Anthropogenic Habitats. Annual Review of Ecology, Evolution, and Systematics 42:1–22.

Woodward CL (2005) Reproductive Success, Genetic Diversity and Gene Flow in Fragmented Populations of Understory Trees in Costa Rica. University of Wisconsin- Madison.

13

Pollinators May Not Limit Native Seed Set at Puget Lowland Prairie Restoration Nurseries

Jennie F. Husby[1],* Carri J. LeRoy[1], Cheryl Fimbel[2]

[1]The Evergreen State College, 2700 Evergreen Parkway NW, Olympia, Washington 98505, USA
[2]Center for Natural Lands Management, 120 E, Union #215, Olympia, Washington 98501, USA

Abstract—Land managers often rely on large-scale production of native seeds in nurseries for replanting into natural environments as part of restoration strategies. Nursery managers question if unmanaged insects can be sufficient to pollinate large increases in native forbs planted into young nurseries in non-native landscapes. This study investigated pollination of deltoid balsamroot (*Balsamorhiza deltoidea* Nutt.) and sicklekeel lupine (*Lupinus albicaulis* Douglas) at a native seed nursery compared to dense patches of native plants at a natural Puget lowland prairie to determine if insect visitation affected viable seed production for those two species. In 2011 and 2012, insect visitation rates were recorded for each plant species at more than 62 plots within two study areas. In 2012, seeds were collected from hand-pollinated and naturally-pollinated inflorescences and tested for viability. Overall visitation rates were significantly higher at the nursery than the prairie for both plant species. However, pollen limitation was not evident for either plant species at either site. Natural pollination by insects and supplemental hand-cross-pollination treatments did not yield different quantities of viable seed. Factors other than pollinator visitation, such as soil nutrients and seed handling practices, may be influencing seed viability, but increasing insect visitation will not likely significantly increase seed viability for these two species at this restoration nursery. Planting dense rows of native plant species may be enough to attract a sufficient amount of unmanaged insects to provide adequate pollination for seed production for some species at even young nurseries.

Keywords: restoration nursery, pollen limitation, insect visitation, prairie restoration, *Balsamorhiza deltoidea*, *Lupinus albicaulis*

Introduction

Pollinators play a key role in the reproduction of wild plants as they are linked to viable seed production and native plant population growth. Pollinators and their activities thus provide an ecosystem-wide service, especially for landscapes with many insect-dependent forb species like prairies (Kremen et al. 2007). Even self-compatible plant species may rely on pollinators to provide conspecific pollen to mix genes, preventing inbreeding depression (Heschel & Page 1995; Price et al. 2008). Native seed from nurseries plays an important role in ecosystem restoration. Ecosystems in need of conservation attention may be stressed by factors such as invading species, fragmentation, and climate change; all of which can suppress a plant species' population size and limit its reproductive ability (McCarty 2001; Vila & Weiner 2004; Fazzino et al. 2011; Tscheulin & Petanidou 2011). Many restoration practitioners depend on native seed grown in nurseries for repopulating plant species in natural areas. Native plant nursery managers strive to produce large quantities of high quality seed to meet restoration demands.

When plants produce fewer viable seeds than maternal resources allow because of insufficient quantity or quality of pollen, they are pollen-limited (Wagenius & Lyon 2010).

Several aspects of pollination can influence the production of viable seeds. Insect visitation rates can positively affect pollen receipt (Engel & Irwin 2003). Differences in pollinator community structure can affect overall pollination effectiveness (Perfectti et al. 2009) and pollinator behaviour, such as order in which a pollinator visits flowers, can affect whether a flower is self-pollinated or cross-pollinated (Kunin 1993). When plants are pollen-limited due to insufficient pollinator activity, they are pollinator-limited (Dieringer 1992).

The Puget lowland prairie ecosystem has been fragmented by coniferous forest encroachment and urban and agricultural development so that now only 2-3% of the original habitat remains (Dunwiddie & Bakker 2011). Re-establishing native flora has been a priority of Puget lowland prairie conservation partners (Stanley et al. 2008), and these partners rely on large-scale production of native seeds in their nurseries for replanting into the Puget lowland prairies as part of their restoration strategies. Some years restoration practitioners have struggled to produce large quantities of viable seeds for certain plant species at their largest nursery, Webster Nursery. These managers have questioned whether or not two critical restoration species, *Balsamorhiza deltoidea* Nutt. (Asteraceae) (deltoid balsamroot) and *Lupinus albicaulis* Douglas (Fabaceae) (sicklekeel lupine), are producing the highest proportion of viable seeds possible at this restoration nursery and if there is a way to increase the proportion. Results of testing by the nursery in 2013 revealed that Webster Nursery produced 1,424 g of *B. deltoidea* seed with 38.8% viability, and 10,299 g of *L. albicaulis* seed with 35.3% viability (S. Smith, Center for Natural Lands Management, pers. comm.). The cause of this problem may be due to seed handling or storage techniques, inadequate environmental conditions where the plants are grown (such as soil nutrients, weather, etc.), seed predation, or pollen limitation. This study addressed the latter by investigating the status of pollination at Webster Nursery and comparing it to dense stands of native plants at a natural Puget lowland prairie to determine if inadequate pollination restricted viable seed production at the nursery.

More specifically, to better understand the role of pollination at a native seed production facility, we investigated the following research questions: (1) Do insect visitation rates to dense floral patches differ between a nursery site and a natural prairie site for two prairie forbs? (2) Is there evidence of pollen limitation for either plant species at either the nursery or prairie sites?

MATERIALS AND METHODS

Study Plant Species

We focused this study on two native prairie plants, *B. deltoidea* and *L. albicaulis*. Both plant species grow along the west coast of the United States and into Canada (USDA Natural Resources Conservation Service 2012). These plants are both found at natural prairie sites and are being produced from seed at Webster Nursery, Tumwater, WA, USA.

Deltoid balsamroot (*B. deltoidea*) is a species of potential concern in Washington State (Washington Natural Heritage Program 2012), its flowers are popular with insect visitors, and it is a valued restoration plant. The federally endangered butterfly, Taylor's checkerspot (*Euphydryas editha taylori*), frequently uses this plant as a nectar resource. *Balsamorhiza deltoidea* bloomed from the last week of May to mid-June in 2011 and from May 7 to June 1 in 2012. This perennial has yellow, compact head inflorescences containing many fertile female ray flowers and bisexual disk flowers. The fruits are achenes, each with a single ovule. Fazzino et al. (2011) documented that this plant species is self-incompatible and does not reallocate resources among flower heads.

Sickle-keel lupine (*L. albicaulis*) provides food for caterpillars and adults of 'blue' butterflies, such as the Puget blue (*Plebejus icarioides blackmorei*), a species of concern in Washington State, and occasionally the federally endangered Fender's blue butterfly (*Icaricia icarioides fenderi* Macy) (Wilson et al. 1997). This plant is a popular floral resource for several species of native bees and provides vertical vegetative structure on the low stature Puget prairies. *Lupinus albicaulis* is a perennial and bloomed from late June to mid-July in 2011 and from May 29 to June 29 in 2012. The blue, papilionaceous flowers develop basally first in racemes. Each flower contains 10 monodelphous stamens and a simple carpel with several ovules (Hitchcock & Cronquist 1998). An average of five ovules was presumed to be in each carpel of the flowers in this study, which was calculated by counting the number of cells in the collected

pods, including those that were empty (likely due to ovule abortion). The *L. albicaulis* inflorescences tested in this study had an average of 48.1 flowers per inflorescence. Little is known about the pollination system of *L. albicaulis*. *Lupinus* species in general are typically self-compatible, though some require a pollinator to trigger autogamy and have increased seed set when cross-pollinated (Kaye 1999, Kittelson & Maron 2000).

Study Areas

Washington Department of Natural Resources (DNR) owns Webster Nursery, a portion of which is leased and managed by conservation partners to produce seed from native plants at a large scale for restoring Puget lowland prairies. The plants are grown outdoors in dense rows. The native seed nursery was first established in 2008 with partial rows (~100 m) of 10 species. The rows planted with *B. deltoidea* and *L. albicaulis* were last fertilized during their installation in 2008, are watered only by rain, and were not sprayed with pesticides or herbicides in 2011 or 2012 (Angela Winter, nursery manager, pers. comm. 2012). Other species grown at Webster Nursery include: hookedspur violet (*Viola adunca*), spring gold (*Lomatium utriculatum*), nine-leaf biscuitroot (*Lomatium triternatum*), shortspur seablush (*Plectritis congesta*), slender cinquefoil (*Potentilla gracilis*), harsh Indian paintbrush (*Castilleja hispida*), golden paintbrush (*Castilleja levisecta*), sea pink (*Armeria maritima*), Nuttall's Larkspur (*Delphinium nuttallii*), woolley sunflower (*Eriophyllum lanatum*), buttercup (*Ranunculus occidentalis*), Pacific lupine (*Lupinus lepidus*), bicolor lupine (*Lupinus bicolor*), farewell to spring (*Clarkia amoena*), and camas (*Camassia quamash*). Farmland, open grassland, residential properties, and forested areas surround the 5.3 km² nursery (Tab. 1, Fig. 1).

The US Department of Defense manages Johnson Prairie, a natural prairie site on Joint Base Lewis-McChord. Johnson is one of the few remaining Puget lowland prairies dominated by semi-native vegetation, and is located near Rainier, WA. *Camassia quamash* is a dominant flowering species and the site includes similar plant species as grown at Webster Nursery in clumped patches throughout the fescue-dominated grassland. This prairie is subject to some recreational activity, though less military training activity than other prairie sites located on the base (Stinson 2005). A portion of this site, including the study area, was burned in August, 2011 for restoration purposes. Coniferous forest and open non-native grasslands border this 7.5 km² prairie site (Tab. 1, Fig. 1).

TABLE I. Percent land-use in a 1 km buffer around the study areas in Thurston County, WA.

Land-use	Webster Nursery	Johnson Prairie
Open Grassland	7.0%	7.2%
Forestland	55.4%	92.8%
Agriculture	14.1%	-----
Residential	23.5%	-----

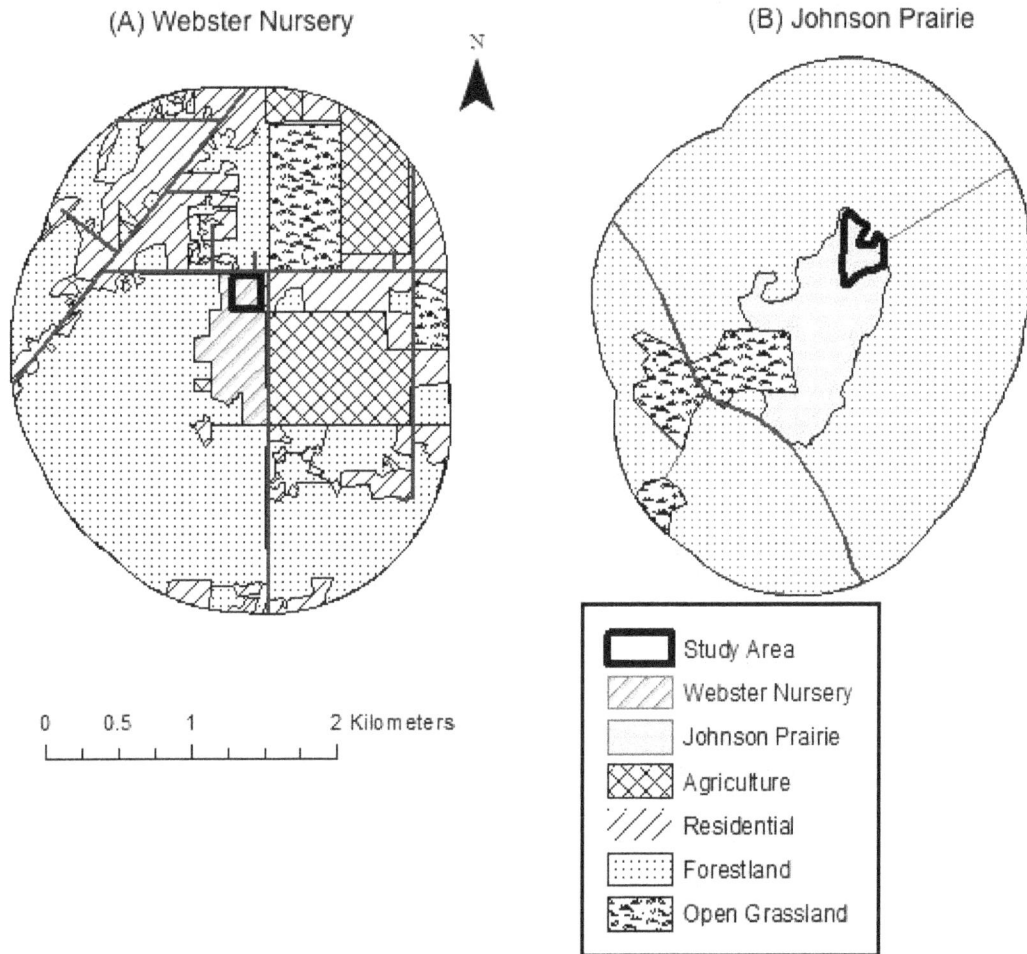

FIGURE 1. Land-use in 1km buffer around: A) Webster Nursery (46.951817 latitude, -122.962126 longitude) and B) Johnson Prairie (46.927746 latitude, -122.732272 longitude). Spatial Reference: NAD27 UTM Zone 10N. Digitized using ESRI® ArcGIS® 10.2 World Imagery basemap Source: ESRI, DigitalGlobe, GeoEye, i-cubed, Earthstar, Geographics, CNES/Airbus DS, USDA USGS, AEX, Getmapping, Aerogrid, IGN, IGP, swisstopo, and the GIS User Community, 2014.

Visitation Rates

The methods used for this study were adapted from Arroyo et al. (1982) who recorded the number of visits to a known number of flowers for a set time interval. Others (Arroyo et al. 1985; Inouye & Pyke 1988; Berry & Calvo 1989; McCall & Primack 1992) replicated this method to allow comparisons among studies (Kearns & Inouye 1993). For this study, plots were selected to collect visitation rate data for both study plant species in 2011 and 2012. At Webster nursery, the target plant species were planted in dense rows. Plot locations were selected at Webster Nursery by breaking each row of the target plant species into two-meter segments. Rows were 1.5 m deep. This plot size was chosen, as it was the largest area that could be observed by one person without missing visits and entirely encompassed most patches of the target species at Johnson prairie. Plots were then randomly selected. If a plot selected was directly adjacent to a plot previously selected, it was thrown-out and a new random number was generated to ensure at least two-meters of distance between plots.

At Johnson prairie, both plant species have a clumped distribution, spread across the site in patches. To reduce the chance of potential confounding factors differentially contributing to visitation and seed set rates, only patches of plants with similar densities to the planted rows at Webster nursery were selected at Johnson prairie for plot locations. The selected patches contained few other flowering species to reduce the chance of small scale floral competition or facilitation happening at the Johnson prairie plots and not at the monoculture plots at Webster nursery. A 7,937 m² macroplot was chosen at Johnson prairie in the Northeast corner where the majority of patches of the target species were. Patches of plants with similar floral densities as found at the nursery were identified within the macroplot, and plot locations were selected randomly from those patches. All plots selected were spaced at least two-meters apart.

At both sites, floral density was calculated for each plot by counting the number of inflorescences of the focal species in bloom and dividing that number by the area of the plot (2 m × 1.5 m). Plot densities varied from 0.3-9.3 inflorescences per square meter for *B. deltoidea*, and from 4.6-82.0 inflorescences per square meter for *L. albicaulis*, but variation in plot density was similar between sites (Appendix 1 and 2). In 2011, six plots were selected for *B. deltoidea* and 16 plots for *L. albicaulis* at each location and observed once. In 2012, 30 plots were selected at each site for *B. deltoidea*, and each observed once ($N = 30$). Ten plots were selected at each site and each sampled three times for *L. albicaulis* in 2012 ($N = 10$). After sampling *B. deltoidea*, we noted that visitation rates varied more than anticipated during the bloom period, likely due to different insect species being more active at certain times of the year. On average, visits on the first day of observations were 2.68 visits per flower per hour higher at Webster nursery and 0.63 visits per flower per hour higher at Johnson prairie, than on the last day of observation. Although *B. deltoidea* plots were sampled relatively evenly throughout the bloom period, the experimental design was changed for *L. albicaulis* in 2012 based on a recent study (Tscheulin & Petanidou 2011), to reduce the influence of variation in visitation rates. Instead of recording each plot only once, we thus recorded visits to plots for three rounds of timed intervals and calculated a mean number of visits per flower per hour. In each of the three rounds, the order in which the ten plots were observed was randomized.

Observations took place during peak flowering times on three days for each plant species between May 20 and July 6 in 2011. In 2012, observations took place between May 8 and June 21 on six days for *B. deltoidea* and five days for *L. albicaulis*. Each observation period lasted 10 minutes per plot, and all observations were made between 1000 and 1530 hours. Sampling dates were chosen to be as close together as possible on days with similar temperature, cloud cover, and wind conditions within an optimal range for insect activity (temperatures ranging from 9 to 27 °C, clear to cloudy skies with shadows present, and still air to light breeze). We assumed all flowers in bloom were receptive to pollen.

The number of visits made by insects was recorded during each 10 min period. A visit was recorded only if the insect landed on the reproductive parts of a flower. If an insect appeared to be "nectar robbing," where there was no potential for pollen transfer, the visit was discounted. Nectar robbing was rarely observed in this study. All visiting insects were identified in flight. Because identification could not accurately be made to a species level and no local guide for pollinator identification for this area was available, the observed insects were grouped into morphotypes: small dark bees (Halictidae, Colletidae: Hylaeinae, Apidae: Xylocopinae, and Andrenidae), large dark bees (*Andrena* sp. and Colletidae), green metallic bees (*Agapostemon* sp.), cuckoo bees (Apidae: Nomadinae), honey bees (*Apis mellifera*), bumblebees (*Bombus* sp.), flies (Diptera), syrphids (Syrphidae), ants (Formicidae), wasps (Hymenoptera: Apocrita), and beetles (Coleoptera) based on Donovall & vanEngelsdorp (2008). It could not be determined whether visiting honeybees came from feral populations or nearby managed hives. Because cuckoo bees, flies, ants, wasps, and beetles were either absent or rare, and potentially did not facilitate pollination, these groups were removed from the statistical analysis. If no insects from a particular morphotype visited, a value of zero was recorded and the trial was not discounted.

Permutative two-way ANOVAs (Manly 2007) were used to compare mean visitation rates at the nursery to the prairie for each plant species in each year, and to detect if there was a year by site interaction. Two tests compared overall visitation rates (one per plant species). For *B. deltoidea* we ran an additional six permutative ANOVAs for the insect morphotypes to determine if there were effects of location, year or their interaction on individual morphotypes. For *L. albicaulis* we ran an additional three permutative ANOVAs for selected insect morphotypes. All analyses were conducted using Resampling Stats for Excel 2007. Alphas were adjusted using Bonferroni corrections to address multiple comparisons. We used an alpha = 0.003 for the two overall visitation rate tests, an alpha = 0.008 for six separate individual insect visitation rate tests for visitors of *B. deltoidea*, and an alpha = 0.017 for three separate individual insect visitation rate tests for visitors of *L. albicaulis*.

Pollen Limitation

Procedures for the pollen limitation experiment were adapted from methods used by Fazzino et al. (2011) who compared seed set from naturally-pollinated *B. deltoidea* inflorescences to hand-cross-pollinated inflorescences to investigate pollen limitation on Puget prairies. In 2012, a subset of 10 plots for *B. deltoidea* at each site was selected randomly from the visitation rate plots, and all plots from the *L. albicaulis* visitation rate observations were used for the pollinator limitation experiment. Two similar inflorescences were chosen within each plot and marked with thread before the styles had matured. One inflorescence was left to be naturally-pollinated, and the second inflorescence was hand-pollinated as well as naturally-pollinated.

Hand-pollination treatments were applied every other day to all flowers of the selected inflorescences until the stigmas shriveled, then all inflorescences in both treatments were covered with a coarse mesh bag to prevent seed predation. When the fruits matured, the inflorescences were collected, and the seeds were extracted and counted.

A tetrazolium assay was used to test the seeds for viability using procedures adapted from the International Seed Testing Association (2012). Ten plump seeds were randomly selected from each inflorescence for *B. deltoidea*, and all seeds from each of the *L. albicaulis* inflorescences were tested. *Balsamorhiza deltoidea* seeds were soaked in warm water for four hours, and *L. albicaulis* seeds were soaked for 24 hours. A 1% aqueous solution of 2,3,5-triphenyltetrazolium chloride was prepared and the pH adjusted to 6.8. All seed coats were pierced before soaking the seeds in the tetrazolium solution. After four hours, the embryos were examined for the red staining that indicates viability.

A permutative two-way ANOVA was also used to compare the percentage of viable seeds produced by the inflorescences of each treatment group for each plant species, and to detect if there was a treatment by site interaction. To determine if there was pollen limitation for either plant species at Webster Nursery or Johnson Prairie, we compared

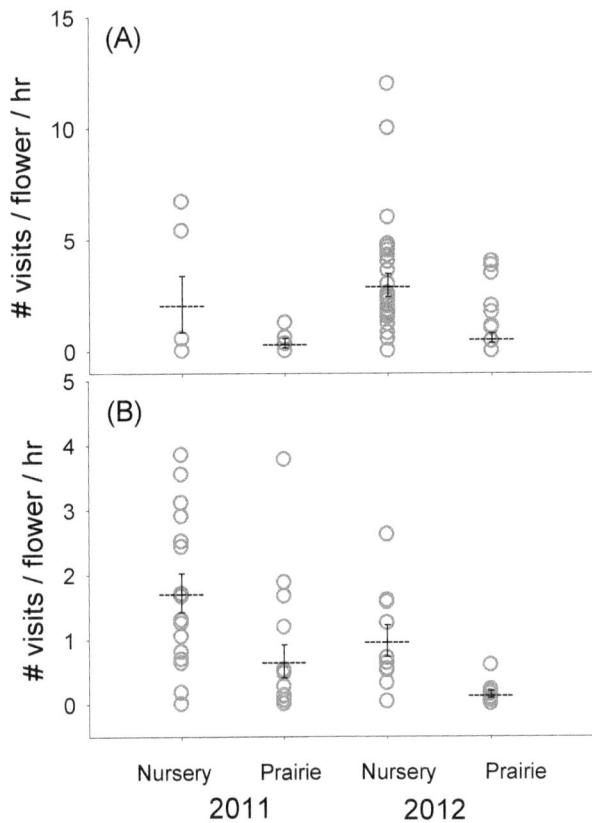

FIGURE 2. Overall insect visitation rates (# visits / plant / hr) at Webster Nursery and Johnson Prairie for: A) *Balsamorhiza deltoidea*, and B) *Lupinus albicaulis* in 2011 and 2012. Open gray circles represent each visitation rate and black horizontal lines represent means +/- 1 standard error. Bonferroni correction for two tests would necessitate an alpha = 0.025.

the percentages of viable seed produced by the hand-pollinated inflorescences to the naturally pollinated inflorescences. All analyses were conducted using Resampling Stats for Excel 2007. Alphas were adjusted using Bonferroni corrections for two comparisons (alpha = 0.025).

RESULTS

Insect visitation rates differed between Webster Nursery and Johnson Prairie, both overall and for many of the insect groups. Overall visitation rates were significantly higher at Webster Nursery than at Johnson Prairie for both *B. deltoidea* (Fig. 2; SSsite: 92.77, $P < 0.001$; SSyear: 7.46, $P = 0.496$; SSsite*year: 2.58, $P = 0.691$) (Appendix I) and *L. albicaulis* (SSsite: 12.32, $P < 0.001$; SSyear; 5.49, $P = 0.039$; SSsite*year: 0.18, $P = 0.715$) (Appendix II).

Webster Nursery also had significantly higher visitation rates than Johnson Prairie for several insect morphotypes visiting each of the plant species, specifically bumblebees and small dark bees (Tab. 2, Appendix I & II). In 2012, there were significantly higher rates of bumblebee visits at both sites than in 2011. In addition, there were higher visitation rates for small dark bees and bumblebees to *L. albicaulis* at Webster Nursery.

Pollen limitation was not evident for either plant species at either site. No significant difference was found between percentage of viable seeds produced by naturally-pollinated inflorescences or hand-cross-pollinated inflorescences for *B. deltoidea* (SSpollen: 3232.69, $P = 0.100$; SSsite: 8566.90, $P = 0.005$; SSpollen*site: 74.13, $P = 0.8021$) (Appendix III) or *L. albicaulis* (SSpollen: 2.35, $P = 0.985$; SSsite: 51212.67, $P < 0.001$; SSpollen*site: 286.09, $P = 0.724$) (Fig. 3, Appendix IV). For both species, the trend in seed viability was significantly higher for seeds from the Nursery site (Fig. 3).

TABLE 2. Results of permutative two-way ANOVAs (SS and *P*-values) comparing insect visitation rates at Webster nursery and Johnson prairie for both *B. deltoidea* and *L. albicaulis* in 2011 and 2012. Significant results are in bold. All significant site effect results indicate higher visitation rates at Webster Nursery than at Johnson Prairie. Bonferroni corrections for six tests (*B. deltoidea*) would necessitate an alpha = 0.008, and for three tests (*L. albicaulis*) would necessitate an alpha = 0.017.

Insect Morphotype	Source	SS	P
B. deltoidea			
Small Dark Bees	Site	1.76	0.010[1]
	Year	1.36	0.206
	Site × Year	0.92	0.284
Large Dark Bees	Site	0.17	0.499
	Year	<0.01	1.000
	Site × Year	<0.01	1.000
Green Metallic Bees	Site	0.60	0.235
	Year	4.83	0.165
	Site × Year	4.83	0.141
Honey Bees	Site	0.62	1.000
	Year	8.12	0.170
	Site × Year	8.12	0.170
Bumblebees	**Site**	**43.00**	**0.002**
	Year	51.96	0.041[1]
	Site × Year	25.51	0.172
Syrphids	Site	0.05	0.117
	Year	**0.39**	**0.006**
	Site × Year	**0.39**	**0.002**
L. albicaulis			
Small Dark Bees	**Site**	**0.01**	**0.013**
	Year	<0.01	0.501
	Site × Year	<0.01	0.698
Large Dark Bees	Site	<0.01	0.757
	Year	0.03	0.136
	Site × Year	<0.01	0.601
Bumblebees	**Site**	**13.09**	**0.001**
	Year	6.50	0.024[2]
	Site × Year	0.12	0.780

[1]No longer significant at Bonferroni-corrected alpha = 0.008
[2]No longer significant at Bonferroni-corrected alpha = 0.017

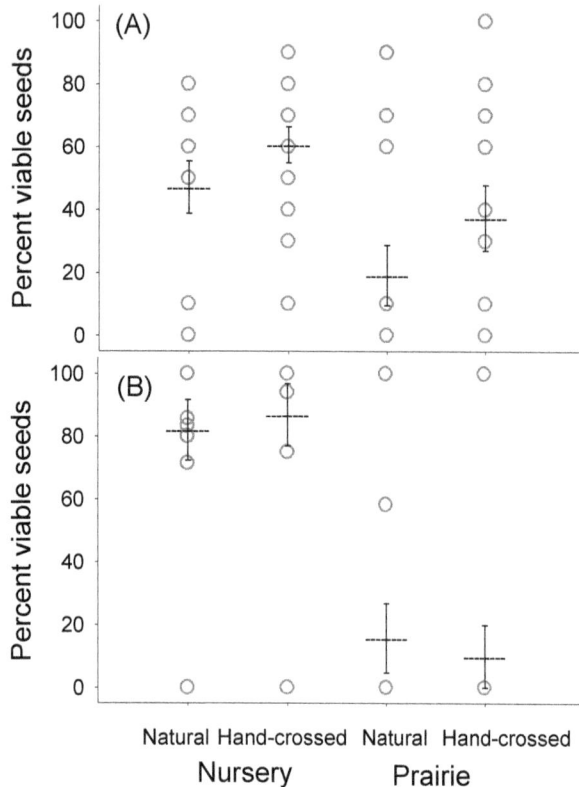

FIGURE 3.   Percentage of viable seeds produced by hand-cross pollination and natural pollination at either Webster Nursery or Johnson Prairie: A) *Balsamorhiza deltoidea* inflorescences, and B) *Lupinus albicaulis* inflorescences. Open gray circles represent percent viability and black horizontal lines represent means +/- 1 standard error. Bonferroni correction for two tests would necessitate an alpha = 0.025.

DISCUSSION

Characteristics of Webster Nursery appear to be facilitating higher rates of insect visitation to these two plant species than at the natural prairie site, Johnson Prairie. The nursery is located in an area with fewer assumed floral resources for native pollinators, though insects appear to be responding to the large influx of floral resources at the relatively new nursery that started in 2008. Although visitation was higher at the nursery site, we found no evidence of pollen limitation and no differences in viability for seeds from either location. Restoration managers at the nursery site were hoping to determine methods for increasing native plant yield and this study suggests that neither increased pollinator populations nor hand pollination would increase seed viability for these two species.

Several studies suggest that human developed landscapes are not necessarily pollinator deprived. Matteson et al. (2012) found it inappropriate to generalize about landscapes created by humans due to high variability in habitat suitability for pollinators within land-use categories. Some researchers found that bee abundance increases in human-constructed landscapes developed with a superabundance of floral resources, and that a combination of natural and developed landscapes can provide a greater diversity of habitat resources (Frankie et al. 2009). Some insects, such as

larger bumblebees, have relatively large foraging areas (Greenleaf et al. 2007). Hagen et al. (2011) found the foraging area for some *Bombus sp.* to be between 0.25-43.53 hectares in one to four days. It is most probable that the high densities of flowers and the many blooming species found in a small area are attracting pollinators to this native plant nursery. Some bees can rapidly produce more offspring in response to an increase in floral resources because greater foraging efficiency means less time they are exposed to predators and parasites (Goodell 2003). For some plant species, rapid accumulation of dense native plant resources at a native plant nursery may attract sufficient unmanaged insects to provide pollination services similar or higher than those found in native landscapes. Restoration nurseries are human-altered landscapes with surpluses of native floral resources, which recent studies have found to be ideal factors in drawing diverse and abundant pollinator responses (Winfree et al. 2011). The visitation rate results of this study support the conclusion that restoration nurseries that are several years old may have sufficient unmanaged insects to pollinate many of their forbs.

Manipulatively increasing insect visitation at Webster nursery may not be a conservation priority given a lack of evidence for pollen limitation for either plant species. In addition, no evidence was found that supplemental pollen increases viable seed production for the plants in this study. An earlier study by Fazzino et al. (2011) found that hand-pollinated inflorescences produced more sprouting seeds than naturally-pollinated inflorescences for *B. deltoidea* in nearby Puget lowland prairies. In contrast, the *B. deltoidea* plants in this study were either not pollen-limited or the hand-pollinated inflorescences did not receive enough supplemental pollen by hand to show a significant difference.

The discrepancy between this earlier study and the results presented here could also be due to differences in weather between study years, as poor flight weather can dampen insect visitation (Vicens & Bosch 2000), or differences in methodology. Weather data were not collected in either study throughout the bloom periods at the study sites, though past local weather reports indicate similar fair weather temperature ranges and precipitation levels in 2009 and 2012 when *B. deltoidea* was in bloom (U.S. Climate Data 2014). Fazzino et al. (2011) examined seed germination and we measured seed viability using lab methods. Due to various factors that can affect germination, and the destructive properties of tetrazolium testing, these two studies are not directly comparable. Future studies could involve both viability and germination testing and should increase the number of replicates to further test for pollen-limitation in both plant species. Though not statistically significant, our results did indicate a trend that hand-pollinated inflorescences produced a higher percentage of viable seeds. Managers may still consider providing supplemental pollen treatments to plants or placing managed bee colonies on site if planning to collect seed during a year with poor weather for insect visitation.

In this experiment, we tested whether or not more pollen would increase viable seed production, though supplemental pollen does not always benefit plant reproduction. When maximum seed production is reached, there are no longer unfertilized ovules for additional pollen to be of benefit (Ashman et al. 2004). There can be a point of pollen saturation on stigmas (Cane & Schiffhauer 2003); too much pollen added too quickly could lead to an underestimation of pollen-limitation (Ashman et al. 2004). In fact, supplemental pollen negatively affected seed weight in a study on pollen limitation at the community level, as plants may reallocate energy and resources in response (Hegland &Totland 2008). Pollen in this study was collected from separate plants on the opposite side of the study area as the plant that was hand-pollinated to ensure cross-pollination. Visiting insects, however, often transfer a combination of pollen from separate plants and pollen that may not be compatible from flowers of the same plant (Wagenius & Lyon 2010), so the effect of increasing insect visitation may not be directly proportional to hand-pollination treatments.

Finally, the percentage of viable seeds was lower at Johnson Prairie compared to Webster nursery for both species, but more dramatically so for *L. albicaulis* (see Fig. 3). Plants at Johnson Prairie appeared to be smaller than at Webster nursery and may have been affected by seed predators, a pathogen, or limited by lack of irrigation. There may have been differences in plant population age, soil nutrients, or microclimate that influenced these differences.

Visitation rate is only one of many factors that may influence the number of viable seeds a plant produces. Availability of resources such as soil nutrients, water, and light can also affect plant reproduction (Stephenson 1981; Corbet 1998; Bos et al. 2007), and seed handling and storage practices can affect seed viability and germination. In addition, changes in light and temperature during germination can affect *L. albicaulis* seed viability (Morey & Bakker 2011). We recommend that land managers turn efforts towards investigating the influence of the above factors on native seed production for these two critical species in future studies. Pollinator visitation and pollen-limitation may not be primary concerns for restoration managers working with these two species in the Puget lowland prairies of western Washington.

ACKNOWLEDGEMENTS

We thank the Center for Natural Lands Management for their support and Joint Base Lewis-McChord for permission to conduct research on their lands. We also thank H. Elizabeth Kirkpatrick, University of Puget Sound, for her suggestions on experimental design for pollinators. We thank Greg Dasso, The Evergreen State College, for helping in the lab. We also thank The Evergreen State College Foundation and the Evergreen Sustainability Fellowship committee for their financial support.

APPENDICES

Additional supporting information may be found in the online version of this article:

Appendix I. Number of insect visits to *Balsamorhiza deltoidea*.

Appendix II. Number of insect visits to *Lupinus albicaulis*.

Appendix III. Number of seeds produced by *Balsamorhiza deltoidea* inflorescences.

Appendix IV. Number of seeds produced by *Lupinus albicaulis* inflorescences.

REFERENCES

Arroyo MTK, Armesto JJ, Primack RB (1985) Community studies in pollination ecology in the high temperate Andes of central Chile II. Effect of temperature on visitation rates and pollination possibilities. Plant Systematics and Evolution 149:187-203.

Arroyo MT, Primack R, Armesto J (1982) Community studies in pollination ecology in the high temperate Andes of central Chile I. Pollination mechanisms and altitudinal variation. American Journal of Botany 69:82-97.

Ashman T, Knight TM, Steets JA, Amarasekare P, Burd M, Campbell DR, Dudash MR, Johnston MO, Mazer SJ, Mitchell RJ, Morgan MT, Wilson WG (2004) Pollen limitation of plant reproduction: Ecological and evolutionary causes and consequences. Ecology 85:2408-2421.

Berry PE, Calvo RN (1989) Wind pollination, self-incompatibility, and altitudinal shifts in pollination systems in the high Andean genus *Espeletia* (Asteraceae). American Journal of Botany 76:1602-1614.

Bos MM, Veddeler D, Bogdanski AK, Klein A, Tscharntke T, Steffan-Dewenter I, Tylainakis JM (2007) Caveats to quantifying ecosystem services: Fruit abortion blurs benefits from crop pollination. Ecological Applications 17:1841-1849.

Cane JH, Schiffhauer D (2003) Dose-response relationships between pollination and fruiting refine pollinator comparisons for cranberry (*Vaccinium macrocarpon* [Ericaceae]). American Journal of Botany 90:1425-1432.

Corbet SA (1998) Fruit and seed production in relation to pollination and resources in Bluebell, *Hyacinthoides non-scripta*. Oecologia 114:349-360.

Dieringer G (1992) Pollinator limitation in populations of *Agalinis strictifolia* (Scrophulariaceae). Bulletin of the Torrey Botanical Club 119:131-136.

Donovall L, vanEngelsdorp D (2008) Pennsylvania Native Bee Survey Citizen Scientist Pollinator Monitoring Guide. The Xerces Society for Invertebrate Conservation. [online] URL: www.xerces.org/download/pdf/PA_Xerces%20Guide.pdf (accessed April 2011).

Dunwiddie PW, Bakker JD (2011) The future of restoration and management of prairie-oak ecosystems in the Pacific Northwest. Northwest Science 85:83-92.

Engel EC, Irwin RE (2003) Linking pollinator visitation rate and pollen receipt. American Journal of Botany 90:1612-1618.

Fazzino L, Kirkpatrick HE, Fimbel C (2011) Comparison of hand-pollinated and naturally-pollinated Puget balsamroot (*Balsamorhiza deltoidea* Nutt.) to determine pollinator limitations on South Puget Sound lowland prairies. Northwest Science 85: 352-360.

Fontaine C, Dajoz I, Meriguet J, Loreau M (2006) Functional diversity of plant-pollinator interaction webs enhances the persistence of plant communities. PLoS Biology 4:0129-0135.

Frankie GW, Thorp RW, Hernandez J, Rizzardi M, Ertter B, Pawelek JC, Witt SL, Schindler M, Coville R, Wojcik VA (2009) Native bees are a rich natural resource in urban California gardens. University of California. California Agriculture [online] URL: http://californiaagriculture.ucanr.edu (accessed Oct 2012).

Goodell K (2003) Food availability affects *Osmia pumila* (Hymenoptera: Megachilidae) foraging, reproduction, and brood parasitism. Oecologia 134:518-527.

Greenleaf SS, Williams NM, Winfree R, Kremen C (2007) Bee foraging ranges and their relationship to body size. Oecologia 153(3):589-596.

Hagen M, Wikelski M, Kissling WD (2011) Space use of bumblebees (*Bombus* spp.) revealed by radio-tracking. PLoS ONE 6(5):e19997.

Hegland SJ, Totland O (2008) Is the magnitude of pollen limitation in a plant community affected by pollinator visitation and plant species specialization levels? Oikos 117:883-891.

Heschel MS, Page KN (1995) Inbreeding depression, environmental stress, and population size variation in scarlet gilia (*Ipomopsis aggregata*). Conservation Biology 9:126-133.

Hitchcock CL, Cronquist A (1998) Flora of the Pacific Northwest. University of Washington Press, Seattle and London.

Inouye DW, Pyke GH (1988) Pollination in the snowy mountains of Australia: Comparisons with montane Colorado, USA. Australian Journal of Ecology 13:191-210.

International Seed Testing Association (2012) International Rules for Seed Testing (2012ed.). Bassersdorf: Switzerland.

Kaye TN (1999) Obligate insect pollination of a rare plant, *Lupinus sulphureus* ssp. *kincaidii*. Northwest Science 73:50-52.

Kearns CA, Inouye DW (1993) Techniques for Pollination Biologists. University Press of Colorado, Niwot.

Kittelson PM, Maron JL (2000) Outcrossing rate and inbreeding depression in the perennial yellow bush lupine, *Lupinus arboreus* (fabaceae). American Journal of Botany 87:652-660.

Kremen C, Williams NM, Aizen MA, Gemmill-Herren B, LeBuhn G, Minckley R, Packer L, Potts SG, Roulston T, Steffan-Dewenter I, Vazquez DP, Winfree R, Adams L, Crone EE, Greenleaf SS, Keitt TH, Klein A, Regetz J, Ricketts TH (2007) Pollination and other ecosystem services produced by mobile organisms: A conceptual framework for effects of land-use change. Ecology Letters 10:299-314.

Kunin WE (1993) Sex and the single mustard: population density and pollinator behavior effects on seed-set. Ecology 74:2145-2160.

Manly BFJ (2007) Randomization, Bootstrap, and Monte Carlo methods in biology (3rd ed.). Chapman & Hall, London.

Matteson KC, Grace JB, Minor ES (2012) Direct and indirect effects of land use on floral resources and flower-visiting insects across an urban landscape. Oikos 116:1588-1598.

Mayer C, Adler L, Armbruster S, Dafni A, Eardley C, Huang S, Kevan PG, Ollerton J, Packer L, Ssymank A, Stout JC, Potts SG (2011) Pollination ecology in the 21st century: Key questions for future research. Journal of Pollination Ecology 3:8-23.

McCall C, Primack RB (1992) Influence of flower characteristics, weather, time of day, and season on insect visitation rates in three plant communities. American Journal of Botany 79:434-442.

McCarty JP (2001) Ecological consequences of recent climate change. Conservation Biology 15:320-331.

Morey M and Bakker J (2011) Effects of light and temperature regimes on germination and viability of native WA prairie species. Undergraduate senior project. School of Forest Resources, University of Washington, Seattle, WA.

USDA Natural Resources Conservation Service (2012) Plants Database. [online] URL: www.plants.usda.gov (accessed Oct 2012).

Perfectti F, Gomez JM, Bosch J (2009) The functional consequences of diversity in plant-pollinator interactions. Oikos 118:1430-1440.

Price MV, Campbell DR, Waser NM, Brody AK (2008) Bridging the generation gap in plants: Pollination, parental fecundity, and offspring demography. Ecology 89:1596-1604.

Stanley AG, Kaye TN, Dunwiddie PW (2008) Regional strategies for restoring invaded prairies: Observations from a multisite, collaborative research project. Native Plants Journal 9:247-254.

Stephenson AG (1981) Flower and fruit abortion: Proximate causes and ultimate functions. Annual Review of Ecology and Systematics 12:253-279.

Stinson DW (2005) Washington State Department of Fish and Wildlife. Washington State status report for the Mazama pocket gopher, streaked horned lark, and Taylor's checker spot. Olympia.

Tscheulin T, Petanidou T (2011) Does spatial population structure affect seed set in pollen-limited *Thymus capitatus*? Apidologie 42:67-77.

U.S. Climate Data (2014) [online] URL: www.usclimatedata.com (accessed November 2014).

Vicens N, Bosch J (2000) Weather-dependent pollinator activity in an apple orchard, with special reference to *Osmia cornuta* and *Apis mellifera* (Hymenoptera: Megachilidae and Apidae). Environmental Entomology 29:413-420.

Vila M, Weiner J (2004) Are invasive plant species better competitors than native plant species? – Evidence from pair-wise experiments. Oikos 105:229-238.

Wagenius S, Lyon SP (2010) Reproduction of *Echinacea angustifolia* in fragmented prairie is pollen-limited but not pollinator-limited. Ecology 91:733-742.

Washington Natural Heritage Program (2012) Washington Department of Natural Resources [online] URL: http://www1.dnr.wa.gov/nhp/refdesk/lists/plantrnk.html (accessed August 2012).

Wilson MV, Hammond PC, Schultz CB (1997) The interdependence of native plants and Fender's blue butterfly. In: Kaye TN, Liston A, Love RM, Luomo D, Meinke RJ, Wilson MV (eds) Conservation management of native flora and fungi. Native Plant Society of Oregon, Corvallis, pp 83-87.

Winfree R, Bartomeus I, Cariveau DP (2011) Native pollinators in anthropogenic habitats. Annual Review of Ecology, Evolution, and Systematics 42:1-22.

14

Innate or Learned Preference for Upward-Facing Flowers?: Implications for the Costs of Pendent Flowers from Experiments on Captive Bumble Bees

Takashi T. Makino[1, 2]* & James D. Thomson[1]

[1]*Department of Ecology and Evolutionary Biology, University of Toronto, 25 Harbord Street, Toronto, Ontario M5S 3G5, Canada*
[2]*Department of Biology, Faculty of Science, Yamagata University, 1-4-12 Kojirakawa, Yamagata, 990-8560, Japan*

Abstract—Pollinator preferences for phenotypic characters, including floral orientation, can affect plant reproductive success. For example, hawkmoths and syrphid flies prefer upward- over downward-facing flowers in field experiments. Although such preferences suggest a cost of pendent flowers in terms of pollinator attraction, we cannot rule out the possibility that the preferences have been affected by prior experience: pollinators might choose the same type of flowers to which they have already become accustomed. To test for innate preference, we observed bumble bees foraging on an array of upward- and downward-facing artificial flowers. Without any prior experience with vertical flowers, 91.7% bees chose an upward-facing flower at the very first visit. In addition to this innate preference, we also found that the preference was strengthened by experience, which suggests that the bees learned upward-facing flowers were easier to handle. Although bumble bees may concentrate on pendent flowers in the field, such learned preferences are evidently imposed on a template of upward-facing preference. Because bee-pollinated pendent flowers face particular difficulties in attracting visits, therefore, we expect them to compensate through other means, such as greater floral rewards.

Keywords: Flower orientation, innate preference, learning, artificial flower, Bombus impatiens

Introduction

In animal-pollinated plants, the number of pollinator visits to a flower is an important determinant of reproductive success because more visits usually mean more pollen transfer (Galen & Stanton 1989; Wilson & Thomson 1991; Jones & Reithel 2001; Engel & Irwin 2003). Many studies on floral traits affecting pollinator visitation have revealed pollinator preferences, for example for larger flowers (Galen & Newport 1987; Johnson et al. 1995; Conner & Rush 1996; Morinaga & Sakai 2006), greater rewards (Pleasants 1981; Thomson 1988; Cartar 2004; Makino & Sakai 2007), and certain colours (Lunau & Maier 1995; Kelber 1997; Weiss 1997; Gumbert 2000), offering insight into the adaptive significance of various floral traits.

Flower orientation is one such trait affecting pollinator visitation. Manipulation of flower orientation has revealed hawkmoths' preference for upward- over downward-facing flowers of *Aquilegia pubescens* (Fulton & Hodges 1999), and syrphid flies' preference for upward-facing or horizontal flowers over downward-facing flowers of *Commelina communis* (Ushimaru & Hyodo 2005, Ushimaru et al. 2009). These findings suggest that a pendent orientation will intrinsically make a negative contribution to pollinator attraction, all else being equal. This would tend to increase

the likelihood of pollen limitation or lower siring success, which in turn might select for countervailing characteristics, such as greater floral rewards in pendent flowers. On the other hand, no preference for flower orientation has been found in hummingbirds (Tadey & Aizen 2001; Castellanos et al. 2004). The different responses among pollinators have some implications for "pollination syndromes", in which flowers show a set of traits that correspond to a particular functional group of pollinators (Fenster et al. 2004). For example, the adaptive function of pendent flowers of the hummingbird-pollinated *Aquilegia formosa* (Fulton & Hodges 1999) may not be to match hummingbirds' preference, but rather to exclude hawkmoths.

Therefore, the role of flower orientation merits study, but we have to be cautious about interpreting observed preferences because we can not rule out the possibility that those preferences have been affected by prior experience, as pointed out by Ushimaru & Hyodo (2005). For example, the hawkmoths might choose upward-facing flowers just because they had already gotten accustomed to upward-facing flowers, and the syrphids might have had experience with upward-facing flowers of other species. To understand how the behavioural ecology of pollinators translates to selection on floral orientation, we wish to determine whether an observed preference is innate or learned; this requires controlling the prior experience of individual pollinators (Thomson & Chittka 2001).

To provide the first test on naïve insects, we performed lab experiments using artificial flowers and captive bumble bees, *Bombus impatiens*. Some bumble bees do specialize in

*Corresponding author; email: mkntkst@gmail.com,

flowers that face down (e.g., Kobayashi et al. 1997; Mahoro 2003), so that researchers tend to think that they do not have any preference against pendent flowers. Indeed, Huang et al. (2002) found no such preference by *Bombus* spp. foraging on *Pulsatilla cernua*, whereas the manipulation of flower orientation by Ushimaru & Hyodo (2005) showed that *Bombus diversus* preferred upward- to downward-facing flowers, though the preference was not statistically significant and might be biased by previous experience. Surprisingly, there are no other studies, and bumble bee preference for flower orientation remains unclear. In this study, we address two questions: 1) do *Bombus impatiens* workers show any preference for upward- or downward-facing flowers?; and 2) is the preference innate, or learned? Then we discuss the costs and benefits of pendent flowers.

MATERIALS AND METHODS

We used two commercial colonies of *Bombus impatiens* (supplied by Biobest, Leamington, Ontario, Canada). A colony was connected to a flight cage with a gated tunnel so that we could control entry of bees into a screen cage, 240 cm long x 220 cm wide x 220 cm high. The nest entrance was located at 100 cm height. We set a table (33 cm long x 55 cm wide x 90 cm high) 180 cm from the entrance, on which we placed artificial flowers.

We made artificial flowers by cutting 1ml pipette tips and mounting a pair of them to a length of styrene plastic tubing (stem) of 6 mm internal diameter, capped at one end (Fig. 1a). The narrow ends of the pipette tips were enlarged to provide a friction-fit over two small tubes that communicated with the interior of the stem. To provide bees with a better grip on flowers, we abraded both inside and outside of floral surfaces with sandpaper. The colours of flowers and tubes were semi-transparent pale blue and

opaque white, respectively. In pre-test and test phases (see below), we offered nectar by inserting a cotton swab into a stem. One end of the swab was dipped into 20% sucrose solution (nectar) before insertion. Bee can drink the nectar from the cotton swab through the small tube at the base of a flower. In order to make a bee leave for another flower, we pulled out the swab about 4 seconds after the bee started probing. The time was too short for a bee to deplete nectar absorbed in the swab tip. We pushed the swab back after the bee landed on the next flower. Note that this method not only makes it easy to offer or withhold nectar, but also solves the difficulty of retaining nectar at the base of an inverted flower. The use of pipette tips as flower cups can be seen, for example, in Ishii (2005) and Makino & Sakai (2007), and 20% sucrose solution was used, for example, in Cnaani et al. (2006) and Worden et al. (2005).

Training phase

To train bees to forage in the cage, we placed six training "plants" (Fig. 1b) randomly on the table and let bees learn to collect nectar from them. Each training plant had a stem and two flowers that were the same as those used in the following test, except that the flowers were oriented horizontally, and nectar was provided continuously by a fixed cotton wick rather than a removable swab. The cotton conveyed nectar upward from a vial, allowing a bee to drink nectar *ad lib*. The gate of the nest was left open to allow bees to forage freely for a few days before testing. Note that bees in this phase did not have any experience with other flower orientations (i.e., upward- and downward-facing flowers), so they were experienced with regard to handling these flowers but naïve with respect to vertical orientation. To identify individual bees, we put correction fluid or numbered tags on the thoraxes of bees that learned flowers. We used those marked bees in the following test.

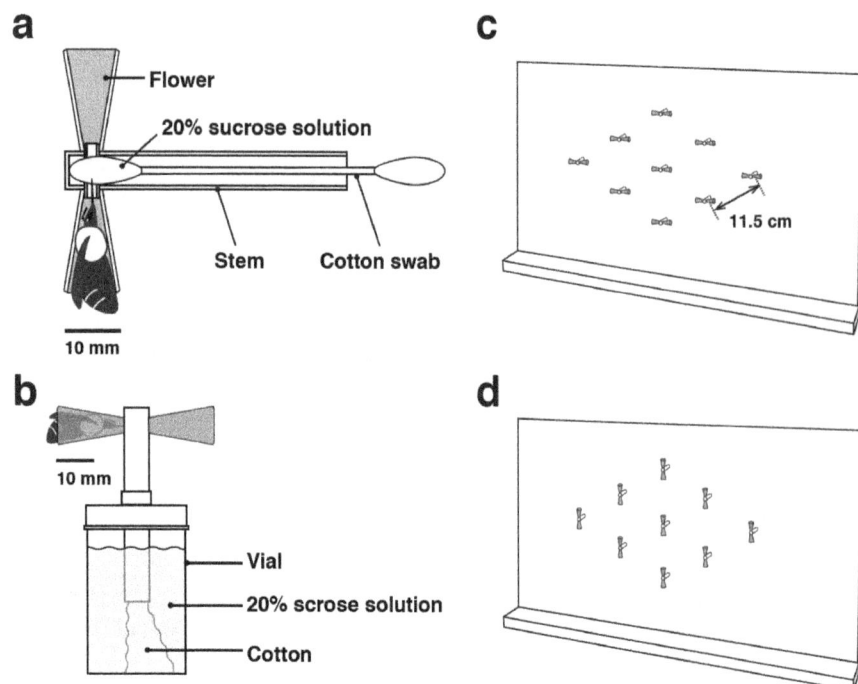

FIG. 1. Schematic views of artificial flowers and experimental arrays. (a) A cross section of upward- and downward-facing flowers on a stem pipe with a cotton swab inside. A bee can drink nectar from the swab until an observer pulls the swab. (b) A training plant with a pair of horizontal flowers. The cotton inside the stem conveyed nectar upward from a vial, allowing a bee to drink nectar unlimitedly. (c) The pre-test array with nine pairs of horizontal flowers. (d) The test array with nine pairs of upward- and downward-facing flowers.

Pre-test phase

Just before the test, we let a focal bee forage in a pre-test array (Fig. 1c) to accustom the bee to swab flowers arranged on a vertical board. The pre-test array comprised nine pairs of horizontal flowers arranged at 11.5 cm intervals in a triangular grid. Flowers were spaced 2.0 cm out from the board, and ranged between 105 cm and 128 cm above the floor. After the bee completed a foraging trip, we rotated each stem 90 degrees to change the pre-test array into the test array of paired upward- and downward-facing flowers (Fig. 1d). We arranged the pairs of flowers at five different heights to eliminate a possible tendency of bees to choose flowers at the same height (when pairs of flowers are placed at the same height, bees could appear to choose only upward- or downward-facing flowers simply by staying at the same height).

Test phase

Each bee was allowed to make three foraging trips in the test array. An observer noted the sequence of the orientations of flowers visited. We re-dipped swabs into nectar between foraging trips in most cases (there were only a few cases in which we did not, but re-dipped the swab of a flower while a bee was foraging another flower). Twenty-four bees were tested (12 bees from each colony). A bee made about 32 flower visits during a single foraging trip on average. We measured the length of the radial cell on the right forewing of each bee as an estimate of body size.

Analysis

To examine experience-dependent changes in the preference for upward-facing flowers, we applied a generalized linear model (GLM) with a logit link function and a binomial error distribution to the ratio of visits to upward-facing flowers to total visits. Foraging trips and individual bees were treated as fixed factors. We applied the same model to every pair of foraging trips for post-hoc comparisons (1st vs. 2nd, 1st vs. 3rd, and 2nd vs. 3rd). Alpha levels were adjusted by the sequential Bonferroni procedure with statistical significance determined at $P = 0.05$.

There were five levels of stem height, and to test whether there was any preference for certain height at the very first visit, we counted the number of visits for each level and performed a chi-square test with expected probabilities of $1/9$, $2/9$, $3/9$, $2/9$ and $1/9$, from the lowest to the uppermost level, respectively. The P value was computed by Monte Carlo simulation with 10,000 replicates. All the analyses were performed using R version 2.13.0 (http://www.r-project.org/).

RESULTS

At the very first visit, 22 of 24 bees chose an upward flower (Fig. 2), a clear deviation from random choice ($n = 24$ bees, $P < 0.0001$, binomial test). There was also a significant tendency to choose the lowest stem at the first visit: the numbers of visits for each of the five levels were 9,

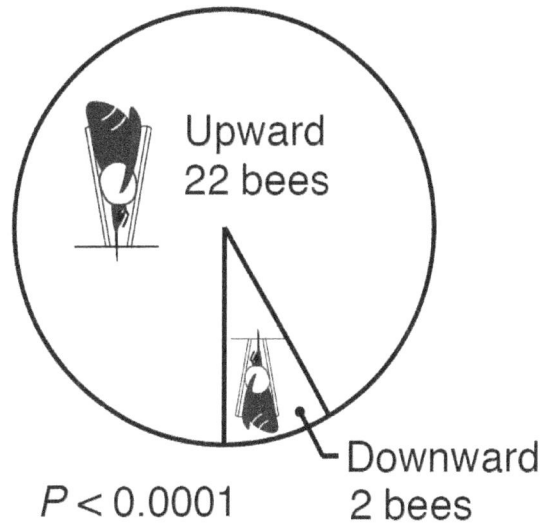

FIG. 2. The number of bees that chose upward- or downward-facing flowers at the very first visit ($n = 24$ bees).

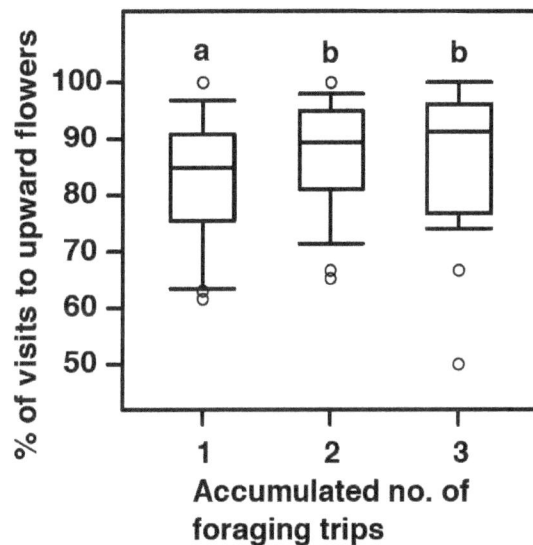

FIG. 3. Box-and-whisker plot of the percentage of visits to upward-facing flowers. The band in a box, the bottom and the top of the box, and the ends of the whiskers represents median, the 25th, 75th, 10th and 90th percentile, respectively ($n = 24$ bees). Small circles are outliers. Different letters indicate significant difference between foraging trips at sequential Bonferroni-adjusted alpha levels with statistical significance determined at $P = 0.05$.

6, 8, 0 and 1, from the lowest to the uppermost level, respectively ($\chi^2 = 21.5$, $P = 0.001$). As shown in Fig. 3, the bees preferred upward- to downward-facing flowers over all three foraging trips (86.7% visits were to upward-facing flowers in total). This preference increased significantly from the first to the second foraging trip (Fig. 3). We also found that the preference significantly differed among bees (Tab. I): the percentage of visits to upward flowers ranged between 61.5% and 100%, 65.2% and 100%, and 50.0% and 100%

TABLE I. Analysis of deviance table on the proportion of visits to upward flowers

	df	Deviance	P
Foraging trip	2	9.7	0.0080
Individual bee	23	177.6	< 0.0001
Residual	24	74.6	

for the first, second and third trips, respectively. However, the variation was not significantly explained by colony ($t = 1.96$, $df = 20.3$, $P = 0.064$, Welch's t-test), nor body size ($n = 24$, Kendall's $\tau = -0.037$, $P = 0.801$).

DISCUSSION

Without any prior experience with upward- and downward-facing flowers, *Bombus impatiens* workers overwhelmingly chose an upward-facing flower at the very first visit (Fig. 2), suggesting that bumble bees are predisposed to prefer upward-facing flowers. This innate preference grew stronger with experience, as evidenced by the significant increase in the proportion of visits to upward-facing flowers (Fig. 3). To our knowledge, this is the first demonstration of innate and learned preference of pollinators for upward-facing flowers. Although *B. impatiens* can certainly learn to handle pendent flowers (e.g., *Vaccinium angustifolium*, Stubbs & Drummond 2001), and may concentrate on them in the field, such learned preferences are evidently imposed on a template of upward-facing preference.

Costs and benefits of pendent flowers in terms of pollinator attraction

The strong preference indicates the possibility that bee-pollinated species with pendent flowers are more prone to pollen limitation (insufficient pollen deposition on a stigma) than those with upward-facing flowers due to their reduced attractiveness. It could be also possible that they invest more in floral traits for pollinator attraction such as petals or nectar to compensate this disadvantage. Future comparisons of those traits among congeneric flowering species with different flower orientations (e.g., *Campanula* and *Rubus*) may reveal some patterns associated with flower orientation.

In contrast, pendent flowers may ensure visits by faithful pollinators by preventing overexploitation of floral resources. All else being equal, reduced attractiveness should increase nectar standing crops of plants, which may let floral visitors specialize on the species (Heinrich 1976, 1979), and also encourage return visits by individual pollinators that learn the locations of beneficial plants (e.g., bumble bees: Burns &

Thomson 2006; Makino & Sakai 2007; hummingbirds: Henderson et al. 2006). The reduced competition may further benefit plants by increasing their chance of being recruited into a pollinator's regular foraging route ("trapline"), which is predicted to increase mating distance and diversity, and also reduce inbreeding of plants (Ohashi & Thomson 2009).

Cause of preference for upward-facing flowers

The exact cause of the preference for upward-facing flowers remains unclear, but it is very likely that bees' tendency to remain upright when flying may predispose them to choose upward-facing flowers. Thus, the preference for upward-facing flowers may simply be a manifestation of general orientation preferences that arise from the basic body plan. The well-known preference for lower positioned flowers at the first visit and working upwards on vertical inflorescences (Waddington & Heinrich 1979; Harder et al. 2004) or experimental arrays (Makino 2008 and this study) may be the same sort of phenomenon, but it is quite possible for a bee to work upward while visiting downward-facing flowers, e.g., on *Digitalis* (Best and Bierzychudek 1982). We should be careful about the possible effect of the prior experience with horizontal flowers in the training phase, but it is hard to think that the experience gave any bias to upward-facing flowers.

The increased preference from the first to the second trips (Fig. 3) suggests that bees learned that upward-facing flowers were easier to handle. Although we did not measure handling time in this study, it is very likely that bees achieved shorter handling times on upward-facing flowers. Indeed, Laverty (1994) showed that naïve *B. rufocinctus* took longer to handle a pendent flower of *Apocynum androsaemifolium* than an upward-facing flower of *A. sibiricum*. In our experiment, we sometimes observed a bee landing on the outer surface of a downward-facing flower keeping its head up and then turning the head down to enter the corolla. We also occasionally saw a bee having difficulty in gripping on the slanting inner surface of a pendent flower, which supports the importance of landing platforms like lower lips of *Digitalis* (Percival & Morgan 1964), or surface structures to provide a better grip (Whitney et al. 2009). It seems worthwhile to see if such facilitations also encourage visits by bees and even invert the innate preference for upward-facing flowers. It would also be interesting to offer more concentrated nectars in downward-facing flowers to see if preferences can be switched.

Interspecific variations

Although *B. impatiens* showed strong preference, we should note that there may be variation among pollinator species, even within *Bombus*. Percival et al. (1968) found that while *B. terrestris* had no difficulty in exploiting pendent flowers of *Digitalis* foxgloves, some other species experienced physical difficulties in entering them. Such difference in handling techniques might lead to interspecific variation in the preference for flower orientation. Like *B. consobrinus*, which easily handle complex flowers of *Aconitum* from their first trials (Laverty 1988), there might be some specialist species for pendent flowers. Clearly, we need more experiments across many pollinator taxa including bees, flies, butterflies, birds, bats, etc., although obtaining naïve subjects will be much harder for vertebrates than for commercially available insects.

Conclusion

We demonstrated that bumble bees, which are known as frequent visitors to pendent flowers, do not necessarily prefer them; *Bombus impatiens* prefers upward-facing flowers if available. This finding indicates the overlooked cost of bearing downward-facing flowers for melittophilous species in terms of pollinator attraction. However, for plants in the presence of better pollinators, hanging flowers might be a good strategy to exclude inferior pollinators (Thomson 2003), as red floral colour reduces bee visitation to bird-pollinated flowers (Schemske and Bradshaw 1999). Pendent flowers are also expected to benefit plant fitness by increasing the precision of pollen transfer (Fenster et al. 2009), by enhancing pollen receipt and removal through increased handling time (Tadey & Aizen 2001), and by protecting nectar and pollen from rain (but see Tadey & Aizen 2001). Examining the balance of those costs and benefits will lead us a better understanding of selective pressures behind the evolution of pendent flowers. We hope our finding stimulates further investigations on other pollinator species. Even hummingbirds, which are usually assumed to have no preference, could have an innate preference for a specific orientation and thereby exert some pressure on floral traits.

ACKNOWLEDGEMENTS

We thank Biobest (Leamington, Ontario) for supplying bees, Jessica Forrest for her useful comment on the experimental design, and Sarah Cao for her help with a part of the experiments. We also thank two anonymous reviewers for their valuable comments. This work was supported by Grant-in-Aid for JSPS (Japan Society of Promotion of Science) overseas fellows to T.T.M. (No. 22.738) and an NSERC (The Natural Sciences and Engineering Research Council of Canada) Discovery Grant to J.D.T.

REFERENCES

Best LS, Bierzychudek P (1982) Pollinator foraging on foxglove (*Digitalis purpurea*): a test of a new model. Evolution 36:70-79.

Burns JG, Thomson JD (2006) A test of spatial memory and movement patterns of bumblebees at multiple spatial and temporal scales. Behavioral Ecology 17:48-55.

Cartar RV (2004) Resource tracking by bumble bees: response to plant-level differences in quality. Ecology 85:2764-2771.

Castellanos MC, Wilson P, Thomson JD (2004) 'Anti-bee' and 'pro-bird' changes during the evolution of hummingbird pollination in *Penstemon* flowers. Journal of Evolutionary Biology 17:876-885.

Cnaani J, Thomson JD, Papaj DR (2006) Flower choice and learning in foraging bumblebees: effects of variation in nectar volume and concentration. Ethology 112:278-285.

Conner JK, Rush S (1996) Effects of flower size and number on pollinator visitation to wild radish, *Raphanus raphanistrum*. Oecologia 105:509-516.

Engel EC, Irwin RE (2003) Linking pollinator visitation rate and pollen receipt. American Journal of Botany 90:1612-1618.

Fenster CB, Armbruster WS, Dudash MR (2009) Specialization of flowers: is floral orientation an overlooked first step? New Phytologist 183:502-506.

Fenster CB, Armbruster WS, Wilson P, Dudash MR, Thomson JD (2004) Pollination syndromes and floral specialization. Annual Review of Ecology Evolution and Systematics 35:375-403.

Fulton M, Hodges SA (1999) Floral isolation between *Aquilegia formosa* and *Aquilegia pubescens*. Proceedings of the Royal Society of London Series B 266:2247-2252.

Galen C, Newport MEA (1987) Bumble bee behavior and selection on flower size in the sky pilot, *Polemonium viscosum*. Oecologia 74:20-23.

Galen C, Stanton ML (1989) Bumble bee pollination and floral morphology: factors influencing pollen dispersal in the alpine sky pilot, *Polemonium viscosum* (Polemoniaceae). American Journal of Botany 76:419-426.

Gumbert A (2000) Color choices by bumble bees (*Bombus terrestris*): Innate preferences and generalization after learning. Behavioral Ecology & Sociobiology 48:36-43.

Harder LD, Jordan CY, Gross WE, Routley MB (2004) Beyond floricentrism: The pollination function of inflorescences. Plant Species Biology 19:137-148.

Heinrich B (1976) The foraging specializations of individual bumblebees. Ecological Monographs 46:105-128.

Heinrich B (1979) "Majoring" and "Minoring" by foraging bumblebees, *Bombus vagans*: an experimental analysis. Ecology 60:245-255.

Henderson J, Hurly TA, Bateson M, Healy SD (2006) Timing in free-living rufous hummingbirds, *Selasphorus rufus*. Current Biology 16:512-515.

Huang S-Q, Takahashi Y, Dafni A (2002) Why does the flower stalk of *Pulsatilla cernua* (Raunculaceae) bend during anthesis? American Journal of Botany 89:1599–1603.

Ishii HS (2005) Analysis of bumblebee visitation sequences within single bouts: implication of the overstrike effect on short-term memory. Behavioral Ecology and Sociobiology 57:599-610.

Johnson SG, Delph LF, Elderkin CL (1995) The effect of petal-size manipulation on pollen removal, seed set, and insect-visitor behavior in *Campanula americana*. Oecologia 102:174-179.

Jones KN, Reithel JS (2001) Pollinator-mediated selection on a flower color polymorphism in experimental populations of *Antirrhinum* (Scrophulariaceae). American Journal of Botany 88:447-454.

Kelber A (1997) Innate preferences for flower features in the hawkmoth *Macroglossum stellatarum*. Journal of Experimental Biology 200:827-836.

Kobayashi S, Inoue K, Kato M (1997) Evidence of pollen transfer efficiency as the natural selection factor favoring a large corolla of *Campanula punctata* pollinated by *Bombus diversus*. Oecologia 111:535-542.

Laverty TM (1994) Bumble bee learning and flower morphology. Animal Behaviour 47:531-545.

Laverty TM, Plowright RC (1988) Flower handling by bumblebees a comparison of specialists and generalists. Animal Behaviour 36:733-740.

Lunau K, Maier EJ (1995) Innate color preferences of flower visitors. Journal of Comparative Physiology A 177:1-19.

Mahoro S (2003) Effects of flower and seed predators and pollinators on fruit production in two sequentially flowering congeners. Plant Ecology 166:37-48.

Makino TT, Sakai S (2007) Experience changes pollinator responses to floral display size: from size-based to reward-based foraging. Functional Ecology 21:854-863.

Makino TT (2008) Bumble bee preference for flowers arranged on a horizontal plane versus inclined planes. Functional Ecology 22:1027-1032.

Morinaga SI, Sakai S (2006) Functional differentiation in pollination processes between the outer and inner perianths in *Iris gracilipes* (Iridaceae). Canadian Journal of Botany 84:164-171.

Ohashi K, Thomson JD (2009) Trapline foraging by pollinators: its ontogeny, economics and possible consequences for plants. Annals of Botany 103:1365-1378.

Percival M, Morgan P (1965) Observations of the floral biology of *Digitalis* species. New Phytologist 61:1-22.

Percival MS, Morgan P, Lewis DE (1968) Behaviour of *Bombus* on *Digitalis* spp. New Phytologist 67:759-769.

Pleasants JM (1981) Bumblebee response to variation in nectar availability. Ecology 62:1648-1661.

Schemske DW, Bradshaw H D, Jr. (1999) Pollinator preference and the evolution of floral traits in monkeyflowers (*Mimulus*).

Proceedings of the National Academy of Sciences of the United States of America 96:11910-11915.

Stubbs CS, Drummond FA (2001) *Bombus impatiens* (Hymenoptera: Apidae): an alternative to *Apis mellifera* (Hymenoptera: Apidae) for lowbush blueberry pollination. Journal of Economic Entomology 94:609-616.

Tadey M, Aizen MA (2001) Why do flowers of a hummingbird-pollinated mistletoe face down? Functional Ecology 15:782–790.

Thomson JD (1988) Effects of variation in inflorescence size and floral rewards on the visitation rates of traplining pollinators of *Aralia hispida*. Evolutionary Ecology 2:65-76.

Thomson JD (2003) When is it mutualism? American Naturalist 162:S1–S9.

Thomson JD, Chittka L (2001) Pollinator individuality: when does it matter? In: Chittka L, Thomson JD (eds) Cognitive Ecology of Pollination. Cambridge University Press, Cambridge, pp 191-213.

Ushimaru A, Dohzono I, Takami Y, Hyodo F (2009) Flower orientation enhances pollen transfer in bilaterally symmetrical flowers. Oecologia 160:667-674.

Ushimaru A, Hyodo F (2005) Why do bilaterally symmetrical flowers orient vertically? Flower orientation influences pollinator landing behaviour Evolutionary Ecology Research 7:151-160.

Waddington KD, Heinrich B (1979) The foraging movements of bumblebees on vertical "Inflorescences": an experimental analysis. Journal of comparative physiology A 134:113-117.

Weiss MR (1997) Innate colour preferences and flexible colour learning in the pipevine swallowtail. Animal Behaviour 53:1043-1052.

Whitney HM, Chittka L, Bruce TJA, Glover BJ (2009) Conical epidermal cells allow bees to grip flowers and increase foraging efficiency. Current Biology 19:948-953.

Wilson P, Thomson JD (1991) Heterogeneity among floral visitors leads to discordance between removal and deposition of pollen. Ecology 72:1503-1507.

Worden BD, Skemp AK, Papaj DR (2005) Learning in two contexts: the effects of interference and body size in bumble bees. Journal of Experimental Biology 208:2045-2053.

15

PRODUCTION OF FLORAL MORPHS IN CLEISTOGAMOUS *RUELLIA BREVIFOLIA* (POHL) C. EZCURRA (ACANTHACEAE) AT DIFFERENT LEVELS OF WATER AVAILABILITY

Amanda Soares Miranda[1,*], Milene Faria Vieira[2]

[1]*Núcleo em Ecologia e Desenvolvimento Ambiental de Macaé. Universidade Federal do Rio de Janeiro. Av. São José Barreto, 764. São José do Barreto, Macaé/ RJ – Brazil; CEP: 27965-045*
[2]*Departamento de Biologia Vegetal. Universidade Federal de Viçosa. Viçosa/MG - Brazil; CEP: 36570-900*

Abstract—In this study we investigated whether the production of cleistogamous (CL) and chasmogamous (CH) floral morphs in *Ruellia brevifolia* is affected by water availability. To this end, the effects of two water levels were tested on plants grown in a greenhouse: soil at 100% water-holding capacity (WHC) (moist soil) and at 50% WHC (water scarcity). Additionally, we investigated fruit and seed production in plants at these two levels of water availability and evaluated whether the drought stress interferes with vegetative growth. The production of floral morphs depended on water availability: plants in moist soil produced only CH morphs and water-stressed plants produced only CL morphs. Fruit production was higher at the higher level of water availability (30.5 ± 28.20 fruits/plant at 100% WHC versus 9 ± 6.04 fruits/plant at 50% WHC; t = 4.384; $P < 0.01$). The mean number of seeds produced by CH and CL morphs were, respectively, 5.93 ± 2.24 and 8.17 ± 2.07 seeds/fruit (t = - 3.304; $P < 0.01$). Although CL morphs produced a greater number of seeds, the total seed production per plant was higher in plants at 100% WHC (180.86 seeds/plant in CH morphs versus 73.53 seeds/plant in CL morphs of plants in soil at 100% and 50% WHC, respectively; t = - 2.759; $P < 0.01$). The plants in soil at 100% WHC were taller (0.48 m ± 0.07) in relation to plants in soil at 50% WHC (0.24 m ± 0.04) (t = 1.781; $P < 0.01$). This study provides new information about the sexual reproductive strategy of *R. brevifolia*, indicating that the main factor inducing cleistogamy is drought stress.

Keywords: chasmogamous floral morph, cleistogamous floral morph, drought stress, floral induction, floral polymorphism, reproductive systems

INTRODUCTION

Ruellia is the largest genus of Acanthaceae, with about 300 species distributed in the tropics (Ezcurra 1993; Tripp 2007). Three species occur at the study location in Viçosa, Minas Gerais, southeastern Brazil (Braz et al. 2002). Two of them, *Ruellia brevifolia* and *Ruellia menthoides*, are typically cleistogamous (Lima et al. 2005; Lima & Vieira 2006), i.e., they produce cleistogamous (CL) and chasmogamous (CH) floral morphs. For the third, *R. subsessilis*, typical cleistogamy has not been reported. This species produces, instead, two distinct chasmogamous morphs ("reduced" and "normal" chasmogamous morphs), depending on the availability of soil moisture (Miranda & Vieira 2014).

Ruellia brevifolia is ornithophilous (bird pollinated; Piovano et al. 1995; Braz et al. 2000; Sigrist & Sazima 2002; Abreu & Vieira 2004), herbaceous, and about 1.0 m tall. The inflorescences are axillary and bear open, potentially outcrossed (CH), red flowers, and closed, obligately self-pollinated (CL), white flowers (Piovano et al. 1995; Sigrist & Sazima 2002; Lima et al. 2005). The pollination tests with the CH morph, performed by Lima & Vieira (2006), obtained similar fruit set without (autofertility, 42.10%) and with pollinators (fruit set of open-pollinated, emasculated flowers, 44.74%). In Viçosa, *R. brevifolia* is found in the forest understory, in shady to partially shaded locations, with year-round flowering and fruiting (Lima et al. 2005; Lima & Vieira 2006).

Observations of Lima et al. (2005) suggest that CH and CL morphs of *R. brevifolia* are produced throughout the year, except in August and September (dry season) for the CH morph and except for January and February (rainy season) for the CL morph. The alternation of chasmogamous and cleistogamous cycles in *R. brevifolia* has been related to ecological and climatic conditions, because the production of CL morphs was mainly observed in the months of low precipitation and temperature (Piovano et al. 1995; Sigrist & Sazima 2002). It is generally thought that the allocation of resources for the production of each floral morph may be influenced by water availability (Lord 1981),

*Corresponding author: asoaresmiranda@gmail.com

and by variations in the photoperiod (Langer & Wilson 1965), temperature (Hexslow 1888), and light intensity (Schemske 1978). However, only a few studies have assessed the role of such factors in *Ruellia* (e.g. *Ruellia nudiflora*, Munguías-Rosas et al. 2012) and or other species (e.g. *Impatiens* sp., Schemske 1978; *Collomia grandiflora*, Minter & Lord 1983; *Calathea micans*, Le Corff 1993).

This study assessed the production of floral morphs in *R. brevifolia* at two levels of water availability in plants grown in a greenhouse: soil at 100% water-holding capacity (WHC) (moist soil) and at 50% WHC (water scarcity). In addition, fruit and seed production in plants at these two levels of water availability and the influence of drought stress on vegetative growth were investigated.

MATERIALS AND METHODS

The experiments were conducted from May 2009 to January 2010 in a greenhouse of the Federal University of Viçosa. *Ruellia brevifolia* plants ($N = 20$) were randomly obtained from previously rooted cuttings (about 25 cm long) from adult plants of a natural population in Viçosa (with about 100 plants), at the Station of Research, Environmental Training and Education Mata do Paraíso, a semideciduous forest reserve. The climate of Viçosa is characterized by hot, humid summers from October to March and cool, dry winters from April to September (Pezzopane 2001).

The plants were grown in pots filled with 3 Kg of substrate consisting of a mixture of soil from the site of plant occurrence with sand (ratio 2:1). The preliminary chemical analysis of this substrate indicated suitability for cultivation. About 50 days after planting (plants in reproductive stage), watering treatments were applied to soil with: 100% of water-holding capacity (WHC) (moist soil, control; $N = 10$ plants) and 50% WHC (drought stress; $N = 10$ plants).

The water-holding capacity was determined in five 100 g samples of the dried substrate at 103 °C for 48 hours, by saturation with water until the percolated water volume became constant (Freire et al. 1980). The values were extrapolated to the amount of soil contained in the pots, corresponding to the control with 100% WHC, and a value of 50% WHC was determined. Irrigation was monitored by the gravimetric method (weighing the pots), adding water until the pots reached the predetermined value for water-holding capacity, based on soil and water weight (Freire et al. 1980).

The 20 plants were checked weekly for the presence of CH and CL morphs. The average numbers of fruits per plant of both water treatments were calculated on two occasions: in August (two months after treatment) and in December 2009 (six months after treatment). In addition, the average number of seeds per fruit ($N = 30$, 15 per treatment) was calculated in December 2009.

To ensure that the plants were drought-stressed (50% WHC), three plants of each treatment (drought stress and control) were evaluated in January 2010, between 8:00 and 10:30 AM, for net photosynthesis (A), stomatal conductance (Gs) and transpiration rates (E). For this purpose, we used an infrared gas analyzer (Irga) - Licor 6400, with steady light sources 700 μmol m^{-2} s^{-1}, indicated as optimum value by the light curve.

The plant height (base to stem apex) was measured at the end of the experiment, to compare the vegetative growth of the plants under water stress compared to those in moist soil ($N = 20$, 10 per treatment). From each value, the initial length of the cuttings was subtracted.

The data were tested for normality and homogeneity (Kolmogorov-Smirnov and Cochran C) and when normal and homogeneous, the Student test was used (Zar 1996).

RESULTS AND DISCUSSION

Plants grown in soil at 100% WHC produced only CH morph and plants in soil at 50% WHC only CL morphs. According to Brown (1952), drought stress may be the trigger of cleistogamy. Indeed, in this study, the monthly means of soil water potential (ψ_w) estimated during the experiment (100% WHC: $\psi_w = 0.15$ MPa; 50% WHC: $\psi_w = -0.36$ MPa) showed that the water scarcity treatment probably induced drought stress. Moreover, the net photosynthesis (A; μmol m^{-2} s^{-1}), transpiration (E; mmol m-2 s-1) and stomatal conductance rates (Gs; mol m^{-2} s^{-1}) were higher in plants grown in soil at 100% WHC (A = 4.8; E = 0.73 and G$_s$ = 0.043) than in plants grown in soil at 50% WHC (A = 0.06; E = 0.15 and G$_s$ = 0.008).

For these reasons, the gene expression of cleistogamy seems to have been selected in stressed plants, possibly due to the lower production and transpiration costs for CL morphs (Galen et al. 1999; Webster & Grey 2008), which are smaller and produce no floral nectar, as seems to be the case of the studied species. Based on this premise, the production of different floral morphs in natural *R. brevifolia* populations should have a seasonal pattern, in response to the water level throughout the year. Thus, plants tend to produce CH morphs in the period of greatest soil water availability (rainy season), and CL morphs in the period of lower availability (dry season). The findings of Sigrist & Sazima (2002) and Lima et al. (2005) for this species demonstrated this trend, reinforcing the results of this study.

Fruit production was higher at the higher level of water availability (30.5 $\pm$ 28.20 fruits/plant at 100% WHC versus 9 $\pm$ 6.04 fruits/plant at 50% WHC; t = 4.384; $P < 0.01$). During the experiment, the fruit set of plants grown in soil at 100% WHC decreased (from 50 $\pm$ 27.34 fruits/plant in August to 11 $\pm$ 8.02 fruits/plant in December; t = 4.384; $P < 0.01$), but no significant variation was observed in plants on 50% WHC soil (from 11 $\pm$ 6.92 fruits/plant in August to 6.7 $\pm$ 4.4 fruits/plant in December).

The reduction in fruit production throughout the experiment in plants at 100% WHC may result from changes in the photoperiod (Langer & Wilson 1965), temperature (Hexslow 1888), light intensity (Schemske

1978), or also from nutrient depletion in the potting soil. Additional studies are needed to confirm these possibilities.

The lower fruit set of plants at 50% WHC may be explained by the reduced photo assimilation, which could reduce the amount of assimilates allocated to fruit production (Garrido et al. 2000). These plants may be resilient to this stress level, since fruiting throughout the experiment varied little, as similarly observed in *R. subsessilis* by Miranda & Vieira (2014).

The average numbers of seeds produced by CH and CL morphs were, respectively, 5.93 ± 2.24 and 8.17 ± 2.07 seeds/fruit (t = - 3.304; $P < 0.01$). The higher seed production by CL morphs was due to the more efficient self-pollination process (95% of fruit set, Lima & Vieira 2006). The smaller number of seeds of CH morphs resulted from self-pollination that was less efficient (42.10% of fruit set, Lima & Vieira 2006). Despite the greater per-flower number of seeds produced by CL morphs (only in plants on 50% WHC soil), the total seed production per plant was higher in CH-morph plants on 100% WHC soil (180.86 ± 167.31 seeds/plant in plants at 100% WHC versus 73.53 ± 49.33 seeds/plant in plants at 50% WHC; t = - 2.759; $P < 0.01$), showing that drought stress can reduce the reproductive success of *R. brevifolia*.

The plants in 100% WHC soil were taller than plants in 50% WHC soil (respectively, $0.48 m \pm 0.07$ and $0.24 m \pm 0.04$; t = 1.781; $P < 0.01$), because the water stress affected vegetative growth, as previsously reported (Larcher 2004). Changes in soil moisture can also alter the availability of nutrients such as nitrogen (Birch 1964), which can also affect plant growth in soil at 50% WHC.

Conclusions

Our greenhouse experiments demonstrated that water availability is a primary factor in the induction of the type of floral morph produced by in *R. brevifolia*: CH morphs are produced in high-moisture soil (100% water-holding capacity) and CL morphs are produced by plants in drier soil (50% WHC). Drought stress reduces vegetative growth and fruit and seed production. All three measures were higher in plants on 100% WHC soil, which produced only CH flowers.

REFERENCES

Abreu CRM, Vieira MF (2004) Os beija-flores e seus recursos florais em um fragmento florestal de Viçosa, sudeste brasileiro. Lundiana 5: 129-134.

Birch HF (1964) Mineralization of plant nitrogen following alternate wet and dry conditions. Plant Soil 20: 43-49.

Braz DM, Carvalho-Okano RM, Kameyama C (2002) Acanthaceae da Reserva Florestal Mata do Paraíso, Viçosa, Minas Gerais. Revista Brasileira de Botânica 25: 495-504.

Braz DM, Vieira MF, Carvalho-Okano RM (2000) Aspectos reprodutivos de espécies de Acanthaceae Juss. de um fragmento florestal do município de Viçosa, Minas Gerais. Revista Ceres 47: 229-239.

Brown WV (1952) The relation of soil moisture to cleistogamy in *Stipa leucotricha*. Botanical Gazette 113: 438-444.

Ezcurra, C (1993) Systematics of *Ruellia* (Acanthaceae) in southern South America. Annals of Missouri Botanical Garden 80: 787-845.

Freire JC, Ribeiro MSV, Bahia VG, Lopes AS, Aquino LH (1980) Resposta do milho cultivado em casa de vegetação a níveis de água em solos da região de Lavras (MG). Revista Brasileira de Ciência do Solo 4: 5-8.

Galen C, Sherry R, Carroll A (1999) Are flowers physiological sinks or faucets? Costs and correlates of water use by flowers of *Polemonium viscosum*. Oecologia 118: 461-470.

Garrido MAT, Del Pino MAIT, Silva AM, Andrade MJB (2000) Crescimento, absorção iônica e produção do feijoeiro sob dois níveis de nitrogênio e três lâminas de irrigação. Ciências Agrotécnicas de Lavras 24: 187-194.

Griffith CJ (1996) Distribution of *Viola blanda* in relation to within-habitat variation in canopy openness, soil phosphorus and magnesium. Bulletin of Torrey Botanical Club 123: 281-285.

Hexslow G (1888) Cleistogamy. Nature 39: 104-105.

Langer RHM, Wilson D (1965) Environmental control of cleistogamy in prairie grass (*Bromus unioloides* H. B. K.). New Phytologist 64: 80-85.

Larcher W (2004) Ecofisiologia Vegetal. Rima, São Carlos.

Le Corff J (1993) Effects of light and nutrient availability on chasmogamy and cleistogamy in an understory tropical herb, *Calathea micans* (Marantaceae). American Journal of Botany 80: 1392–1399.

Lima NAS, Vieira MF (2006) Fenologia de floração e sistema reprodutivo de três espécies de *Ruellia* (Acanthaceae) em fragmento florestal de Viçosa, Sudeste brasileiro. Revista Brasileira de Botânica 29: 681-687.

Lima NAS, Vieira MF, Carvalho-Okano RM, Azevedo AA (2005) Cleistogamia em *Ruellia menthoides* (Nees) Hiern e *Ruellia brevifolia* (Pohl) C. Ezcurra (Acanthaceae) em fragmento florestal do Sudeste brasileiro. Acta Botanica Brasilica 19: 443-449.

Lord EM (1981) Cleistogamy: a tool for the study of floral morphogenesis, function and evolution. The Botanical Review 47: 421-449.

Minter TC, Lord EM (1983) Effects of water stress, abscisic acid, and gibberellic acid on flower production and differentiation in the cleistogamous species *Collomia grandiflora* Dougl. ex Lindl. (Polemoniaceae). American Journal of Botany 70: 618–624.

Miranda AS, Vieira MF (2014) *Ruellia subsessilis* (Nees) Lindau (Acanthaceae): a species with a sexual reproductive system that responds to different water availability levels. Flora 209: 711-717.

Munguías-Rosas MA, Parra-Tabla V, Ollerton J, Cervera C (2012) Environmental control of reproductive phenology and the effect of pollen supplementation on resource allocation in the cleistogamous weed, *Ruellia nudiflora* (Acanthaceae). Annals of Botany 109: 343-350.

Pezzopane JEM (2001) Caracterização microclimática, ecofisiológica e fitossociológica em uma floresta estacional semidecidual secundária, em Viçosa, MG. Universidade Federal de Viçosa, Viçosa (Dr tese).

Piovano M, Galetto L, Bernardello L (1995) Floral morphology, nectar features and breeding system in *Ruellia brevifolia* (Acanthaceae). Revista Brasileira de Biologia 55: 409-418.

Schemske DW (1978) Evolution of reproductive characteristics in *Impatiens* (Balsaminaceae): the significance of cleistogamy and chasmogamy. Ecology 59: 596-613.

Sigrist MR, Sazima M (2002) *Ruellia brevifolia* (Pohl) Ezcurra (Acanthaceae): fenologia da floração, biologia da polinização e reprodução. Revista Brasileira de Botânica 25: 35-42.

Tripp EA (2007) Evolutionary relationships within the species-rich genus *Ruellia* (Acanthaceae). Systematic Botany 32: 628-649.

Webster TM & Grey TL (2008) Growth and reproduction of benghal dayflower (*Commelina benghalensis*) in response to drought stress. Weed Science 56: 561-566.

Zar JH (1996) Biostatistical analysis. Prentice-Hall International, New Jersey.

16

POLLINATION ECOLOGY OF THE INVASIVE TREE TOBACCO *NICOTIANA GLAUCA*: COMPARISONS ACROSS NATIVE AND NON-NATIVE RANGES

Jeff Ollerton[1]*, Stella Watts[1,7], Shawn Connerty[1], Julia Lock[1], Leah Parker[1], Ian Wilson[1], Sheila K. Schueller[2], Julieta Nattero[3,8], Andrea A. Cocucci[3], Ido Izhaki[4], Sjirk Geerts[5,9], Anton Pauw[5] and Jane C. Stout[6]

[1]*Landscape and Biodiversity Research Group, School of Science and Technology, University of Northampton, Avenue Campus, Northampton, NN2 6JD, UK.*
[2]*School of Natural Resources and Environment, University of Michigan 440 Church Street Ann Arbor, MI 48109-1115, USA.*
[3]*Instituto Multidisciplinario de Biología Vegetal (IMBIV). Conicet-Universidad Nacional de Córdoba. Casilla de Correo 495. 5000, Córdoba. Argentina.*
[4]*Department of Evolutionary and Environmental Biology, Faculty of Science and Science Education, University of Haifa, 31905 Haifa, Israel.*
[5]*Dept of Botany and Zoology, Stellenbosch Univ., Private Bag X1, Matieland, 7602, South Africa.*
[6]*Trinity Centre for Biodiversity Research and School of Natural Sciences, Trinity College Dublin, Dublin 2, Republic of Ireland.*
[7]*Current address: Laboratory of Pollination Ecology, Institute of Evolution, University of Haifa, Haifa 31905, Israel.*
[8]*Current address: Cátedra de Introducción a la Biología, Facultad de Ciencias Exactas, Físicas y Naturales, Universidad Nacional de Córdoba. X5000JJC, Córdoba, Argentina.*
[9]*Current Address: South African National Biodiversity Institute, Kirstenbosch National Botanical Gardens, Claremont, South Africa*

Abstract—Interactions with pollinators are thought to play a significant role in determining whether plant species become invasive, and ecologically generalised species are predicted to be more likely to invade than more specialised species. Using published and unpublished data we assessed the floral biology and pollination ecology of the South American native *Nicotiana glauca* (Solanaceae) which has become a significant invasive of semi-arid parts of the world. In regions where specialised bird pollinators are available, for example hummingbirds in California and sunbirds in South Africa and Israel, *N. glauca* interacts with these local pollinators and sets seed by both out-crossing and selfing. In areas where there are no such birds, such as the Canary Islands and Greece, abundant viable seed is set by selfing, facilitated by the shorter stigma-anther distance compared to plants in native populations. Surprisingly, in these areas without pollinating birds, the considerable nectar resources are only rarely exploited by other flower visitors such as bees or butterflies, either legitimately or by nectar robbing. We conclude that *Nicotiana glauca* is a successful invasive species outside of its native range, despite its functionally specialised hummingbird pollination system, because it has evolved to become more frequently self pollinating in areas where it is introduced. Its invasion success is not predictable from what is known of its interactions with pollinators in its home range.

Key words: Argentina, California, Canary Islands, Greece, hummingbird, invasive species, Israel, mutualism, Peru, pollination, Solanaceae, South Africa, sunbird

INTRODUCTION

Plant-flower visitor relationships evolve and are maintained within a fluctuating ecological context in which populations of pollinating, pollen collecting and nectar robbing animals can change significantly from year to year (e.g. Herrera 1988; Fishbein & Venable 1996; Ollerton 1996; Lamborn & Ollerton 2000; Alarcón et al. 2008). This is particularly relevant to introduced invasive plant species which lack ecological or functional pollinator specificity and are therefore ecological generalists (*sensu* Waser et al. 1996; Fenster et al. 2004; Ollerton et al. 2007). Such plants can form relationships with pollinators and maintain viable populations following human dispersal

beyond their native range, negatively affecting local habitats by monopolising space and soil resources, and in the process out-competing native species (Theoharides & Dukes 2007).

Invasive plants have also been shown to have more subtle, but still potentially important, detrimental effects on the native flora by becoming integrated into local pollination interaction webs (*sensu* Memmott & Waser 2002; Vilà et al. 2009; Padrón et al. 2009) and influencing patterns of flower visitation and pollen flow, resulting in lower seed set and quality, and reduced pollinator abundance (Chittka & Schürkens 2001; Schürkens & Chittka 2001; Moragues & Traveset 2005; Traveset & Richardson 2006; Bjerknes et al. 2007; Aizen et al. 2008; Morales & Traveset 2009; Stout & Morales 2009). Other studies, however, have found no negative effects (e.g. Aigner 2004) indicating that the outcomes of such indirect interactions are likely to be species and/or community specific (Moragues & Traveset

2005) and also to depend on intensity of invasion (Dietzsch et al. 2011) and spatial scale (Cariveau & Norton 2009). It is therefore important for us to understand why some plant species are more likely than others to become a threat to local plants, and particularly whether such plants can be predicted from their floral traits (Rodger et al. 2010, though see the recent exchange of views in Trends in Ecology and Evolution stimulated by Thompson & Davis 2011). For example, Chittka & Schürkens' (2001) study suggests that introduced plants with very high rates of nectar production may draw pollinators away from native plants, reducing their reproductive success. High rates of nectar production may therefore be a predictive trait for such negative indirect effects (although see Nienhuis et al. 2009).

Other than self-pollinating species, plants with ecologically generalized pollination systems, which can attract, reward and therefore utilise a wide range of pollinators, have been considered the most probable invasive species (Baker 1974; Richardson et al. 2000; Olesen et al. 2002; Vilá et al. 2009). Such plants are theoretically more likely to co-opt native or introduced flower visitors as pollinators, ensuring their reproductive success and subsequent invasiveness, and there is growing evidence that this is the case (e.g. Forster 1994; Stout 2007; Bartomeus et al. 2008; Vilá et al. 2009; Harmon-Threatt et al. 2009). Nevertheless more studies are required to test the generality of this idea. In particular we should compare the pollination ecologies of invasive plants within their normal distributional range and within the areas of invasion. The only such published study that we know to exist is of *Rhododendron ponticum* (Stout et al. 2006, though see Rodger et al. 2010) and this is a gap in the knowledge of the ecology of invasive species generally (Tillberg et al. 2007).

This paper focuses on the invasive tree tobacco *Nicotiana glauca* Graham (Solanaceae), a native of central and north west Argentina and Bolivia (Goodspeed 1954) which has been widely introduced to the subtropics as a garden ornamental, only to escape and densely colonise native habitats across the globe, including other parts of South America (Moraes et al. 2009; Cocucci, Watts, pers. obs.); Australia (Florentine & Westbrook 2005; Florentine et al 2006); California (Schueller 2004); Hawaii (Izhaki, pers. obs.); the north and east Mediterranean region (Tadmor-Melamed et al. 2004, Bogdanović et al. 2006) including Israel where *N. glauca* was first observed in 1890 (Bornmuller 1898); Mexico (Hernández, 1981); North Africa (Ollerton., pers. obs.); Southern Africa (Geerts & Pauw 2009; Henderson 1991; R. Raguso, pers comm.); and the Canary Islands (Kunkel 1976; Ollerton pers. obs.; Stout pers. obs.). The species is listed in the Global Invasive Species Database (http://www.issg.org/ase/welcome/), and a number of regional organisations consider it invasive, for example in Hawai'i (http://www.hear.org/pier/species/nicotiana_glauca.htm), Europe (http://www.europe-aliens.org/index.jsp) and South Africa (http://www.agis.agric.za).

In its native range, *N. glauca* is strictly hummingbird pollinated (Nattero & Cocucci 2007) although bees and other insects may pierce the base of the corolla tube to rob nectar. Our study therefore addresses the following two questions:

(1) Is *N. glauca*, with its apparently functionally specialised pollination system and abundant nectar resources, pollinated by functionally equivalent pollinators (i.e. flower-feeding birds) throughout its native and non-native range?

(2) Are these interactions with local pollinators (i.e. integration into the local pollination web) a prerequisite for reproductive success in this highly invasive species?

DATA COLLECTION AND SYNTHESIS

We have synthesised published and unpublished data from studies of the species in north western Argentina and Bolivia (Nattero & Cocucci 2007, Nattero et al. 2010, and unpublished data), where the species is native, with research from areas where the species is introduced, including South America (Peru – Watts unpublished data); other populations of Argentina outside the native range (Nattero & Cocucci 2007 and unpublished data); North America (México – Hernández 1981; California – Schueller 2004 and 2007 and unpublished data); and the Old World, including Tenerife (Ollerton et al. unpublished data), Gran Canaria (Stout unpublished data), Greece (Schueller 2002 and unpublished data) and Crete (Ollerton unpublished data); Israel (Tadmor-Melamed 2004; Tadmor-Melamed et al. 2004; Izhaki unpublished data); and South Africa (Skead 1967; Knuth 1898-1905; Marloth 1901; Geerts & Pauw 2009 and unpublished data).

The methods for the published data collection can be found in the relevant papers; methods for the unpublished data are summarised only briefly and more details can be obtained via the corresponding author. Measurements of floral traits, including nectar production, followed standard protocols (Kearns & Inouye 1993; Dafni et al. 2005). Unless otherwise stated, the authors cited above were responsible for the data collected in specific geographical regions. Data were analysed using SPSS 17.0: all data fulfilled assumptions of normality (one-sample Kolmogorov-Smirnov Test) and mean values are presented as $\pm$ SD unless otherwise stated.

DISTRIBUTION, REPRODUCTION AND ABUNDANCE OF *NICOTIANA GLAUCA*

Within its native range *N. glauca* is an occasional plant of dry, naturally and anthropogenically disturbed areas such as river banks, track sides and abandoned quarries. It is found mainly in semi-arid environments from low to high altitudes (0-3500 m), but never at wet localities. The plant is rarely abundant and is mainly found as scattered, usually multi-stemmed individuals, though stem densities on anthropogenically disturbed sites can range from 3.0 to 12.5 m^{-2} (Nattero & Cocucci 2007). Mean population fruit set per plant ranges from about 28.0% to 67.0% (grand mean = 42.4 $\pm$ 13.1% - Table I).

TABLE 1: Reproductive output and mean stigma-anther distances of populations of *Nicotiana glauca*. Stigma-anther distances were measured to the nearest 0.1 mm using a digital calliper. Sample sizes vary considerably and are available from the corresponding author on request. All means are ± SD. Status: N = Native, I = Introduced

Region	Locality	Status	Mean fruit set (%)	Mean minimum stigma-anther distance (mm)
Argentina and Bolivia (Nattero and Cocucci 2007, Nattero et al. unpublished data)	Tupiza	N	36 ± 25	2.2 ± 0.8
	Cuesta de Miranda	N	39 ± 16	1.2 ± 0.6
	Cochabamba	N	32 ± 14	3.0 ± 0.8
	Dique Los Sauces	N	28 ± 17	2.6 ± 0.8
	Potosí	N	47 ± 10	3.6 ± 1.1
	Sanagasta	N	48 ± 26	1.3 ± 0.8
	Sucre	N	67 ± 16	2.6 ± 1.5
	Paraná	I	67 ± 7	2.5 ± 0.9
	Costa Azul	I	41 ± 18	1.2 ± 0.6
	Bella Vista	I	56 ± 8	1.2 ± 0.8
Peru (Watts, unpublished data)	Urubamba	I	-	1.7 ± 0.4
California (Schueller 2004)	Santa Cruz Island	I	41 ± 49	1.6 ± 0.3
	Santa Catalina Island	I	75 ± 20	1.8 ± 0.2
	Sedgwick Reserve	I	26 ± 44	1.9 ± 0.5
	Starr Ranch	I	70 ± 46	2.1 ± 0.2
Israel (Izhaki, unpublished data)	Jezreel Valley	I	55 ± 8	1.8 ± 0.7
Tenerife (Ollerton et al., unpublished data)	South West	I	80 ± 21	1.4 ± 0.4
Greece (Schueller, 2002 and unpublished data)	Athens (Ano Illioupolis)	I	Data un-quantified, but fruit set high and seeds viable.	1.5 ± 0.5
	South-central Peloponese (Gerolimenas and Gythio)	I	ditto	1.6 ± 0.3
	Crete, Agia Galini	I	ditto	1.4 ± 0.9
	Crete, Tympaki	I	ditto	1.9 ± 0.5
Ollerton (unpublished data)	Crete, Agios Nikolaos	I	42 ± 23	1.1 ± 0.2
South Africa (Geerts and Pauw unpublished data)	Buffelsrivier	I	61 ± 10	1.4 ± 0.3
	Leipoldtville	I	74 ± 9	

In its non-native range *N. glauca* is a conspicuous, profusely blooming invasive species growing predominantly along roadsides and on disturbed land in semi-arid regions.

It can be extremely abundant; for example, in Tenerife we have recorded stem densities of 0.7 ± 0.3 m⁻², covering hundreds of square metres, and in an extensive population in South Africa (Buffelsrivier) we recorded stem densities of 2.0 ± 0.4 m⁻². Similarly, in Israel it forms relatively dense scrub in both mesic and semi-arid regions. These high densities are achieved mainly from seed production; there is no clonal growth, though broken stems can re-sprout and there may be some rooting from horizontal branches in contact with the soil.

Fruit set in populations can be high (Table 1) and each fruit contains hundreds of tiny, dry seeds with viabilities of about 90% (Table 2). Seedlings are common in non-native populations.

THE FLORAL BIOLOGY OF *N. GLAUCA*

Within its native range, the flowers of *N. glauca* are typically yellow and tubular, ranging from on average 32.0 ± 2.2 mm to 41.9 ± 4.9 mm in length (n = 6 populations – see Nattero et al. 2009). The mouth of the corolla is green when the flower first opens, but changes to yellow over several days, until the flower is a single hue (Fig. 1). In scattered

TABLE 2. Seed production and viability in introduced population of *Nicotiana glauca*. Seed germination was assessed by sowing 20 or 25 (in Israel) seeds on damp filter paper in each of 10 Petri dishes.

Region	Mean seeds per fruit ± SD	Number of fruit scored	Seed viability (%)	Number of seeds
Israel	1122.7 ± 655.8	12	92.7 ± 5.2%	250
California	655 ± 247	16	-	-
South Africa	1435.8 ± 1063.6	7	87.5% ± 10.6%	200
Tenerife	-	-	85.5% ± 6.4%	200

FIGURE 1: Stages of flower development in *Nicotiana glauca*. From left to right: closed late stage bud; newly opened flower; older flower showing colour change of mouth of corolla tube from green to yellow. Photograph by Jeff Ollerton in an invasive population on Tenerife.

populations of northwest Argentina, a flower colour polymorphism is present which includes dark red, reddish yellow and yellow morphs.

All non-native populations studied to date possess only the typical yellow flower colour variant which may reflect the introduction of a limited set of genotypes into the alien range (Fig. 1). Corolla length also tends to be shorter in non-native populations than the maximum observed in native populations (up to 57 mm); for example, flowers on Tenerife are on average 37.6 ± 1.7 mm in length (n = 21 flowers from 5 plants); South Africa (Buffelsrivier) = 33.7 ± 0.5 mm (n = 10 flowers on each of 16 plants); California = 35.5 ± 1.8 mm (n = 10 flowers per plant on 85 plants across 4 sites) though island populations (more recently colonized and containing shorter billed hummingbird visitors) have slightly shorter corollas than the mainland plants (Schueller 2007); northern Israel = 34.8 ± 2.0 mm (n = 10 flowers on each of 10 plants); Peru = 33.3 ± 1.5 mm (n = 10 flowers on each of 4 plants); finally, Greek

populations have the shortest recorded corolla lengths with an average of 31.8 ± 2.5 mm (n = 95 flowers on 9 plants).

Nectar is abundantly produced and of moderate sugar concentration in both native and introduced populations (Table 3). Sugar composition has been analysed as 48.6%: 38.9%: 13.2% (sucrose: fructose: glucose, Galetto & Bernardello 1993b). Data from non-native populations were obtained using a variety of protocols, e.g. bagged for various periods versus standing crop from open flowers, at various times of the day, making direct comparisons problematic. But they largely agree with the results from the native populations in that they show that *N. glauca* flowers produce moderate to substantial quantities of moderately concentrated nectar (Table 3). A daily rhythm of nectar volume was detected in some populations but this varied. In Israel the lowest volumes were found in the morning and rose in the afternoon, whilst in Gran Canaria the peak was at midday. Perhaps more expected for a bird pollinated plant is the observation of peak nectar volumes in the early morning in a population in South Africa (Table 3).

Although, as we mentioned, the nectar data have been collected using a range of protocols and are therefore not directly comparable, nonetheless these results emphasise our main point that *N. glauca* produces abundant nectar in all populations, even those that are predominantly selfing (see below).

The population mean minimum stigma-anther (S-A) distance is a measure of the average ability of flowers to autogamously self pollinate (e.g. Armbruster 1988). In *N. glauca* in California, island populations have shorter S-A distances than mainland populations. This is probably a result of the initial colonising plants being predominantly selfing rather than a result of natural selection favouring self pollinating genotypes, as the island populations (contrary to expectation) did not experience lower pollinator visitation rates compared to mainland populations (Schueller 2004). However, S-A distances vary greatly among populations (Table 1) and there is a trend of smaller S-A distances when one compares native populations, with non-native populations where specialised bird pollinators are present and populations with no pollinators (Fig. 3). The difference between mean S-A of plants in their native range (2.4 ± 0.9 mm, n = 7 populations) versus those from introduced populations where there are no pollinators (1.5 ± 0.3 mm, n = 6 populations) is small in absolute terms (only 0.9 mm on

TABLE 3. Nectar production in flowers from native and introduced population of *Nicotiana glauca*. All means are ± SD. Status: N = Native, I = Introduced. Unless otherwise cited, see text for details of authors responsible for data collection. Sample sizes, duration of bagging and times of collection vary considerably; details available from first author on request.

	Status	Mean nectar volume (μl)	Mean nectar concentration (%)	Notes
Argentina	N	20.0 ± 8.1	25.2 ± 3.7	Galetto & Bernardello (1993a)
California	I	25.4 ± 16.8	25.1 ± 6.0	
Mexico	I	2.2 ± 5.8	36.0 ± 1.7	Hernández (1981)
Peru	I	12.7 ± 12.1	20.2 ± 5.8	
South Africa	I	15.5 ± 14.4	26.9 ± 4.0	at 08h30
		2.8 ± 4.7	31.8 ± 6.4	at 14h30
Tenerife	I	5.7 ± 4.7	26.8 ± 7.4	
Gran Canaria	I	4.9 ± 2.8	34.4 ± 8.4	at 09h45
		9.0 ± 8.9	44.0 ± 14.6	at 12h00
		1.5 ± 2.2	27.1 ± 27.6	at 15h45
Greece	I	23.5 ± 8.6	-	
Israel	I	5.7 ± 3.4	20.4 ± 1.0	at 08h00
		9.8 ± 3.8	19.9 ± 3.7	at 14h00

average). But in proportional terms this represents a decrease in stigma-anther distance of over one third from plants in the native ancestral range to the introduced invasive populations.

The small S-A distances of invasive compared to native populations may play a role in the ability of invasive populations to produce greater proportional fruit set (Table 1) as on average introduced populations have statistically significantly (at $P = 0.06$) greater reproductive output than native populations [mean fruit set: Native = 42.4 ± 13.1% (n = 7 populations); Introduced = 57.3 ± 16.9% (n = 12 populations); independent samples t-test: t = -2.0, df = 17, $P = 0.06$]. However, there is some geographical variation to this pattern; for example, self pollination in Israel is rare and occurs in only 6% of bagged *Nicotiana glauca* flowers (Tadmor-Melamed 2004), bagged flowers in a Mexican population studied by Hernández (1981) did not set fruit, whilst within California populations, mean fruit set of bagged flowers varied from 6 to 29% (Schueller 2004).

FLOWER VISITORS TO *N. GLAUCA*

In its native range in South America, *N. glauca* is pollinated by several species of hummingbirds and nectar robbed by *Xylocopa* carpenter bees (Table 4). In addition, the hummingbird *Chlorostilbon aureoventris* behaved as a secondary nectar robber in a population from northern Argentina and as a legitimate pollinator in others (Table 4). None of these hummingbirds, nor the *Xylocopa*, are *Nicotiana* specialists: all visit the flowers of other plants for nectar.

Outside of its native range, two distinct patterns emerge, depending upon whether or not the populations establish within the range of specialist flower visiting birds. In Argentina, Peru, the south western USA and Mexico, hummingbirds once again act as pollinators, with bees and flies also making occasional legitimate visits to flowers (Table 4). In addition, the flowers are nectar robbed by native bees, honeybees (Fig. 2), hoverflies (Syrphidae) and flower piercers of the genus *Diglossa*. In Israel, sunbirds (*Nectarinia osea*) are likely to be the main pollinators: 60% of their visits were legitimate, with nectar being accessed from the front of the flower; in the other 40% of visits the birds pierced the corolla base, and thus acted as nectar robbers. The Hummingbird hawkmoth (*Macroglossum stellatarum*) was also observed as an occasional legitimate visitor in Israel, with *Xylocopa* and also several species of ants acting as nectar robbers (Cohen et al., pers. obs.; see Table 4). In South Africa, three species of sunbirds, the Malachite sunbird (*Nectarinia famosa*), the Dusky sunbird (*Cinnyris fuscus*) and the Southern double-collared sunbird (*C. chalybea*) have been confirmed as pollinators (Geerts & Pauw 2009). The former species is the most effective and frequent pollinator, while the latter two species also rob during 7% and 61% of visits respectively. There are also records of flower visitation of honeyeaters for Australia but their role in pollination is not clear (Table 4).

In the northern Mediterranean and Tenerife, where there are no specialist flower visiting birds, flower visitors have never been observed in any populations, despite extensive observations (Table 4). For example, five contrasting populations in the arid south west of Tenerife were surveyed during peak *N. glauca* flowering in April 2006. These populations had different abundances and densities of plants, and ranged from suburban post-demolition sites to rural, goat-grazed habitats. Despite the presence in all of these habitats of potential flower visiting

FIGURE 2: (A) Flowers of *Nicotiana glauca* being nectar robbed by introduced honeybees (*Apis mellifera*) in a non-native population in the high Andes of Peru. Photograph by Stella Watts. (B) A Canary Island blue tit (*Parus teneriffae*) observed robbing flowers for nectar on Gran Canaria in 2012. Photograph by Jane Stout

insects (including bees and butterflies) and birds (principally the Canary chiffchaff *Phylloscopus canariensis* a generalist bird that opportunistically visits other flowers for nectar – see below) visits to flowers were never observed. In addition we checked over 1600 flowers (on average 330 per population) and found no evidence of nectar robbing.

Finally, one population was surveyed for nocturnal visitors, particularly large night-flying moths, on three evenings. This population was chosen because of the presence of larvae of the Barbary spurge hawkmoth (*Hyles tithymali tithymali*) feeding on *Euphorbia broussonetii*, indicating that these potential pollinators were present in that community. As well as checking flowers with flashlights, we added fluorescent dye powder (see Kearns &

Inouye 1993; Dafni et al. 2005) to 10 flowers on each of 4 trees on one evening. On the two subsequent evenings we checked for dye transfer to nearby flowers but none was observed. These results confirmed previous observations by Ollerton et al. in 2003, 2004 and 2005 that *N. glauca* flowers on Tenerife are rarely, if ever, visited by nectar-feeding animals. These results strongly suggest that Tenerife populations are wholly selfing.

In contrast, observations in 2012 in Gran Canaria revealed a range of insects and passerine birds visiting the flowers of *N. glauca* at one site (Table 4, Fig. 2). This is surprising given the observations made on the nearby island of Tenerife. However, it is worth noting that the winter of 2011-12 was one of the driest experienced by the Canary Islands since records began. Some plant species failed to flower and others flowered much later than usual, resulting in flowers being visited by insects that are not normally seen on them (Ollerton, unpublished data). It is possible that the Gran Canaria observations were of visitors utilising less favoured flowers as a nectar/water source.

DISCUSSION

In its native range, *Nicotiana glauca* usually forms dense stands only in disturbed sites with recent soil exposure, for example dry river beds and road sides (Nattero & Cocucci 2007). Outside of its native range *N. glauca* is clearly a successful invasive weed of disturbed areas where it forms dense, monodominant colonies because of the high rate of fruit and seed set, the viability of seeds and the frequent recruitment of seedlings into the population.

The pollination system of *Nicotiana glauca* in its native range can be best described using the terminology of Fenster et al. (2004) and Ollerton et al. (2007) as functionally specialized for hummingbird pollination, but ecologically generalized in that a range of hummingbird species can act as pollinators (Nattero et al. 2010). In this respect it seems to be an unlikely candidate as an ecologically invasive species, in relation to its ability to co-opt other non-hummingbird pollinators (Richardson et al. 2000; Olesen et al. 2002). In parts of Argentina, Peru, Mexico, California, South Africa and Israel, where *N. glauca* has been introduced for at least 100 years, the species is clearly well integrated into the local pollination web via its interactions with specialist flower feeding birds including hummingbirds (Hernández 1981; Schueller 2004, 2007) and sunbirds (Tadmor-Melamed et al. 2004; Geerts & Pauw 2009).

The successful pollination of *N. glauca* by sunbirds in the Old World comes as a surprise. Like many other hummingbird pollinated New World flowers, the flowers of *N. glauca* are oriented towards open space, an adaptation for pollination by birds that hover while feeding. According to conventional wisdom, Old World birds perch while feeding, so Old World flowers need to be oriented towards a perch in order to receive

TABLE 4: Flower visitors to *Nicotiana glauca* within its native range and in areas where it is introduced. "Legitimate flower visitors" are those which enter from the front of the flower and are the most likely pollinators; flower robbing visitors pierce holes at the base of the corolla to access the nectar, or make secondary use of previously excavated holes.

Range and locality	Legitimate flower visitors	Flower robbing visitors
Native – within the range of specialist flower visiting birds		
Argentina and Bolivia (6 sites - Nattero & Cocucci 2007, Nattero et al. 2010)	Hummingbirds (4 spp.)	*Xylocopa ordinaria* *Chlorostilbon aureoventris*
Introduced – within the range of specialist flower visiting birds		
Peru (3 sites within the Sacred Valley during February, June and August 2002 - SW, unpublished data)	Hummingbirds (5 spp.)	*Bombus* sp., *Xylocopa* sp. and other native bees, *Apis mellifera*, Syrphidae, *Diglossopis cyanea*
USA California (4 sites - Schueller 2004) California, Sonara and Sinaloa (Stiles 1973, 1976)	Hummingbirds (3 spp.) Bees and Diptera (very infrequently) Hummingbirds	House finches and white-crowned sparrows observed pecking at flowers and usually destroying them or ripping corolla; also occasional holes at base of corolla made by unknown bee and frequently find ants in flowers that consume a lot of the nectar, but do not act as pollinators.
Israel (Tadmor-Melamed 2004, Tadmor-Melamed et al. 2004, Cohen 2007)	Palestine Sunbirds (60% of 274 visits were legitimate Hummingbird hawkmoth (*Macroglossum stellatarum*)	Palestine Sunbirds (40% of 274 visits were nectar robbery) *Xylocopa pubescens* *Apis mellifera* (secondary nectar robber) Seven ant species (Formicidae)
South Africa (Skead 1967, Knuth 1898-1905; Marloth 1901; Geerts & Pauw 2009; SG, unpublished data)	Malachite sunbirds (*Nectarinia famosa*) Dusky sunbirds (*Cinnyris fuscus*) Southern double-collared sunbirds (*Cinnyris chalybea*)	*C. chalybea* and *C. fuscus* Weavers (*Ploceus capensis* and *P. velatus*) destroy flowers to access nectar
Australia (Hobbs 1961)		White-fronted honeyeater (*Phylidonyris albifrons*)*
México (I site Hernández 1981)	Hummingbirds (4 spp.)	*Diglossa baritula* and *Xylocopa* sp.
Argentina (3 sites - Nattero & Cocucci 2007)	Hummingbirds (3 spp.)	*Xylocopa ordinaria*
Introduced – outside of the range of specialist flower visiting birds		
Tenerife (5 sites in the arid south west - Ollerton et al. unpublished data)	None observed	None observed > 1000 flowers checked
Gran Canaria (one site in the Arguineguin Valley – Stout unpublished data)	Solitary bees (2 spp?) Neuroptera	Common whitethroat (*Sylvia communis*) Canary Islands blue tit (*Parus teneriffae*) Both bird species were observed removing flowers from plants, then pecking a hole in the base of the corolla to drink nectar before dropping flowers to the ground.
Crete, Agios Nikolaos (Ollerton unpublished data)	None observed	None observed on c. 200 flowers
Greece (Schueller unpublished data)	None observed	None observed

*Hobbs (1961) does not indicate if visits were legitimate. The short, broad bill of this species suggests that it acts as a robber (B. Lamont pers. com.).

pollination (Westerkamp 1990). Unexpectedly, Old World sunbirds were found to adapt their behaviour and hover for extended periods of time while feeding from the hummingbird adapted flowers of *N. glauca* (Geerts & Pauw 2009) though it is possible that feeding whilst hovering is under documented (see Janecek et al. 2011). It remains to be determined whether Australian honeyeaters are also able to adopt this novel behaviour and act as pollinators of *N. glauca*.

In the northern Mediterranean and the Canary Islands, in contrast, *N. glauca* has not become integrated into the local flower visitation web, either via pollinators or nectar robbers. There are a restricted number of native Canarian and European taxa which could potentially pollinate *N. glauca*, for example long tongued bees such as *Xylocopa* and *Bombus* (which could also act as nectar robbers), and the larger Lepidoptera, including various hawkmoths (Sphingidae). Non-flower specialist passerine birds, particularly chiffchaffs (*Phylloscopus canariensis* and *P. collybita*), are known to pollinate a number of native Canarian plants (Vogel et al. 1984; Valido et al. 2004, Ollerton et al. 2008), at least one continental European species (Ortega-Olivencia et al. 2005) and are opportunistic feeders at the flowers of other non-native plants (Clement 1995; Ollerton pers. obs.). Their beak and tongue lengths are too short for them to legitimately access the nectar of *N. glauca*; however they are known to nectar rob other plants in the Canary Islands, for example *Aloe* spp. (Bramwell 1982). Extensive observation of populations of *N. glauca* on Tenerife and in Greece revealed no instances of nectar robbery, however. This is despite the presence of chiffchaffs in all populations on Tenerife, some of which were observed to perch in the larger *N. glauca* trees.

The nectar available in flowers of *N. glauca* is a significant energy and water resource for animals in semi-arid habitats. Multiplying the nectar values obtained in Tenerife (see above) by the mean number of open flowers per inflorescence and the mean number of inflorescences per stem, suggests that on Tenerife, each stem on average maintains a standing crop of $374.8 \pm 820.7\mu l$ of relatively sugar-rich nectar. Using the data for flowering stem densities (above), the nectar resources available to animals that can exploit these flowers would be of the order of $277.6\mu l \ m^{-2}$ in low density areas to $832.9\mu l \ m^{-2}$ in high density areas. We do not know the rate of replenishment of nectar in these flowers in Tenerife (though for an Argentinean population it was 0.2 ± 0.2 ml/h - Galetto & Bernardello 1993a); nonetheless this standing crop represents a large potential resource of energy and water to any flower visiting animals within the semi-arid zone of Tenerife. It far exceeds the standing crops of most native species, with the exception of some of the specialised passerine-pollinated endemics (Ollerton et al. 2008) which are mainly restricted to the wetter laurel forest communities of the island. Why this resource is not utilised, resulting in the subsequent integration of the species into the local flower visitation web, is unclear. It is possible that the alkaloid content of the nectar of *N. glauca* deters animals that might otherwise exploit the nectar (Tadmor-Melamed et al. 2004) which would suggest that pollinators and nectar robbers within the native range of the plant, as well as in California, Israel and South Africa, have digestive strategies adapted to cope with these compounds. The relatively high fraction of sucrose, which can only be digested by specialized nectarivores, i.e. hummingbirds and sunbirds, might additionally deter generalist passerines such as chiffchaffs. These areas deserve further research.

Despite the absence of pollinators in some parts of its modern range, *N. glauca* is a plant which is reproductively successful to the point of being a problematical invasive. High fruit and seed set, and relatively small S-A distances, suggest that these populations are largely selfing; apomixis is unlikely as emasculated and bagged flowers of plants in California never resulted in fruit or seed set (Schueller 2002). The difference in average S-A distances in native versus non-native habitats implies that populations in the native range are less frequently selfing. In native populations in Bolivia, where S-A distance is greatest, and presumably with a long history of interaction with the giant hummingbird (*Patagona gigas*), fruit set is relatively low (Nattero et al. 2010; Loayza et al. 1999). The populations with a small S-A distance therefore have pre-adapted the species to be a successful invader and fits with the ecology of the plant as a weedy colonising small tree of disturbed soil in South America. Nevertheless, despite small S-A distances and high levels of selfing, plants in South Africa that receive visits from sunbirds set significantly more fruit and seeds than pollinator-excluded controls (Geerts and Pauw 2009). The trend of decreasing S-A distances from native populations, to invasive populations that are within the range of specialised flower visiting birds, to those where no birds are present (Fig. 3) is precisely what we would expect if initial founder events by largely self pollinating, isolated individuals are important prior to the establishment of larger populations that then subsequently attract significant numbers of native bird pollinators (if available) or remain as selfing populations if no suitable pollinators exist in the locality.

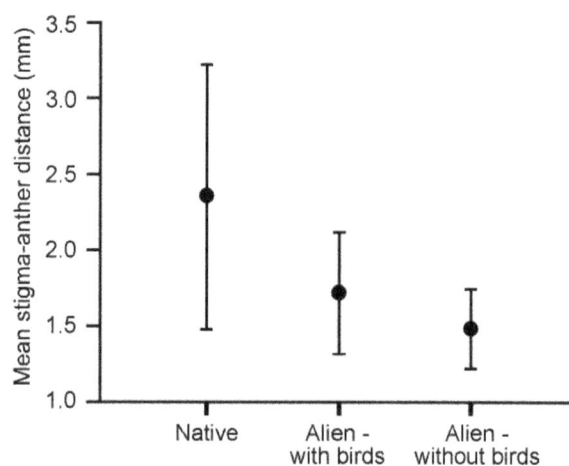

FIGURE 3: Mean ($\pm$ 95% confidence interval) stigma-anther distance for native populations of *Nicotiana glauca* ("Native" – n = 7 populations) compared to non-native populations within ("Alien-with birds" – n = 11 populations) and outside ("Alien-without birds" – n = 6 populations) of the range of specialised flower visiting birds (hummingbirds and sunbirds). One-way ANOVA: $F_{2,20} = 4.4$, $P = 0.027$. Only the contrasts between Native vs. Alien-with birds (LSD post-hoc test: $P = 0.03$) and Native vs. Alien-without birds (LSD post-hoc test: $P = 0.01$) are significantly different, but there is an apparent trend and the latter has only a small sample size.

Invasive plants with a high rate of nectar production almost invariably have a high rate of pollinator visits to flowers, for example *Buddleja davidii* and *Impatiens glandulifera* in Europe and *Lantana camara* and *Melaleuca quinquenervia* in subtropical North America (Chittka & Schürkens 2001; Koptur 2006). The assumption is that many of these species are likely to be ecological and/or functional generalists in their native habitats (reviewed by Corbet 2006 and Traveset & Richardson 2006; see also Harmon-Threatt 2009; Rodger et al. 2010). For instance, in Europe *Impatiens glandulifera* is a functional specialist but an ecological generalist (it is pollinated by a range of bumblebees *Bombus* spp. – Chittka & Schürkens 2001; Lopezaraiza-Mikel et al. 2007; Nienhuis et al. 2009; Nienhuis & Stout 2009). In its native Himalayan habitats it is also *Bombus* pollinated (Saini & Ghattor 2007) which suggests that this plant was pre-adapted to be a successful invader in terms of its ability to integrate into existing pollination webs containing *Bombus* spp. or functionally equivalent medium to large sized bees.

Invasive species may therefore re-establish their pollination niche only if they can appropriately interact with suitable local pollinators. Intuitively we might consider this to be less likely if the species is an ecological, functional or phenotypic specialist, but the data here are contradictory: functionally and phenotypically specialised ("euphilous") introduced plants were shown by Corbet (2006) to be the most successful group of plants (as measured by range increase) within the British and Irish flora. This may be due to the tendency of gardeners to introduce plants with large, showy flowers. *Nicotiana glauca* was introduced to Tenerife as a garden ornamental in the early 19th Century (Kunkel 1976) and so there would have been ample time for the species to establish relationships with pollinators or nectar robbers if suitable species were available. However in that time *N. glauca* has largely forsaken outcrossing for a successful selfing reproductive strategy. Island ecosystems are especially vulnerable to plant invasions (Simberloff 1995; Olesen et al. 2002) and *N. glauca* is only one of a large number of introduced plant species which may be negatively impacting on the flora and fauna of the Canary Islands and two of the California Channel Islands studied by Schueller (2002). *N. glauca* has some degree of frost tolerance and, whilst adult plants may be killed by very low temperatures, seeds survive to germinate the next year (Ollerton, pers. obs.). Therefore the increasingly arid environments predicted for some regions by future climate change may result in the species spreading even further north in Europe and North America.

In order to test our hypothesis that shorter S-A distances have evolved in invasive populations of *N. glauca* that lack pollinators, further research is required. In particular we would like to know why it is that seeds produced from self pollination show such high viability with no apparent cost from inbreeding depression. In addition, we would like to understand the role of other mechanisms that could prevent or facilitate selfing (such as reduced incompatibility of self-pollen, and synchronization of pollen viability and stigma receptivity) in native and invasive populations with and without pollinators. Finally we require data on the extent to which native populations of *Nicotiana glauca* are pollen limited, the degree of self compatibility amongst individuals and the natural rate of self pollination.

In conclusion, we can state that *Nicotiana glauca* is a successful invasive species outside of its native range, despite its functionally specialised pollination system. In areas where suitable bird pollinators are available, for example hummingbirds in California and sunbirds in South Africa and Israel, *N. glauca* becomes integrated into the local pollination web and sets seed by both out-crossing and selfing. In regions where there are no specialised flower visiting birds, such as the Canary Islands and Greece, abundant seed set is maintained by selfing, and the considerable nectar resources are hardly utilised by native flower visitors, either legitimately or by nectar robbing. In the case of *N. glauca*, invasion success is therefore not predictable from its interactions with pollinators in its native range.

ACKOWLEDGEMENTS

The authors thank the various funding agencies who have supported different aspects of this work, as well as two anonymous referees whose comments greatly improved this paper. JO would especially like to acknowledge the assistance of students on the University of Northampton Tenerife Field Course who helped with data collection over the years and Rob Raguso for discussion. JCS thanks Anne Doyle, Kathleen Fitzgibbon, Clare Guy and Brian Kennedy from Trinity College Dublin for help with the Gran Canaria data collection.

REFERENCES

Aigner PA (2004) Ecological and genetic effects on demographic processes: pollination, clonality and seed production in *Dithyrea maritima*. Biological Conservation 116: 27-34.

Aizen MA, Morales CL, Morales JM (2008) Invasive mutualists erode native pollination webs. PLOS Biology 6: 396-403.

Alarcón R, Waser NM, Ollerton J (2008) Year-to-year variation in the topology of a plant-pollinator interaction network. Oikos 117: 1796-1807.

Armbruster, WS (1988) Multilevel comparative analysis of the morphology, function, and evolution of *Dalechampia* blossoms. Ecology 69:1746–1761.

Baker, HG (1974) The evolution of weeds. Annual Review of Ecology and Systematics 5: 1-24.

Bartomeus I, Vila M, Santamaria L (2008) Contrasting effects of invasive plants in plant-pollinator networks. Oecologia 155: 761-770.

Bjerknes A-L, Totland Ø, Hegland SJ, Nielsen A (2007) Do alien plant invasions really affect pollination success in native plant species? Biological Conservation 138: 1-12.

Bogdanović S, Mitić B, Ruščić M, Dolina K (2006) *Nicotiana glauca* Graham (Solanaceae), a new invasive plant in Croatia. Acta Botanica Croatica 65:203-209.

Bornmuller J (1898) Ein Beitrag zur Kenntnis der Flora von Syrien und Palastina. Verhandlungen der Zoologisch-Botanischen Gesellschaft in Wien 48: 544–653.

Bramwell D (1982) Conserving biodiversity in the Canary Islands. Annals of the Missouri Botanical Gardens 77: 28-37.

Cariveau DP, Norton AP (2009) Spatially contingent interactions between an exotic and native plant mediated through flower visitors. Oikos 118:107-114.

Chittka L, Schürkens S (2001) Successful invasion of a floral market. Nature 411:653-653.

Clement P (1995) The chiffchaff. Hamlyn, London.

Corbet SA (2006) A typology of pollination systems: implications for crop management and the conservation of wild plants. In: Waser NM, Ollerton J (eds.) Plant-pollinator interactions: from specialization to generalization, University of Chicago Press, Chicago.

Cohen N (2007) The effects of alkaloids in the floral nectar on the relationships between *Nicotiana glauca* and ants. M.Sc. Thesis. University of Haifa, Israel.

Dafni A, Kevan PG, Husband BC (2005) Practical pollination biology. Enviroquest, Cambridge, Ontario.

Dietzsch AC, Stanley DA, Stout JC (2011) Relative abundance of an invasive alien plant affects native pollination processes. Oecologia 167:469-479.

Fenster CB, Armbruster WS, Wilson P, Dudash MR, Thomson JT (2004) Pollination syndromes and floral specialization. Annual Review of Ecology, Evolution and Systematics 35:375-403.

Fishbein M, Venable DL (1996) Diversity and temporal change in the effective pollinators of *Asclepias tuberosa*. Ecology 77: 1061-1073.

Florentine SK, Westbrooke ME (2005) Invasion of the noxious weed *Nicotiana glauca* R. Graham after an episodic flooding event in the arid zone of Australia. Journal of Arid Environments 60: 531-545.

Florentine SK, Westbrooke ME, Gosney K, Ambrose G, O'Keefe M (2006) The arid land invasive weed *Nicotiana glauca* R. Graham (Solanaceae): Population and soil seed bank dynamics, seed germination patterns and seedling response to flood and drought. Journal of Arid Environments 66: 218-230.

Forster PI (1994) Diurnal insects associated with the flowers of *Gomphocarpus physocarpus* E Mey (Asclepiadaceae), an introduced weed in Australia. Biotropica 26: 214-217.

Galetto L, Bernardello L (1993a) Nectar sugar composition in angiosperms from Chaco and Patagonia (Argentina): an animal visitor's matter? Plant Systematics and Evolution 238: 69–86.

Galetto L, Bernardello L (1993b) Nectar secretion pattern and removal effects in 3 species of Solanaceae Canadian Journal of Botany 71: 1394-1398.

Geerts S, Pauw A (2009) African sunbirds hover to pollinate an invasive, hummingbird-pollinated plant. Oikos 118: 573-579.

Harmon-Threatt AN, Burns JH, Shemyakina LA, Knight TM (2009) Breeding system and pollination ecology of introduced plants compared to their native relatives. American Journal of Botany 96: 1544–1550.

Henderson L (1991) Invasive alien woody plants of the Northern Cape. Bothalia 21: 177-189.

Hernández HM (1981) Sobre la biología reproductiva de *Nicotiana glauca* Graham: una maleza de distribución cosmopolita. Boletin de la Sociedad de Botánica de México 41: 47-73.

Herrera CM (1988) Variation in mutualisms - the spatio-temporal mosaic of a pollinator assemblage. Biological Journal of the Linnean Society 35:95-125.

Hobbs JN (1961) The birds of Southwest New South Wales. Emu 61: 21-55.

Janecek S, Patacova E, Bartos M, Padysakova E, Spitzer L, Tropek R (2011) Hovering sunbirds in the Old World: occasional behaviour or evolutionary trend? Oikos 120: 178-183.

Kearns CA, Inouye DW (1993) Techniques for pollination biologists. University Press of Colorado, Niwot.

Knuth P (1898-1905) Handbuch der Blütenbiologie. Engelmann, Leipzig.

Koptur SA (2006) The conservation of specialized and generalized pollination systems in subtropical ecosystems: a case study. In: Waser NM, Ollerton J (eds.) Plant-pollinator interactions: from specialization to generalization, University of Chicago Press, Chicago.

Kunkel G (1976) Notes on the introduced elements in the Canary Islands' flora. In: Kunkel G (ed) Biogeography and ecology in the Canary Islands. Dr W. Junk, The Hague, pp249-266.

Lamborn E, Ollerton J (2000) Experimental assessment of the functional morphology of inflorescences of *Daucus carota* (Apiaceae): testing the "fly catcher effect". Functional Ecology 14:445-454.

Loayza A, Rios R, Castillo C (1999) Actividad alimenticia de *Patagona gigas* en arbustos de *Nicotiana glauca* (Solanaceae). Ecología en Bolivia 33:75-79.

Lopezaraiza-Mikel ME, Hayes RB, Whalley MR, Memmott J (2007) The impact of an alien plant on a native plant-pollinator network: an experimental approach. Ecology Letters 10: 539-550.

Marloth R (1901) Die Ornithophilie der Flora Süd-Afrikas. Berichte der Deutschen Botanischen Gesellschaft 19: 176-179.

Memmott J, Waser NM (2002) Integration of alien plants into a native flower-pollinator visitation web. Proceedings of the Royal Society of London B 269: 2395-2399.

Moraes AD, de Melo E, Agra MD, Franca F (2009) The Solanaceae family from the inselbergs of semi-arid Bahia, Brazil. Iheringia Serie Botanica 64: 109-122.

Moragues E, Traveset A (2005) Effect of *Carpobrotus* spp. on the pollination success of native plant species of the Balearic Islands. Biological Conservation 122: 611-619.

Morales CL, Traveset A (2009) A meta-analysis of impacts of alien vs. native plants on pollinator visitation and reproductive success of co-flowering native plants. Ecology Letters 12: 716-728.

Nattero J, Cocucci AA (2007) Geographic variation in floral traits of the tree tobacco in relation to its hummingbird pollinator fauna. Biological Journal Linnean Society 90: 657-667.

Nattero J, Sérsic AN, Cocucci AA (2010) Patterns of contemporary phenotypic selection and flower integration in the hummingbird-pollinated *Nicotiana glauca* between populations with different flower-pollinator combinations. Oikos: 119: 852-863.

Nienhuis CM, Dietzsch AC, Stout JC (2009) The impacts of an invasive alien plant and its removal on native bees. Apidologie 40: 450-463.

Nienhuis CM, Stout JC (2009) Effectiveness of native bumblebees as pollinators of the alien invasive plant *Impatiens glandulifera* (Balsaminaceae) in Ireland. Journal of Pollination Ecology 1: 1-11.

Olesen JM, Eskildsen LI, Venkatasamy S (2002) Invasion of pollination networks on oceanic islands: importance of invader complexes and endemic super generalists. Diversity and Distributions 8: 181-192.

Ollerton J, Killick A, Lamborn E, Watts S, Whiston M (2007) Multiple meanings and modes: on the many ways to be a generalist flower. Taxon 56: 717-728.

Ollerton J, Cranmer L, Stelzer R, Sullivan S, Chittka L (2008) Bird pollination of Canary Island endemic plants. Naturwissenschaften 96: 221-232.

Ortega-Olivencia A, Rodríguez-Riaño T, Valtueña FJ, López J, Devesa JA (2005) First confirmation of a native bird-pollinated plant in Europe. Oikos 110:578-590.

Padrón B, Traveset A, Biedenweg T, Díaz D, Nogales M, Olesen JM. (2009) Impact of alien plant invaders on pollination networks in two archipelagos. PLoS One 4: e6275.

Richardson DM, Allsopp N, D'Antonio CM, Milton SJ, Rejmanek M (2000) Plant invasions - the role of mutualisms. Biological Reviews 75: 65-93.

Rodger JG, van Kleunen M, Johnson SD (2010) Does specialized pollination impede plant invasions? International Journal of Plant Sciences 171: 382-391.

Saini MS, Ghattor HS (2007) Taxonomy and food plants of some bumble bee species of Lahaul and Spiti valley of Himachal Pradesh. Zoos' Print Journal 22: 2648-2657.

Schürkens S, Chittka L (2001) The significance of the invasive crucifer species Bunias orientalis (Brassicaceae) as a nectar source for central European insects. Entomologia Generalis 25: 115-120.

Schueller SK (2002) Hummingbird pollination and floral evolution of introduced Nicotiana glauca and native Epilobium canum: California island-mainland comparisons. Ph.D. dissertation, University of Michigan, Ann Arbor, Michigan, USA.

Schueller SK (2004) Self-pollination in island and mainland populations of the introduced hummingbird-pollinated plant, Nicotiana glauca (Solanaceae). American Journal of Botany 91: 672-681.

Schueller SK (2007) Island-mainland difference in Nicotiana glauca (Solanaceae) corolla length: a product of pollinator-mediated selection? Evolutionary Ecology 21: 81-98.

Simberloff D (1995) Why do introduced species appear to devastate islands more than mainland areas? Pacific Science 49: 87-97.

Skead CJ (1967) The sunbirds of southern Africa. Balkema, Cape Town.

Stiles FG (1973) Food supply and the annual cycle of the Anna Hummingbird. University of California Publications in Zoology 97: 1-116.

Stiles FG (1976) On taste preferences, color preferences, and flower choice in hummingbirds. The Condor 78: 10-26.

Stout JC (2007) Reproductive biology of the invasive exotic shrub, Rhododendron ponticum L. (Ericaceae). Botanical Journal of the Linnean Society 155: 373-381.

Stout JC, Parnell JAN, Arroyo J, Crowe TP (2006) Pollination ecology and seed production of Rhododendron ponticum in native and exotic habitats. Biodiversity and Conservation 15:755-777.

Stout JC, Morales CL (2009) Ecological impacts of invasive alien species on bees. Apidologie 40: 388-409.

Tadmor-Melamed H (2004) The ecological role of secondary metabolites in floral nectar governing the interaction between Nicotiana glauca and Nectarinia osea. M.Sc. Thesis. The Hebrew University of Jerusalem, Israel.

Tadmor-Melamed H, Markman S, Arieli A, Distl M, Wink M, Izhaki I (2004) Limited ability of Palestine Sunbirds Nectarinia osea to cope with pyridine alkaloids in nectar of Tree Tobacco Nicotiana glauca. Functional Ecology 18:844-850.

Theoharides KA, Dukes JS (2007) Plant invasion across space and time: factors affecting nonindiginous species success during four stages of invasion. New Phytologist 176: 256-273.

Thompson K, Davis MA (2011) Why research on traits of invasive plants tells us very little. Trends in Ecology and Evolution 26: 155–156.

Tillberg CV, Holway DA, LeBrun EG, Suarez AV (2007) Trophic ecology of invasive Argentine ants in their native and introduced ranges. Proceedings of the National Academy of Sciences USA 104: 20856–20861.

Valido A, Dupont YL, Olesen JM (2004) Bird-flower interactions in the Macaronesian islands. Journal of Biogeography 31: 1945-1953.

Vilá M, Bartomeus I, Dietzsch AC, Petanidou T, Steffan-Dewenter I, Stout JC, Tscheulin T (2009) Invasive plant integration into native plant–pollinator networks across Europe. Proceedings of the Royal Society series B 276: 3887-3893.

Vogel S, Westerkamp C, Thiel B, Gessner K (1984) Ornithophilie auf den Canarischen Inseln. Plant Systematics and Evolution 146: 225-248.

Westerkamp C (1990) Bird-flowers - hovering versus perching exploitation. Botanica Acta 103: 366-371.

17

CITIZEN SCIENTISTS DOCUMENT GEOGRAPHIC PATTERNS IN POLLINATOR COMMUNITIES

Alison J. Parker* and James D. Thomson

University of Toronto, Department of Ecology and Evolutionary Biology, 25 Harbord Street, Toronto, Ontario M5S 3G5 Canada

Abstract—It is widely recognized that plants are visited by a diverse community of pollinators that are highly variable in space and time, but biologists are often unable to investigate the pollinator climate across species' entire ranges. To study the community of pollinators visiting the spring ephemerals *Claytonia virginica* and *Claytonia caroliniana*, we assembled a team of citizen scientists to monitor pollinator visitation to plants throughout the species' ranges. Citizen scientists documented some interesting differences in pollinator communities; specifically, that western *C. virginica* and *C. caroliniana* populations are visited more often by the pollen specialist bee *Andrena erigeniae* and southern populations are visited more often by the bombyliid fly *Bombylius major*. Differences in pollinator communities throughout the plants' range will have implications for the ecology and evolution of a plant species, including that differences may affect the male fitness of individual plants or the reproductive success of plant populations, or both.

Keywords: citizen science, plant-pollinator interactions

INTRODUCTION

A rich history of research has explored the role of a pollinator species in determining the reproductive success of a plant, selecting for plant traits, and in some cases influencing reproductive isolation (van der Niet et al. 2014). We know that most plants are visited by a diverse community of pollinators (Waser et al. 1996), and that the diverse community of pollinators can be highly variable geographically (Herrera et al. 2006). Yet the vast majority of studies of plant-pollinator interactions are conducted in one geographic location; Herrera et al. (2006) calculates that 88.4% of plant-pollinator studies look at only one site.

The diversity and abundance of pollinators visiting a plant population in any given location and time form the "pollinator climate" (Grant & Grant 1965). A number of studies have documented variation in the pollinator climate in different plant populations within the same plant species. Most studies focus on a small number of sites (e.g. Miller 1981; Robertson & Wyatt 1990; Arroyo & Dafni 1995; Johnson & Steiner 1997; Price et al. 2005), and many are confined to a relatively small geographic area (e.g. Miller 1981; Robertson & Wyatt 1990; Arroyo & Dafni 1995; Johnson & Steiner 1997; Price et al. 2005; Gomez et al. 2008). These studies have contributed to the understanding that pollinator communities are variable. However, in the majority of studies there is no obvious pattern or process that explains the documented variation in pollinator communities, so the conclusions to these studies are limited to the specific populations studied.

Looking for and documenting large-scale patterns in pollinator communities requires a great deal of observational data. Studies are often limited to just one or a few plant populations (Herrera et al. 2006). Some limitations are specific to plant-pollinator studies; often the flowering season of a study species will limit the time available for traveling throughout the plant range, and pollinator identification can be very difficult for novice research assistants (Lye et al. 2011; Kremen et al. 2011). Recently, more and more biologists have begun to employ the efforts of amateur naturalists and volunteers in their research efforts (Dickinson et al. 2010; Silvertown 2009). Citizen scientists can benefit research in ecology and evolution in many ways, including expanding the scope of data collected over space and time, filling gaps in natural history knowledge, and increasing access to otherwise inaccessible spaces, such as private land (e.g., Lye et al. 2011 recruited volunteers to document bumblebee nest sites in their private gardens). Citizen science also provides an opportunity for scientists to connect with the public, hear valuable observations from residents of an area, and build public support for science and conservation (Dickinson et al. 2010; Cooper et al. 2007; Toomey & Domroese 2013; Lewandowski & Oberhauser 2017; McKinley et al. 2017; Ballard et al. 2017). In pollination research especially, there is great potential for harnessing the enthusiasm of amateur naturalists to support large-scale data collection and fill in gaps in our understanding of the basic natural history and biogeography of plants and pollinators. Citizen scientists are a great resource for pollinator monitoring; often, amateur naturalists have practice in plant and insect identification, are eager to spend time outdoors to contribute to monitoring, and are already located throughout the range of the plant and pollinator species of interest (Kremen et al. 2011). Citizen science for the biology and conservation of *Danaus plexippus* (monarch) butterflies is a model for demonstrating the

*Corresponding author: alison.parker@alum.utoronto.ca

potential impact of citizen science (Ries & Oberhauser 2015) and pollination citizen science projects such as the Great Sunflower Project and Bumble Bee Watch have made strong contributions to research in pollination and pollinator biology (Acorn 2017, Lye et al. 2011, Birkin et al. 2015, Roy et al. 2016, Deguines et al. 2012). Except for these efforts, citizen science is relatively underused in pollination research and has potential to add greatly to our understanding of pollination and pollinator biology. Kremen et al. (2011) compared data sets of pollination observations collected by citizen scientists to data sets collected by experts, and found that although the citizen scientists missed some taxonomic diversity, the qualitative results were comparable. Citizen science data can help identify how the pollinator climate varies in space and time, and help document large-scale geographic patterns in a plant species' pollinator climate (Dickinson et al. 2010).

To better understand patterns in pollinator climates, we recruited citizen scientists to do pollinator observations across the range of two plant species, *Claytonia virginica* and *Claytonia caroliniana*. Specifically, we ask: 1) Do pollinator climates vary along large-scale patterns like latitude, longitude, and altitude? 2) By conducting observations over an entire species' range, can we uncover patterns in variation in pollinator diversity and abundance? Previous observations caused us to predict that the pollinator climate of *C. virginica* and *C. caroliniana* would vary latitudinally, with higher fly visitation in Southern populations (Parker et al. 2017).

MATERIALS AND METHODS

Claytonia virginica and *Claytonia caroliniana* (Portulacaceae), collectively known as "spring beauty", are spring ephemeral wildflowers native to North American eastern woodlands, where they are visited by a variety of insects, among them the oligolectic solitary bee *Andrena erigeniae*, which collects pollen exclusively from these two species (Fig. 1). Female bees may eat some pollen, but most pollen is used to provision *A. erigeniae* larvae. These two species of *Claytonia* and *A. erigeniae* have overlapping geographic ranges and are phenologically matched (Davis & LaBerge 1975). A number of generalist species also visit *C. virginica* and *C. caroliniana*, collecting pollen or nectar from these plants and other sources. The generalist bee species are from many genera and include both pollen-foragers and social parasites that do not amass pollen provisions. The other most frequent visitor, the bee fly *Bombylius major*, is focused on nectar-collecting and mostly ignores pollen. *Bombylius major* is a parasite of solitary bee species, probably including *Claytonia*'s oligolege *A. erigeniae*. The distribution of these *Claytonia* species ranges from Georgia to Ontario, and from the East Coast west to Kansas and Nebraska. These species are protandrous; pollen and nectar are offered on the first day, in the male phase, while only nectar is produced in the succeeding female phase (Fig. 1).

To recruit volunteers, we advertised the project on established email listservs, including Native Plant Societies and Master Gardener lists. Hundreds of volunteers responded with interest and ultimately, 27 people submitted usable data. The instructions and learning materials for participating in the project were compiled onto a website. Before participating, we asked volunteers to visit the website, familiarize themselves with the project protocol, and study the identification of the bees and flies that they were likely to see in the field. Volunteers were also responsible for locating a patch of *C. virginica* or *C. caroliniana* in their local area; these patches could be in any habitat type. Volunteers were able to ask questions via email and also on the project website, where they could view our responses to questions as well as respond to one another.

Volunteers throughout the range of the plants located a patch of *C. virginica* or *C. caroliniana* in their area, recorded general information about the site and patch, and conducted observations a few times throughout March, April and May of 2011, 2012, and 2013. We asked volunteers to conduct observations three times throughout the season, but many volunteers were only able to do two, and some conducted many more than three. Each set of observation periods included six five-minute observation periods, each focusing on a defined number of focal flowers. During observations, volunteers recorded the identity of visiting insects and the number of visits that each insect made to male- and female-phase flowers. To facilitate identification, we organized the floral visitors into groups according to taxon, morphology, and behaviour. We provided volunteers with an information sheet with photographs and distinguishing characteristics of these pollinator groups, including size, colour, and body shape; we asked that they refer to this sheet during observations (Fig. 2). We encouraged volunteers to use these groups but also allowed identifications at any level or descriptions of the visitor (e.g., "unknown", "unknown bee", "small black bee with yellow stripes"). After conducting observations, volunteers submitted their data by mailing in their original data sheets, by entering and emailing data on a spreadsheet that we provided, or by entering and emailing data on a fillable PDF of the data sheets. Following submission of data by volunteers, we reviewed the data submitted and removed submissions that did not follow the data collection protocol.

We encouraged volunteers to conduct observations on three different days throughout the season, and as much as possible on sunny days between 10 am and 12:30 pm. On each day of observations, volunteers recorded information on the date, site, and plants, including the location, the plant species observed, and the phenology of the plant individuals. Before each set of observation periods, volunteers recorded the temperature and provided a general rating of the amount of wind and cloud cover. Before each observation period, volunteers defined an observation area that included a number of flowers; volunteers chose the number of flowers that they observed during each observation period. Volunteers observed different flowers in each observation period, though in small patches the flowers may have been very close to one another. They defined their area of observation by using a hula hoop or other square or circular perimeter. Volunteers identified how many of their focal flowers were male-phase and how many were female-phase, using an information sheet that contained detailed photographs and outlined the morphological differences between male- and female-phase flowers. Before beginning their observation period, volunteers set a stopwatch for five minutes.

FIGURE 1.   Photos of the *Claytonia virginica* pollination system. (A) A *C. virginica* female-phase flower. (B) A *C. virginica* male-phase flower. (C) The bee-fly *Bombylius major* visiting *C. virginica*. (D) The oligolectic bee *Andrena erigeniae* visiting *C. virginica*. (Reprinted from Parker et al. 2016)

During each five-minute observation period, volunteers observed their focal flowers. When an insect visited, volunteers recorded the identity of that visitor to the best of their ability and recorded the number of male- and female-phase visits that visitor made. Volunteers counted the number of total visits regardless of whether they were made by the same pollinator individual or different individuals. When identifying insects, we encouraged volunteers to use the functional groups that we provided; however, if the volunteer was not sure of the identification, the volunteer identified the insect as "unknown".

All statistical analyses were done using generalized linear mixed models (GLMMs) in *R 3.0.1* (R Core Team 2013). The function *glmmADMB* in the R library *glmmABMB* (Fournier et al. 2012) allowed us to account for highly variable observation times, include random effects, and account for overdispersion in the response variables. We analysed the total number of visits from a visitor group in thirty minutes, including re-visits. The response variable was the number of visits by any individual in a particular floral visitor group in 30 minutes; groups of floral visitors included *A. erigeniae*, *B. major*, small dark bees, parasitic bees, other bees, other flies, all bees, and all flies. At the start, we included the following predictor variables in each model: latitude; longitude; the interaction between latitude and longitude; elevation; whether observations were conducted in a designated natural area/park or a residential area; in natural areas, the approximate size of the park where observations

were conducted; the plant species observed (*C. virginica* or *C. caroliniana*); the temperature; the approximate level of wind (windy, light breeze, or still); the approximate degree of cloud cover (sunny, partly cloudy, or overcast); and the proportion of flowers observed that were male-phase flowers. When one of these factors did not improve the model fit, as indicated by log likelihood ratios, we removed it from the final model. Of these response variables, all final models included only latitude, longitude, and whether observations were conducted in a designated natural area/park or a residential area, because the remaining response variables did not improve model fit. To account for the high variation in the number of flowers observed and number of observation periods conducted, we included the total number of flowers observed as an offset. Because visits observed by a volunteer on a particular day are not independent, we included this as a random effect; as such, each day of observations for a particular volunteer represents the level of replication in each model. To account for overdispersed data, we used a negative binomial error distribution. Because weather influences pollinator visitation, we also ran each of the models with significant results on a subset of the data that included only those observations that occurred when the temperature was above 15°C. These models produced qualitatively similar results to those presented here.

RESULTS

We received usable data from 27 volunteers over three years, who together conducted 655 observation periods (95

FIGURE 2. Volunteer information sheet with photographs and distinguishing characteristics of these pollinator groups, including size, colour, and body shape; we asked that they refer to this sheet during observations

sets of six five-minute observation periods) and observed a total of 14,159 flowers (Fig. 3). Observations were submitted from 24 locations ranging from Vermont to Wisconsin to Kansas, with high representation from Maryland and Virginia. Within these data (1,328 insect visits), 46.46% of insect visits were reported as *A. erigeniae*, 10.24% as *Bombylius major*, 23.87% small dark bees, 3.6% parasitic bees, 3.92% other bees, 9.49 other flies, and 2.41% unspecified. All categories of visitors were included in analyses (as follows), with the exception of unspecified insects, or entries marked as "unknown"; these were excluded.

The number of visits by the pollen specialist bee *A. erigeniae* in 30 minutes (one set of observation periods) varied significantly with longitude, with more visits in the western part of the range than the eastern (Tab. I, Fig. 3C, $Z = 2.62$, $P = 0.0088$). The number of visits by the bombyliid fly *B. major* in 30 minutes varied significantly with latitude, with more visits in the southern part of the range than the northern (Tab. I, Fig. 3B, $Z = 3.25$, $P = 0.0011$). There was no significant effect of latitude on *A. erigeniae* visitation (Tab. I, Fig. 3D, $Z = 0.40$, $P = 0.687$, and no significant effect of longitude on *B. major* visitation (Tab. I, Fig. 3A, $Z = 0.29$, $P = 0.772$). The type of land use in the local area of the observations (whether the observations were done in a residential area or in a natural area) had a significant effect on

the model fit, with higher *A. erigeniae* visitation in natural areas than residential areas.

The number of visits by small dark bees, parasitic bees, other bees, other flies, all bees, and all flies did not vary significantly with latitude or longitude, and for the most part none of the other predictor variables included in the model had a significant effect on visitation by these insects. The exception is that there were more visits by other flies when there was a higher proportion of male-phase flowers observed ($Z = 2.67$, $P = 0.0076$).

DISCUSSION

Data collected by citizen scientists revealed a significant effect of large-scale geographic parameters on the number of visits by two important pollinators of *C. virginica* and *C. caroliniana* in thirty minutes. In other words, the pollinator climate – i.e., the diversity and abundance of *A. erigeniae* and *B. major* specifically – changes fairly predictably along both a latitudinal gradient and a longitudinal one. In lower latitudes, *Claytonia* populations are visited by more bombyliid flies; in western populations, *Claytonia* populations are visited by more *A. erigeniae* bees. These patterns are consistent with the high visitation rates of *B. major* to *C. virginica* documented by Motten et al. (1981) in North Carolina, as well as with our

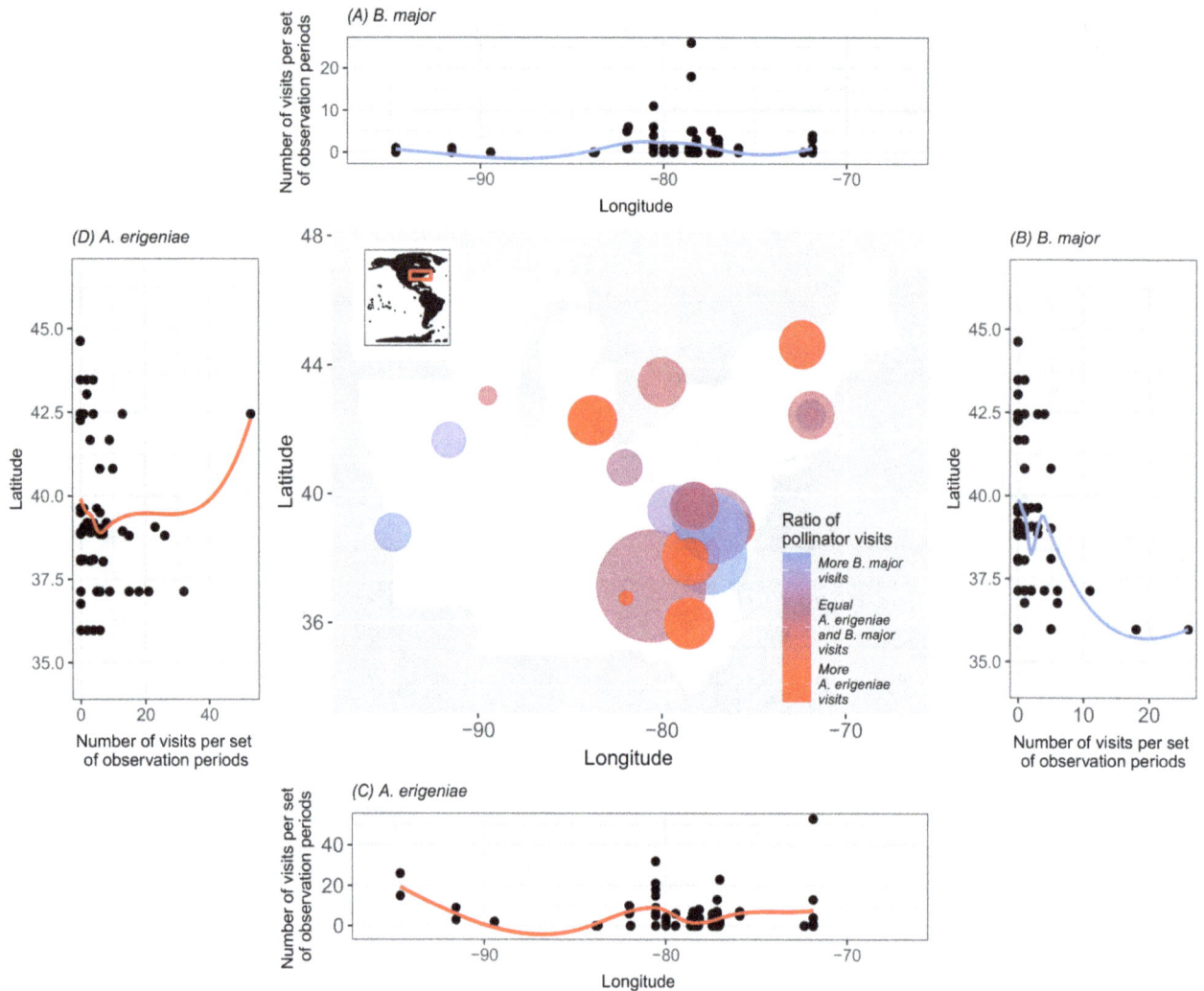

FIGURE 3.   Patterns in *A. erigeniae* and *B. major* visitation to *C. virginica* and *C. caroliniana*. (A) The relationship between longitude and the number of visits by *B. major* per set of observation periods. Curves were fit using locally weighted least squares (LOWESS). (B) The relationship between latitude and the number of visits by *B. major* per set of observation periods. Curves were fit using locally weighted least squares (LOWESS). (C) The relationship between longitude and the number of visits by *A. erigeniae* per set of observation periods. Curves were fit using locally weighted least squares (LOWESS). (D) The relationship between latitude and the number of visits by *B. major* per set of observation periods. Curves were fit using locally weighted least squares (LOWESS). (E) Ratio of pollinator visits at each site of pollinator observations. Ratios are calculated as the number of *A. erigeniae* visits over the number of *A. erigeniae* and *B. major* visits. The colour represents the ratio of pollinator visits, with more blue circles representing higher ratios of *A. erigeniae* visits, and more red circles representing lower ratios of *A. erigeniae* visits, which corresponds with higher ratios of *B. major* visits. The size of the circles represents the number of sets of observation periods conducted at each site. Axes in (A), (B), (C), and (D) correspond with the direction of latitude and longitude in (E).

own observations of visitation rates in Pennsylvania, Maryland, and North Carolina (Parker et al. 2017). We cannot explain the more specific factors that may be impacting these patterns. They may result from variation in the abundance of insect populations. Despite its importance, we do not understand the factors that underlie bee and fly population dynamics very well (Bischoff 2003; Franzen & Nilsson 2013) and we may under-estimate the value of flies as pollinators (Kearns 2001). The results of this study may be driven by biology; for example, flies are more common in cool, moist habitat and have lower energy requirements (Kearns 2001). Another driving factor may be the relationship of these species to one another. *Bombylius major* is a parasite of solitary bees, including *Andrena*, and therefore may be

parasitizing *A. erigeniae* populations. It is hard to say how this relationship impacts the differences observed here, as one would expect that their abundances would be correlated, but we are not aware of *B. major* visiting other plants in northern regions and *A. erigeniae* does not appear to visit plants besides *C. virginica* when *B. major* is abundant (pers. obs.)

Instead, the patterns in our data may result from changes in relative numbers of visits due to variation in floral attractiveness, the composition of plant communities, or some other factor. In general, our results indicate that large-scale geographic gradients – those patterns that change with latitude and longitude – are likely to be important. Climate, day length, and the range limits of interacting species are some

TABLE 1. Results of generalized linear mixed models. The response variable is the number of visits by any individual in a particular floral visitor group in 30 minutes ($N = 95$). *$P < 0.05$, **$P < 0.01$, ***$P < 0.001$.

Andrena erigeniae	Estimate	Standard	Z value	P value
Latitude	0.0387	0.0960	0.40	0.687
Longitude	-0.1048	0.04	2.62	0.0088**
Residential or Natural Area	-1.529	0.449	3.40	0.00066***
Bombylius major				
Latitude	-0.3310	0.1018	3.25	0.0011**
Longitude	0.0158	0.0544	0.29	0.7715
Residential or Natural Area	0.0643	0.4415	0.15	0.8842

intriguing possibilities that deserve attention in future research.

Because we documented gradients in the number of visits by two significant pollinators of *Claytonia*, we have a better idea of the pollinator climate experienced by plant populations in different parts of the species' range; western populations are visited more often by the oligolege *A. erigeniae*, while southern populations see *B. major* more often. We can now make predictions about how plant populations may respond these differences. The pollinators *A. erigeniae* and *B. major* are very different; the pollen specialist *A. erigeniae* is an adept and systematic forager that collects a great deal of pollen, while the bombyliid fly *B. major* is a haphazard forager that collects and deposits pollen only passively (Motten et al. 1981). Although the two pollinators deliver similar numbers of grains, *A. erigeniae* removes substantially more pollen during visitation than *B. major* (Parker et al. 2016). These results help identify traits that vary geographically that otherwise would be overlooked; in fact, Parker et al. (2017) describe ecotypic variation in pollen-related plant traits in *C. virginica* that corresponds with the pollinator climate.

Our approach was to recruit as many volunteers as possible, train volunteers remotely to the best of our ability using visual guides and specific protocols, gather as many data as possible, and identify and remove those data that did not follow protocol. We targeted recruitment efforts to organizations already familiar with plant and insect species identification (e.g. master naturalists) and targeted the more rural areas of the plant species' geographic range to promote geographic coverage. Beyond those initial efforts, we did not attempt to control participants' qualifications or balance geographic coverage in the resulting dataset. The purpose of this project was to provide a sense of the pollinator climate over a larger spatial and temporal scale and generate hypotheses about spatial and temporal patterns for future study. We supplemented these observations by examining patterns in the pollinator climate and pollen-related plant traits in more detail in specific locations (Parker et al. 2017). In this way, the value of these data collected are "fit for purpose"; other citizen science projects designed for other purposes may employ additional data validation methods, such as the validation of identification through photographs (Kremen et al. 2011, Wiggins et al. 2011, Lye et al. 2011), or employ strategies for increasing or balancing geographic coverage and potential bias.

The use of citizen science provided a landscape-level view of variation in the pollinator climate, which otherwise would not have been possible. Aspects of this project made it especially conducive to citizen scientist participation. The two pollinators of greatest interest in this study (*A. erigeniae* and *B. major*) are very different morphologically and behaviourally and are relatively easy to distinguish from one another. *Claytonia* is abundant in many areas across its geographic range, and many participants already had a connection to local *Claytonia* populations from local parks or even their own property. Although some volunteers sent in images with their observations, we did not specifically test the accuracy of citizen scientists' differentiation among the pollinator groups; instead, we relied on the citizen scientists to follow protocols for species identification or indicate that they were unable to do so (e.g. by recording visitors as "unknown"). We reviewed the data submitted and removed those data that were not collected using the project protocol. This type of "expert review" is employed by 77% of citizen science projects, often accompanied by other data validation methods (Wiggins et al. 2011).

Citizen science is a novel approach for contributing to the foundation of knowledge in the biogeography of pollination systems. Without citizen science, we could not have obtained these data – or this coverage of *Claytonia*'s range – during *Claytonia*'s limited flowering period. Future studies that employ citizen science methods for studies of plants and pollinators will further demonstrate the value of citizen science for a variety of uses in pollination biology. Perhaps most importantly, future work should study the value of citizen science for broader outcomes in pollination biology, including providing an opportunity for pollination biologists to connect with the public and enhancing public support for pollination biology and pollinator conservation (e.g. Lewandowski & Oberhauser 2017), just as the value of citizen science continues to be demonstrated for science generally (Stepenuck & Green 2015, Newman et al. 2017; McKinley et al. 2017; Ballard et al. 2017; Toomey & Domroese 2013).

ACKNOWLEDGEMENTS

Our sincere thanks to the many citizen scientists who collected data with skill and enthusiasm. Thanks also to the many naturalist groups and listservs that promoted the project and connected us with volunteers. We thank Teresa Tufts for help with data compilation. We are very grateful to Neal Williams, Sam Droege, Claire Brittain, Katharina Ullman, Megan Frederickson, Jessica Forrest, and Jane Ogilvie for providing advice and comments on data collection materials and protocols. Natural Sciences and Engineering Research Council of Canada (NSERC) funded this research (Discovery Grant to J.D.T.).

REFERENCES

Armbruster WS (1985) Patterns of character divergence and the evolution of reproductive ecotypes of *Dalechampia scandens* (Euphorbiaceae). Evolution 39:733-752.

Arroyo J, Dafni A (1995) Variations in habitat, season, flower traits and pollinators in dimorphic *Narcissus tazetta* L. (Amaryllidaceae) in Israel. New Phytologist 129:135–145.

Ballard HL, Robinson LD, Young AN, Pauly GB, Higgins LM, Johnson RF, Tweddle JC (2017) Contributions to conservation outcomes by natural history museum-led citizen science: Examining evidence and next steps. Biological Conservation 208: 87-97.

Birkin L, Goulson D (2015) Using citizen science to monitor pollination services. Ecological Entomology (2015) 40: 3-11.

Bischoff I (2003) Population dynamics of the solitary digger bee *Andrena vaga* Panzer (Hymenoptera, Andrenidae) studied using mark-recapture and nest counts. Population Ecology 45:197–204.

Boyd A (2002) Morphological analysis of Sky Island populations of *Macromeria viridiflora* (Boraginaceae). Systematic Botany 27:116–126.

Boyd AE (2004) Breeding system of *Macromeria viridiflora* (Boraginaceae) and geographic variation in pollinator assemblages. American Journal of Botany 91:1809–1813.

Cane JH, Sipes S (2006) Floral specialization by bees: analytical methodologies and a revised lexicon for oligolecty. In: Waser N, Ollerton J (eds.) Plant-Pollinator Interactions: From Specialization to Generalization. University of Chicago Press.

Cooper CB, Dickinson J, Phillips T, Bonney R (2007) Citizen science as a tool for conservation in residential ecosystems. Ecology and Society 12:11.

Darwin C (1877) On the various contrivances by which British and foreign orchids are fertilised by insects. J. Murray.

Davis LR, LaBerge WE (1975) The nest biology of the bee *Andrena (Ptilandrena) erigeniae* Robertson (Hymenoptera: Andrenidae). Illinois Natural History Survey Biological Notes 95: 1–24.

Deguines N, Julliard R, de Flores M, Fontaine C (2012) The whereabouts of flower visitors: Contrasting land-use preferences revealed by a country-wide survey based on citizen science. PLOS One 7(9):e45822.

Dickinson JL, Zuckerberg B, Bonter DN (2010) Citizen science as an ecological research tool: challenges and benefits. Annual Review of Ecology, Evolution, and Systematics 41:149-172.

Fournier DA, Skaug HJ, Ancheta J, Ianelli J, Magnusson A, Maunder MN, Nielsen A, Sibert J (2012) AD Model Builder: using automatic differentiation for statistical inference of highly parameterized complex nonlinear models. Optimization Methods Software 27:233–249.

Franzen M, Nilsson SG (2013) High population variability and source–sink dynamics in a solitary bee species. Ecology 94:1400–1408.

Galen C (1996) Rates of floral evolution: adaptation to bumblebee pollination in an alpine wildflower, *Polemonium viscosum*. Evolution 50:120-125.

Gomez JM, Bosch J, Perfectti F, Fernandez JD, Abdelaziz M, Camacho JPM (2008) Spatial variation in selection on corolla shape in a generalist plant is promoted by the preference patterns of its local pollinators. Proceedings of the Royal Society B 275:2241–2249.

Grant V, Grant KA (1965) Flower pollination in the phlox family. Columbia University Press.

Harder LD, Thomson JD (1989) Evolutionary options for maximizing pollen dispersal of animal-pollinated plants. American Naturalist 133:323–344.

Herrera CM (1995) Microclimate and individual variation in pollinators: flowering plants are more than their flowers. Ecology 76:1516-1524.

Herrera CM, Castellanos MC, Medrano M (2006) Geographical context of floral evolution: towards an improved research programme in floral diversification. In: Harder LD and Barrett SCH (eds), Ecology and evolution of flowers, Oxford University Press.

Hillebrand H (2004) On the generality of the latitudinal diversity gradient. American Naturalist 163:192–211.

Inoue K, Maki M, Masuda M (1996) Evolution of *Campanula* flowers in relation to insect pollinators on islands. In: Lloyd D and Barrett SCH (eds.) Floral Biology. Springer US, Boston, MA, pp 377-400.

Johnson SD (1997) Pollination ecotypes of *Satyrium hallackii* (Orchidaceae) in South Africa. Botanical Journal of the Linnean Society 123:225–235.

Johnson SD, Steiner KE (1997) Long-tongued fly pollination and evolution of floral spur length in the *Disa draconis* complex (Orchidaceae). Evolution 51:45-53.

Kearns C (2001) North American dipteran pollinators: assessing their value and conservation status. Ecology and Society 5(1):5.

Kremen C, Ullman KS, Thorp RW (2011) Evaluating the quality of citizen-scientist data on pollinator communities. Conservation Biology 25:607–617.

Lewandowski EJ and Oberhauser KS (2017) Butterfly citizen scientists in the United States increase their engagement in conservation. Biological Conservation 208: 106-112.

Lye GC, Osborne JL, Park KJ, Goulson D (2012) Using citizen science to monitor *Bombus* populations in the UK: nesting ecology and relative abundance in the urban environment. Journal of Insect Conservation 16:697-707.

MacArthur R (1984) Geographical Ecology. Princeton University Press.

Malo JE, Baonza J (2002) Are there predictable clines in plant–pollinator interactions along altitudinal gradients? The example of *Cytisus scoparius* (L.) Link in the Sierra de Guadarrama (Central Spain). Diversity and Distributions 8:365–371.

McKinley DC, Miller-Rushing A, Ballard HL, Bonney R, Brown H, Cook-Patton SC, Evans DM, French RA, Parrish JK, Phillips TB, Ryan SF, Shanley LA, Shirk JL, Stepenuck KF, Weltzin JF, Wiggins A, Boyle OD, Briggs RD, Soukup MA (2017) Citizen science can improve conservation science, natural resource management, and environmental protection. Biological Conservation 208: 15-28.

Miller RB (1981) Hawkmoths and the geographic patterns of floral variation in *Aquilegia caerulea*. Evolution 35:763-774.

Motten A, Campbell D, Alexander D, and Miller H (1981) Pollination effectiveness of specialist and generalist visitors to a North Carolina population of *Claytonia virginica*. Ecology 62:1278–1287.

Newman E, Manning J, Anderson B (2014) Matching floral and pollinator traits through guild convergence and pollinator ecotype formation. Annals of Botany 113:373-384.

Newman G, Chandler M, Clyde M, McGreavy B, Haklay M, Ballard H, Gray S, Scarpino R, Hauptfeld R, Mellor D, Gallo J (2017) Leveraging the power of place in citizen science for effective conservation decision making. Biological Conservation 208:55-64.

Parker AJ, Williams NM, Thomson JD (2016) Specialist pollinators deplete pollen in the spring ephemeral wildflower *Claytonia virginica*. Ecology and Evolution. doi:10.1002/ece3.2252

Parker AJ, Williams NM, Thomson JD (2017) Geographic patterns and pollination ecotypes in Claytonia virginica. Evolution doi:10.1111/evo.13381

Pérez-Barrales R, Arroyo J Armbruster WS (2007) Differences in pollinator faunas may generate geographic differences in floral morphology and integration in *Narcissus papyraceus* (Amaryllidaceae). Oikos 116: 1904-1918.

Price MV, Waser NM, Irwin RE, Campbell DR, Brody AK (2005) Temporal and spatial variation in pollination of a montane herb: A seven-year study. Ecology 86:2106–2116.

R Core Team (2013) R: A language and environment for statistical computing. Vienna, Austria.

Richards S, Williams N, Harder L (2009) Variation in pollination: causes and consequences for plant reproduction. American Naturalist 174:382–398.

Ries L, Oberhauser K (2015) A citizen army for science: Quantifying the contributions of citizen scientists to our understanding of monarch butterfly biology. BioScience 65: 419-430.

Robertson JL, Wyatt R (1990) Evidence for pollination ecotypes in the yellow-fringed orchid, *Platanthera ciliaris*. Evolution 44:121-133.

Roy HE, Baxter E, Saunders A, Pocock MJO (2016) Focal plant observations as a standardised method for pollinator monitoring: Opportunities and limitations for mass participation citizen science. PLOS One 11(3):e0150794.

Stepenuck KF, Green LT (2015) Individual- and community-level impacts of volunteer environmental monitoring: a synthesis of peer-reviewed literature. Ecology and Society 20(3): 19.

Toomey AH, Domroese, MC (2013) Can citizen science lead to positive conservation attitudes and behaviors? Human Ecology Review 20(1): 50-62.

van der Niet T, Peakall R, Johnson SD (2014) Pollinator-driven ecological speciation in plants: new evidence and future perspectives. Annals of Botany 113:199–211.

van der Niet T, Pirie MD, Shuttleworth A, Johnson SD, Midgley JJ (2014) Do pollinator distributions underlie the evolution of pollination ecotypes in the Cape shrub *Erica plukenetii*? Annals of Botany 113:301-315.

Williams NM, Winfree R (2013) Local habitat characteristics but not landscape urbanization drive pollinator visitation and native plant pollination in forest remnants. Biological Conservation 160:10–18.

Willig MR, Kaufman DM, Stevens RD (2003) Latitudinal gradients of biodiversity: Pattern, process, scale, and synthesis. Annual Review of Ecology, Evolution, and Systematics 34:273–309.

Willmer PG (1983) Thermal constraints on activity patterns in nectar-feeding insects. Ecological Entomology 8:455–469.

Wiggins A, Newman G, Stevenson R, Crowston K (2011) Mechanisms for data quality and validation in citizen science. Proceedings of the 2011 IEEE Seventh International Conference on e-Science Workshops 14-19.

Winfree R, Bartomeus I, Cariveau DP (2011) Native pollinators in anthropogenic habitats. Annual Review of Ecology, Evolution, and Systematics 42:1–22.

18

Experimental Evidence for Predominant Nocturnal Pollination Despite More Frequent Diurnal Visitation in *Abronia umbellata* (Nyctaginaceae)

Laura A. D. Doubleday*[1] and Christopher G. Eckert

Department of Biology, Queen's University, Kingston, Ontario, K7L 3N6 Canada.
[1]*Current Address: Graduate Programs in Organismic and Evolutionary Biology and Entomology, University of Massachusetts, Amherst, Massachusetts, 01003 USA.*

Abstract—Different suites of floral traits are associated with historical selection by particular functional groups of pollinators, but contemporary floral phenotypes are not necessarily good predictors of a plant's effective pollinators. To determine the extent to which plant species specialize on particular functional groups of pollinators, it is important to quantify visitation rates for the full spectrum of flower visitors as well as to experimentally assess the contributions of each functional group to plant reproduction. We assessed whether attracting both diurnal and nocturnal flower visitors corresponded to pollination generalization or specialization in the Pacific coastal dune endemic *Abronia umbellata* var. *umbellata*. In multiple populations over two years, we observed flower visitors during the day and at night to assess visitation rates by different insect groups and conducted pollinator exclusion experiments to assess the contributions of diurnal and nocturnal visitors to seed production.

Flower visitation rates were 8.67 times higher during the day than at night, but nocturnal visitation resulted in significantly higher seed set, suggesting that nocturnal noctuid and sphingid moths are the chief pollinators. Most diurnal visitors were honey bees, with tongues too short to reach *A. umbellata* nectar or contact stigmas and effect pollination. The prevalence of honey bees, combined with the lack of successful seed production resulting from diurnal pollination, suggests that honey bees are pollen thieves that collect pollen but do not deposit it on stigmas. Our results underscore the need to experimentally assess the contributions of different groups of flower visitors to plant reproduction.

Keywords: *Abronia umbellata, moth pollination, Nyctaginaceae, pollination ecology, pollination syndromes, seed set, specialization*

Introduction

The remarkable phenotypic diversity of flowering plants is thought to be due largely to selection exerted on flower morphology and development by pollen vectors (Fenster et al. 2004). One of the most compelling pieces of evidence for the importance of pollinator-mediated selection is the nonrandom association of floral traits among species that differ in the types of animals that mediate cross-pollination, a phenomenon generally referred to as pollination syndromes (van der Pijl 1961). For example, plants that have historically experienced selection exerted predominantly by nocturnal moths typically have medium to long narrow corolla tubes that contain relatively dilute nectar, are white or pale in colour, lack nectar guides, and emit a sweet fragrance in the evening (Willmer 2011).

While pollination syndromes can provide insight into the historical pollinator-mediated selection that helped to shape contemporary floral phenotypes, they do not necessarily predict which flower visitors are currently effective pollinators (Ollerton et al. 2009), nor can they be used to infer the degree of specialization of a plant's current pollinator fauna (Waser et al. 1996). The number of functional groups observed visiting a plant species is often a better indicator of that plant's degree of specialization than the number of visiting taxa alone (Fenster et al. 2004; Armbruster 2017), but it remains critical to distinguish between visitors that effect pollination and those that do not, especially because many animals that visit flowers perform little to no pollination (Hargreaves et al. 2009; Irwin et al. 2010).

The important distinction between floral visitation and effective pollination makes experimental approaches critical for determining the importance of particular visitors or visitor guilds to pollination and plant fitness. Different visitor guilds often visit specific plant taxa at different times of day, thus temporal pollinator exclusion can be used to determine the relative importance of temporally-divergent visitors (Brunet & Holmquist 2009; Walter 2010; Bustamante et al. 2010). Combining such experiments with observations of flower visitors can determine which functional groups are the most effective pollinators. Such experiments should be replicated across populations and years, given that many studies have demonstrated pronounced spatiotemporal variation in pollinator faunas (Herrera 1988; Wolfe & Barrett 1988;

*Corresponding author: ldoubled@psis.umass.edu

TABLE 1.   Locations of *Abronia umbellata* study sites in California, USA, activities conducted at each site, and dates visited. "Coll" = pollinator collections, "Exp" = pollinator exclusion experiments, and "Obs" = pollinator observations.

Site	Code	County	Latitude (°N)	Longitude (°W)	Activities	Dates Visited
(1) McGrath State Beach	CMGA	Ventura	34.21876	119.25853	Coll, Exp, Obs	2010: May 23, July 9–10; 2011: May 22–27, June 11, July 6–7
(2) San Buenaventura State Beach	CBVA	Ventura	34.26788	119.27815	Coll	2010: July 10
(3) Coal Oil Point Reserve	CCOA	Santa Barbara	34.40824	119.87909	Coll, Exp, Obs	2010: May 24, June 1–7; 2011: June 2–8, July 3–4
(4) Coreopsis Hill	CGN2A	San Luis Obispo	35.02181	120.62203	Coll	2010: July 13
(5) Montaña de Oro State Park	CSPA	San Luis Obispo	35.30072	120.87560	Coll, Exp, Obs	2010: July 2–8; 2011: June 26–30, July 28–30
(6) Manresa Uplands State Beach	CMNA	Santa Cruz	36.91531	121.85155	Coll, Exp, Obs	2010: June 11–17; 2011: June 14–17, July 24–25
(7) Seacliff State Beach	CSEA	Santa Cruz	36.96854	121.90492	Coll	2011: July 25

Fishbein & Venable 1996; Eckert 2002). However, spatiotemporal replication is logistically demanding and difficult to achieve in many systems. We are aware of only four studies that include multiple years of study at more than one site (Morse & Fritz 1983; Fleming et al. 2001; Holland & Fleming 2002; Bustamante et al. 2010), just three of which combined temporal exclusion experiments with pollinator observations.

The Pacific coast dune plant *Abronia umbellata* Lam. var. *umbellata* (Nyctaginaceae) possesses some floral traits seemingly specialized for moth attraction at night, but other traits that are attractive to diurnal insects. It provides an excellent opportunity to study the extent to which floral traits associated with attracting visitors from multiple functional groups translate into a generalized vs. specialized pollinator fauna. *Abronia umbellata* var. *umbellata* is self-incompatible (SI) and obligately outcrossing, in contrast to the self-compatible and highly autogamous var. *breviflora* (Doubleday et al. 2013). *Abronia umbellata* var. *umbellata* (hereafter SI *A. umbellata*) exhibits some traits that typify the moth-pollination syndrome: it has reverse herkogamous narrow, tubular flowers with anthers close to the mouth of the floral tube and stigmas recessed deeply within (Fig. 1B) and, in the evening, emits a sweet fragrance containing benzenoid compounds typical of moth-pollinated species (Doubleday et al. 2013). However, unlike canonically white "moth flowers," the flowers are pink-purple with contrasting white "eyespots" encircling the floral tube opening (Fig. 1A), suggesting attraction of diurnal visitors. The flowers are uniovulate and borne on umbellate inflorescences, and fruits are tough, winged diclesia, a type of anthocarp that appears to be

FIGURE 1.   Morphology of inflorescences (A) and flowers (B) of *Abronia umbellata* var. *umbellata* (Nyctaginaceae).

TABLE 2.   Mean flower visitation rates (visits (v) per flower (f) per hour (h)), standard errors of the means, number of observation periods (n_{obs}), total observation time (Time, in h) at each *Abronia umbellata* study site for pollinator observations during the day and night pooled across two years (2010 and 2011). Pollinator observations consisted of recording visitation to all of the flowers in ~ 2 m diameter patch for 3 or 5 min. Sampling effort is the product of number of flowers observed during each observation period and the length of that observation period in hours (f*h). Populations are listed by increasing latitude. Site codes are as in Table 1.

| Site | Day (0600–1800h) | | | | | Night (1800–0600h) | | | | |
	Visit rate (v/f/h)	SE	n_{obs}	Time (h)	Effort (f*h)	Visit rate (v/f/h)	SE	n_{obs}	Time (h)	Effort (f*h)
(1) CMGA	0.0000	0.0000	15	0.75	293.5	0.0012	0.0012	143	9.62	1918.9
(3) CCOA	0.0519	0.0109	159	7.95	7394.8	0.0033	0.0015	220	13.27	6144.5
(6) CSPA	0.0228	0.0072	202	12.47	4494.6	0.0105	0.0075	119	7.02	2670.1
(7) CMNA	0.0013	0.0008	116	5.80	3794.7	0.0020	0.0009	389	22.12	8955.3
Mean	0.026	0.003	–	–	–	0.003	0.001	–	–	–
Total	–	–	492	26.97	15977.60	–	–	871	52.03	19688.80

dispersed by wind (Darling et al. 2008). An umbel's flowers open individually over the course of 1–4 days, each umbel bears apparently functional flowers for 7–10 d, and individual flowers remain open day and night (L. A. D. Doubleday, pers. observation). The timing of pollen presentation and stigma receptivity are unknown. The chief pollinators of SI *A. umbellata* have not been determined, although Tillett (1967) speculated that crepuscular and nocturnal noctuid and sphingid moths are likely the most important pollinators.

This study has two main objectives. First, we document patterns of visitation by different insect guilds to flowers during the day and at night. Second, we use temporal pollinator exclusion experiments to determine which visitors are effective pollinators by directly quantifying the relative importance of nocturnal vs. diurnal flower visitors to seed set. We assess SI *A. umbellata*'s pollination ecology in multiple natural populations in two years.

MATERIALS AND METHODS

Pollinator observations

At each of four sites on the Pacific coast of California, USA (Tab. 1) we estimated the rates of diurnal and nocturnal insect visitation to individual SI *A. umbellata* flowers during standardized 3- or 5-min observation periods. For each period, we randomly selected a patch of inflorescences within a circular plot of ~ 2 m diameter and simultaneously observed all flowers without regard to individual plants (plants are prostrate and intermingle such that individuals that cannot be separated without damaging them). We recorded each flower visitor's identity (to the lowest taxonomic level possible in the field), the number of umbels visited by each flower visitor, and the number of flowers probed on each umbel visited. Our presence, standing ~ 1 m from the plot, did not seem to influence flower visitor behaviour. We used red headlamps (Petzl® TIKKA PLUS2®) for evening (sunset – 0200 h) observations because nocturnal insects are relatively insensitive to red light (Briscoe & Chittka 2001). We conducted observations across all times of day and night except between midnight and 0500 h (Fig. 2). We saw insects probing flowers only between 1000 and 2200 h, with peaks in flower visitation between 1100 and 1600 h and 2000 and 2200 h (Fig. 2). The timing of these flower visitation peaks, combined with a lull in visitation between 1600 and 2000 h (Fig. 2), sunrises between 0545 and 0630 h, and sunsets between 2000 and 2030 h made 0600 and 1800 h reasonable dividing times

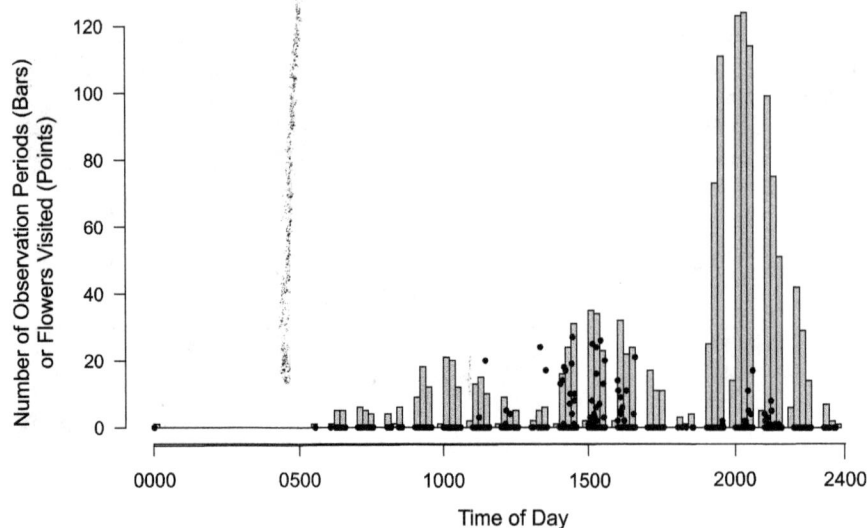

FIGURE 2. Sampling effort and temporal patterns of flower visitation to *Abronia umbellata* var. *umbellata* flowers in natural populations. Bars represent the number of pollinator observation periods conducted at a given time and points represent the total number of flowers visited during an observation period at a given time. Data is pooled across days and sites.

TABLE 3.   Insects collected while visiting flowers of *Abronia umbellata* at sites in California, USA. Site numbers are in Table 1.

Order	Family	Species	Day- or	Year	Sites	Number of specimens
Hymenoptera	Apidae	*Apis mellifera*	Day	2010	1, 3, 5, 6	16, 12, 3, 2
				2011	1, 2, 6	1, 1, 2
		Bombus spp.	Day	2010	3, 7	1, 1
	Unknown	Small, long-tongued solitary bee	Day	2011	6	1
Lepidoptera	Hesperiidae	*Hylephila phyleus*	Day	2010	1	2
	Nymphalidae	*Vanessa cardui*	Day	2010	1	1
				2011	8	1
	Unknown	Small brown and orange butterfly	Day	2011	6	1
	Geometridae	*Euphyia* sp.	Night	2010	2, 7	2, 1
	Noctuidae	*Trichoplusia ni*	Night	2010	7	2
		Copablepharon robertsoni	Night	2010	6	1
		Copablepharon sanctaemonicae	Night	2010	2, 6	3, 6
				2011	6	1
		Autoplusia egenoides	Night	2010	7	3
	Pyralidae	*Phobus funerellus*	Night	2010	7	11
	Sphingidae	*Hyles lineata*	Night	2010	1, 7	1, 1
Diptera	Acroceridae	*Eulonchus* sp.	Day	2010	6	1

between day and night. We made 492 observation periods (26.97 h) between 0600 and 1800 h ("day") and 871 (52.03 h) between 1800 and 0600 h ("night"). There were two reasons why we conducted more observations at night: daylight was important for other sampling activities, so we could not devote as much daytime effort to pollinator observations, and we were determined to learn the taxonomic identities of nocturnal moth pollinators, even though they were relatively rare. Because of the logistic challenges involved in studying multiple sites separated by hundreds of km, while conducting experiments lasting 7–10 d at some sites, sampling effort was unbalanced across sites and sites were visited on different dates (Tab. 1 & 2). Pollinator faunas often fluctuate seasonally (Herrera 1988), and sampling date may have affected which flower visitors were present at a given site. Whenever possible, we collected a sample of flower-visiting insects and identified these to the lowest taxonomic level possible given the expertise available to us (Tab. 3). We calculated total visitation for each insect group as visits per flower per hour.

Temporal pollinator exclusion experiment

At four sites over two years we quantified the relative importance of diurnal vs. nocturnal pollination by excluding pollinators from individual umbels at different times of day. The experiment involved one site in 2010 (CMNA) two others in 2011 (CMGA, CSPA) and a third in both years (CCOA; Tab. 1). Individual umbels were allocated to one of three treatments: (i) exposed to pollination at night but enclosed in a wire cage covered with fine bridal veil during the

day (0600–1800 h); (ii) exposed to pollination during the day but enclosed at night (1800–0600 h); or (iii) exposed to pollination always (i.e. never enclosed). In 2011, we added an additional treatment: (iv) enclosed always. Any seed produced by these umbels would have to result from either autonomous selfing and/or pollinators accessing enclosed flowers. Only a very small number of continuously enclosed umbels set seed: one at CCOA (mean proportion of flowers setting seed ± 1SE = 0.0018 ± 0.0033 seeds/ovule), two at CMGA (0.010 ± 0.0081) and seven at CSPA (0.029 ± 0.012) suggesting that autonomous selfing and/or the failure of the enclosures to exclude pollinators were infrequent. It is possible that stigmas frequently received self-pollen, but fertilization rarely occurred because of strong genetic self-incompatibility (Doubleday et al. 2013). Experimental crosses on both enclosed and exposed plants did not reveal any negative effects of enclosure on seed set (Doubleday 2012).

We randomly assigned ≥ 30 umbels (only one umbel per plant) just at anthesis (1–2 flowers beginning to open) to each of the treatment groups. Because *A. umbellata* infructescences shatter at maturity, we bagged them with bridal veil to capture all mature diclesia. For each umbel, we assessed whether each flower produced a seed by counting the number of flowers on each umbel and the number of filled seeds produced. Failure to set seed included flowers from which a diclesium never developed, and those from which a diclesium developed but did not contain a filled seed. At our study sites, only about 30% of developed diclesia contain seeds (mean ± 1SD: 29.52 ± 30.00%; L. A. D. Doubleday and C. G. Eckert, unpublished data). Diclesia routinely expand but contain no seeds under pollinator-free glasshouse conditions (L. A. D. Doubleday and C. G. Eckert, unpublished data), indicating that successful pollination is not a prerequisite for fruit expansion.

Statistical analyses

We used R (version 3.1.3, R Core Team 2017) for all statistical analyses. Because flower visitation occurred in only 6% of observation periods, visitation rates were zero-inflated. Accordingly, we tested for a difference in visitation rate between night and day using randomization tests. We computed the mean difference between day and night visitation rates, randomized the data without respect to when an observation was conducted using the "sample()" function in R, and calculated mean visitation rates for day and night for each of 10,000 randomizations. The proportion of differences calculated from randomized data that were equal to or greater than the observed difference is equivalent to a P value (P_{rand}). We report these as approximate because each randomization run returns a slightly different value.

We tested for differences in seed set among temporal enclosure treatments by fitting generalized linear models (GLMs, glm() function in R) with binomial error structure (logit link function) to variation in seed set (the number of flowers making a seed vs. the number not making a seed). Because we studied different populations and performed a different set of treatments in each year, data for 2010 and 2011 were analyzed separately. Hence, we evaluated site and enclosure treatment and their interaction as potential predictors. We used quasi-likelihood estimation because data were overdispersed, and performed likelihood ratio tests to evaluate the significance of each term in the model using F tests following Buckley (2015). We used the lsmeans() and contrast() functions in the lsmeans R package (version 2.25, Lenth 2016) to perform post-hoc, pairwise contrasts among enclosure treatments.

RESULTS

Flower visitation

Visitation was 8.67-times higher during day than night observation periods for populations and years combined (Tab. 2; P_{rand} < 0.0001), 5.09-times higher in 2010 (P_{rand} < 0.0001), and 13.10-times higher in 2011 (P_{rand} ~ 0.0008). When we separately analysed each of three sites with more than 100 observation periods for each time period, day visitation was greater at CCOA (P_{rand} < 0.0001), but not CSPA (P_{rand} ~ 0.13) or CMNA (P_{rand} ~ 0.50). Most diurnal visitors were introduced honey bees (*Apis mellifera*: Apidae, Tab. 3). Other diurnal visitors included bumble bees, butterflies, flies (Tab. 3), and, on one occasion, a diurnal sphingid (possibly *Hemaris* sp.) (L. A. D. Doubleday, personal observation), but these other visitors were infrequent. When we excluded honey bee visits from the analysis (pooling years and populations), mean night visitation (mean ± 1 SE: 0.0030 ± 0.0011) was higher than day visitation (0.0008 ± 0.0009) but not significantly so (P_{rand} ~ 0.13). The differences for individual sites analysed separately were also not significant (all P_{rand} > 0.11). When observations were pooled by year with honey bee visits excluded, night visitation

was more frequent than day visitation for 2010 (P_{rand} ~ 0.037) but not 2011 (P_{rand} ~ 0.18).

We caught 81 visitors that probed SI *A. umbellata* flowers and identified 70 to species, nine to genus, and two to order (Tab. 3). The 49 day visitors represented at least seven species (though 33 visitors were honey bees) and three orders. The 32 night visitors, all of which were Lepidoptera, represented at least seven species from four families. Noctuid moths were most common (50% of night visitors), with three of four species collected from only one population and the fourth species collected from two populations.

Temporal pollinator exclusion

Seed set was generally very low, even among open-pollinated umbels always exposed to pollinators (mean ± 1 SE: 0.071 ± 0.012 seeds/ovule, Fig. 3). In 2011, umbels exposed to pollinators only at night set 3.8-times more seed than those exposed only during the day and did not differ from umbels exposed all the time (Fig. 3, Tab. 4). Seed set was somewhat higher at sites CSPA and CMGA than CCOA, but there was no difference in treatment effects between sites. Results from 2010 suggest 3.4-times higher seed set by night-exposed than day-exposed umbels (Fig. 3), but the difference among treatments was not quite significant ($P = 0.073$, Tab. 4). Again, there was variation among sites in mean seed set (CCOA > CMNA), but not in the effects of enclosure treatment on seed set. At CCOA for 2010 and 2011 combined, seed set did not vary among treatments or between years, nor did the treatment effect vary between years (Tab. 4).

DISCUSSION

Self-incompatible, outcrossing *A. umbellata* var. *umbellata* possesses some traits suggestive of historical selection by nocturnal moths but does not completely conform to the classic moth-pollination syndrome. Our pollinator observations suggest that several functional groups of insects visit the plant during the day and at night, with much higher rates of flower visitation during the day in some populations. However, our temporal pollinator exclusion experiments suggest that nocturnal visitors are more effective pollinators than diurnal visitors. In 2011, night-pollinated umbels had significantly higher seed set (3.8-times higher) than day-pollinated umbels, and the difference in 2010 was of similar magnitude, but not statistically significant (Fig. 3). Taken together, these results suggest predominant nocturnal pollination despite more frequent visitation during the day.

Of the taxa we observed visiting flowers at night, noctuid moths were most common, but we also observed sphingid, pyralid, and geometrid moths. Average tongue lengths were available for members of three of the four moth families we collected from SI *A. umbellata* flowers: 25–33mm for *Hyles lineata*, the sphingid species we observed; 10–20 mm for most temperate noctuids; and 4–9 mm for a different pyralid species than the one we observed (Willmer 2011). Stigmas are recessed 14.07 ± 1.01 mm (mean ± 1 SD, C. G. Eckert, unpublished data) from the floral face in SI *A. umbellata*, suggesting that sphingids and many noctuids would be

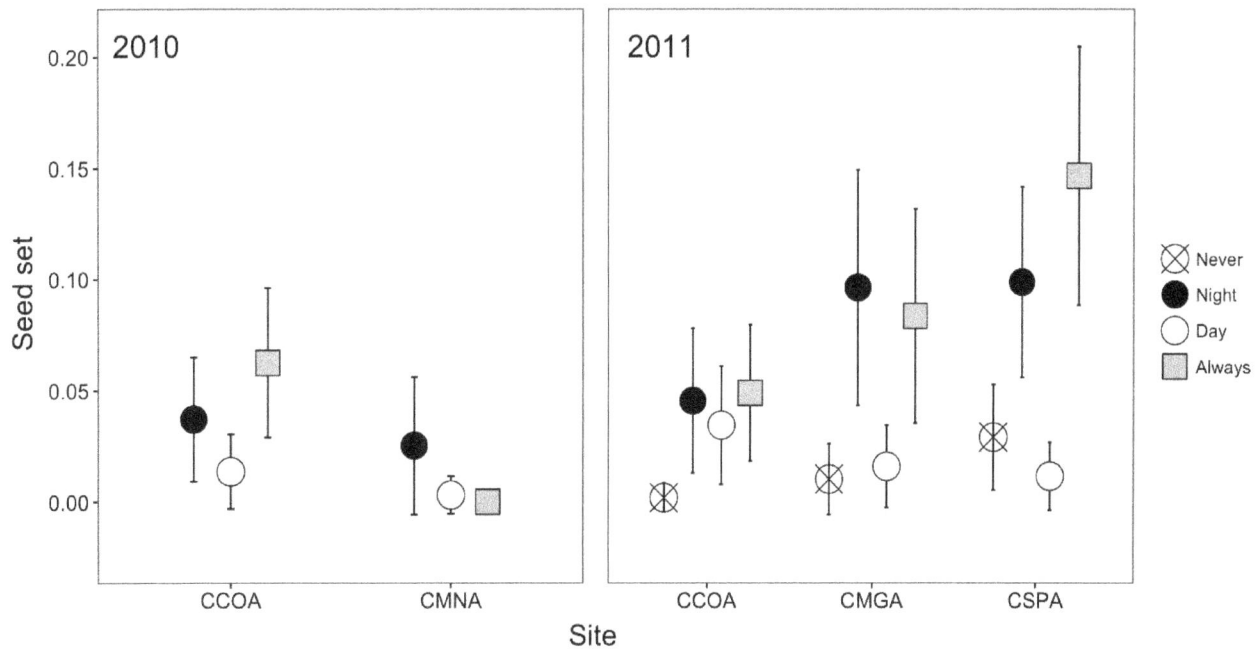

FIGURE 3.   Mean seed set of umbels subjected to different pollinator exclusion treatments in natural populations of *Abronia umbellata* var. *umbellata*. Umbels were available to pollinators all the time ("Always"), from 1800 – 0600h ("Night"), from 0600 – 1800h ("Day") or always enclosed ("Never"). Sample sizes in 2010: CCOA Night = 18, Day = 19, Always = 19; CMNA Night = 10, Day = 16, Always = 11; 2011: CCOA Never = 26, Night = 24, Day = 28, Always = 30; CMGA Never = 27, Night = 22, Day = 28, Always = 22; CSPA Never = 28, Night = 29, Day = 29, Always = 21. Each point is a back-transformed least-squares treatment mean, and error bars are 95% confidence intervals.

effective pollinators because their tongues are long enough to successfully contact the stigma.

Many species of *Abronia* may be predominantly moth pollinated, as the floral display of most species is typified by umbels of fragrant, tubular flowers with recessed stigmas, requiring pollinators with long, narrow tongues. Jabis et al. (2011) suggested that the sphingid moth *Hyles lineata* was the most effective potential pollinator of *A. alpina*, but that diurnal taxa also contributed substantially to pollination. Saunders and Sipes (2006) suggested that several species of noctuid and sphingid moths were likely the most important pollinators of *A. ammophila*, but that butterflies were also likely to affect some pollination. Similarly, Williamson et al. (1994) speculated that noctuid and sphingid moths were key

pollinators of *A. macrocarpa* as multiple moth species were observed visiting flowers and bearing *A. macrocarpa* pollen in the field.

During the day, we observed much higher frequencies of visitation by honey bees than by other taxa and, and when performing pollen supplementations as part of another experiment, we found it increasingly difficult to obtain sufficient donor pollen from open flowers, which we suspect was due to pollen theft by honey bees (L. A. D. Doubleday, personal observation). The combination of recessed stigmas in flowers of *A. umbellata*, the relatively short glossae of *A. mellifera* (mean ± 1 SD = 5.15 ± 0.47mm, $n = 28$; L. A. D. Doubleday, unpublished data), and our observation of low

TABLE 4.   Analyses of variation in seed production by umbels of *Abronia umbellata* after experimental isolation from pollinators. Cells present *F* and *P* values from likelihood ratio tests comparing generalized linear models fit with binomial errors to the number of flowers setting seed vs. not setting seed. Contrasts among treatment means are shown when the effect of treatment was significant (A = umbels always exposed to pollinators, N = exposed only during the night, D = exposed only during the day, X = never exposed). All four treatments were applied in 2011, whereas only three were used in 2010 (A, N, D). For the analysis of CCOA including both years, the full model did not fit the data better than a null model ($F_{5,132}$ = 1.08, $P = 0.38$). NS = not significant at $P < 0.05$. Least squares means by treatment and site are in Fig. 2.

Sites	Years	Treatment (T)	Site (S)	T x S	Contrasts
CCOA, CMNA	2010	$F_{2,89} = 2.70, P = 0.073$	$F_{1,89} = 6.86,$ $P = 0.010$	$F_{2,87} = 2.52,$ $P = 0.086$	NS
CCOA, CMGA, CSPA	2011	$F_{3,308} = 15.89,$ $P < 0.0001$	$F_{2,308} = 4.64,$ $P = 0.010$	$F_{6,302} = 1.90,$ $P = 0.080$	A = N > D = X
		Treatment (T)	Year (Y)	T x Y	
CCOA	2010, 2011	$F_{2,132} = 1.70, P = 0.18$	$F_{2,132} = 0.13,$ $P = 0.72$	$F_{2,132} = 0.90,$ $P = 0.42$	NS

seed set among day-pollinated umbels makes it extremely unlikely that honey bees vector substantial outcross pollen. It is also unlikely that they are successfully extracting nectar from flowers, because available nectar is deeply recessed within the flower at the base of the floral tube (L. A. D. Doubleday, personal observation). This, and the close proximity of dehiscing anthers to the mouth of the floral tube suggests that these visitors are gathering pollen as pollen thieves, but further study would be required to confirm this as we were unable to assess pollen loads on honey bees foraging on SI *A. umbellata* flowers. It is likely that pollen theft has negative effects on plant fitness: for example, Hargreaves et al. (2010) demonstrated that pollen theft by bees decreases reproductive success in *Aloe maculata*. Bees are the most commonly reported pollen thieves (Hargreaves et al. 2009).

Introduced pollinators like honey bees have significant, but varied, effects on native ecosystems (Goulson 2003). It would be premature to generalize about the effects of honey bees across diverse native ecosystems, but honey bees have been shown to dominate the spectrum of flower-visiting taxa in certain fragmented habitats in the Neotropics (Aizen & Feinsinger 1994), and meta-analysis has shown that habitat loss has negative effects on the abundance of unmanaged bee species but not on honey bee abundance (Winfree et al. 2009). Honey bees are ineffective pollinators of some native plants but effective pollinators of others (reviewed by Butz Huryn 1997) and introduced honey bees may compete for nesting cavities with native bees, birds, and mammals, but definitive studies are lacking (reviewed by Goulson 2003). Adding hives of Africanized honey bees to habitats in French Guiana reduced visitation to certain plants by native stingless bees, due to honey bees outcompeting native bees for limited floral resources, and removing the honey bee hives reversed the trend (Roubik 1978). Most studies of the effects of honey bees on native pollinators focus on introduced vs. native bees, and less is known about the effects of honey bees on moth pollinators and moth-pollinated plants. The presence of honey bees in Pacific coastal dune systems is likely to have a negative effect on SI *A. umbellata*'s reproductive success, because pervasive pollen theft would increase pollen limitation and *A. umbellata*'s seed set is partly limited by pollen (L. A. D. Doubleday & C. G. Eckert, unpublished data).

ACKNOWLEDGEMENTS

We thank A. Boag, D. Denley, S. Lynskey, and B. Gresik for help in the field; N. Derieg, J.E. Dugan, B. Gresik, S. Hodges, M. King, M. Jacobs, H.M. Page, G. Rochon-Terry, C. Willis, and J. Viengkone for lab assistance; D. Hubbard, J.E. Dugan, and C. Sandoval for support in the field; D. Lafontaine, J.-F. Landry, and C. Schmidt of the Canadian National Collection of Insects, Arachnids and Nematodes for insect identifications; California State Parks, the University of California Natural Reserve System, and the United States Fish and Wildlife Service for research permits and access to study sites. This work was supported by the Government of Ontario (Ontario Graduate Scholarship to L.A.D.D.) and the Natural Sciences and Engineering Research Council of Canada (Discovery Grant to C.G.E. and Canada Graduate Scholarship to L.A.D.D.).

REFERENCES

Aizen MA, Feinsinger P (1994) Habitat fragmentation, native insect pollinators, and feral honey bees in Argentine "Chaco Serrano." Ecological Applications 4:378–392.

Armbruster WS (2017) The specialization continuum in pollination systems: diversity of concepts and implications for ecology, evolution and conservation. Functional Ecology 31:88–100.

Briscoe AD, Chittka L (2001) The evolution of color vision in insects. Annual Review of Entomology 46:471–510.

Brunet J, Holmquist KGA (2009) The influence of distinct pollinators on female and male reproductive success in the Rocky Mountain columbine. Molecular Ecology 18:3745–3758.

Buckley Y (2015) Generalized linear models. In: Fox GA, Negrete-Yankelevich S, Sosa VJ (eds) Ecological Statistics. Contemporary Theory and Application. Oxford University Press, Oxford, UK, pp 131–148.

Bustamante E, Casas A, Búrquez A (2010) Geographic variation in reproductive success of *Stenocereus thurberi* (Cactaceae): effects of pollination timing and pollinator guild. American Journal of Botany 97:2020–2030.

Butz Huryn VM (1997) Ecological impacts of introduced honey bees. The Quarterly Review of Biology 72:275–297.

Darling E, Samis KE, Eckert CG (2008) Increased seed dispersal potential towards geographic range limits in a Pacific coast dune plant. New Phytologist 178:424–435.

Doubleday LAD (2012) Moth pollination, low seed set, and vestigialization of attractive floral traits in *Abronia umbellata* (Nyctaginaceae). Queen's University (Canada) [online] URL: http://search.proquest.com/openview/ab5665e7a8d700341932 5747424Ifb61/1?pq-origsite=gscholar&cbl=18750&diss=y

Doubleday LAD, Raguso RA, Eckert CG (2013) Dramatic vestigialization of floral fragrance across a transition from outcrossing to selfing in *Abronia umbellata* (Nyctaginaceae). American Journal of Botany 100:2280–2292.

Eckert C G (2002) Effect of geographical variation in pollinator fauna on the mating system of *Decodon verticillatus* (Lythraceae). International Journal of Plant Sciences 163:123–132.

Fenster CB, Armbruster WS, Wilson P, Dudash MR, Thomson JD (2004) Pollination syndromes and floral specialization. Annual Review of Ecology, Evolution, and Systematics 35:375–403.

Fishbein M, Venable DL (1996) Diversity and temporal change in the effective pollinators of *Asclepias tuberosa*. Ecology 77:1061–1073.

Fleming TH, Sahley CT, Holland JN, Nason JD, Hamrick JL (2001) Sonoran desert columnar cacti and the evolution of generalized pollination systems. Ecological Monographs 71:511–530.

Goulson D (2003) Effects of introduced bees on native ecosystems. Annual Review of Ecology, Evolution, and Systematics 34:1–26.

Hargreaves AL, Harder LD, Johnson SD (2009) Consumptive emasculation: the ecological and evolutionary consequences of pollen theft. Biological Reviews 84:259–276.

Hargreaves AL, Harder LD, Johnson SD (2010) Native pollen thieves reduce the reproductive success of a hermaphroditic plant, *Aloe maculata*. Ecology 91:1693–1703.

Herrera CM (1988) Variation in mutualisms: the spatiotemporal mosaic of a pollinator assemblage. Biological Journal of the Linnean Society 35:95–125.

Holland NJ, Fleming TH (2002) Co-pollinators and specialization in the pollinating seed-consumer mutualism between senita cacti and senita moths. Oecologia 133:534–540.

Irwin RE, Bronstein JL, Manson JS, Richardson L (2010) Nectar robbing: ecological and evolutionary perspectives. Annual Review of Ecology, Evolution, and Systematics 41:271–292.

Jabis MD, Ayers TJ, Allan GJ (2011) Pollinator-mediated gene flow fosters genetic variability in a narrow alpine endemic, *Abronia alpina* (Nyctaginaceae). American Journal of Botany 98:1583–1594.

Lenth RV (2016) Least-squares means: the R package lsmeans. Journal of Statistical Software 69:1–33.

Morse DH, Fritz RS (1983) Contributions of diurnal and nocturnal insects to the pollination of common milkweed (*Asclepias syriaca* L.) in a pollen-limited system. Oecologia 60:190–197.

Ollerton J, Alarcón R, Waser NM, Price MV, Watts S, Cranmer L, Hingston A, Peter CI, Rotenberry J (2009) A global test of the pollination syndrome hypothesis. Annals of Botany 103:1471–1480.

van der Pijl L (1961) Ecological aspects of flower evolution. II. Zoophilous flower classes. Evolution 15:44–59.

R Core Team (2017) R: a language and environment for statistical computing. R Foundation for Statistical Computing, Vienna, Austria. [online] URL: http://www.R-project.org/

Roubik DW (1978) Competitive interactions between neotropical pollinators and Africanized honey bees. Science 201:1030–1032.

Saunders NE, Sipes SD (2006) Reproductive biology and pollination ecology of the rare Yellowstone Park endemic *Abronia ammophila* (Nyctaginaceae). Plant Species Biology 21:75–84.

Tillett SS (1967) The maritime species of *Abronia* (Nyctaginaceae). Brittonia 19:299–327.

Walter HE (2010) Floral biology of *Echinopsis chiloensis* ssp. *chiloensis* (Cactaceae): evidence for a mixed pollination syndrome. Flora - Morphology, Distribution, Functional Ecology of Plants 205:757–763.

Waser NM, Chittka L, Price MV, Williams NM, Ollerton J (1996) Generalization in pollination systems, and why it matters. Ecology 77:1043–1060.

Williamson PS, Muliani L, Janssen GK (1994) Pollination biology of *Abronia macrocarpa* (Nyctaginaceae), an endangered Texas species. The Southwestern Naturalist 39:336–341.

Willmer P (2011) Pollination and Floral Ecology. Princeton University Press, Princeton. [online] URL: http://public. eblib.com/choice/publicfullrecord.aspx?p=793219 (accessed 16 May 2015).

Winfree R, Aguilar R, Vázquez DP, LeBuhn G, Aizen MA (2009) A meta-analysis of bees' responses to anthropogenic disturbance. Ecology 90:2068–2076.

Wolfe LM, Barrett SCH (1988) Temporal changes in the pollinator fauna of tristylous *Pontederia cordata*, an aquatic plant. Canadian Journal of Zoology 66:1421–1424.

19

IDENTIFICATION OF PLANT SPECIES FOR POLLINATOR RESTORATION IN THE NORTHERN PRAIRIES

Diana B. Robson[1],*, Cary Hamel[2] and Rebekah Neufeld[3]

[1]*Manitoba Museum, 190 Rupert Avenue, Winnipeg, MB Canada R3B 0N2*
[2]*Nature Conservancy of Manitoba, 611 Corydon Ave., Winnipeg, MB, Canada, R3L 0P3*
[3]*Nature Conservancy of Manitoba, 207-1570 18th Street, Brandon, MB, Canada R7A 5C5*

Abstract—Research on diurnal plant–pollinator interactions indicates that a small number of generalist plants provide a disproportionately high amount of floral resources to pollinating insects. Identifying these generalist plants would help prairie restoration specialists select species that will provide forage for the majority of pollinator taxa. Field research in three Canadian fescue (*Festuca hallii*) prairie preserves that were at most 3.3 km away from each other was conducted in 2014 and 2015 to create pooled, weighted, plant–insect visitor matrices for each site. Using these matrices, generalization (G) scores were calculated for each plant species to help assess their importance to wild insect visitors as this method controls for differences in insect abundances over the year. The three species with the highest average generalization scores were *Solidago rigida*, *Erigeron glabellus* and *Symphyotrichum laeve*. Species accumulation curves were created to determine how many plant species would need to be present before most pollinator taxa would have at least one acceptable forage species. This research indicates that the 16 plant species (33% of the total) with the highest average generalization scores were visited by 90% of the observed pollinator taxa. To detect exceptionally attractive plant species while accounting for natural differences in abundance, we calculated the insect, bee and fly visitation rates per inflorescence. There was several specialized plant species that were visited frequently by bees. Most of these specialized plants had purple or yellow, tubular flowers, and bloomed in mid to late summer when bee populations were most numerous.

Keywords: Fescue prairie, generalists, insect visitors, pollination, restoration, seed mix

INTRODUCTION

Fescue prairie is a unique North American grassland formation dominated by *Festuca hallii* (Coupland & Brayshaw 1953; Coupland 1961). Interest in conserving and restoring Canada's fescue prairies is increasing due to the relative rarity of this ecosystem; agricultural conversion, industrial activity and exotic species encroachment have drastically reduced the amount of native prairie remaining. Complete restoration of all species that were native to an ecosystem is the ideal end goal as literature suggests that the more species present, the greater the resiliency of the community (LaBar et al. 2014). Due to the short growing season and potential for both late spring and early fall frosts in the northern prairies, using locally-adapted seeds is recommended to avoid phenological mismatching (Chuine 2010). Currently, the seeds of native grasses are the ones most commonly available from commercial seed growers in western Canada. Unfortunately, grasses are less important to pollinators than plants that offer large amounts of nectar. The addition of insect-pollinated herbs or shrubs to seed mixes is needed to ensure that pollinators will be attracted to restored areas (Menz et al. 2011). Some plants may be exceptionally attractive to pollinators; research is needed to identify them.

Research on plant–pollinator interactions has found that some plants are super-generalists that act as the skeleton of an ecosystem (Jordano et al. 2003; Bascompte & Jordano 2007). Essentially, super-generalists enable less common species to persist (Memmott et al. 2004; Saavedra et al. 2011). Usually these super-generalist plants are visited by a wide diversity of insect taxa due to their large floral displays, radially symmetrical flowers and easily accessible nectar (Elle et al. 2012; Koski et al. 2015). Ecological theory suggests that including super-generalist plants in a seed mix will result in a more diverse and stable plant–pollinator community than adding species at random (Montoya et al. 2012; LaBar et al. 2014; Harmon-Threatt & Hendrix 2015). As a result, plant–pollinator interaction matrices are increasingly being used to help identify the most important plants for restoration purposes (Forup et al. 2008; Devoto et al. 2012; Montoya et al. 2012; Russo et al. 2013). Indeed, including the most generalized plants to restoration seed mixes does increase pollinating insect diversity and abundance over randomly chosen mixes (LaBar et al. 2014; Kremen & M'Gonigle 2015; M'Gonigle et al. 2015).

Following Isaacs et al. (2009) note that regional research efforts are needed to identify the best native plant species to support the pollinator communities in each region, the goal of this study was to identify plant species most attractive to pollinators to use for restoration of Canada's fescue prairies. To conduct our evaluation we collected plant–pollinator interaction data from three fescue prairie preserves in south

*Corresponding author: drobson@manitobamuseum.ca

western Manitoba during the summers of 2014 and 2015. To identify the super-generalist plant species, and make it easier to compare species with a different number of available insect pollinators, we calculated a generalization score (G) for each species developed by Medan et al. (2006). This generalization score takes into account not just the number of insect-visiting taxa (S), but their resource usage (RU) (i.e. the proportion of all available insect taxa that visit the plant), and evenness (E). Plants with high generalization scores are visited equally by a large number of the available insect taxa compared to plants with low generalization scores, which tend to be visited by only a few insect taxa and thus are more specialized. We identified the species with the highest generalization scores in these fescue prairies and considered them to be top candidates for inclusion in fescue prairie seed mixes. However, Harmon-Threatt and Hendrix (2015) and Russo et al. (2013) noted that there are certain specialist plants (often legumes) that are visited very frequently by bees and may be important to include in a seed mix to support these insects. Further, it is possible that plant density will affect the G score simply by increasing the likeliness that an encounter will be observed or that insects will visit that plant species. Therefore, we also determined the visitation rate per inflorescence (instead of the rate per plot) by all the pollinators, by just bee taxa, and by just fly taxa. We created species accumulation curves (Ebeling et al. 2008) to help determine how many of the most generalized plants would have to be added to a seed mix before most insect taxa (90%) had at least one plant species to forage on. Lastly, we graphed the phenology of all plant species observed to determine the flowering sequence and detect any gaps.

MATERIALS AND METHODS

Study Sites

Three prairie conservation sites in the Aspen Parkland Ecoregion south of Riding Mountain National Park, Manitoba were surveyed: Elk Glen (50.849444°N, 100.819417°W), Cleland (50.830693°N, 100.788083°W) and Crown (50.834333°N, 100.787750°W). The plots at Elk Glen were 2.8 to 3.3 km away from the plots at the Cleland and Crown sites; plots at the latter two sites were all within 600 m of each other. The soils in this area are moderately calcareous consisting of mainly sand and silt derived from glacial till. The sites contain mainly mixed wood forests with some fescue prairie patches; past grazing practices are thought to have reduced the fescue grass component in the Riding Mountain region (Trottier 1986). As the prairies nearest the roads were heavily invaded by non-native species such as *Bromus inermis* and *Poa pratensis*, our choices regarding plot locations were limited to areas that were accessible only by foot, which constrained the number of plots that could be surveyed in a day. There are at least 155 vascular plant species that occur in the area, but many of these species are found in the wetlands or forested parts of the sites. The close proximity of the sites means that the climate and landscapes were similar. The Elk Glen and Cleland sites are owned and managed as nature preserves by the Nature Conservancy of Canada (NCC) but both were

previously grazed by cattle, the latter as recently as 2013. The Crown land is managed by the Manitoba government for wildlife and agricultural values and may have been historically grazed by cattle several decades ago. As there were some differences in the plant communities due to their grazing history, they were all considered separate sampling sites.

Vegetation Surveys

In 2014, we established six 4 m² permanent plots in each of the three sites on flat to gently sloping land. Small plots rather than transects were chosen to enable: (1) future comparison of these data with other Manitoba timed pollinator surveys in tall-grass prairie (Robson 2008, 2013), (2) observations of interactions between less common plants and insects (Gibson et al. 2011), and (3) calculation of visits/inflorescence in a reliable way. The plots were randomly selected and at least 10 m apart. In 2014, sampling was conducted for four consecutive days per site, which was repeated four times in mid-June, -July, -August and -September (16 days total). In 2015, sampling was conducted on the previously established plots for four consecutive days per site, repeated four times at the beginning and at the end of June, in early July and in late August (16 days total). In total 15 more plant species were observed in 2015 than in 2014 as a result of conducting the surveys on different calendar days. Plant richness and number of inflorescences of each species in the plots was recorded each sampling day.

Floral Visitor Surveys

Flower-visiting insect sampling occurred on the same days and in the same plots as the vegetation surveys: four consecutive days for four months in each of 2014 and 2015 for a total of 32 days. Each plot was surveyed for 10 minutes each sampling day thus the total time spent surveying was 96 hrs. Due to natural variability in abundance, some plant species were present in more plots than others and thus were observed for a longer period of time. Surveys were conducted between 09:30 and 17:00 when insect foraging activity is at a maximum (Kevan & Baker 1983) thus nocturnal insects were not observed. During the survey periods the mean temperature recorded at the nearest Environment Canada weather station (i.e. Brandon, MB) was 17.7°C (± 0.9 SE), the mean relative humidity 53.0% (± 1.9 SE) and the mean wind speed 10.0 km/hr (± 0.62 SE). The order in which the plots were visited was randomized each day. Some of the flower-visiting insects (e.g. crabronid wasps in the genus *Ectemnius*) may have been predators of other flower-visiting insects. Regardless of whether the wasps were foraging for pollen, nectar or prey, they were considered potential pollinators and their visitations recorded. Ambush bugs (*Phymata* spp.) and crab spider (*Misumena* spp.) visits were not recorded as they remain stationary on one inflorescence for a long time and thus were unlikely to act as pollinators.

All insect visitations to any inflorescence in the plot were recorded but the quality of the visit was not assessed and therefore may not have resulted in pollination of the plant. However, even if an insect visit did not fulfill the reproductive goal of the plant, the plant was probably important for the insect's survival; this is an aspect of plant–

pollinator interaction that we wanted to capture. The first time an insect taxon was observed a specimen was obtained and given a unique collection number. When the same (or what appeared to be a very similar) taxon was observed later on, the collection number was used to link that insect visit to the plant. Although this technique does not allow for complete identification "on the wing" (resulting in an underestimate of insect taxa) it does enable evaluation of insect visitation frequency, which was then used to determine the visitation rate for each plant species by all insect taxa, by bees only and by flies only (Parachnowitsch & Elle 2005). All insect voucher specimens were identified by zoologists using reference specimens at the Manitoba Museum (MM) and the Wallis Roughley Entomology Museum in Winnipeg, Manitoba; the specimens were deposited in MM's zoology collection.

Data Analysis

The daily inflorescence density at each site was calculated by counting the number of inflorescences of each species in each plot, dividing by the plot area (4 m²) and averaging the values of all plots. The average inflorescence density at all three sites over the length of the study (32 days) for each species was then calculated. Using the density data we created a phenology graph. To do this, we determined the calendar day (from pooled 2014 and 2015 data) when each plant species began and ceased blooming in our plots. Differences in flowering times from year to year are known to vary in this region due to differences in snow cover, spring temperatures and late frosts, thus bloom times should only be considered approximate. Further, as we did not conduct the surveys every single day, the length of the flowering period is likely slightly longer than indicated.

For each of the three sites the data were used to create two year cumulative, weighted (i.e. by the number of visits recorded) plant–insect visitor matrices. These matrices were used to calculate a generalization (G) score (Medan et al. 2006) for each plant species using pooled data over two years (Appendix I) to facilitate comparison between them as:

$$G = RU \times E$$

Where

RU (Resource use) is:

$$RU = \frac{\# \ of \ visiting \ insect \ taxa}{\# \ of \ available \ insect \ taxa}$$

E (evenness) is:

$$E = \frac{-\sum_i p_i \cdot \ln p_i}{\ln S}$$

where p_i is the proportion of all interactions corresponding to the ith individual visiting insect of a given plant, and S is the total number of individual insects visiting it. When evenness is maximal, E is equal to 1 and when it approaches 0, interaction frequencies are very unequal. When a plant had only one insect visitor, the index could not be computed because it required dividing by zero; these plants were excluded from the analysis. The available insect taxa were all the taxa observed at that site at any time over the

two year study. The G scores from each site were then averaged to get a single G score for each plant species. Although pooling the data over two years obscures some of the variation, we believe the results reflect the core of the most common insect visitors and interactions in this area. When a plant flowered at more than one site, the overall G score was the average of all the scores. The G, RU and E scores varied between 0 and 1. Low G scores indicated specialists while higher ones indicated generalists. All plants were ranked according to their G scores. If two species had the same G score, the plant that was visited at a higher rate by all insects was listed first. We then graphed species accumulation curves using Microsoft Excel where the plants were arranged in decreasing order of their G scores. The cumulative number of insect taxa that visited the accumulative number of plant species was determined using the matrix data. These data were used to determine the minimum number of plant species that would need to be added to an ecosystem to ensure that most (~90%) insect taxa had at least one preferred forage plant species.

The density of each plant species at the sites was different, which obscured evaluation of their importance to the insect visitors. To address this, we calculated the insect visitation rate per inflorescence to reduce the chance that density was affecting the results. The visitation rates are the averages over the number of days the plant was in flower. We also calculated the bee and fly visitation rates per inflorescence to determine which plant species were particularly attractive to them. We used linear regression analysis to determine the relationship between plant density and G, and plant density and insect visitation rate per inflorescence by all insects, by bees only and by flies only. Logarithmic transformations to homogenize variances were applied to the data. These statistical tests were performed using Analyze-it software (Analyze-it Software Ltd. 2009). The percentage of all visits each insect taxon made at each site was determined using pooled data from 2014 and 2015, and then averaged when the taxon occurred at more than one site. Lastly, we obtained lists of species available from local Canadian seed suppliers to determine which plants are currently available for use in restorations.

RESULTS

During the study, a total of 110 insect taxa were observed visiting the flowers of 48 forbs and/or shrubs found within the study plots (Appendix II). The average generalization score of all 48 plant species was 0.07 with the highest value 0.19 and the lowest 0.01 (Tab. 1). *Solidago rigida* had the highest G score and received the greatest number of insect visitor taxa (38). *Erigeron glabellus* has the second highest G score even though it was visited by fewer insect taxa than *Symphyotrichum laeve* and *Solidago nemoralis* due to a greater evenness. Three of the species with the highest G scores also had high insect visitation rates per inflorescence: *Solidago rigida*, *S. canadensis* and *Symphyotrichum laeve*. *Solidago rigida* and *Symphyotrichum laeve* were also visited frequently by bees. However, many of the plant species with the highest visitation rates had relatively low G scores (≤ 0.06) including: *Cirsium*

TABLE I.   Average generalization scores (G), insect visitation rates, number of insect visitor taxa (S), and inflorescence (infl.) density and flowering period data for vascular plant species in three fescue prairie sites in southern Manitoba (sorted by G score then insect visitation rate). Plant species with only one insect visitor were excluded. Numbers in bold are the ten highest values for that variable.

Scientific name	G (mean ±SE)[1]	Insect[2] visitation rate (mean visits infl.[-1]day[-1]±SE)	Bee[3] visitation rate (mean visits infl.[-1]day[-1]±SE)	Diptera visitation rate (mean visits infl.[-1]day[-1]±SE)	S (total # of insect visitor taxa)	Infl. density (mean #/m²±SE)	Flowering period[4]
Solidago rigida	0.19±0.03	**0.50±0.11**	**0.35±0.08**	0.13±0.04	**38**	0.34±0.11	M-L
Erigeron glabellus	0.15±0.06	0.19±0.07	0.07±0.02	0.09±0.05	22	0.16±0.04	E-M-L
Symphyotrichum laeve	0.14±0.01	**0.41±0.09**	**0.30±0.08**	0.10±0.03	**31**	**0.41±0.11**	L
Solidago nemoralis	0.14±0.08	0.23±0.09	0.14±0.06	0.08±0.04	**32**	**0.51±0.14**	M-L
Linum lewisii	0.14±0.0	0.19±0.08	0.08±0.04	0.03±0.02	11	0.07±0.02	E-M
Solidago canadensis	0.12±0.01	0.31±0.09	0.14±0.05	0.09±0.04	19	0.05±0.02	L
Campanula rotundifolia	0.12±0.04	0.07±0.01	0.06±0.01	0.0±0.0	20	0.36±0.10	M-L
Achillea millefolium	0.12±0.02	0.06±0.01	0.02±0.01	0.04±0.01	20	0.32±0.08	E-M-L
Symphoricarpos occidentalis	0.12±0.03	0.03±0.02	0.01±0.01	0.02±0.02	26	0.29±0.09	M-L
Apocynum androsaemifolium	0.10±0.0	0.20±0.10	0.09±0.04	0.07±0.05	7	0.03±0.01	M
Astragalus laxmannii	0.10±0.04	0.09±0.06	0.07±0.06	0.0±0.0	14	0.36±0.14	M
Prunus virginiana	0.09±0.0	**0.50±0.29**	**0.40±0.31**	0.04±0.02	9	0.07±0.03	E
Drymocallis arguta	0.09±0.03	0.23±0.05	0.18±0.05	0.02±0.01	18	0.12±0.03	M
Rudbeckia hirta	0.09±0.03	0.22±0.06	0.06±0.04	0.16±0.06	18	0.12±0.03	M-L
Lathyrus venosus	0.09±0.0	0.12±0.07	0.11±0.06	0.0±0.0	6	0.03±0.01	E-M
Cerastium arvense	0.09±0.01	0.06±0.02	0.02±0.01	0.03±0.02	11	0.19±0.06	E
Symphyotrichum ericoides	0.08±0.0	0.01±0.00	0.0±0.0	0.01±0.01	6	**0.71±0.31**	L
Liatris ligulistylis	0.07±0.04	0.24±0.12	0.18±0.09	0.02±0.02	8	0.03±0.01	L
Erigeron strigosus	0.07±0.0	0.15±0.09	0.0±0.0	**0.15±0.08**	4	0.04±0.01	M
Vicia americana	0.07±0.0	0.11±0.04	0.10±0.04	0.02±0.02	6	0.10±0.03	E-M
Cirsium drummondii	0.06±0.05	**1.18±0.41**	**1.17±0.41**	0.0±0.0	10	0.02±0.01	E-M
Agastache foeniculum	0.06±0.03	**0.46±0.16**	**0.45±0.16**	0.0±0.0	10	0.08±0.02	M-L
Monarda fistulosa	0.06±0.02	**0.37±0.14**	**0.32±0.12**	0.0±0.0	12	0.16±0.05	M-L
Liatris punctata	0.05±0.0	0.22±0.08	0.11±0.06	0.04±0.04	4	0.04±0.02	L
Helianthus pauciflorus	0.05±0.0	0.19±0.08	0.10±0.07	0.16±0.06	4	0.04±0.02	L
Hedysarum boreale	0.05±0.01	0.18±0.05	0.18±0.05	0.0±0.0	5	0.23±0.07	M
Astragalus crassicarpus	0.05±0.04	0.15±0.04	0.04±0.03	0.0±0.0	7	0.17±0.06	E-M
Geum triflorum	0.05±0.02	0.05±0.02	0.05±0.02	0.0±0.0	7	**0.59±0.15**	E
Zizia aptera	0.04±0.03	0.19±0.06	0.06±0.02	0.11±0.04	13	0.12±0.05	E
Agoseris glauca	0.04±0.01	0.22±0.08	0.13±0.07	0.06±0.05	5	0.02±0.01	M-L
Allium stellatum	0.04±0.02	0.04±0.03	0.04±0.02	0.01±0.01	4	0.04±0.02	L
Lithospermum canescens	0.04±0.02	0.02±0.01	0.01±0.0	0.0±0.0	10	**0.68±0.22**	E
Dalea purpurea	0.03±0.01	**0.79±0.35**	**0.78±0.35**	0.01±0.0	8	0.13±0.05	L

Table I continued.

Scientific name	G (mean ±SE)[1]	Insect[2] visitation rate (mean visits infl.⁻¹day⁻¹±SE)	Bee[3] visitation rate (mean visits infl.⁻¹day⁻¹±SE)	Diptera visitation rate (mean visits infl.⁻¹day⁻¹±SE)	S (total # of insect visitor taxa)	Infl. density (mean #/m²±SE)	Flowering period[4]
Tragopogon dubius	0.03±0.0	0.13±0.12	0.08±0.08	0.04±0.04	2	0.01±0.01	E
Gaillardia aristata	0.03±0.02	0.12±0.06	0.08±0.05	0.03±0.03	4	0.02±0.01	M-L
Oxytropis campestris	0.03±0.0	0.11±0.02	0.09±0.02	0.0±0.0	4	0.19±0.06	M
Galium boreale	0.03±0.02	0.04±0.03	0.01±0.0	0.01±0.01	5	0.27±0.09	M
Rosa acicularis	0.02±0.01	**0.55±0.26**	0.02±0.01	**0.53±0.26**	6	0.03±0.01	E-M
Astragalus agrestis	0.02±0.01	0.19±0.05	0.16±0.02	0.0±0.0	3	0.04±0.01	E
Penstemon gracilis	0.02±0.01	0.14±0.11	0.14±0.10	0.0±0.0	3	0.02±0.01	M
Heuchera richardsonii	0.02±0.01	0.11±0.04	0.11±0.04	0.0±0.0	4	0.02±0.01	E-M
Pediomelum esculentum	0.02±0.01	0.11±0.05	0.10±0.05	0.0±0.0	3	0.04±0.01	E-M
Polygala senega	0.02±0.02	0.01±0.01	0.0±0.0	0.0±0.0	4	0.21±0.08	E
Fragaria virginiana	0.01±0.01	0.01±0.01	0.01±0.0	0.0±0.0	2	0.06±0.02	E

[1]G = generalization score. A G score (see Methods section for the formula) was calculated for each of the three sites using pooled data over two years (Appendix I) and averaged if the plant occurred at more than one site.
[2]Includes the visits by all insect taxa pooled over two years and averaged from all three sites.
[3]Includes only visits by bees (Hymenoptera) in the Andrenidae, Apidae, Colletidae, Halictidae and Megachilidae pooled over two years and averaged from all three sites.
[4]E= Early (June), M=Mid (July), L= Late (August-September). Determined using pooled data over two years from all three sites.

drummondii, *Dalea purpurea*, *Agastache foeniculum* and *Monarda fistulosa*. This is because these plants were visited frequently by two of the most frequently observed insect taxa - *Bombus ternarius* and *B. sandersoni* (Tab. 2) - which greatly decreased the evenness and consequently the G score. *Rosa acicularis* was among the most frequently visited plants but its visitors were mainly root maggot flies (*Drymeia* sp.) not bumblebees (*Bombus* spp.).

The average density of each plant varied from a high of 0.7 inflorescences/m² for *Symphyotrichum ericoides* to < 0.1 inflorescences/m² for *Tragopogon dubius*. Differences in inflorescence density may have influenced the G scores as these two variables were strongly positively correlated ($y = 0.3393x + 2.202$, $R^2 = 0.2186$, $P < 0.001$). This may be in part because species with low densities were observed for less time than more abundant plants, potentially decreasing the number of insect visitor taxa observed and influencing the G score. However, inflorescence density was not correlated with the insect visitation rate per inflorescence ($y = -0.1839x + 2.6452$, $R^2 = 0.0287$, $P = 0.25$), bee visitation rate per inflorescence ($y = -0.0224x + 2.4753$, $R^2 = 0.0004$, $P = 0.25$) or fly visitation rate per inflorescence ($y = -0.0017x + 2.3504$, $R^2 = 6E^{-06}$, $P = 0.99$) suggesting that factors other than abundance are being used by some insects to select plants. Indeed some species with low densities had some of the highest visitation rates per inflorescence (e.g. *Prunus virginiana* and *Solidago canadensis*).

The insect taxa observed in this study belonged to five orders: 7 beetle species (Coleoptera), 39 fly species (Diptera), 3 bug species (Hemiptera), 39 bee, ant and wasp species (Hymenoptera) and 22 butterfly and moth species (Lepidoptera) (Tab. 2). The Hymenoptera consisted of mostly bees (29 species), seven species of wasps, two species of ants and one sawfly. On average, *Bombus* species together made 54.7% of the visits at each site, with *Bombus ternarius* alone responsible for over 32.0%. *Bombus* species also visited the greatest number of plant species: 12.4% over the whole course of the study on average compared to 3.7% for all other taxa. The root maggot fly *Drymeia* sp. was the most frequently observed fly taxon making on average 3.8% of all insect visits, mainly to *Rosa acicularis*. Of the fly species observed, the tachinid *Peleteria* sp. visited the highest number of plant species (12). Most insect taxa (56.0%) were observed in only one time period (e.g. early, mid or late summer) but 28.0% were observed in two periods and 15.0% in all three. The taxa most frequently observed were also the ones observed over the longest period of time.

The species accumulation curve for all insect visitor taxa indicates that most (90%) would have at least one appropriate forage plant once 16 of the most generalized plants were present (Fig. 1). Fly taxa reached the 90% point after nine of the most generalized plant species were added, and bees after 13 species. Figure 2 shows the flowering range of all the plants observed being visited by insects during the survey period. Of the 16 most generalized plant species (see Table I) seven began flowering in June, six in July and three in August.

TABLE 2. The time period when 20 insect visitor taxa with the highest percentage of visits and number of plant species visited (S) in three fescue prairie sites in southern Manitoba were observed.

Order	Family	Common Name	Scientific Name	Time period[1]	Visits (mean %/site±SE)[2]	S (total # of plant species visited)
Hymenoptera	Apidae	Tri-colored bumblebee	*Bombus ternarius*	E-M-L	32.4±5.3	18
Hymenoptera	Apidae	Sanderson bumblebee	*Bombus sandersoni*	E-M-L	9.5±1.7	21
Hymenoptera	Apidae	Nevada bumblebee	*Bombus nevadensis*	E-M-L	4.8±1.5	18
Diptera	Muscidae	Root maggot fly	*Drymeia* sp.	E-M	3.8±1.9	3
Diptera	Tachinidae	Tachinid fly	*Tachina* sp.	M-L	3.2±0.2	7
Hymenoptera	Halictidae	Sweat bee	*Augochlorella aurata*	E-M-L	2.5±0.99	14
Hymenoptera	Apidae	Northern amber bumblebee	*Bombus borealis*	E-M-L	2.5±0.86	13
Diptera	Tachinidae	Tachinid fly	*Chaetogaedia* sp.	L	2.5±0.95	7
Hymenoptera	Apidae	Yellow-banded bumblebee	*Bombus terricola*	E-M-L	2.2±0.49	7
Diptera	Bombyliidae	Bee fly	*Poecilanthrax alcyon*	L	1.9±0.60	8
Hymenoptera	Apidae	Confusing bumblebee	*Bombus perplexus*	L	1.8±0.90	6
Diptera	Tachinidae	Tachinid fly	*Peleteria* sp.	L	1.6±0.20	12
Hymenoptera	Megachilidae	Unarmed leaf-cutter bee	*Megachile inermis*	E-M-L	1.6±0.51	10
Hymenoptera	Apidae	Red-belted bumblebee	*Bombus rufocinctus*	L	1.5±0.73	4
Diptera	Syrphidae	Hover fly	*Toxomerus geminatus*	E-M-L	1.3±0.08	11
Hymenoptera	Halictidae	Sweat bee	*Lasioglossum succinipenne*	E-M-L	1.3±0.58	10
Diptera	Bombyliidae	Bee fly	*Villa nigra*	L	1.3±0.43	3
Hymenoptera	Megachilidae	Small-handed leaf-cutter bee	*Megachile gemula*	E-M-L	1.2±0.33	10
Lepidoptera	Lycaenidae	Silvery blue	*Glaucopsyche lygdamus*	E-M	1.2±1.10	8
Hymenoptera	Andrenidae	Miserable andrena	*Andrena miserabilis*	E-M-L	1.2±0.39	6
Other Diptera	n/a	Flies	32 spp.	E-M-L	11.6	27
Other Hymenoptera (bees)	n/a	Bees	17 spp.	E-M-L	8.0	29
Other Lepidoptera	n/a	Butterflies/moths	21 spp.	E-M-L	7.6	25
Other Hymenoptera (ants & wasps)	n/a	Ants/wasps	10 spp.	E-M-L	2.6	12
Coleoptera/ Hemiptera	n/a	Beetles/bugs	10 spp.	E-M-L	2.6	9

[1]E= Early (June), M=Mid (July), L= Late (August-September). Determined using pooled data over two years from all three sites.

[2]The % of visits was calculated for each taxa at each of three sites using pooled data over two years and averaged if the insect was seen at more than one site.

DISCUSSION

The results of this study and many others (Forup et al. 2008; Devoto et al. 2012; Montoya et al. 2012; Russo et al. 2013) suggest that just a small number of the most generalized entomophilous plant species supply most of the insect pollinator taxa in a community with appropriate forage. This likely has to do with the functional redundancy of some plant species for the pollinator community (Goldstein & Zych 2016). For example, many of the legumes (e.g. *Astragalus* spp., *Vicia americana*, *Hedysarum*

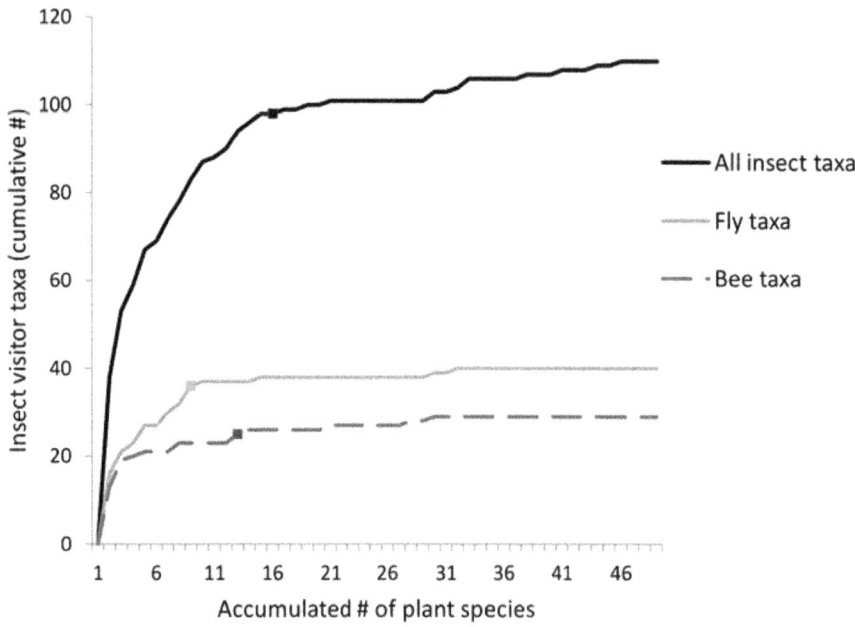

FIGURE 1. Species accumulation curves for all insect visitor taxa recorded (110), also broken down for the taxa of flies (39) and bees (29) as the most to the least generalized plants are added. Squares represent the points at which the plant species present were visited by 90% of all insect taxa in that group.

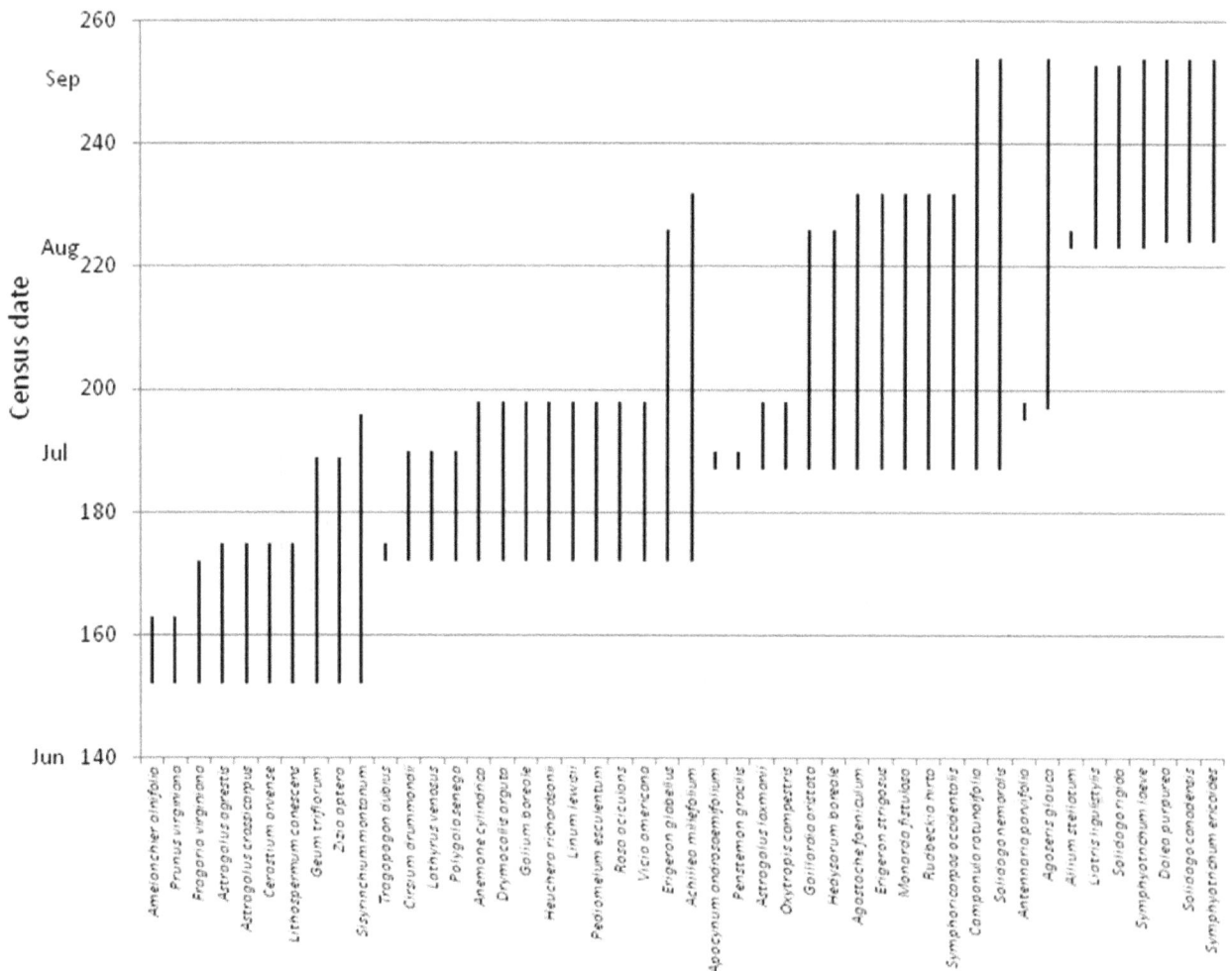

FIGURE 2. Flowering phenology of 48 plant species in south western Manitoba observed over 102 calendar days ordered by earliest to latest flowering period. The line indicates the flowering duration in the research plots using merged 2014 and 2015 survey data.

boreale and *Dalea purpurea*) were visited by the exact same suite of insect taxa. However, the aforementioned legumes flower at different times so at least three species would be needed to provide the insect community with consistent leguminous forage over the year. This is important to remember as restoration ecologists are often under pressure to keep costs down. Wilkerson et al. (2014) found that although using a seed mix with nine species had the highest germination, cover and floral resources, it was not as cost effective as a mix with only five species due to the higher price. However, the actual diversity of pollinators was not assessed in the Wilkerson study. Although adding fewer species may be more cost effective, doing so may result in inadequate floral resources: in our study less than 1% of all early summer insect visits would have occurred if only the top five plant species (in terms of their G scores) were present. A lack of early blooming flowers inhibits the development of healthy bumble bee colonies, which start out in spring with a single queen (Goulson 2003; Goulson et al. 2015; Isaacs et al. 2009). By determining G scores and preparing species accumulation curves with our data, we were able to identify the minimum number of species that would be needed to provide forage for most insect taxa over the entire growing season.

In addition to species with high generalization scores, it may be beneficial to add more specialized plant species to a seed mix if providing consistent forage for bees is desirable. Several studies have shown that *Bombus* spp. preferred plants with relatively deep nectar tubes (Stang et al. 2006; Forup et al. 2008; Elle et al. 2012), and researchers have found that adding such plants (particularly legumes) to restored areas results in increased richness and abundance of this genus (M'Gonigle et al. 2015; Carvell et al. 2011). Although four of the top ten plant species most frequently visited by bees were among the 16 most generalized plants, six species were not. These six plants had purple flowers and short to long nectar tubes that could only be accessed by relatively longer-tongued insects: *Cirsium drummondii*, *Dalea purpurea*, *Agastache foeniculum*, *Monarda fistulosa*, *Hedysarum boreale* and *Liatris ligulistylis*. The differences in the species with the highest G scores and insect visitation rates per inflorescence are due to bee preferences for the aforementioned flowers. In general, plants with the highest visitation rates were visited mainly by bees but one species was frequently visited by flies: *Rosa acicularis*. This rose had a low G score but a high visitation rate mainly due a species of *Drymeia* that swarmed the flowers once they bloomed, quickly removing all pollen. Therefore, specialized plant species should not be discounted for inclusion in wildflower plantings, if they are frequently visited. The visitation rate per inflorescence was thus a valuable metric to identify popular species.

Unfortunately, two of the plant species with the highest generalization scores appear to be currently unavailable from Canadian native seed supply companies (NPSS 2013): *Astragalus laxmannii* and *Cerastium arvense*. Seeds of these species are likely unavailable because they are difficult to harvest or germinate (Baskin & Baskin 1998). Further, seeds from the plant with the highest visitation rate per inflorescence (*Cirsium drummondii*) appear to be unavailable as the plant is uncommon and, due to its prickly leaves, not generally considered a desirable plant (although its popularity with bees may encourage restoration specialists to reconsider it). In the event these plants cannot be acquired, potential substitutes that flower in the same season with slightly lower G scores or visitation rates, and are visited by a similar subset of pollinators, could be selected instead. For example, *Geum triflorum* or *Astragalus crassicarpus* could be substituted for *Cerastium arvense* as they all flower at the same time. Possible substitutes for the mid-summer species -*Astragalus laxmannii* and *Cirsium drummondii*- are two leguminous species: *Vicia americana* and/or *Hedysarum boreale*. Spreading native hay is another possible strategy to introduce propagules for a restoration (Donath et al. 2007), but this may be difficult due the high number of invasive plants present in most fescue prairies.

There are several limitations of this study that need to be considered. Although moths are important pollinators in northern ecosystems (MacGregor et al. 2015), nocturnal insect activity was not monitored and, as a result, the importance of some plant species was likely underestimated. In addition, the 48 plant species were not observed for an equal length of time due to their natural variability in abundance. The possibility for modification to the G score to adjust for differences plant abundance was not addressed in the original reference (Medan et al. 2006), so additional observations are needed for the rarer species. An assumption was made that if these plants are grown as part of a prairie restoration, the insect taxa that were observed under natural conditions would eventually begin to inhabit the site. However, some studies comparing plant–pollinator communities in restored habitats have found that although they eventually function in a similar way, species composition between sites was somewhat different, possibly due to habitat differences (Forup and Memmott 2005; Forup et al. 2008; Menz et al. 2011; Tarrant et al. 2012). Further, there are strong temporal variations in insect pollinator composition from year to year (Dupont et al. 2009; MacLeod et al. 2016; CaraDonna et al. 2017). Thus, a restored prairie may not have the exact same composition as a native reference community. However, Forup et al. (2008) argues that restoration success should be based less on pollinator composition and more on functional similarity. This is because increased species richness may not be strongly correlated with functional diversity (Mayfield et al. 2010; Cadotte et al. 2011). Assessing the functional diversity of an ecosystem is increasingly being used to help set restoration goals (Thorpe & Stanley 2011; Giannini et al. 2016) and assess community health (Cadotte et al. 2011).

Considering all our data and the aforementioned studies, we propose one possible species list for fescue prairie restoration (Tab. 3). It includes 16 of the most generalized plants, plus four that are more specialized but frequently visited: *Agastache foeniculum*, *Dalea purpurea*, *Monarda fistulosa* and *Rosa acicularis*. All of the species in Tab. 3 were also among the top ten most generalized plants at the individual sites except one (*Liatris ligulistylis*). The plants on this list are in eight families: Asteraceae (7 species), Rosaceae (4), Fabaceae (3), Lamiaceae (2), and one species each in the Apocynaceae, Campanulaceae, Caprifoliaceae and Linaceae.

TABLE 3.   Suggested minimum fescue prairie seed mix, and data regarding the expected insect community that would be supported by it.

Flowering period	Species	Insect taxa (cumulative #)[1]	Insect visits (cumulative %)[1]	Most attracted insect taxa[1]
Early	*Geum triflorum*[2]	7	2.0	Eusocial & solitary bees
	Prunus virginiana	13	3.0	Beetles & bees
	Rosa acicularis	18	4.2	Flies & solitary bees
	Lathyrus venosus	21	4.5	Butterflies & eusocial bees
	Linum lewisii	30	5.5	Flies & butterflies
Early-mid	*Apocynum androsaemifolium*	35	5.9	Solitary bees & flies
	Drymocallis arguta	41	7.6	Solitary bees & beetles
	Vicia americana[3]	42	9.1	Butterflies & eusocial bees
Mid	*Rudbeckia hirta*	53	10.9	Flies & butterflies
	Achillea millefolium	61	12.4	Beetles & flies
	Campanula rotundifolia	67	14.5	Solitary & eusocial bees
	Erigeron glabellus	68	16.5	Flies & butterflies
	Monarda fistulosa	69	20.1	Eusocial bees & butterflies
	Symphoricarpos occidentalis	74	25.8	Flies & eusocial bees
Late	*Solidago rigida*	86	43.3	Flies & beetles
	Agastache foeniculum	86	45.6	Eusocial & solitary bees
	Solidago canadensis	90	47.2	Beetles & solitary bees
	Solidago nemoralis	96	55.5	Beetles & flies
	Symphyotrichum laeve	98	69.7	Eusocial bees & flies
	Dalea purpurea	98	77.6	Eusocial bees & butterflies

[1]Based on 2014-2015 insect visitation observations in fescue prairie.

[2]This species is a suggested substitute if seeds of *Cerastium arvense* are unavailable from commercial Canadian seed suppliers

[3]This species is a suggested substitute if seeds of *Astragalus laxmannii* are unavailable from commercial Canadian seed suppliers

Three of the species were shrubs (i.e. *Prunus virginiana*, *Symphoricarpos occidentalis* and *Rosa acicularis*) and the remainder are herbaceous. Almost 90% of all insect taxa observed visited at least one of the 20 plants on this list. All of the major insect flower visitor groups (i.e. eusocial & solitary bees, beetles, butterflies and flies) would have at least one plant they favoured in each flowering period (i.e. early, early-mid, mid and late summer). Many of the species we recommended were previously noted as being important pollen and nectar sources for pollinators in North America including: *Agastache foeniculum*, *Campanula rotundifolia*, *Dalea purpurea*, *Monarda fistulosa*, *Prunus virginiana*, *Solidago* spp., *Symphoricarpos occidentalis* and *Symphyotrichum* spp. (Isaacs et al. 2009; Mader et al. 2011; Evans 2013). Substitutions could, of course, be made to accommodate different environmental conditions but care should be taken to select species that flower at the same time and support a similar insect community. For example, *Liatris ligulistylis* could be added instead of *Dalea purpurea* where the soils are relatively moist as the latter species appears to be less tolerant of such conditions (Anderson & Schelfhout 1980). Differences in seed availability may also affect the precise mix. This seed mix should be considered the minimum number of forb/shrub species for any fescue prairie restoration. The addition of other plant species may be important to meet other restoration goals, such as increasing the total plant richness or providing appropriate floral resources for rare pollinators such as oligolectic bees (Sheffield et al. 2014) or butterflies such as *Danaus plexippus* (Landis 2014). Additional research on pollination in fescue prairie is needed to: (1) identify pollinators of plant species we missed, (2) determine the plant species required by rarer pollinators, (3) determine if there is regional variability in the pollinator community, and (4) determine what role landscape variability may have on pollinator nesting.

ACKNOWLEDGEMENTS

The authors gratefully acknowledge the financial support of the Nature Conservancy of Canada and The Manitoba Museum Foundation Inc. Thanks to the Government of Manitoba for permitting research on their lands. We are grateful for Manitoba Museum staff and volunteer assistance with specimen processing. Special thanks to Sarah Semmler for preparation and identification of insect specimens. We are also grateful to the reviewers of this manuscript for their thoughtful comments, which greatly improved our paper.

APPENDICES

Additional supporting information may be found in the online version of this article:

APPENDIX I.   Generalization scores (G) for vascular plant species in three fescue prairie sites in southern Manitoba.

APPENDIX II. Two-year pooled plant–insect visitor matrix in three fescue prairie sites, Manitoba.

References

Analyze-it Software, Ltd. (2009) http://www.analyse-it.com/.

Anderson RC, Schelfhout S (1980) Phenological patterns among tallgrass prairie plants and their implication for pollinator competition. The American Midland Naturalist 104:253-263.

Bascompte J, Jordano P (2007) Plant-animal mutualistic networks: the architecture of biodiversity. Annual Review of Ecology, Evolution and Systematics 38:567-593.

Baskin CC, Baskin JM (1998) Seeds: ecology, biogeography, and, evolution of dormancy and germination. Elsevier

Cadotte MW, Carscadden K, Mirotchnick N (2011). Beyond species: functional diversity and the maintenance of ecological processes and services. Journal of Applied Ecology 48:1079-1087.

CaraDonna PJ, Petry WK, Brennan RM, Cunningham JL, Bronstein JL, Waser NM, Sanders NJ (2017) Interaction rewiring and the rapid turnover of plant–pollinator networks. Ecology Letters 20:385-94.

Carvell C, Osborne JL, Bourke AFG, Freeman SN, Pywell RF, Heard MS (2011) Bumble bee species' responses to a targeted conservation measure depend on landscape context and habitat quality. Ecological Applications 21:1760-1771.

Chuine I (2010) Why does phenology drive species distribution? Philosophical Transactions B Royal Society of London 365:3149-3160.

Coupland RT (1961) A reconsideration of grassland classification in the northern great plains of North America. Journal of Ecology 49:135-167.

Coupland RT, Brayshaw TC (1953) The fescue grassland in Saskatchewan. Ecology 34:386-405.

Devoto M, Bailey S, Craze P, Memmott J (2012) Understanding and planning ecological restoration of plant–pollinator networks. Ecology Letters 15:319-328.

Donath TW, Bissels S, Hölzel N, Otte A (2007) Large scale application of diaspore transfer with plant material in restoration practice–Impact of seed and microsite limitation. Biological Conservation 138:224-234.

Dupont YL, Padrón B, Olesen JM, Petanidou T (2009) Spatio-temporal variation in the structure of pollination networks. Oikos 118:1261-9.]

Ebeling A, Klein AM, Schumacher J, Weisser WW, Tscharntke T (2008) How does plant richness affect pollinator richness and temporal stability of flower visits? Oikos 117:1808-15.

Elle E, Elwell SL, Gielens GA (2012) The use of pollination networks in conservation. Botany 90:525-534.

Evans MM (2013) Influences of grazing and landscape on bee pollinators and their floral resources in rough fescue grassland. M.Sc. Dissertation, University of Calgary, Calgary, Alberta.

Forup ML, Memmott J (2005) The restoration of plant–pollinator interactions in hay meadows. Restoration Ecology 13:265-274.

Forup ML, Henson KS, Craze PG, Memmott J (2008) The restoration of ecological interactions: plant–pollinator networks on ancient and restored heathlands. Journal of Applied Ecology 45:742-752.

Giannini TC, Giulietti AM, Harley RM, Viana PL, Jaffe R, Alves R, Pinto CE, Mota NF, Caldeira CF, Imperatriz-Fonseca VL, Furtini AE (2016) Selecting plant species for practical restoration of degraded lands using a multiple-trait approach. Austral Ecology.

Gibson RH, Knott B, Eberlein T, Memmott J (2011) Sampling method influences the structure of plant–pollinator networks. Oikos 120:822-831.

Goldstein J, Zych M (2016). What if we lose a hub? Experimental testing of pollination network resilience to removal of keystone floral resources. Arthropod-Plant Interactions 10:263-271.

Goulson D (2003) Bumblebees: behaviour and ecology. Oxford Univ. Press.

Goulson D, Nicholls E, Botías C, Rotheray EL (2015) Bee declines driven by combined stress from parasites, pesticides, and lack of flowers. Science 347:1255957.

Harmon-Threatt AN, Hendrix SD (2015) Prairie restorations and bees: the potential ability of seed mixes to foster native bee communities. Basic and Applied Ecology 6:64-72.

Isaacs R, Tuell J, Fiedler A, Gardiner M, Landis D (2009) Maximizing arthropod-mediated ecosystem services in agricultural landscapes: the role of native plants. Frontiers in Ecology and the Environment 7:196-203.

Jordano P, Bascompte J, Olesen JM (2003) Invariant properties in coevolutionary networks of plant-animal interactions. Ecology Letters 6:69-81.

Kevan PG, Baker HG (1983) Insects as flower visitors and pollinators. Annual Review of Entomology 28:407-453.

Kremen C, M'Gonigle LK (2015). Small-scale restoration in intensive agricultural landscapes supports more specialized and less mobile pollinator species. Journal of Applied Ecology 52:602-610.

Koski MH, Meindl GA, Arceo-Gómez G, Wolowski M, LeCroy KA, Ashman TL (2015) Plant–flower visitor networks in a serpentine metacommunity: assessing traits associated with keystone plant species. Arthropod-Plant Interactions 9:9-21.

LaBar T, Campbell C, Yang S, Albert R, Shea K (2014) Restoration of plant–pollinator interaction networks via species translocation. Theoretical Ecology 7:209-220.

Landis TD (2014) Monarch waystations: propagating native plants to create travel corridors for migrating monarch butterflies. Native Plants Journal 15:5-16.

MacGregor CJ, Pocock MJ, Fox R, Evans DM (2015) Pollination by nocturnal Lepidoptera, and the effects of light pollution: a review. Ecological Entomology 40:187-198.

MacLeod M, Genung MA, Ascher JS and Winfree R (2016) Measuring partner choice in plant–pollinator networks: using null models to separate rewiring and fidelity from chance. Ecology 97:2925-2931.

Mader E, Shepard M, Vaughan M, Hoffman Black S, LeBuhn G (2011) Attracting native pollinators: Protecting North America's bees and butterflies. The Xerces Society, North Adams, Massachusetts.

Mayfield MM, Bonser SP, Morgan JW, Aubin I, McNamara S, Vesk PA (2010) What does species richness tell us about functional trait diversity? Predictions and evidence for responses of species and functional trait diversity to land-use change. Global Ecology and Biogeography 19:423-431.

Medan D, Basilio AM, Devoto M, Bartoloni NJ, Torretta JP, Petanidou T (2006) Measuring generalization and connectance in temperate, year-long active systems. In: Waser N, Ollerton J (eds) Plant–pollinator interactions: from specialization to generalization. University of Chicago Press, Chicago, Illinois, pp 245-259.

Memmott J, Waser NM, Price MV (2004) Tolerance of pollination networks to species extinctions. Proceedings of the Royal Society of London Series B: Biological Sciences 271:2605-2611.

Menz MH, Phillips RD, Winfree R, Kreme C, Aizen MA, Johnson SD, Dixon KW (2011) Reconnecting plants and pollinators: challenges in the restoration of pollination mutualisms. Trends in Plant Science 16:4-12.

M'Gonigle LK, Ponisio LC, Cutler K, Kremen C (2015) Habitat restoration promotes pollinator persistence and colonization in intensively managed agriculture. Ecological Applications 25:1557-1565.

Montoya D, Rogers L, Memmott J (2012) Emerging perspectives in the restoration of biodiversity-based ecosystem services. Trends in Ecology and Evolution 27:666-672.

NPSS (Native Plant Society of Saskatchewan) (2013) Native plant material and services supplier list [online] URL: http://www.npss.sk.ca/docs/2_pdf/Native_Plant_Source_List_2013_-_revised.pdf (accessed 28 September 2016).

Parachnowitsch AL, Elle E (2005) Insect visitation to wildflowers in the endangered Garry Oak, *Quercus garryana*, ecosystem of British Columbia. The Canadian Field Naturalist 119:245-253.

Robson DB (2008) The structure of the flower-insect visitor system in tall-grass prairie. Botany 86:1226-1278.

Robson DB (2013) An assessment of the potential for pollination facilitation of a rare plants by common plants: *Symphyotrichum sericeum* (Asteraceae) as a case study. Botany 91:34-42.

Russo L, DeBarros N, Yang S, Shea K, Mortensen D (2013) Supporting crop pollinators with floral resources: network-based phenological matching. Ecology and Evolution 3:3125-3140.

Saavedra S, Stouffer DB, Uzzi B, Bascompte J (2011) Strong contributors to network persistence are the most vulnerable to extinction. Nature 478:233-235.

Sheffield CS, Frier SD, Dumesh S (2014) The bees (Hymenoptera: Apoidea, Apiformes) of the Prairie Ecozone, with comparisons to other grasslands of Canada. In: Giberson J, Carcamo HA (eds) Arthropods of Canadian grasslands volume 4: Biodiversity and systematics part 2, Biological Survey of Canada, pp 427-467.

Stang M, Klinkhamer PG, Van Der Meijden E (2006) Size constraints and flower abundance determine the number of interactions in a plant–flower visitor web. Oikos 112:111-121.

Tarrant S, Ollerton J, Rahman ML, Tarrant J McCollin D (2012) Grassland restoration on landfill sites in the east midlands, United Kingdom: an evaluation of floral resources and pollinating insects. Restoration Ecology 21:560-568.

Thorpe AS, Stanley AG (2011) Determining appropriate goals for restoration of imperilled communities and species. Journal of Applied Ecology 48:275-279.

Trottier GC (1986) Disruption of rough fescue, *Festuca hallii*, grassland by livestock grazing in Riding Mountain National Park, Manitoba. Canadian Field-Naturalist 100:488-495.

Wilkerson ML, Ward KL, Williams NM, Ullmann KS, Young TP (2014) Diminishing returns from higher density restoration seedings suggest trade-offs in pollinator seed mixes. Restoration Ecology 22:782-789.

20

Does lack of pollination extend flower life?

Hannah F. Fung[1] and James D. Thomson[1,2]*

[1]Department of Ecology and Evolutionary Biology, University of Toronto, 25 Harbord Street, Toronto, Ontario, M5S 3G5, Canada
[2]Rocky Mountain Biological Laboratory, Post Office Box 519, Crested Butte, CO 81224-0519, USA

Abstract—Across angiosperm species, the longevity of individual flowers can range from fixed to highly plastic. The orchid family is noteworthy for frequent reports of species in which flower lifespans are greatly prolonged if flowers are not pollinated. Less dramatic cases of pollination-induced senescence of anthesis have been reported for various species in other families, but such reports are scattered. Frequently, such findings are peripheral components of more general pollination studies. Because pollination-dependent plasticity can ameliorate phenological dislocations between plants and pollinators, it is worthwhile to conduct systematic surveys of its magnitude and taxonomic distribution. As a start, we report a set of experiments comparing the active lifespans of pollinated flowers to those of unpollinated controls in a set of nine species from a local subalpine flora. In all species, unpollinated flowers had longer mean times of receptiveness than pollinated ones, although the differences in means were often small. Three species exhibited significantly extended floral longevity in the absence of pollination.

Keywords: Plasticity, anthesis, floral longevity, pollination-induced senescence, Rocky Mountain Biological Laboratory

Introduction

The longevity of individual flowers varies across plant species; some of the variation is species-specific and genetically determined (Primack 1985; Stratton 1989), and some of it arises from plastic responses to immediate conditions. Here we are concerned with one potential plastic response, the extension of flower life as a response to lack of pollination. Such extension is particularly relevant in the context of recent concerns that climate change can cause phenological mismatches between plants and their pollinators (Kudo et al. 2004; Memmott et al. 2007; McKinney et al. 2012). As environmental conditions change, plants and pollinators may respond to different cues and emerge at different times, leading to pollination deficits for plants and food shortages for pollinators. The detrimental effects of phenological mismatches can be ameliorated, however, if plants can prolong the lifespans of flowers that have not been pollinated. In Colorado subalpine meadow communities, for example, Forrest & Thomson (2011, p. 487) argued that plant-pollinator dislocations in time are likely to be "quantitative effects… rather than… complete decoupling of formerly interacting organisms." In such situations, the ability of flowers to prolong their lifespans by even a few days might substantially increase the probability of receiving pollinating visits.

Floral longevity, or the length of time that a flower is open and functional (Ashman & Schoen 1994), has been treated as a resource allocation strategy in which the maintenance of flowers diverts resources from other reproductive or vegetative functions (Schoen & Ashman 1995; Ashman & Schoen 1997). Increased floral longevity is likely to increase plant reproductive success through prolonged pollen and stigma presentation (Thomson & Barrett 1981; Lloyd & Yates 1982; Galen et al. 1986; Harder & Thomson 1989), but it also exacts significant carbon and water costs through nectar production, respiration, and transpiration (Ashman & Schoen 1997). Thus, floral longevity can be viewed as a trade-off between fitness accrual through reproduction and the costs of floral maintenance.

Consequences of this trade-off could be particularly dramatic in many orchid species, where pollination can trigger rapid senescence of flowers that would stay receptive much longer if unpollinated (van Doorn 1997). Pollination-induced senescence has been reported in diverse genera such as *Leporella* and *Caladenia* (Peakall 1989), *Encyclia* (Ackerman 1989), *Cypripedium* (Primack & Hall 1990), *Cleistes* (Gregg 1991), *Calypso* (Proctor & Harder 1995), and *Myrmecophila* (Parra-Tabla et al. 2009). The effects of reproductive activity on floral longevity are not restricted to female function: in *Chloraea alpina*, pollinia removal, in addition to deposition, shortened the longevity of unpollinated flowers (Clayton & Aizen 1996). Similarly, both pollinia removal and deposition induced senescence in *Cattleya porcia* in an ethylene-dependent process (Strauss & Arditti 1984).

Orchids are an extreme case. Nevertheless, in other families, floral longevity does respond plastically to the completion of male and/or female function. Effects of pollen deposition on floral lifespan have been reported in numerous families, including Onagraceae (Addicott & Lynch 1955; Ashman & Schoen 1997), Caryophyllaceae (Nichols 1971; Motten 1986), Solanaceae (Gilissen 1976, 1977), Plantaginaceae (Stead & Moore 1979, 1983), Campanulaceae (Devlin & Stephenson 1984, 1985; Richardson & Stephenson 1989; Evanhoe & Galloway 2002), Liliaceae (Schemske et al. 1978; Motten 1983, 1986; Ishii & Sakai 2000), Portulacaceae (Motten 1986; Aizen

*Corresponding author: james.thomson@utoronto.ca

FIGURE 1. Photos of nine animal-pollinated species surveyed for pollination-induced senescence in and near the Rocky Mountain Biological Laboratory, Gothic, Colorado. Top row, left to right: *Mertensia fusiformis* (Boraginaceae), *Delphinium barbeyi*, *Aconitum columbianum* (Ranunculaceae); second row, left to right: *Ipomopsis aggregata* (Polemoniaceae), *Vicia americana* (Fabaceae), *Chamerion angustifolium* (Onagraceae); third row, left to right: *Sidalcea candida* (Malvaceae), *Gentianopsis detonsa* (Gentianaceae), *Campanula rotundifolia* (Campanulaceae). All photos by H. F. Fung.

1993), Brassicaceae (Motten 1986; Preston 1991), Gentianaceae (Webb & Littleton 1987), and Ericaceae (Rathcke 1988a, 1988b; Blair & Wolfe 2007). In *Lobelia cardinalis* and *Campanula rapunculoides*, pollen removal and deposition shortened the duration of the staminate and pistillate phases respectively (Devlin & Stephenson 1984; Richardson & Stephenson 1989). Interestingly, pollen removal, but not deposition, accelerated senescence in *Brassica napus* (Bell & Cresswell 1998).

To broaden the study of pollination-induced senescence and to further explore its implications for phenological mismatch, we surveyed nine plant species (Fig. 1) at or near the Rocky Mountain Biological Laboratory (RMBL) in Gothic, Colorado, USA. A great deal of pollination research, including the previously cited study of phenological

dislocation by Forrest and Thomson (2011), has been conducted near this field station. In addition, the mating systems and pollinator faunas of most species at Gothic are well-characterized. The species in this study were chosen for convenience: most are abundant, and they produce large, tractable flowers that are characteristically visited by bees or hummingbirds. Using a series of controlled hand-pollinations, we examined whether plants in this subalpine community can prolong the lifespan of unpollinated flowers.

MATERIALS AND METHODS

We sampled the nine species at sites within and near the RMBL (38.96° N, 106.99° W, 2900 m asl) from June to August 2016. To minimize environmental variation, individuals of each species were sampled at one site. Prior to flower opening, we covered plants with polyolefin drawstring bags or sand-bag style exclusion bags (Thomson et al. 2011) to exclude flower visitors.

We marked the pedicels of two flowers, paired for size, floral age, and position on plant, with felt-tipped markers. We chose species-specific markers of floral age that could be scored by gross inspection, such as bud break, stylar exsertion, wilting, colour changes, and abscission (see Appendix I). These easily scored characteristics may not precisely delimit the onset and cessation of flower functions, but we consider them appropriate for detecting differences in floral lifespan in our paired design.

Within pairs, flowers were randomly assigned to experimental or control treatments. Once mature and receptive, experimental flowers were hand-pollinated with Microbrush® applicators (Microbrush International; Grafton, Wisconsin, USA) bearing a mixture of fresh pollen from other plants in the vicinity. We attempted to apply as much pollen as the stigmas could retain. Control flowers were not pollinated, but were manipulated in the same way as experimental flowers with clean applicators. To ensure successful hand pollination, we pollinated each flower twice, on consecutive days. An exception was *Campanula rotundifolia*, which was hand pollinated three times. Criteria for stigma receptivity are provided in the Supplementary Data.

We checked individuals only once a day. Although more frequent checks could have produced finer-grained data, we compromised so as to be able to score more species and more replicates. We scored flowers for a range of floral age indicators, including extent of anther dehiscence, perianth colour, openness or accessibility to interior (Olesen et al. 2007), degree of wilting, and corolla abscission (details in Supplementary Data). Our response variable, floral lifespan, was the number of days to senescence from first hand-pollination.

Data analysis

We used one-tailed exact Wilcoxon signed rank tests to determine whether anthesis was prolonged in unpollinated flowers. Analyses were performed using R v. 3.1.2 (R Foundation for Statistical Computing 2014). Data are expressed as means ± standard deviation.

RESULTS

The extension of flower life in unpollinated flowers varied across species (Fig. 2). Based on single Wilcoxon tests, the following species showed evidence of pollination-induced senescence (mean paired differences, control - treatment): *Mertensia fusiformis* (0.67 ± 0.87 days, $V = 77$, $P = 0.011$), *Chamerion angustifolium* (0.58 ± 1.82 days, $V = 165$, $P = 0.039$), and *Gentianopsis detonsa* (2.71 ± 1.93 days, $V = 253$, $P < 0.0001$). The remaining species did not: *Delphinium barbeyi* (0.07 ± 1.36 days, $V = 150$, $P = 0.58$), *Ipomopsis aggregata* (2.64 ± 4.36 days, $V = 161$, $P = 0.14$), *Vicia americana* (0.07 ± 2.71 days, $V = 90$, $P = 0.31$), *Aconitum columbianum* (0.56 ± 2.17 days, $V = 147$,

$P = 0.058$), *Sidalcea candida* (0.26 ± 0.62 days, $V = 52$, $P = 0.19$), and *Campanula rotundifolia* (0.33 ± 2.09 days, $V = 39$, $P = 0.61$). Despite the lack of significance for the majority of the species, it is worth noting that control flowers had longer estimated mean lifetimes in all nine species, which is itself a highly significant pattern by sign test ($P = 0.0039$).

DISCUSSION

This study represents one of the few attempts to explore pollination-mediated senescence across species within a local community. Of the nine species surveyed in and near the RMBL, three showed individually significant evidence of pollination-induced senescence (Fig. 2). Overall, unpollinated flowers did tend to last longer, but for the majority of species the differences were too small and inconsistent to overcome the substantial variation within treatments. In further studies, larger sample sizes would be desirable. Our results suggest that species vary in their ability to prolong the lifespan of unpollinated flowers, a finding that is consistent with other broad surveys of floral longevity. In a study of the spring wildflower community in North Carolina, pollen deposition triggered floral senescence in five of eight species (Motten 1986). Likewise, pollination-induced senescence was observed in four of six shrub species in The Great Swamp, Rhode Island, USA (Rathcke 1988b). Evidently, meaningful plasticity in floral lifespan is common but not universal.

Flowers of *Gentianopsis detonsa* responded to pollination in a similar fashion to two closely related species, *Gentiana saxosa* and *G. serotina*. In addition to changes in colour and turgor, pollination caused the corollas of these species to close (H. F. Fung, pers. obs.; Webb & Littleton 1987), which may protect the developing ovary from predators and subsequent pollinations (Webb & Littleton 1987). Thus, corolla closure may be a common response to pollination among the gentians. Interestingly, we did not find evidence of pollination-induced senescence in *Campanula rotundifolia* (Fig. 2), in contrast to studies of *C. rapunculoides* (Richardson & Stephenson 1989) and *C. americana* (Evanhoe & Galloway 2002).

From a resource allocation perspective, pollination-induced senescence offers a way in which plants can plastically optimize the trade-off between reproductive success and the costs of maintaining flowers (Primack 1985; Harrison & Arditti 1976; Ishii & Sakai 2000). Through pollination-induced senescence, plants can direct their resources toward maintaining unpollinated flowers, and in doing so, reduce the likelihood that plants and their pollinators are phenologically mismatched. The findings presented here and elsewhere indicate that certain plants can prolong the lifespans of unpollinated flowers, but it is unclear that this is prevalent enough to ameliorate phenological asynchrony substantially. After all, pollination had no effect on floral longevity in the majority of plant species surveyed in the present study.

There are several reasons why floral longevity may not evolve to respond plastically to pollination. First, the trade-

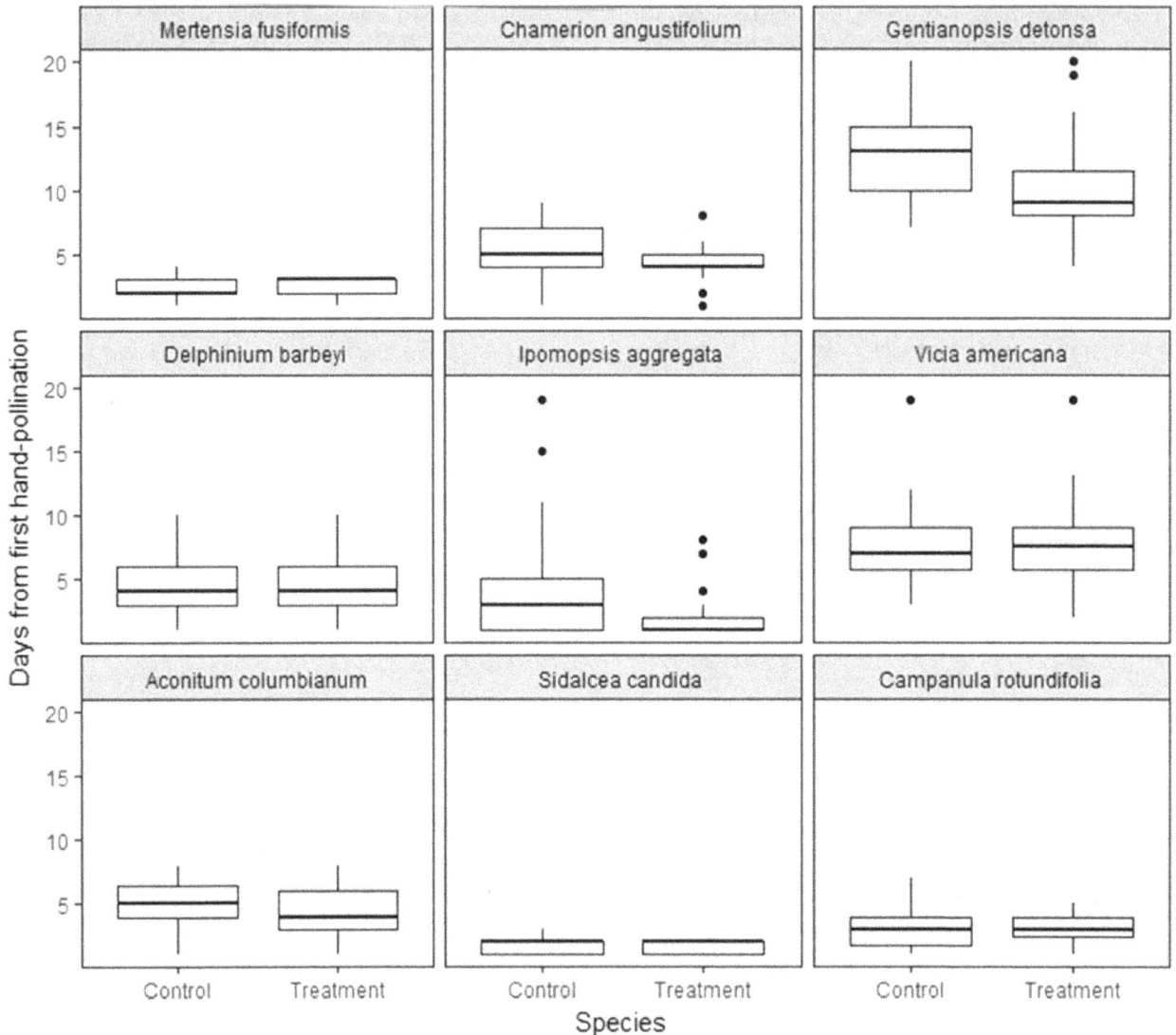

FIGURE 2. Floral longevity of nine plant species in or near the Rocky Mountain Biological Laboratory, Gothic, Colorado, USA. Experimental flowers were hand-pollinated once receptive, while control flowers were not pollinated. Sample sizes, in pairs, for the species, in order from left to right, top to bottom: 13, 25, 19, 29, 28, 28, 27, 26, 18.

off between reproductive success and maintenance costs may not be as pronounced in plants that continue to grow and photosynthesize after reproduction (Ashman & Schoen 1997). In the snow buttercup *Ranunculus adoneus*, for example, flowers accounted for a significant proportion of carbon assimilation (Galen et al. 1993). Similarly, in *Ambrosia trifida* L., reproductive structures contributed 41% and 57% of the carbon required to construct male and female inflorescences respectively (Bazzaz & Carlson 1979). Moreover, increased photosynthesis in adjacent leaves may compensate for some of the costs of floral maintenance (Gifford & Evans 1981; Lehtilä & Syrjänen 1995; Ashman & Schoen 1997). Such cases are more complicated than the simple concept of allocation from a fixed pool of resources.

Second, plants may continue to maintain pollinated flowers to increase floral display size and to facilitate long-distance attraction of pollinators (Primack 1985; reviewed by Snow et al. 1996; Evanhoe & Galloway 2002). As display size increases, pollinator visitation tends to increase,

increasing pollen export and receipt. Large displays, however, can represent a significant drain on resources (Evanhoe & Galloway 2002) and can increase the rate of geitonogamy (Barrett & Harder 1996; Snow et al. 1996).

Finally, flowers can contribute to male reproductive success even after they are pollinated (Primack 1985; Ishii & Sakai 2000). In *Erythronium japonicum*, for example, floral longevity was nearly constant in flowers pollinated between days one and 12 of anthesis (Ishii & Sakai 2000). This 12-day period was subsequently shown to be necessary for *E. japonicum* flowers to shed the majority of their pollen (Ishii & Sakai 2000). Based on these results, Ishii & Sakai (2000) proposed that flowers have a genetically determined minimum longevity that functions to facilitate male reproduction.

This 'minimum longevity' hypothesis yields the following prediction: pollination-induced senescence should be more common in species in which senescence in response to pollination does not interfere with pollen dispersal, as in protandrous species (Ashman & Schoen 1996; Ishii & Sakai 2000; Evanhoe & Galloway 2002). There is some evidence that protandrous species are more likely to show pollination-induced senescence. In a survey of spring wildflowers in North Carolina, Motten (1986) found that weakly protogynous and protandrous species were more likely to senesce in response to pollen deposition than protogynous species. In the present study, however, pollination significantly accelerated senescence in only two of seven protandrous species (Fig. 2).

To address this apparent discrepancy, it is important to recognize that floral longevity may be governed by the pollination status at the plant level, as opposed to the flower level. In other words, plants in which the majority of ovules have been fertilized may present shorter-lived flowers than their unpollinated counterparts. As a result, among-plant comparisons may reveal instances of pollination-induced senescence that were not detected at the within-plant level, as in the cases of *Delphinium*, *Ipomopsis*, *Vicia*, *Aconitum*, *Sidalcea*, and *Campanula* (Fig. 2).

To conclude, we observed significant pollination-induced senescence in three out of nine study species in and near the RMBL, but the magnitudes of those effects were small-half a day to a few days. Regarding our motivating question of phenological dislocation, plastic extensions of floral lifespan may frequently have trivial effects. Still, species of particular interest should be examined individually.

APPENDICES

Additional supporting information may be found at the end of this article:

APPENDIX I. Criteria for measuring floral age, stigma receptivity, and senescence.

REFERENCES

References

Ackerman JD (1989) Limitations to sexual reproduction in *Encyclia krugii* (Orchidaceae). Systematic Botany 14:101-109.

Addicott FT, Lynch RS (1955) Physiology of abscission. Annual Review of Plant Physiology 6:211-238.

Aizen MA (1993) Self-pollination shortens flower lifespan in *Portulaca umbraticola* H.B.K. (Portulacaceae). International Journal of Plant Sciences 154:412-415.

Ashman T-L, Schoen DJ (1994) How long should flowers live? Nature 371:788-791.

Ashman T-L, Schoen DJ (1996) Floral longevity: fitness consequences and resource costs. In: Lloyd DG, Barrett SCH (eds) Floral biology. Chapman and Hall, New York, pp 112-139.

Ashman T-L, Schoen DJ (1997) The cost of floral longevity in *Clarkia tembloriensis*: An experimental investigation. Evolutionary Ecology 11:289-300.

Barrett SCH, Harder LD (1996) Ecology and evolution of plant mating. Trends in Ecology and Evolution 11:73-79.

Bazzaz FA, Carlson RW (1979) Photosynthetic contribution of flowers and seeds to reproductive effort of an annual colonizer. New Phytologist 82:223-232.

Bell SA, Cresswell JE (1998) The phenology of gender in homogamous flowers: temporal change in the residual sex function of flowers of oil-seed rape (*Brassicus napus*). Functional Ecology 12:298-306.

Blair AC, Wolfe LM (2007) The association between floral longevity and pollen removal, pollen receipt, and fruit production in flame azalea (*Rhododendron calendulaceum*). Canadian Journal of Botany 85:414-419.

Clayton S, Aizen MA (1996) Effects of pollinia removal and insertion on flower longevity in *Chloraea alpina* (Orchidaceae). Evolutionary Ecology 10:653-660.

Devlin B, Stephenson AG (1984) Factors that influence the duration of the staminate and pistillate phases of *Lobelia cardinalis* flowers. Botanical Gazette 145:323-328.

Devlin B, Stephenson AG (1985) Sex differential floral longevity, nectar secretion, and pollinator foraging in a protandrous species. American Journal of Botany 72:303-310.

Evanhoe L, Galloway LF (2002) Floral longevity in *Campanula americana* (Campanulaceae): A comparison of morphological and functional gender phases. American Journal of Botany 89:587-591.

Forrest JRK, Thomson JD (2011) An examination of synchrony between insect emergence and flowering in Rocky Mountain meadows. Ecological Monographs 81:469-491.

Galen C, Dawson TE, Stanton ML (1993) Carpels as leaves: Meeting the carbon cost of reproduction in an alpine buttercup. *Oecologia* 95:187-193.

Galen C, Shykoff JA, Plowright RC (1986) Consequences of stigma receptivity schedules for sexual selection in flowering plants. The American Naturalist 127:462-476.

Gifford RM, Evans LT (1981) Photosynthesis, carbon partitioning and yield. Annual Review of Plant Physiology 32:485-509.

Gilissen LJW (1976) The role of the style as a sense organ in relation to the wilting of the flower. Planta 131:201-202.

Gilissen LJW (1977) Style controlled wilting of the flower. Planta 133:275-280.

Gregg KB (1991) Reproductive strategy of *Cleistes divaricata* (Orchidaceae). American Journal of Botany 78:350-360.

Harder LD, Thomson JD (1989) Evolutionary options for maximizing pollen dispersal of animal pollinated plants. The American Naturalist 133:323-344.

Harrison CR, Arditti J (1976) Post-pollination phenomena in orchid flowers. VII. Phosphate movement among floral segments. American Journal of Botany 63:911-918.

Ishii HS, Sakai S (2000) Optimal timing of corolla abscission: experimental study on *Erythronium japonicum* (Liliaceae). Functional Ecology 14:122-128.

Kudo G, Nishikawa Y, Kasagi T, Kosuge S (2004) Does seed production of spring ephemerals decrease when spring comes early? Ecological Research 19:255-259.

Lehtilä K, Syrjänen K (1995) Positive effects of pollination on subsequent size, reproduction and survival of *Primula veris*. Ecology 76:1084-1098.

Lloyd DG, Yates JMA (1982) Intersexual selection and the segregation of pollen and stigmas in hermaphroditic plants, exemplified by *Wahlenbergia albomarginata* (Campanulaceae). Evolution 36:903-915.

McKinney AM, CaraDonna PJ, Inouye DW, Barr B, Bertelsen CD, Waser NM (2012) Asynchronous changes in phenology of migrating broad-tailed hummingbirds and their early-season nectar resources. Ecology 93:1987-1993.

Memmott J, Craze PG, Waser NM, Price MV (2007) Global warming and the disruption of plant-pollinator interactions. Ecology Letters 10:710-717.

Motten AF (1983) Reproduction of *Erythronium umbilicatum* (Liliaceae): Pollination success and pollinator effectiveness. Oecologia 59:351-359.

Motten AF (1986) Pollination ecology of the spring wildflower community of a temperate deciduous forest. Ecological Monographs 56:21-42.

Nichols R (1971) Induction of flower senescence and gynaecium development in the carnation (*Dianthus caryophyllus*) by ethylene and 2-chloroethylphosphonic acid. Journal of Horticultural Science and Biotechnology 46:323-332.

Olesen JM, Dupont YL, Ehlers BK, Hansen DM (2007) The openness of a flower and its number of flower-visitor species. Taxon 56(3):729-736.

Parra-Tabla V, Abdala-Roberts L, Rojas JC, Navarro J, Salinas-Peba L (2009) Floral longevity and scent respond to pollen manipulation and resource status in the tropical orchid *Myrmecophila christinae*. Plant Systematics and Evolution 282:1-11.

Peakall R (1989) The unique pollination of *Leporella fimbriata* (Orchidaceae): pollination by pseudocopulating male ants (*Myrmecia urens*, Formicidae). Plant Systematics and Evolution 167:137-148.

Preston RE (1991) The intrafloral phenology of *Streptanthus tortuosus* (Brassicaceae). American Journal of Botany 78:1044-1053.

Primack RB (1985) Longevity of individual flowers. Annual Review of Ecology and Systematics 16:15-37.

Primack RB, Hall P (1990) Costs of reproduction in the pink lady's slipper orchid: A four-year experimental study. The American Naturalist 136:638-656.

Proctor HC, Harder LD (1995) Effect of pollination success on floral longevity in the orchid *Calypso bulbosa* (Orchidaceae). American Journal of Botany 82:1131-1136.

Rathcke B (1988a) Interactions for pollination among coflowering shrubs. Ecology 69:446-457.

Rathcke B (1988b) Flowering phenologies in a shrub community: Competition and constraints. Journal of Ecology 76:975-994.

Richardson TE, Stephenson AG (1989) Pollen removal and pollen deposition affect the duration of the staminate and pistillate phases in *Campanula rapunculoides*. American Journal of Botany 76:532-538.

Schemske DW, Willson MF, Melampy MN, Miller LJ, Verner L, Schemske KM, Best LB (1978) Flowering ecology of some spring woodland herbs. Ecology 59:351-366.

Schoen DJ, Ashman T-L (1995) The evolution of floral longevity: Resource allocation to maintenance versus construction of repeated parts in modular organisms. Evolution 49:131-139.

Snow AA, Spira TP, Simpson R, Klips RA (1996) The ecology of geitonogamous pollination. In: Lloyd DG, Barrett SCH (eds) Floral biology. Chapman and Hall, New York, pp 191-216.

Stead AD, Moore KG (1979) Studies on flower longevity in *Digitalis*: Pollination induced corolla abscission in *Digitalis* flowers. Planta 146:409-414.

Stead AD, Moore KG (1983) Studies on flower longevity in *Digitalis*: The role of ethylene in corolla abscission. Planta 157:15-21.

Stratton DA (1989) Longevity of individual flowers in a Costa Rican cloud forest: Ecological correlates and phylogenetic constraints. Biotropica 21:308-318.

Strauss MS, Arditti J (1984) Postpollination phenomena in orchid flowers. XII. Effects of pollination, emasculation, and auxin treatment on flowers of *Cattleya porcia* 'Cannizaro' and the rostellum of Phalaenopsis. Botanical Gazette 145:43-49.

Thomson JD, Barrett SCH (1981) Selection for outcrossing, sexual selection and the evolution of dioecy in plants. The American Naturalist 118:443-449.

Thomson JD, Forrest JRK, Ogilvie JE (2011) Pollinator exclusion devices permitting easy access to flowers. Journal of Pollination Ecology 4:24-25.

Thomson JD, Thomson BA (1992) Pollen presentation and viability schedules in animal-pollinated plants: Consequences for reproductive success. In: Wyatt R (ed) Ecology and evolution of plant reproduction: New approaches. Chapman and Hall, New York, pp 1-24.

van Doorn WG (1997) Effects of pollination on floral attraction and longevity. Journal of Experimental Botany 48:1615-1622.

Webb CJ, Littleton J (1987) Flower longevity and protandry in two species of *Gentiana* (Gentianaceae). Annals of the Missouri Botanical Garden 74:51-57.

21

Differences in Pollination Syndromes and the Frequency of Autonomous Delayed Selfing Between Co-Flowering *Hibiscus Aponeurus* (Sprague and Hutch) and *H. Flavifolius* (Ulbr) from Kenya

Juan Carlos Ruiz-Guajardo[1,7]*, Andrew Schnabel[2], Britnie McCallum[3], Adriana Otero Arnaiz[4], Katherine C. R. Baldock[5], and Graham N. Stone[6]

[1]*Department of Evolution and Ecology, University of California at Davis, Davis CA 95618 USA.*
[2]*Department of Biological Sciences, Indiana University South Bend, South Bend IN 46634 USA.*
[3]*Belk College of Business, University of North Carolina, at Charlotte, Charlotte NC 28223 USA.*
[4]*Office of Agricultural Affairs, USDA, U.S. Embassy, Mexico City*
[5]*School of Biological Sciences, University of Bristol, Bristol BS8 1TQ UK*
[6]*Institute of Evolutionary Biology, University of Edinburgh, Edinburgh EH9 3JT UK*
[7]*Mpala Research Centre, Laikipia, Kenya*

Abstract—Delayed autonomous selfing offers a mechanism for seed production when pollination levels are low or unpredictable. At Mpala Research Centre (MRC) in Kenya, we examined the relationships between floral attraction, insect visitation, and delayed autonomous selfing through backwards stylar curvature in the co-flowering *Hibiscus aponeurus* and *H. flavifolius*. Despite producing similar pollen and nectar rewards, visitation rates and the composition of floral visitor guilds varied significantly between these species. Across four years of observations, floral visitation in *H. flavifolius* was dominated by bees, and in *H. aponeurus* by a mixture of bees, butterflies and beetles. Visitation rates to *H. flavifolius* flowers (range 0.17 - 2.1 visits flr⁻¹hr⁻¹) were two times greater than to *H. aponeurus* flowers (range 0 - 2.7 visits flr⁻¹hr⁻¹), which resulted in significantly higher pollen deposition and removal rates in *H. flavifolius* than in *H. aponeurus*. Field crosses demonstrated little pollen limitation in either species. In open-pollinated flowers, *H. aponeurus* displayed significantly greater stylar curvature and apparent self-pollination than did *H. flavifolius*. Floral attributes in *H. aponeurus*, such as a smaller corolla size and a downwards orientation of the stylar column, also suggest that delayed selfing is a more important mechanism of reproductive assurance in this species than in *H. flavifolius*. Determining whether these differences in insect visitation and stylar curvature are characteristic for these species or are unique to MRC will require comparison with populations located in other parts of the ranges, genetic tests of selfing rates, and chemical analyses of nectar, pollen, and floral volatiles.

Keywords: Delayed autonomous selfing, Hibiscus, pollination, stylar curvature, Mpala Research Centre

Introduction

Human activities that destroy natural habitats and fragment populations are reducing the size of pollinator populations and potentially decreasing the reproductive success of many plant species (Aguilar et al. 2006; Eckert et al. 2010; Thomann et al. 2013; Vanbergen 2013; Somme et ting systems, habitat fragmentation can increase reliance on self-pollination (e.g. Brys & Jacquemyn 2012) and over time potentially can result in inbreeding depression, reduced genetic variability, and increased risk of local extinction due to the loss of adaptive potential (Stebbins 1957; Goodwillie et al. 2005; Eckert et al. 2010). As reviewed by Wright et al. (2013), however, selfing also has potential benefits, including

a 50 per cent transmission advantage over outcrossing. Further, selfing can sometimes provide reproductive assurance when potential mates are present at low densities, such as during colonization events (e.g. Pannell & Barrett 1998; Rambuda & Johnson 2004), or when pollination rates are low or unpredictable, such as in the decline or absence of suitable pollinators (Zhang et al. 2014).

From a functional standpoint, selfing can be mediated by pollinators (geitonogamy and facilitated autogamy) or it can be autonomous and occur spontaneously within flowers without the aid of a pollinator (Lloyd & Schoen 1992). Of the three modes of autonomous selfing described by Lloyd & Schoen (1992), delayed autonomous selfing is favoured under the widest range of ecological conditions (Morgan & Wilson 2005; Morgan et al. 2005), because it does not occur until opportunities for outcrossing have passed, thereby eliminating seed and pollen discounting. The frequency and importance of autonomous selfing in plant populations show considerable

*Corresponding author: jcruizguajardo@ucdavis.edu

variation among individuals (Kalisz et al. 1999), among populations (Klips & Snow 1997), and among closely related species (Brys & Jacquemyn 2011), most likely because selfing rates are strongly influenced by local biotic and abiotic conditions (Kalisz et al. 2004; Qu et al. 2007; Ruan et al. 2009; Vaughton & Ramsey 2010; Jorgensen & Arathi 2013).

Delayed autonomous selfing occurs in several genera of the Malvaceae (Ruan et al. 2010, 2011), but detailed exploration of its ecological importance has been limited to *Hibiscus laevis* (Klips & Snow 1997), *Hibiscus trionum* (Ramsey et al. 2003; Seed et al. 2006), and *Kosteletzkya virginica* (Ruan et al. 2005, 2008a, 2008b, 2009). Most Malvaceae capable of delayed selfing have flowers with styles that are surrounded by and extend beyond monadelphus stamens. In these species, the styles curve out and backwards as flowers age until the stigmas contact the pollen located in the upper anthers (see images in Ruan et al. 2010; Kumar et al. 2014). Studies in *Hibiscus* (e.g. Klips & Snow 1997) suggest that this backward bending can be stopped by prior pollination, although the number of pollen grains required to halt curvature and the physiological mechanism behind that interaction are not well understood (Buttrose et al. 1977; Klips & Snow 1997; Seed et al. 2006; Ruan et al. 2008a).

Here, we report results of a multiple-year study examining the pollination biology, breeding system, and mechanisms for reproductive assurance in *Hibiscus aponeurus* (Sprague & Hutch.) and *Hibiscus flavifolius* (Ulbr). These species of semiarid Africa are morphologically and ecologically similar, and they co-flower following seasonal rains at Mpala Research Centre (MRC) in Laikipia, Kenya, where we conducted our study. Our initial observations revealed that these species produce abundant pollen and nectar rewards and attract a broad array of insect visitors, but visitation frequency to *H. aponeurus* appeared to be lower than to *H. flavifolius*. Further, at the end of the day when flowers were closing, both species were capable of autonomous selfing through stylar curvature, but this action appeared to be more common in *H. aponeurus*. To examine the links between floral attraction, insect visitation, fruit and seed production, and delayed selfing in these species, we performed a series of field observations and greenhouse manipulations between 2004 and 2013. Specifically, we asked: 1) How do *H. aponeurus* and *H. flavifolius* differ in floral display and in the rewards offered to potential pollinators? 2) Are the insect visitor assemblages of these *Hibiscus* species similar and consistent across years? 3) How do insect visitation rates, levels of pollen deposition, and levels of pollen removal compare between species? 4) Is fruit set and seed production limited by low rates of pollination? 5) How common and effective is delayed selfing as a mechanism of reproductive assurance?

Based on our preliminary observations, we hypothesized that *H. aponeurus* at MRC will more heavily rely on delayed selfing as a mechanism to ensure pollination, and we predicted that for open-pollinated flowers, stylar curvature would be significantly greater in *H. aponeurus* than in *H. flavifolius*. Further, because pollination in other malvaceous species can prevent the curving of stigmas, we predicted that manually pollinated flowers of both species would exhibit less stylar curvature than would unpollinated, intact flowers allowed to perform autonomous selfing at the end of the day.

MATERIALS AND METHODS

Study site and species

Our fieldwork was conducted at Mpala Research Centre (MRC) in Central Kenya (37°52'E, 0°17'N) from 2004-2008 and in 2013. The vegetation at MRC is predominantly savannah woodland, with alternating areas dominated by grasses and forbs (see Baldock et al. 2011; Ruiz-Guajardo 2008, for a more detailed description of vegetation). Throughout MRC, *Hibiscus aponeurus* is less abundant and more patchily distributed than *H. flavifolius*. We concentrated our efforts in two approximately 0.5 Ha plots where both species commonly coflowered in close proximity to one another. For some variables, additional data were collected in 2008 from plants in a greenhouse located at Indiana University South Bend, USA.

Both *Hibiscus* species are erect perennials found in wooded grasslands, but differ subtly in growth habit, microhabitat, flower shape, and more distinctly in flower colour (Agnew & Agnew 1994). *Hibiscus aponeurus* grows mostly underneath and scandently up through acacia trees. In contrast, *H. flavifolius* has sturdier stems, is a better colonizer of open areas grazed by cattle and native herbivores, and is therefore often found growing in grassy glades away from trees. Flowers of both species are solitary in the leaf axils; they open before 0800 hrs, close before 1800 hrs, and do not reopen thereafter. The petals in both species open to produce a nearly flat landing surface for insects that completely exposes the stylar column with its monadelphous stamens (Fig. 1A). Both species produce abundant bright orange pollen, but petals in *H. aponeurus* are bright crimson, while in *H. flavifolius* are pure white. As is typical in *Hibiscus* (Pfeil et al. 2002), the stylar columns of both species have five branches, each of which ends in a capitate stigma. Ovaries of both species contain approximately 30 ovules.

Floral display

We measured corolla diameters, length of stylar column, and the angle of the stylar column with respect to the nearly vertical plane of the corolla for 22 *H. aponeurus* flowers and 35 *H. flavifolius* flowers. A measurement of 90° indicated that the stylar column was orthogonal to the plane of the corolla, whereas angles greater than 90° indicated that it pointed downwards. To ensure that flowers were fully opened, we conducted all floral measurements between 0830 – 1100 hrs. We compared floral measurements between species using t-tests. All statistical analyses were conducted in JMP v.8.0.1 software (SAS Institute Inc., Cary, North Carolina, USA), and all means are shown ± 1 S.E.

Floral abundance and floral rewards

Higher floral abundance often increases visitation rates and may affect pollen transfer, because pollinators do not have to search long before finding the next flower (Elliot & Irwin 2009; Scriven et al. 2013). To estimate seasonal variation in floral abundances, we conducted a total of 18 single-day

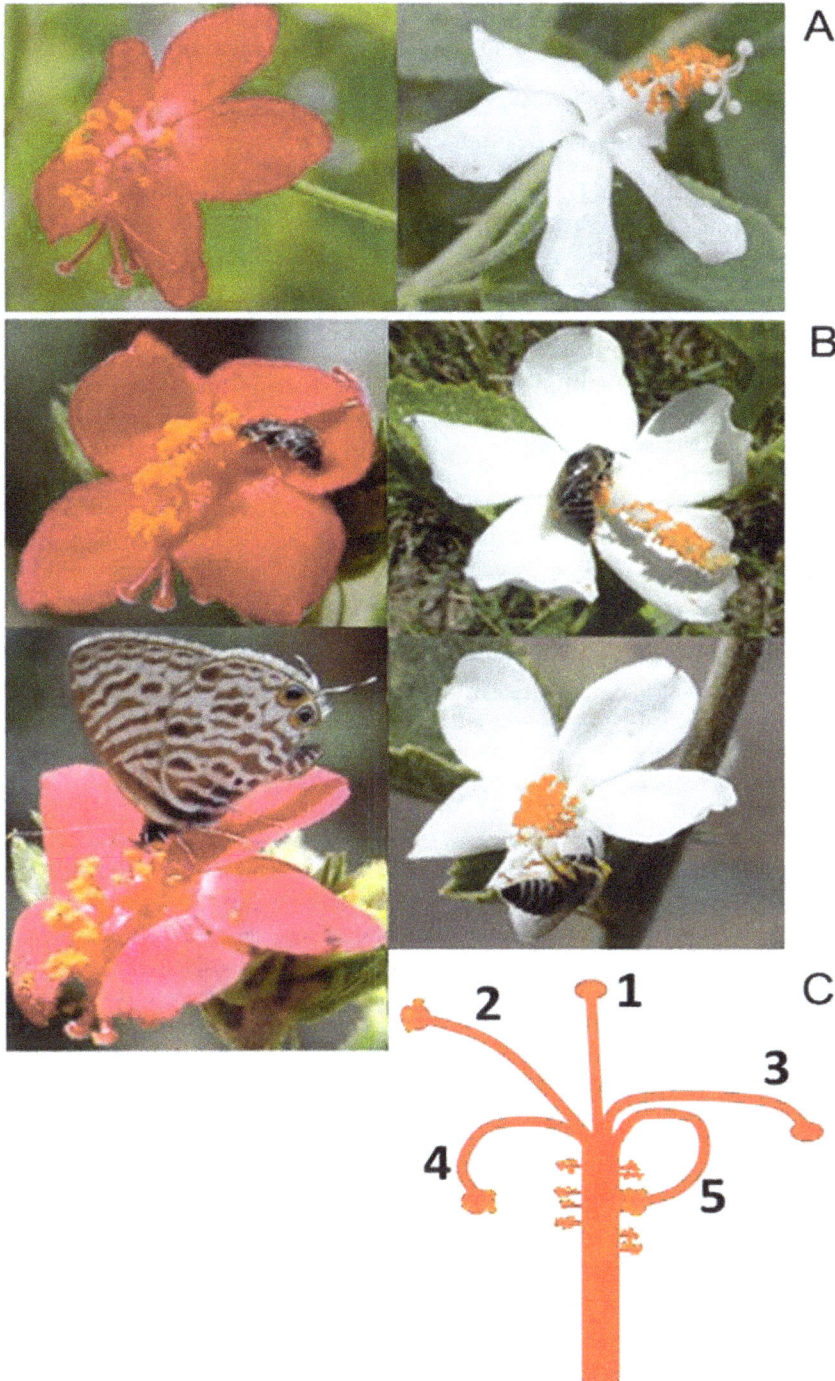

FIGURE 1. Floral morphology, examples of insect visitation behaviour, and stigma behaviour for *Hibiscus aponeurus*, and *Hibiscus flavifolius*. A) Differences between *H. aponeurus* (left) and *H. flavifolius* (right) in staminal tube and stylar orientation relative to the plane of fully opened petals. Note also the somewhat inaccessible location of *H. aponeurus* stigmas. B) Examples of backward facing visitation to *H. aponereus* by a bee (top left) and a butterfly (bottom left). A contrasting illustration of a backward facing visitation by a bee to an *H. falvifolius* flower (top right), and a forward facing visitation where the bee directly lands on the stigma (bottom right). C) Scoring scheme for stylar curvature showing monadelphus stamens with orange pollen grains, and red styles and stigmas.

surveys between 2004 and 2006. In 2004, we counted the number of open *H. aponeurus* and *H. flavifolius* flowers on one day during each of three weeks in May, one day during each of two weeks in June, July, and August, and on one day in October. Additional single-day counts were made 2005 (March, April, May, October) and 2006 (May, June, July, October).

To examine daily variation in standing crops, we sampled nectar and pollen from unbagged flowers in 2005 and 2006 once each hour in the following 2-hour intervals: 0800 - 0900, 1200 -1300, and 1600 - 1700 hrs. At each collection,

we sampled 4 - 12 flowers / species from plants not included in the patch used for observations of insect visitation. To measure nectar standing crop volume (μL / flower), we used glass microcapillaries of known total volumes (Camlab, UK) to probe at the base of the corolla, carefully avoiding to clog the tubes (see Stone et al. 1998). We measured sucrose concentration (% sucrose g / 100 g solution) directly with hand held pocket refractometers modified by the makers (Bellingham & Stanley, UK) for volumes down to 0.15 µl. Sucrose concentrations were later converted into g sucrose / L of nectar using values from tables published in Kearns (1993). For further details about sampling methods, see Stone et al. (1996, 1998) and Raine et al. (2007). The percentage of pollen remaining in anthers was scored on a scale of 0 - 4 (0 = < 1%; 1 = 1 - 25%; 2 = 26 - 50%; 3 = 51 - 75%; 4 = 76 - 100%). To minimize scoring bias, we spent multiple days calibrating pollen estimates. Each crew member scored several flowers throughout the day, using a fully dehisced, unvisited flower as the basis for a pollen score of 4 and an emptied flower as the basis for a score of 0. We performed these exercises until sampling was consistent by each crew member. We compared between-species differences in pollen, nectar volume, and sugar standing crops for each time interval and year of observation using nested analysis of variance (ANOVA). We fitted day of observation as a random variable nested within species. Because sampling floral resources often requires the partial destruction of corollas, during 2008 when flowering densities for both species were much lower, we did not sample nectar standing crop and only scored percentage of pollen remaining on the anthers at the end of the day (1630 - 1730 hrs) for the flowers that had been observed for insect visitation. We used nested ANOVA to compare end-of-day pollen scores between the two species.

Insect visitation

Based on the strong daily structuring of floral visitation and high seasonality among plant-visitor interactions reported by Baldock et al. (2011) for our study area, we conducted floral observations that spanned multiple daily time intervals and seasons. In 2004, each species was observed for 20 minutes during each of four time intervals (0600 - 0900, 0900 - 1200, 1200 - 1500, and 1500 - 1800 hrs) on two different days within a two-week period (see Baldock et al. 2011). To obtain an even more detailed representation of the visitation patterns of these plant species, in May and June of 2005, 2006, and 2008, flowers were observed for 40 min each hour from anthesis to closure (0800 - 1800 hrs). Observational patches in 2004, 2005 and 2006 varied from 8 - 15 flowers / species, but in 2008 due to lower flower abundances, patches were smaller (mean of 4 flowers / patch in *H. aponeurus* and 8 flowers / patch in *H. flavifolius*). A visit was recorded whenever an insect landed anywhere on the corolla, or approached the flower and contacted anthers or stigmas. For each *Hibiscus* species, we monitored insect visitation for at least 2 days in 2004 and 2005, 5 days in 2006, and 9 days in 2008. We noticed that many visitors landed on the corolla in an orientation that did not allow contact with the stigmas. Thus, in 2008 we recorded the orientation of the insect on the flower as either frontward facing (body of insect could contact anthers and possibly also stigmas) or backward facing (head near floral tube to allow probing for nectar, but body oriented at some angle away from the distal end of the style, so that contact with stigmas was not possible and even contact with anthers was minimized; see Fig. 1B). Our "backward" category includes approach angles that are similar to, but more extreme than, side-working as understood for apples and similar flowers (e.g. Delaplane et al. 2013).

We grouped insect visitors by order (Hymenoptera, Lepidoptera, Coleoptera, Diptera, and Hemiptera) and for each year independently (2004 - 2008), calculating the proportion of visits accounted by each order and the hourly visitation rates on each day of observation (number of visits $\text{flr}^{-1}\text{hr}^{-1}$). To compare the composition of insect visitor

assemblages between *Hibiscus* species within years, and the proportions of frontward *versus* backward facing visits recorded in 2008, we used chi-square tests or Fisher's exact tests if sample sizes were small. We used t-tests for comparisons of visitations rates between species for each year independently.

Tests for apomixis, pollen limitation, autofertility, and reproductive assurance

We performed floral manipulations in wild populations at MRC during 2005 - 2007 and 2013, and in the greenhouse at Indiana University South Bend in 2008 (see Supplemental Information Appendix I (A) for seed germination and (B) for growing conditions). Rates of fruit set (percentage of total flowers pollinated that developed a fruit) and seed production (number of seeds per mature fruit) were examined for five treatments, although not all treatments were repeated across all years of observation: a) apomixis (seed production without fertilization), b) manual outcrossing (pollen transferred manually from one plant to another), c) manual selfing (pollen transferred manually from the anthers to the stigmas of the same flower), d) autonomous selfing (seeds produced in bagged flowers through delayed selfing), and e) open pollination (seeds produced either through pollen transferred between plants by wild pollinators, through facilitated selfing, or through delayed autonomous selfing). In both species, flower production per plant was low (usually < 3 flowers per day), and flowers were open for a single day only. For the autonomous selfing treatment, we covered flowers with small bags made from fine mosquito net and threaded with cotton cords, which were used to close the sacks tightly and to tie them to the stems. Although great care was exercised to avoid contacting sexual organs, it is possible that some stigmas were accidentally pollinated during bagging or later in the day if wind movement knocked the bags against the stylar column. All treatments were completed shortly after flowers opened in the morning, and bags were removed at the end of the day after the flowers had closed. For all manual pollinations, we used small brushes to collect pollen and gently dab it onto stigmas until all five lobes were completely covered. In 2007, we eliminated facilitated autogamy as a potential mode of pollination by emasculating open-pollinated and manually pollinated flowers.

Regular monitoring of fruit development in the field and in the greenhouse indicated that if pollination was not successful and fruit set was not going to occur, then the ovary did not begin to swell, and flower abscission occurred within a week following treatment. Therefore, we considered fertilization to have occurred successfully, and fruit to have been set, if we recorded the presence of a swelling fruit at least one week after pollination. Although this interpretation of fruit set is not as accurate as following all fruits to dehiscence, it was a necessary aspect of the fieldwork, because complete fruit maturation can take several weeks and during that time many fruits, and sometimes whole plants, were eaten by mammals. For this reason also, sample sizes (number of flowers per treatment) for seed set counts were lower than for the level of fruit set. As a consequence of herbivory and low flower production per plant, statistical analyses for field experiments did not attempt to account for variation among maternal plants. To assess differences in fruit set among treatments, depending on sample sizes, we used chi-square or Fisher's exact tests, and for seed numbers, we used ANOVA with *post hoc* Tukey-Kramer comparisons.

Treatment comparisons for estimating various pollination parameters followed Eckert et al. (2010). First, we verified that fruits could not be produced by apomixis by either removing all stigma lobes from mature undehisced flowers (MRC in 2005 and greenhouse in 2008) or brushing away all pollen from anthers and bagging the flowers to exclude pollinators (MRC in 2007). Second, we tested for pollen limitation by comparing levels of fruit and seed production between open-pollinated and manually pollinated flowers (both outcrossed and selfed to help account for variation in pollen quality; see Thomson 2001; Aizen & Harder 2007). Third, we tested for autofertility (the proportion of maximum seed production that can potentially be achieved through autonomous selfing) in the field by comparing fruit and seed production between autonomously selfed flowers with manually outcrossed and manually selfed flowers. We performed an equivalent test for autofertility in the greenhouse, where we eliminated the potential effect of facilitated autogamy by comparing un-manipulated flowers with manually outcrossed and manually selfed flowers. The greenhouse was free of pollinating insects, so delayed selfing was the only possible means of pollination for unmanipulated flowers. Fourth, we conducted a small field test of reproductive assurance in 2007 by comparing fruit set between open-pollinated intact flowers and open-pollinated emasculated flowers.

Use of stylar curvature for delayed selfing

To examine whether the frequency of stylar curvature for delayed selfing differed between species, we surveyed open-pollinated late afternoon flowers at MRC in 2005, 2006, 2007, and 2008. We scored the degree of stylar curvature using a scale of 1 - 5, allocating an independent score to each of the five stylar branches of the flowers and then taking the average (see Fig. 1C). Our scores represent: (1) upright with little or no curvature, (2) outwards curvature (~ 45°) showing clearly that stigma lobes were spreading apart, (3) curvature to a position close to, but not exceeding 90° to the long axis of stylar column, (4) curvature > 90° but the stigma not touching the anthers, (5) stylar branch fully recurved so that the stigma was in contact with the anthers. We used Welch's t-tests to compare mean curvature scores between the two species in each year. In 2005 and 2008, we also recorded whether each stigma was pollinated or not and calculated the proportion of partially recurved stigmas (i.e. those with scores of 1 - 4) that had been pollinated during that day. Additionally, in 2006 and 2007 at MRC and in 2008 for the greenhouse populations, we examined whether, as has been reported for other *Hibiscus* species, prior pollination slowed or completely prevented stylar curvature. We compared levels of style curvature between flowers that had received manual selfed pollen or manual outcrossed pollen against flowers that had been allocated to the autonomous selfing treatment. To analyse these data, we used the curvature score for each flower as the dependent variable and one-way ANOVAs with *post hoc* Tukey-Kramer comparisons.

RESULTS

Floral display

Mean corolla diameter was smaller in *H. aponeurus* (5.6 ± 0.12 cm) than in *H. flavifolius* (6.2 ± 0.06 cm; t_{33} = 4.9, $P < 0.0001$), but the length of the stylar column was longer in *H. aponeurus* (2.9 ± 0.06 cm) than in *H. flavifolius* (2.1 ± 0.04 cm; t_{40} = 10.4, $P < 0.0001$). Stylar columns were much more downwardly oriented with respect to the plane of the corolla in *H. aponeurus* (142.2 ± 3.1 degrees) than in *H. flavifolius* (11.7 ± 2.3 degrees; t_{55} = 8.1, $P < 0.0001$), giving *H. aponeurus* flowers a mildly zygomorphic shape when compared to *H. flavifolius* flowers (Fig. 1A).

Floral abundance and floral rewards

Floral abundance varied across years for both species, but pooled across all surveys, we recorded approximately 6.5 times more *H. flavifolius* than *H. aponeurus* flowers (694 *versus* 106 respectively). The greatest difference in floral abundance was recorded in 2004 (566 *H. flavifolius versus* 77 *H. aponeurus* flowers), but strong differences remained in 2005 (53 *versus* 25) and 2006 (75 *versus* 4). Detailed comparisons between species for nectar volume, sugar concentration, and pollen scores during each time interval and year of observation are shown in supplementary information (Appendix II, Tab. S1).

i) Nectar volume — Relative nectar volumes in the two species varied across years of observation. In 2005, nectar levels early in the morning were similar ($F_{1,3}$ = 1.7, $P = 0.280$; Fig. 2A), but by midday nectar volume in *H. flavifolius* was significantly lower ($F_{1,3}$ = 633.6, $P = 0.00014$; Fig. 2A), and by the end of the day half of the flowers of this species contained no measurable nectar (< 0.1 μl). In contrast, in 2005 nectar standing crop in *H. aponeurus* flowers averaged above 0.6 μl / flower in all three time intervals (Fig. 2A). In 2006, early morning standing crop nectar volume for *H. flavifolius* flowers was on average more than twice that recorded in *H. aponeurus* ($F_{1,8}$ = 7.6, $P = 0.025$), but later in the day volumes decreased in both species and did not differ significantly (Fig. 3A).

ii) Sugar concentration — Nectar sugar concentration did not differ significantly between the two species in any time interval for either of the two years of observation (Figs. 2B and 3B). In *H. aponeurus,* sugar concentration ranged from 29% - 48% in 2005 and 38% - 41% in 2006, whereas in *H. flavifolius,* respective estimates were 20% - 46% and 32% - 39%.

iii) Pollen availability — In 2005 and 2006, pollen availability scores for *H. flavifolius* flowers declined strongly through time on all sampling days, with nearly all pollen removed by the end of the day (mean scores 0.03; Figs. 2C and 3C). In contrast, mean pollen scores for *H. aponeurus* remained above 3.0 throughout the day on all days and years of observation, indicating very low pollen removal by floral visitors. Consequently, end-of-day pollen scores for 2005 and 2006 differed significantly between species ($P < 0.0001$). Differences were less pronounced in 2008, but end-of-day pollen availability was still greater in *H. aponeurus* (2.42 ±

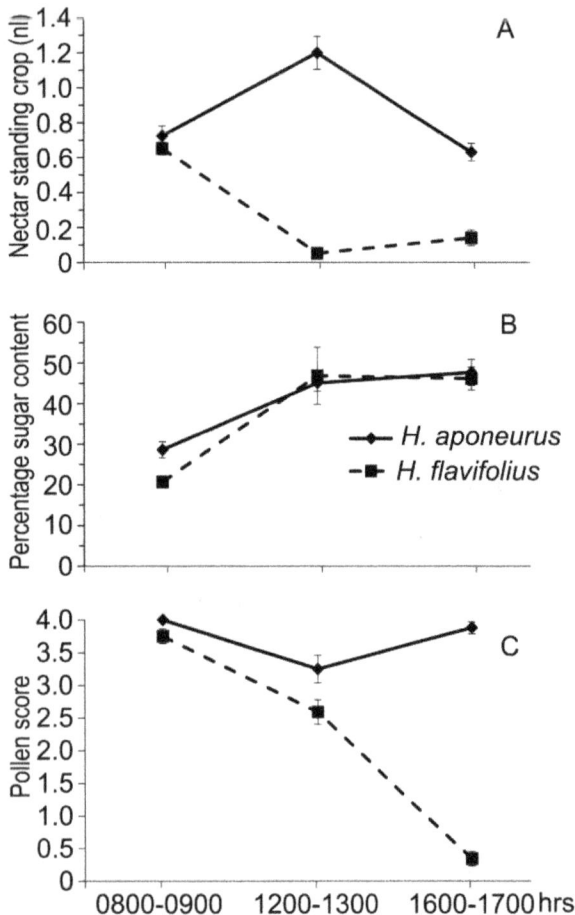

FIGURE 2. Change in floral resources for *H. aponeurus* and *H. flavifolius* as recorded in 2005 during three daily time periods at Mpala Research Centre. Graphs show A) nectar standing crop (μL / flower); B) nectar sugar concentration (% sugar); and C) pollen availability (as pollen score). Data points are grand means ± S.E. across all flowers (25 - 49 per species) and sampling dates (2 days for *H. aponeurus* and 3 days for *H. flavifolius*). The percentage of pollen remaining in anthers was scored on a scale of 0 - 4 (0 = < 1%; I = I - 25%; 2 = 26 - 50%; 3 = 5I - 75%; 4 = 76 - 100%).

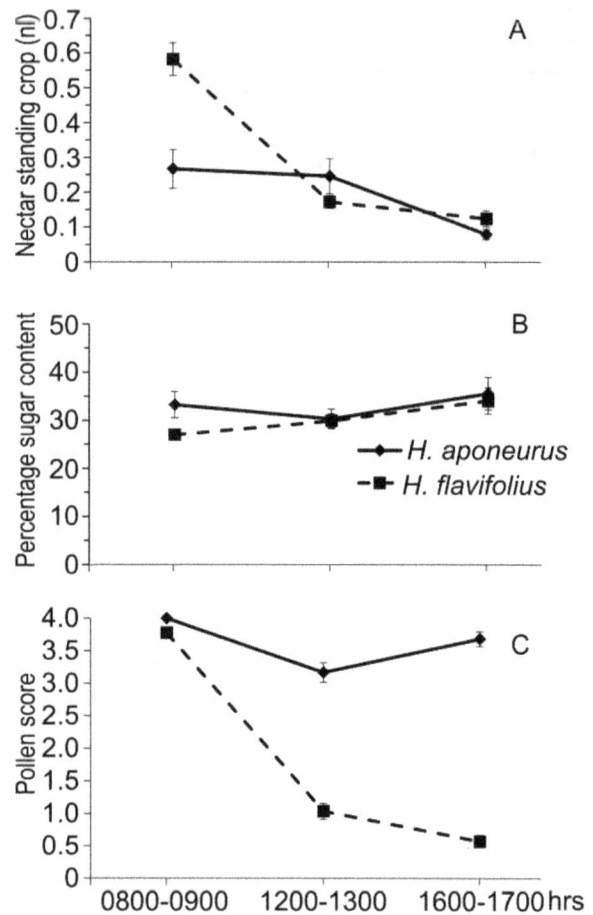

FIGURE 3. Change in floral resources for *H. aponeurus* and *H. flavifolius* as recorded in 2006 during three daily time periods at Mpala Research Centre. Graphs show A) nectar standing crop (μL / flower); B) nectar sugar concentration (% sugar); and C) pollen availability (as pollen score). Data points are grand means ± S.E. across all flowers (46 - 91 per species) and sampling dates (5 days for each species). The percentage of pollen remaining in anthers was scored on a scale of 0 - 4 (0 = < 1%; I = I - 25%; 2 = 26 - 50%; 3 = 5I - 75%; 4 = 76 - 100%).

0.15; N = 36) flowers than in *H. flavifolius* (2.00 ± 0.09; N = 75) flowers ($F_{1.16}$ = I I.0; P = 0.00I).

Insect visitation

Across four years of observation (2004 - 2008), we recorded 245 visits to *H. aponeurus* flowers, and 663 visits to *H. flavifolius* flowers (see Appendix III, Tab. S2 for a taxonomic checklist). However, our 2008 data revealed that only a minority of the visits to either species involved insects landing in the forward-facing position that facilitates pollen transfer onto the stigmas. These potentially effective visits were significantly more frequent (Fisher's exact test, P = 0.00I) for *H. flavifolius* (21%, N = I07) than for *H. aponeurus* (0%, N = 36).

i) Diversity of visitor assemblages — Across all four years, the composition of visitor assemblages and the relative abundance of the visiting taxa differed significantly between *Hibiscus* species (examining four categories: bees, butterflies,

beetles, other insects: χ^2 = 25I.7, df = 3, P < 0.000I; Fig. 4). In both plants, the most common visitor taxa were bees, and the rarest were bugs and flies. Five and six bee genera, respectively, visited *H. aponeurus* and *H. flavifolius*, and four of those genera (*Apis mellifera*, *Braunsapis* sp., *Lasioglossum* sp., and *Tetralonia* sp) were recorded visiting flowers of both species (Appendix III, Tab. S2). Bee visits accounted for 87.5% of the total to *H. flavifolius* and 46.I% of the total to *H. aponeurus*. By contrast, butterflies were both more diverse and more frequently recorded visiting *H. aponeurus* flowers (nine genera; 4I.2% of visits) than *H. flavifolius* flowers (three genera; 2.7% of visits). Two main beetle genera, *Coryna* and *Meloidea*, were observed consuming the flowers of both species (II.6% of visits to *H. aponeurus*; 6.3% to *H. flavifolius*), and it is possible that in a few cases they facilitated autogamy by knocking pollen onto the stigmas.

□ Hemiptera   ■ Hymenoptera
□ Coleoptera   ◪ Hemiptera
▨ Lepidoptera   & Diptera

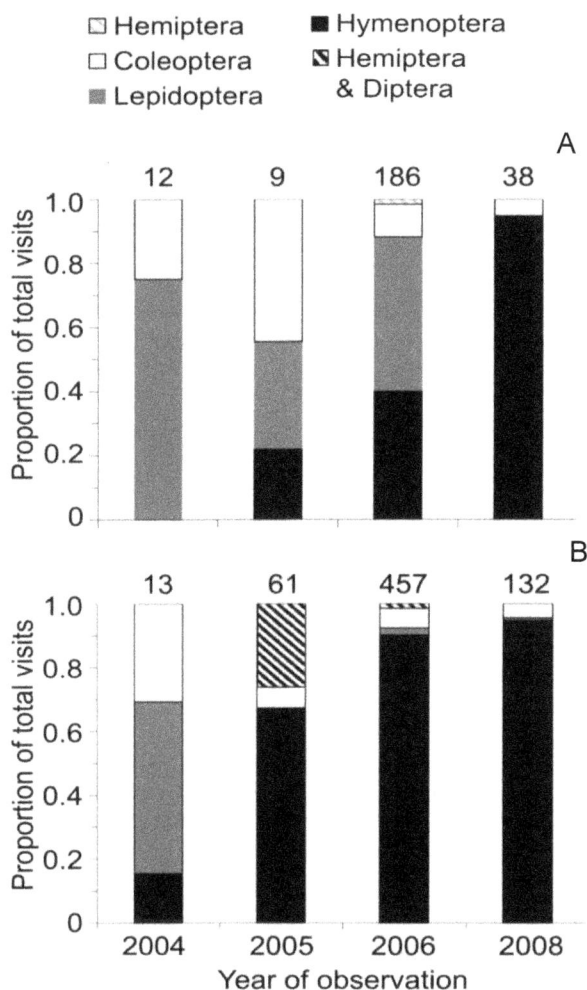

FIGURE 4. Diversity of insect taxa grouped by order that were observed visiting A) *H. aponeurus* and B) *H. flavifolius* flowers in 2004, 2005, 2006, and 2008 at Mpala Research Centre. Total numbers of recorded visits are given above each column.

Patterns of visitation differed significantly between *Hibiscus* species in 2005 (bees *versus* other insects; Fisher's

exact test, $P = 0.022$) and 2006 (bees *versus* butterflies *versus* other insects; $\chi^2 = 214.3$, df $= 2$, $P < 0.0001$). However, visitor frequencies did not differ significantly between plant species during 2004, when most visitors were butterflies, or in 2008, when over 90% of all visits were by bees (mostly honeybees, *Apis mellifera*, a species not recorded in previous years).

ii) Floral visitation rates per hour — Across all visitor taxa, *H. flavifolius* flowers received five times as many visits per hour as did *H. aponeurus* in 2005, and about twice as many in both 2006 and 2008 (Tab. 1). Visitation rates by bee visitors alone were 16 times greater to *H. flavifolius* than to *H. aponeurus* in 2005, and nearly five times greater in 2006 (Tab. 1). Across all surveys in 2008, *H. aponeurus* flowers that had received higher numbers of visits had significantly less pollen remaining in their anthers at the end of the day ($R^2 = 0.73$, $P = 0.003$). Although we observed a similar trend in *H. flavifolius*, this relationship was not statistically significant ($R^2 = 0.23$, $P = 0.194$).

Tests for apomixis, pollen limitation, autofertility, and reproductive assurance

None of the apomixis treatments produced fruits in either species, whether at MRC in 2005 or 2007 (54 and 11 *H. aponeurus* flowers; 66 and 6 *H. flavifolius* flowers, respectively), or in the greenhouse (26 *H. aponeurus* and 3 *H. flavifolius* flowers). Both species appear to be highly self-compatible and to experience negligible levels of early acting inbreeding depression. Fruit set from manual selfing was 61 - 98% at MRC (Tab. 2) and 91- 92% in the greenhouse (Tab. 3) These levels did not differ significantly from fruit set from manual outcrossing, except for *H. aponeurus* in 2005 (Tab. 2), when fruit set was much higher in the manually selfed flowers than for other treatments ($\chi^2 = 11.3$, df $= 3$, $P = 0.01$). Likewise, seed number per fruit for manually selfed flowers generally matched that of manually outcrossed flowers (Tabs. 2 and 3), except for the *H. flavifolius* crosses in 2005, when seed number differed significantly among treatments ($F_{3, 54} = 4.0$, $P = 0.013$), and manually selfed flowers produced about 64% as many seeds as did manually outcrossed flowers (Tukey-Kramer test; $P = 0.032$).

TABLE 1.   Visitation rates (number of visits/flower/hour) to *H. flavifolius* and *H. aponeurus* in 2005, 2006, and 2008. Statistical comparisons use two-sample t-tests for 2005 and 2006 and paired t-tests for 2008.

Visitors - Year	*H. flavifolius*	*H. aponeurus*	$P\|t\|$
All insects			
2005	0.25 ± 0.03	0.05 ± 0.02	$t_3 = 4.2, P = 0.025$
2006	1.55 ± 0.16	0.69 ± 0.51	$t_8 = 1.6, P = 0.148$
2008	0.32 ± 0.04	0.18 ± 0.04	$t_8 = 4.2, P = 0.003$
Bees only			
2005	0.16 ± 0.03	0.01 ± 0.01	$t_8 = 3.8, P = 0.032$
2006	1.39 ± 0.14	0.28 ± 0.25	$t_8 = 3.9, P = 0.004$
2008	0.29 ± 0.05	0.17 ± 0.05	$t_8 = 4.8, P = 0.001$

TABLE 2. Fruit set (percentage of total flowers pollinated that developed a fruit) and mean ± S.E. number of seeds / fruit for four pollination treatments from *H. aponeurus* and *H. flavifolius* populations at Mpala Research Centre. Numbers of maternal plants used for crosses in 2005, 2006, 2007, and 2013, respectively, were 32, 49, 33, and 32 in *H. aponeurus* and 31, 57, 35, and 27 in *H. flavifolius*. In 2007 only, we emasculated the open pollinated, manually outcrossed, and manually selfed flowers. Values in parentheses represent numbers of flowers pollinated (for fruit set) and numbers of fruits sampled (for seed number).

Species - Treatment	2005 Fruit set	2005 Seed number	2006 Fruit set	2007 Fruit set	2013 Fruit set	2013 Seed number
Hibiscus aponeurus						
Open pollination	79.0 (62)	22.2 ± 1.7 (13)	69.5 (131)	40.9 (22)	0.75 (36)	17.6 ± 1.3 (27)
Autonomous selfing	84.2 (57)	17.6 ± 3.2 (8)	72.4 (105)	61.1 (18)	---	---
Manual outcrossing	86.3 (51)	20.2 ± 2.8 (9)	74.8 (103)	65.6 (32)	81.6 (38)	16.9 ± 1.5 (31)
Manual selfing	98.0 (50)	19.2 ± 1.9 (10)	---	60.7 (28)	86.7 (15)	18.8 ± 2.7 (13)
Hibiscus flavifolius						
Open pollination	79.0 (100)	17.7 ± 1.8 (19)	70.3 (91)	76.9 (13)	85.7 (35)	18.5 ± 1.5 (30)
Autonomous selfing	75.0 (80)	22.6 ± 1.9 (11)	78.5 (79)	85.7 (14)	---	---
Manual outcrossing	84.5 (71)	23.1 ± 2.5 (18)	80.0 (95)	84.2 (19)	90.4 (21)	24.6 ± 1.2 (19)
Manual selfing	78.8 (85)	14.8 ± 1.7 (18)	---	88.2 (17)	87.5 (16)	22.8 ± 2.2 (14)

i) Pollen limitation — We found little evidence of pollen limitation in either species. The relative levels of fruit set for open pollinated flowers were 6 - 12% lower than for manually pollinated flowers in our 2005 and 2006 pollinations at MRC (Tab. 2), and those differences were never significant (χ^2 tests, df = 3; $P > 0.05$ in all cases). For seed number, only the *H. flavifolius* data for 2013 suggested pollen limitation ($F_{2, 60} = 4.26$; $P = 0.019$; Tab. 3), with approximately a 30% reduction for open pollination relative to manual outcrossing (Tukey-Kramer test; $P = 0.019$). When we eliminated the possibility of facilitated selfing in 2007 by emasculating the flowers, open-pollinated fruit set was reduced by 33 - 35% relative to all other treatments in *H. aponeurus* ($\chi^2 = 3.53$, df = 3, $P = 0.32$; Tab. 2), but was reduced by only 9 - 13% relative to other treatments in *H. flavifolius* ($\chi^2 = 0.71$, df = 3, $P = 0.87$; Tab. 2).

ii) Autofertility — Field and greenhouse pollinations produced contrasting results for autofertility. At MRC, fruit set for autonomously selfed bagged flowers was lower than but not significantly different from fruit set for manually outcrossed flowers (Tab. 2). Likewise, in 2005, seed number did not differ between autonomously selfed flowers and manually outcrossed flowers (Tab. 2). In the greenhouse, where we could eliminate the chances of accidental pollination due to bagging, autonomous selfing of unbagged *H. aponeurus* flowers produced roughly half the fruits recorded for the manually pollinated flowers ($\chi^2 = 47.5$, df = 2, $P < 0.0001$; Tab. 3). An even larger difference between these two treatments was recorded for *H. flavifolius*, where autonomous selfing resulted in less than a third of fruit set produced by manually pollinated flowers ($\chi^2 = 36.8$, df = 2, $P < 0.0001$; Tab. 3). Seed number in autonomously selfed fruits of *H. aponeurus* was 36% lower than in manually outcrossed and manually selfed fruits ($F_{2, 121} = 20.7$; $P < 0.0001$; Tab. 3). Autonomously selfed fruits of *H. flavifolius* also contained fewer seeds than did fruits from manually pollinated flowers, but this difference was not significant (Welch's ANOVA; $F_{2, 18} = 1.5$; $P = 0.248$).

TABLE 3. Fruit set (percentage of total flowers pollinated that developed a fruit) and mean ± S.E. number of seeds/fruit for three pollination treatments from *H. aponeurus* and *H. flavifolius*, performed in the greenhouse at Indiana University South Bend. Values in parentheses represent numbers of flowers pollinated (for fruit set) and numbers of fruits sampled (for seed set).

Species - Treatment	Fruit set	Seed number
Hibiscus aponeurus		
Autonomous selfing	48.3 (60)	14.5 ± 1.1 (21)
Manual outcrossing	94.6 (56)	22.6 ± 0.8 (45)
Manual selfing	92.5 (67)	22.6 ± 0.7 (58)
Hibiscus flavifolius		
Autonomous selfing	30.0 (30)	20.0 ± 1.2 (9)
Manual outcrossing	100 (15)	23.5 ± 0.9 (14)
Manual selfing	91.3 (23)	23.6 ± 0.8 (20)

iii) Reproductive assurance — In our 2007 test for reproductive assurance, fruit set for *H. aponeurus* open-pollinated intact flowers was almost twice as great as for open-pollinated emasculated flowers (78% *versus* 41%), but that difference was not significant due to small sample sizes (9 intact and 22 emasculated flowers; Fisher's exact test, $P = 0.11$). For *H. flavifolius*, fruit set from open-pollinated intact flowers (83%; $N = 6$ flowers) was very similar to fruit set from open-pollinated emasculated flowers (77%; $N = 13$ flowers).

Use of stylar curvature for delayed selfing

As predicted, our end-of-day surveys of open-pollinated flowers showed that *H. aponeurus* flowers displayed significantly greater stylar curvature than did *H. flavifolius* flowers in all four years of observation (Fig. 5). Average curvature for *H. aponeurus* consistently exceeded 90° to the long axis of stylar column (mean score always > 3.0; see Fig. 1C), whereas this degree of curvature rarely occurred in *H. flavifolius*. For *H. aponeurus*, 75% of the styles in 2005 had

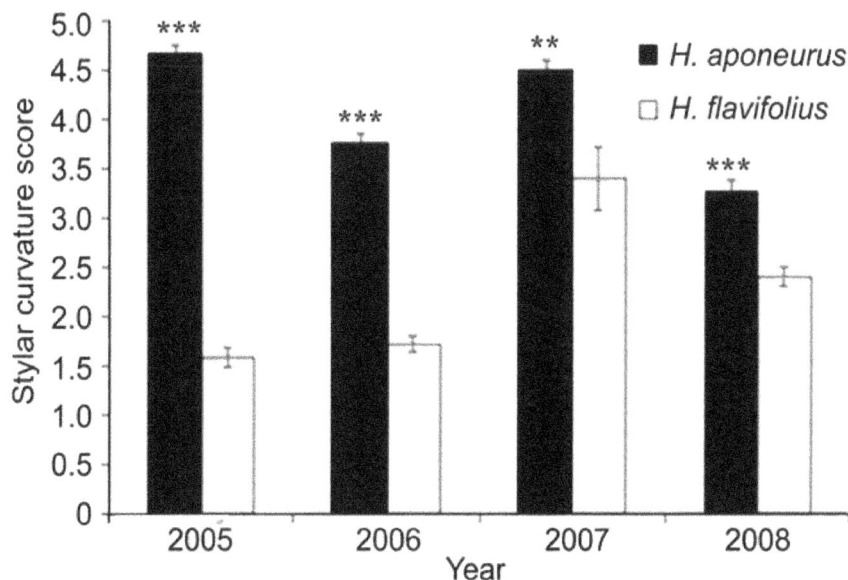

FIGURE 5. Comparison of stylar curvature scores (mean ± S.E.) for end-of-day, open-pollinated flowers of *H. aponeurus* and *H. flavifolius* at Mpala Research Centre. See Figure 1C for an illustration of curvature scores. In 2005, $N = 27$ and 41 flowers, respectively ($t_{60} = 22.5$, $P < 0.0001$); in 2006, $N = 94$ and 79 flowers ($t_{157} = 14.7$, $P < 0.0001$); in 2007, $N = 40$ and 13 flowers ($t_{15} = 3.2$, $P = 0.0062$); and in 2008, $N = 35$ and 75 flowers ($t_{88} = 5.8$, $P < 0.0001$).

TABLE 4. Comparison of stylar curvature scores (mean ± S.E.) for autonomously selfed and manually pollinated flowers from 2006 and 2007 at Mpala Research Centre and for 2008 in the greenhouse at Indiana University South Bend. See Figure 1C for an illustration of curvature scores. Within years and species, treatments not sharing the same superscript are significantly different from one another (ANOVA with Tukey-Kramer tests).

Year-Pollination treatment	*Hibiscus aponeurus*		*Hibiscus flavifolius*	
	No. flowers	Curvature score	No. flowers	Curvature score
MRC 2006				
Autonomous selfing	50	4.53 ± 0.16	36	2.78 ± 0.17
Manual outcrossing	47	2.47 ± 0.17	32	1.64 ± 0.18
	$F_{1,95} = 76.7$; P < 0.0001		$F_{1,66} = 21.4$; P < 0.0001	
MRC 2007				
Autonomous selfing	21	4.33 ± 0.25[A]	14	3.43 ± 0.21[A]
Manual selfing	37	2.75 ± 0.19[B]	26	2.43 ± 0.16[B]
Manual outcrossing	36	2.65 ± 0.19[B]	24	2.11 ± 0.16[B]
	$F_{2,91} = 16.9$; P < 0.0001		$F_{2,61} = 12.4$; P < 0.0001	
IUSB greenhouse 2008				
Autonomous selfing	56	4.60 ± 0.10[A]	49	4.40 ± 0.12[A]
Manual selfing	66	2.48 ± 0.13[B]	43	2.20 ± 0.13[B]
Manual outcrossing	50	2.77 ± 0.16[B]	23	2.04 ± 0.17[B]
	$F_{2,169} = 76.3$; P < 0.0001		$F_{2,112} = 103.8$; P < 0.0001	

completely recurved to effect self-pollination, and of those that were not fully recurved (scores 1 - 4), only 35% of the stigmas had been pollinated. In 2008, however, only 7.4% of styles were fully recurved, and the rate of pollination in the remaining styles was 67%. By contrast, *H. flavifolius* showed greater similarity between years, with low percentages of fully recurved styles (14% in 2005 and 5.1% in 2008) and high rates of pollination on stigmas with partially recurved styles (86% in 2005 and 82% in 2008).

In agreement with our prediction that pollination early in the day would delay or stop stylar curvature, styles in manually pollinated flowers always curved less than styles in autonomously selfed flowers (Tab. 4). In *H. aponeurus*, whether in the field or in the greenhouse, curvature for the autonomously selfed treatment consistently neared completion (mean scores 4.3 - 4.6, Tab. 4; Fig. 1C), and curvature for manually pollinated flowers always stopped well before the styles reached 90° (mean scores 2.5 - 2.8, Tab. 4; Fig. 1C). In *H. flavifolius*, the results for the greenhouse crosses mirrored those for *H. aponeurus*, but in the field, curvature stopped well before stigmas could contact anthers in all treatments (mean scores 1.6 - 3.4, Tab. 4).

DISCUSSION

The sympatric co-flowering *H. aponeurus* and *H. flavifolius* offer abundant floral rewards that attract a diversity of visitor taxa, but floral visitor arrays and visitation rates strongly differ between species and across years. As a result, much less pollen is removed and fewer stigmas are pollinated in *H. aponeurus* than in *H. flavifolius*. Despite lower levels of visitation and pollination, *H. aponeurus* appears not to be pollen limited with regard to fruit or seed set, a result that can be attributed to a combination of facilitated and autonomous selfing along with a small number of ovules that probably can be fully fertilized with deposition of only 20-30 self or outcross pollen grains. Both species clearly possess mixed mating systems that include facultative delayed autonomous selfing as a potential means of reproductive assurance, but this mechanism was observed much more frequently in *H. aponeurus* than in *H. flavifolius*. The differences in levels of delayed selfing between species correlate well with differential rates of insect visitation and the distinct floral morphology of *H. aponeurus* that appears to reduce chances of outcross pollen deposition while increasing chances for facilitated autogamy.

Visitor assemblages and potential for pollinator competition

Hibiscus aponeurus and *H. flavifolius* flowers were visited by substantially different sets of insects from year to year, with *H. aponeurus* generally attracting a higher proportion of butterflies and beetles, and *H. flavifolius* attracting more bees. Although we observed no interspecific flights by bees or butterflies as we monitored visitation, competition for pollination services may still occur. Flowers of both species are open at the same time, and some visitor taxa were recorded on both species in the same year (e.g. honeybees in 2008). To corroborate such competition, we would need to analyze pollen loads and make careful observations to determine if insects contacted the stigmas of both species in ways which could facilitate heterospecific pollen transfer. Future surveys should also attempt to examine the potential role of top-down (e.g. climate) influences in insect activity patterns, which may also help explain the significant year-to-year differences in visitor assemblages recorded in our study. For example, while in 2004 a mixture of bees, butterflies, and beetles were observed visiting both *Hibiscus* species, in 2008 visitation was strongly dominated by bee taxa. Days of observation in 2008 were cooler and windier than in other years, with mean hourly temperatures in May and June of 25.8 °C and 24.2 °C, in comparison to 31.4 °C and 27.4 °C in 2004. Lower temperatures may have favoured the flight physiology of bees, which, even in small species, can maintain higher activity under cooler conditions than can other insects (Willmer & Stone 2004). Moreover, additional observations of pollen deposition in relation to floral morphology may help us determine the pollination effectiveness of certain visitor taxa. Beetles, for example, were observed to be mostly florivores, but they could have facilitated autogamy by knocking pollen onto stigmas. Even in *Hibiscus* species with strong separation between anthers and stigmas, the movements of floral visitors within a flower can sometimes knock pollen about and increase rates of deposition (Sampson et al. 2016). This seems particularly likely in *H. aponeurus* flowers that very often have stigmas located below the anthers due to the downward curving shape of the stigmatic column.

Hibiscus aponeurus and *H. flavifolius* flowers offer abundant pollen and similar volumes of nectar with comparable sugar concentrations. Several lines of evidence, however, demonstrate that *H. aponeurus* receives less outcross pollen than does *H. flavifolius*. First, in 2005, 2006, and 2008, visitation rates to *H. aponeurus* were lower than to *H. flavifolius*. Except for a single unusual day in 2006, each *H. aponeurus* flower received on average no more than three visits over the course of an entire day (10-hr period of anthesis), whereas *H. flavifolius* flowers received as many as 16 visits in a day. Second, the amount of pollen remaining in anthers at the end of the day was always greater in *H. aponeurus* than in *H. flavifolius*, a pattern consistent with lower rates of pollen removal by flower visitors and a potential reduction in male function for *H. aponeurus*. The 2008 data also showed a significant negative correlation between visitation rates and the amount of pollen remaining in the anthers of *H. aponeurus* flowers, a relationship not observed in *H. flavifolius*, presumably because rates of visitation were sufficiently high or because visitors were highly efficient in harvesting pollen, so at the end of the day most flowers were nearly empty. Third, in 2005 and 2008 our end-of-day surveys showed that significantly fewer stigmas were pollinated in *H. aponeurus* than in *H. flavifolius* flowers. Fourth, data from 2008 showed that when bees landed on *H. aponeurus* flowers, all did so in a manner that minimized contact with the stigma. Fifth, *H. aponeurus* flowers often had stigmas located below the opened petal lobes, whereas *H. flavifolius* flowers always have stigmas located above the petals. The combination of a sometimes inaccessible stigma location, a downwardly curved stylar column, and the backwards landing orientation of most insect visitors in *H. aponeurus* would seem to make insect-mediated cross-pollination more difficult than in *H. flavifolius*.

Differences between the *Hibiscus* species in visitor assemblages and visitation rates could have several causes. For example, *H. flavifolius* flowers were consistently much more abundant than *H. aponeurus* flowers. As reviewed by Knight et al. (2005), smaller plant populations are expected to show lower pollinator visitation and pollen deposition than large plant populations. Bees especially tend to show floral constancy (Waser 1986; Goulson 1999; Willmer & Stone 2004), so in a particular foraging bout, we might expect them to focus on collecting nectar or pollen from *H. flavifolius*, as choosing this species over *H. aponeurus* would likely reduce flight time between flowers and result in collection of greater reward. This advantage to *H. flavifolius* might be compounded by the strong colour difference between species. The crimson red of *H. aponeurus* flowers may not be as easily seen by bees as the white *H. flavifolius* flowers, possibly because the *H. aponeurus* flowers are often set against a complex background of green foliage under acacia trees (Chittka & Waser 1997; Spaethe et al. 2001; Rodríguez-Gironés & Santamaría 2004). Thus, it is possible that *H. aponeurus* is competing for many of the same bee pollinators as *H. flavifolius*, with the outcome that the more frequently encountered, and possibly more easily detected, *H. flavifolius* receives the great majority of visits.

An alternative explanation is that despite offering similar quantities of pollen, nectar and sugar, the rewards offered by *H. aponeurus* may be perceived as having a lower nutritional value by the array of local bees. Floral scents and other secondary metabolites may also play an important role in attracting or deterring visitors (Baker & Baker 1982; González-Teuber & Heil 2009, Byers et al. 2014). From this perspective, it is possible that different floral scents are produced by the two *Hibiscus* species, which attract different visitor taxa. Little is known about how variable floral traits are among populations of *H. aponeurus,* but studies in other species show that the chemical composition of volatiles and other floral traits can be subject to pollinator-mediated selection (see Azuma et al. 2001; Whitehead & Peakall 2009). Records from herbaria are scarce, but *H. aponeurus* has a wide geographical range that spans parts of Ethiopia, Sudan, southern Somalia, western and central Kenya, Uganda, Tanzania, Rwanda, Burundi, and the Democratic Republic of Congo (Exell 1961; Agnew & Agnew 1994; Barkhadle et al. 1994; Friis & Vollesen 1998; African Plant Database 2017). The populations at MRC thus represent some of the eastern most recorded localities for *H. aponeurus*, and it is possible that these populations arrived recently from more central parts of its geographical range, and that their floral traits may have evolved to attract pollinator guilds that are largely absent from MRC. Examining the plausibility of this explanation would require detailed observations of floral visitation and floral traits from *H. aponeurus* populations located in other areas of its geographical distribution, chemical analyses of the composition and nutritional value of its floral rewards, and feeding preference assays under controlled conditions.

Pollen limitation, autofertility, and the importance of delayed selfing

Consistent with studies of other Malvaceae (e.g. Klips & Snow 1997; Ruan et al. 2008b), both *H. aponeurus* and *H. flavifolius* are fully self-compatible. Across three years of crosses in the field and one set of greenhouse crosses, fruit set and seed number for manually selfed flowers were almost never lower than for manually outcrossed flowers. In addition, neither species showed strong or consistent evidence of pollen limitation, a result that agrees with published findings that low levels of pollen limitation are associated with intrinsic factors, such as self-compatibility and actinomorphy, and with extrinsic factors, such as high abundance, low plant diversity, and high pollinator diversity (Burd 1994; Larson & Barrett 2000; Knight et al. 2005; Vamosi et al. 2006, 2013; Alonso et al. 2010). Except for the low abundance of *H. aponeurus* at MRC and its slight tendency towards zygomorphy, our populations possess all of these traits. Moreover, both *H. aponeurus* and *H. flavifolius* have the potential to use stylar curvature as a mechanism of delayed autonomous selfing to compensate for low levels of insect-mediated pollination. However, our data are unclear on the degree to which delayed selfing provides reproductive assurance. Field crosses using intact flowers showed 79% - 100% autofertility in *H. aponeurus* and 87 - 100% autofertility in *H. flavifolius*, but in the greenhouse these rates were much lower, with only 33% autofertility in *H. aponeurus* and 27% in *H. flavifolius*. Eckert et al. (2010) pointed out that high autofertility indicates only the potential for reproductive assurance, and that in field experiments, it is often not well correlated with reproductive assurance, which can be measured as the fraction of reproduction due to autonomous selfing. Our small experiment to test directly for reproductive assurance was

ambiguous due to low sample sizes, but it showed that stylar curvature is more important in *H. aponeurus*, where 47% of fruit set could be attributed to delayed selfing, compared to only 7% that could be ascribed to delayed selfing in *H. flavifolius*.

Regardless of its effectiveness at offsetting low pollination rates, stylar curvature is clearly much more common in *H. aponeurus* flowers than in *H. flavifolius*, an event that appears to be driven by differences in insect visitation rates. In *H. aponeurus*, the extent of stylar curvature in open-pollinated flowers, which at MRC regularly received very low rates of visitation, was always similar to the curvature observed in bagged, autonomously selfing flowers, and always significantly greater than in manually pollinated flowers. In contrast, with the exception of 2007, stylar curvature for open-pollinated *H. flavifolius* flowers was similar to manually pollinated flowers, indicating that open-pollinated flowers were being pollinated at high enough rates to stop curvature. Thus, although we found both species to have approximately equal capacity to use autonomous selfing through stylar curvature as a means of reproductive assurance, low visitation rates in *H. aponeurus* appeared to force this species to resort to this mechanism more frequently than the more regularly visited *H. flavifolius*.

Future directions

Two complementary hypotheses could potentially explain the differences we found between *Hibiscus* species in the use of delayed selfing. *Hibiscus aponeurus* populations at MRC could be locally adapted to a largely selfing reproductive strategy or they could more simply be showing a phenotypically plastic response to low visitation levels by inadequate pollinators. Under either hypothesis, we expect that open-pollinated progeny tests using genetic markers (perhaps in artificial populations in the field where floral densities could be controlled) would find higher rates of selfing in *H. aponeurus* than in *H. flavifolius*. If the local adaptation hypothesis were true, then comparisons of selfed and outcrossed progeny should find less inbreeding depression in *H. aponeurus* than in *H. flavifolius*. Conditions within *H. aponeurus* populations, in fact appear to be favourable to the evolution of increased selfing: small population sizes, low levels of insect visitation, and high variability in pollinator composition across seasons (Morgan et al. 2005; Morgan & Wilson 2005; Knight et al. 2005). Consistent with this hypothesis are also the reduced size of the corolla in *H. aponeurus* compared to *H. flavifolius* and the placement of stigmas in a position that appears to decrease chances for outcrossing and to facilitate autogamy.

Additional evidence that MRC populations of *H. aponeurus* are evolving towards increased selfing could come from comparisons with populations situated in other parts of this species' geographical range. Klips & Snow (1997), for example, showed that in the United States, northern populations of *Hibiscus laevis* are capable of selfing, whereas southern populations are not, because styles in the southern populations were too long to allow stigmas to touch anthers even when fully recurved. They argued that the southern populations may experience more reliable bee pollination than the northern populations, resulting in the evolution of selfing only in the northern populations. Also, the "abundant centre" model predicts that population sizes and densities will be greatest near the central region of a species' range and will decline towards the peripheries of the range (Brown 1984). A possible consequence of low population size and density is lower levels of outcrossing and selection for traits that increase autogamous self-fertilization (Jain 1976; Herlihy & Eckert 2005). As discussed above, because *H. aponeurus* populations at MRC represent the eastern periphery of the range, an analysis of more central populations could reveal evidence of higher pollination rates and less reliance on delayed selfing for seed production. Our future work contemplates visiting populations of these two species in other areas across their geographical ranges to perform floral visitation observations and seed collections. Because some of the differences in visitation observed between the two species may be due to differences in nectar and pollen composition, we are currently performing chemical analyses to examine sugar, amino acid, and potentially secondary metabolite contents. We are also assessing whether floral volatiles are released during anthesis or dehiscence that may help explaining attraction of floral visitors. To examine potential genetic consequences of autonomous selfing, we are developing genetic markers, which we will use to measure outcrossing levels of maternal families collected in the wild.

ACKNOWLEDGEMENTS

The authors thank the Kenyan government for permission to conduct this research (National Council of Science and Technology Permit number MOEST 13/001/33C 116), and Mpala Research Centre for excellent logistical support. We are deeply grateful for the hard work in the field and the constructive intellectual input provided by P. Daly, A. De LaPaz, H. Dueck, R. Erengay, P. Lenguya, J. Lima, K. Relos, and Prof. P. Willmer. This research was conducted in close collaborations with the Invertebrate Section at The National Museums of Kenya, and the East African Herbarium. In particular we thank Dr. Wanja Kinuthia, Dr. Siro Masinde, and Dr. Itambo Malombe for their help in curating and exporting insect and plant materials. We also thank Prof. Connal Eardley in Pretoria South Africa, for his invaluable taxonomic help to identify all bee taxa collected during visitation observations. This work was supported by the U.S. National Science Foundation under Grant No. DEB-0344519 (AS), Consejo Nacional de Ciencia y Tecnología in México under Scholarship CONACyT-168410 (JCRG), and a NERC quota studentship to Edinburgh University (KB).

APPENDICES

Additional supporting information may be found in the online version of this article:

APPENDIX I.   Information about seed germination and growth in greenhouses.

APPENDIX II.   Significance level of the differences in resource production between species across time intervals and years of observation.

APPENDIX III.   Floral visitors recorded for each plant species and each year of observation.

REFERENCES

African Plant Database (version 3.0.4). Conservatoire et Jardin Botaniques de la Ville de Genève and South African National Biodiversity Institute, Pretoria, South Africa. [online] URL:http://www.ville- ge.ch/musinfo/bd/cjb/africa/details.php?id=81504   (accessed March 2017).

Agnew ADQ, Agnew S (1994) Upland Kenya wildflowers: a flora of the ferns and herbaceous flowering plants of upland Kenya, 2nd ed. East Africa Natural History Society, Nairobi, Kenya.

Aguilar R, Ashworth L, Galetto L, Aizen MA (2006) Plant reproductive susceptibility to habitat fragmentation: review and synthesis through a meta-analysis. Ecology Letters 9:968–980.

Aizen MA, Harder LD (2007) Expanding the limits of the pollen-limitation concept: effects of pollen quantity and quality. Ecology 88:271-281.

Alonso C, Vamosi JC, Knight TM, Steets JA, Ashman TL (2010) Is reproduction of endemic plant species particularly pollen limited in biodiversity hotspots? Oikos 119:1192–1200.

Azuma, H, Toyota M, Asakawa Y (2001) Intraspecific variation of floral scent chemistry in *Magnolia kobus* DC. (Magnoliaceae). Journal of Plant Research 114:411–422.

Baker HG, Baker I (1982) Chemical constituents of nectar in relation to pollination mechanisms and phylogeny. In: Nitecki M. (ed.) Biochemical aspects of evolutionary biology. University of Chicago Press, Chicago, pp.1311-1371.

Baldock KCR, Memmott J, Ruiz-Guajardo JC, Roze D, Stone GN (2011) Daily temporal structure in African savanna flower visitation networks and consequences for network sampling. Ecology 92:687–698.

Barkhadle AMI, Ongaro L, Pignatti S (1994) Pastoralism and plant cover in the lower Shabelle region, Southern Somalia. Landscape Ecology 9:79-88.

Brown JH (1984) On the relationship between abundance and distribution of species. The American Naturalist 124:255-279.

Brys R, Jacquemyn J (2011) Variation in the functioning of autonomous self-pollination, pollinator services and floral traits in three *Centaurium* species. Annals of Botany 107:917-925.

Brys R, Jacquemyn J (2012) Effects of human-mediated pollinator impoverishment on floral traits and mating patterns in a short-lived herb: an experimental approach. Functional Ecology 26:189-197.

Burd M (1994) Bateman's principle and plant reproduction: the role of pollen limitation in fruit and seed set. The Botanical Review 60:83-139.

Buttrose MS, Grant WJR, Lott JNA (1977) Reversible curvature of style branches of *Hibiscus trionum* L., a pollination mechanism. Australian Journal of Botany 25:567-570.

Byers KJRP, Bradshaw Jr HD, Riffell JA (2014) Three floral volatiles contribute to differential pollinator attraction in monkey flowers (*Mimulus*). Journal of Experimental Biology 217:614-623.

Chittka L, Wasser N (1997) Why red flowers are not invisible to bees. Israel Journal of Plant Sciences 45:169-183.

Eckert CG, Kalisz S, Geber MA, Sargent R, Elle E, Cheptou PO, Goodwillie C, Johntson MO, Kelly JK, Moeller DA, Porcher E, Ree RH, Vallejo-Marín M, Winn AA (2010) Plant mating systems in a changing world. Trends in Ecology and Evolution 25:35-43.

Elliot SE, Irwin RE (2009) Effects of flowering plant density on pollinator visitation, pollen receipt, and seed production in *Delphinium barbeyi* (Ranunculaceae). American Journal of Botany 96:912–919.

Exell AW (1961) Malvaceae In: Exell AW, Wild H (eds) Flora Zambesiaca: Mozambique, Federation of Rhodesia and Nyasaland, Bechuanal and Protectorate, vol. I, part 2, p. 420. [online] http://apps.kew.org/efloras/search.do. (Accessed 3 November 2016). Information used with permission of the Trustees of the Royal Botanic Gardens, Kew.

Friis I, Vollensen K (1998) Flora of the Sudan-Uganda border area east of the Nile. Catalogue of vascular plants, ser. I, part I. The Royal Danish Academy of Sciences and Letters, Munksgaard, Copenhagen.

González-Teuber M, Heil M (2009) Nectar chemistry is tailored for both attraction of mutualists and protection from exploiters. Plant Signaling & Behavior 9:809-813.

Goodwillie C, Kalisz S, Eckert CG (2005) The evolutionary enigma of mixed mating systems in plants: occurrence, theoretical explanations, and empirical evidence. Annual Review of Ecology, Evolution, and Systematics 36:47-79.

Goulson D (1999) Foraging strategies of insects for gathering nectar and pollen, and implications for plant ecology and evolution. Perspectives in Plant Ecology, Evolution and Systematics 2:185-209.

Herlihy CR, Eckert CG (2005) Evolution of self-fertilization at geographical range margins? A comparison of demographic, floral, and mating system variables in central vs. peripheral populations of *Aquilegia canadensis* (Ranunculaceae). American Journal of Botany 92:744-751.

Jain SK (1976) The evolution of inbreeding in plants. Annual Review of Ecology and Systematics 7:469–495.

Jorgensen R, Arathi HS (2013) Floral longevity and autonomous selfing are altered by pollination and water availability in *Collinsia heterophylla*. Annals of Botany 112:821-828.

Kalisz S, Vogler DW, Hanley KM (2004) Context-dependent autonomous self-fertilization yields reproductive assurance and mixed mating. Nature 430:884-887.

Kalisz S, Vogler D, Fails B, Finer M, Shepard E, Herman T, Gonzales R (1999) The mechanism of delayed selfing in *Collinsia verna* (Scrophulariaceae). America Journal of Botany 86:1239-1247.

Klips RA, Snow AA (1997) Delayed autonomous self-pollination in *Hibiscus laevis* (Malvaceae). American Journal of Botany 84:48-53.

Knight TM, Steets JA, Vamosi JC, Mazer SJ, Burd M, Campbell DR, Dudash MR, Johnston MO, Mitchell RJ, Ashman TL (2005) Pollen limitation of plant reproduction: pattern and process. Annual Review of Ecology, Evolution, and Systematics 36:467-497.

Kumar G, Kadam GB, Saha TN, Girish KS, Tiwari AK, Kumar R (2014) Studies on floral biology of *Malva sylvestris* L. Indian Journal of Horticulture 71:295-297.

Larson BMH, Barrett SCH (2000) A comparative analysis of pollen limitation in flowering plants. Biological Journal of the Linnean Society 69:503-520.

Lloyd DG, Schoen DJ (1992) Self- and cross-fertilization in plants. I. Functional dimensions. International Journal of Plant Sciences 153:358-369.

Morgan MT, Wilson WG (2005) Self-fertilization and the escape from pollen limitation in variable pollination environments. Evolution 59:1143-1148.

Morgan MT, Wilson WG, Knight TM (2005) Plant population dynamics, pollinator foraging, and the selection of self-fertilization. The American Naturalist 166:169-183.

Pannell JR, Barrett SCH (1998) Baker's law revisited: reproductive assurance in a metapopulation. Evolution 52:657-668.

Pfeil BE, Brubaker CL, Craven LA, Crisp MD (2002) Phylogeny of *Hibiscus* and the tribe Hibisceae (Malvaceae) using chloroplast DNA sequences of *ndhF* and the *rpl16* intron. Systematic Botany 27:333-350.

Qu R, Li X, Luo Y, Dong M, Xu H, Chen X, Dafni A (2007) Wind-dragged corolla enhances self-pollination: a new mechanism of delayed self-pollination. Annals of Botany 100:1155-1164.

Raine NE, Pierson AS, Stone GN (2007) Plant-pollinator interactions in a Mexican *Acacia* community. Arthropod-Plant Interactions 1:101-117.

Rambuda TN, Johnson SD (2004) Breeding systems of invasive alien plants in South Africa: does Baker's rule apply? Diversity and Distributions 10:409-416.

Ramsey ML, Seed L, Vaughton G (2003) Delayed selfing and low levels of inbreeding depression in *Hibiscus trionum* (Malvaceae). Australian Journal of Botany 51:275-281.

Rodríguez-Gironés MA, Santamaría L (2004) Why are so many bird flowers red? Public Library of Science Biology 2:1515-1519.

Ruan CJ, Chen SC, Qun L, Texeira Da Silva JA (2011) Adaptive evolution of context-dependent style curvature in some species of the Malvaceae: a molecular phylogenetic approach. Plant Systematics and Evolution 297:57-74.

Ruan CJ, Li H, Mopper S (2008a) The impact of pollen tube growth on stigma lobe curvature in *Kosteletzkya virginica*: the best of both worlds. South African Journal of Botany 74:65-70.

Ruan CJ, Mopper S, Texeira Da Silva JA, Qin P, Zhang QX, Shan Y (2009) Context-dependent style curvature in *Kosteletzkya virginica* (Malvaceae) offers reproductive assurance under unpredictable pollinator environments. Plant Systematics and Evolution 277:207-215.

Ruan CJ, Qin P, Xi Y (2005) Floral traits and pollination modes in *Kosteletzkya virginica* (Malvaceae). Belgian Journal of Botany 138:39-46.

Ruan CJ, Mopper S, Texeira Da Silva JA, Qin P (2010) Style curvature and its adaptive significance in the Malvaceae. Plant Systematics and Evolution 288:13-23.

Ruan CJ, Zhou L, Zeng F, Han R, Qin Q, Luttis S, Saad L, Mahy G. (2008b) Contribution of delayed autonomous selfing to reproductive success in *Kosteletzkya virginica*. Belgian Journal of Botany 141:3-13.

Ruiz-Guajardo JC (2008) Community plant-pollinator interactions in a Kenyan savannah. Ph.D. dissertation, University of Edinburgh, Edinburgh, UK.

Sampson BJ, Pounders CT, Werle CT, Mallette TR, Larsen D, Chetelain L, Lee KC (2016) Aggression between floral specialist bees enhances pollination of *Hibiscus* (section *Trionum*: Malvaceae). Journal of Pollination Ecology 18:7-12.

Scriven LA, Sweet MJ, Port GR (2013) Flower density is more important than habitat type for increasing flower visiting insect diversity. International Journal of Ecology 2013, Article ID 237457, 12 pages.

Seed L, Vaughton G, Ramsey M (2006) Delayed autonomous selfing and inbreeding depression in the Australian annual *Hibiscus trionum* var. *vesicarius* (Malvaceae). Australian Journal of Botany 54:27-34.

Somme L, Mayer C, Jacquemart AL (2014). Multilevel spatial structure impacts on the pollination services of *Comarum palustre* (Rosaceae). PLoS One 9:1-10.

Spaethe J, Tautz J, Chittka L (2001) Visual constraints in foraging bumblebees: flower size and color affect search time and flight behavior. Proceedings of the National Academy of Sciences USA. 98:3898-3903.

Stebbins GL (1957) Self-fertilization and population variability in higher plants. The American Naturalist 91:337-354.

Stone G, Willmer P, Nee S (1996) Daily partitioning of pollinators in an African *Acacia* community. Proceedings of the Royal Society of London Series B-Biological Sciences 263:1389-1393.

Stone GN, Willmer P, Rowe JA (1998) Partitioning of pollinators during flowering in an African *Acacia* community. Ecology 79:2808-2827.

Thomann ME, Imbert E, Devaux C, Cheptou PO (2013) Flowering plants under global pollinator decline. Trends in Plant Science 18:353-359.

Thomson JD (2001) Using pollination deficits to infer pollinator declines: can theory guide us? Conservation Ecology 5:6. [online] URL: http://www.consecol.org/vol5/iss1/art6/.

Vamosi JC, Knight TM, Steets JA, Mazer SJ, Burd M, Ashman TL (2006) Pollination decays in biodiversity hotspots. Proceedings of the National Academy of Sciences USA 103:956-961.

Vamosi JC, Steets JC, Ashman TL (2013) Drivers of pollen limitation: macroecological interactions between breeding system, rarity, and diversity. Plant Ecology & Diversity 6:171-180.

Vanbergen, AJ (2013) Threats to an ecosystem service: pressures on pollinators. Frontiers in Ecology & the Environment 11: 251–259.

Vaughton G, Ramsey M (2010) Pollinator-mediated selfing erodes the flexibility of the best-of-both-worlds mating stragey in *Bulbine vagans*. Functional Ecology 24:374-382.

Waser NM (1986) Flower constancy: definition, cause, and measurement. The American Naturalist 127:593-603.

Watanabe ME (2014) Pollinators at risk: human activities threaten key species. BioScience 64:5-10.

Whitehead MR, Peakall R (2009) Integrating floral scent, pollination ecology and population genetics. Functional Ecology 23:863-874.

Willmer PG, Stone GN (2004) Behavioral, ecological and physiological determinants of the activity patterns of bees. Advances in the Study of Behavior 34:347-466.

Wright SI, Kalisz S, Slotte T (2013) Evolutionary consequences of self-fertilization in plants. Proceedings of the Royal Society B 280:20130133.

Zhang C, Zhou GY, Yang YP, Duan YW (2014) Better than nothing: evolution of autonomous selfing under strong inbreeding depression in an alpine annual from Qinghai-Tibet Plateau. Journal of Systematics and Evolution 52:363-367.

22

Contrasting pollination efficiency and effectiveness among flower visitors of *Malva sylvestris, Borago officinalis* and *Onobrychis viciifolia*

Anna Gorenflo*[1], Tim Diekötter[2], Mark van Kleunen[3,4], Volkmar Wolters[5], Frank Jauker[5]

[1]*Applied Entomology, Department of Insect Biotechnology, Justus Liebig University Giessen, Heinrich-Buff-Ring 26-32, 35392 Giessen, Germany*
[2]*Department of Landscape Ecology, Institute for Natural Resource Conservation, Kiel University, Olshausenstrasse 75, 24118 Kiel, Germany*
[3]*Ecology, Department of Biology, University of Konstanz, Universitätsstrasse 10, 78457 Konstanz, Germany*
[4]*Zhejiang Provincial Key Laboratory of Plant Evolutionary Ecology and Conservation, Taizhou University, Taizhou 318000, China*
[5]*Department of Animal Ecology, Justus Liebig University Giessen, Heinrich-Buff-Ring 26-32, 35392 Giessen, Germany*

Abstract—Biotic pollination is an important factor for ecosystem functioning and provides a substantial ecosystem service to human food security. Not all flower visitors are pollinators, however, and pollinators differ in their pollination performances. In this study, we determined the efficiencies of flower visitors to the plant species *Malva sylvestris, Borago officinalis* and *Onobrychis viciifolia* by analysing stigmatic pollen deposition. We further calculated pollinator effectiveness by scaling up single-visit pollen deposition using visitation frequency. Flower-visitor groups differed in their efficiencies at the single-visit level and not all of them deposited more pollen compared to unvisited stigmas. Bumblebees tended to be most efficient in depositing pollen per single visit across the three plant species. Due to the by far highest visitation frequencies, *Apis mellifera* showed the highest effectiveness in depositing pollen per hour for *M. sylvestris* and *B. officinalis*, but not for *O. viciifolia*, for which the *Bombus lapidarius* complex was both the most frequent and the most effective pollinator group. Hence, the most frequent flower visitors were most effective in our study. For non-dominant pollinator groups, however, visitation frequencies contributed disproportionally to pollinator effectiveness. Thus, combining pollen deposition per single-visit with visitation frequency is necessary to reveal true pollinator performance and to better understand flower-visitor interactions.

Keywords: *Apoidae, pollen analysis, pollination service, pollinator importance, solitary bees*

Introduction

Pollen transfer is an essential process for seed and fruit production in sexually reproducing plants and around 90% of all angiosperm species are animal pollinated (Ollerton et al. 2011). Biotic pollination is therefore an important factor for wild plant reproduction (Ollerton et al. 2011) and an important ecosystem service to human food security (Gallai et al. 2009; Calderone et al. 2012). Despite the importance of biotic pollination, many obscurities about plant-pollinator interactions still need to be solved (Mayer et al. 2011). For example, the terms "flower visitor" and "pollinator" are often used synonymously, without actual proof of pollen transfer (Ne'eman et al. 2010). In addition, performances as pollinators have been shown to differ among flower visitors (Ne'eman et al. 2010; Jędrzejewska-Szmek & Zych 2013; Popic et al. 2013; Ballantyne et al. 2015). This general gap in knowledge complicates finding appropriate pollinators for agricultural systems (Slaa et al. 2006), interpreting specialisation and generalisation (Jędrzejewska-Szmek &

Zych 2013; Popic et al. 2013; Ballantyne et al. 2015) and predicting ecological and economic consequences of pollinator loss or invasions (Goulson 2003).

Insects, and above all bees, are considered the most important pollinators (Buchmann & Nabhan 1996). Pollen is transferred passively while insects utilise flowers for foraging, as shelter or mating site, among other reasons (Inouye et al. 1994). Not all flower visitors, however, are involved in pollination. Reasons are mismatches in morphology or behaviour or illegitimate exploitations of floral resources (i.e. nectar and pollen thieves or robbers) (Inouye 1980). Therefore, pollination services of a given flower visitor may differ among plant species and pollination success for a given plant species differs among pollinators (Fenster et al. 2004; King et al. 2013; Popic et al. 2013; Ballantyne et al. 2015).

Both indirect (e.g. visit duration, visitor frequency, pollen removal, stigma receptivity) and direct measurements (e.g. stigmatic pollen deposition, seed set) have been used to analyse pollination performances of flower visitors (summarised by Ne'eman et al. 2010), but previous research suggests that direct measurements are more reliable (Johnson & Steiner 2000; Adler & Irwin 2006; King et al. 2013;

---
*Corresponding author: annagorenflo@googlemail.com

Ballantyne et al. 2015). Therefore, to analyse pollination performances, distinguish flower visitors from true pollinators and determine the strength of plant-visitor interactions, it seems crucial to investigate the transfer of pollen by flower visitors more closely for multiple plant species.

In this study, we analysed the performance of flower visitors to the three plant species *Malva sylvestris, Borago officinalis* and *Onobrychis viciifolia* directly by determining stigmatic pollen deposition. The three plant species are visited by a variety of insect species. Yet, it is not known, whether all flower visitors are pollinators and whether pollinators differ in their pollination performances when direct measurements are used. We first compared pollen loads on unvisited and visited stigmas to evaluate pollination performance of abiotic factors versus flower visitors overall. Then, pollination performances of flower visitors to *M. sylvestris, B. officinalis* and *O. viciifolia* were analysed in a hierarchical approach: (1) A flower-visitor group was identified as efficient pollinator when more pollen grains were deposited on stigmas per single visit compared to pollen loads on unvisited stigmas. (2) Pollinators were ranked regarding their pollination efficiencies by comparing single-visit pollen depositions. (3) Pollinators were ranked regarding their pollination effectiveness by comparing their pollen depositions per hour, including visitation frequencies (Fumero-Cabán & Meléndez-Ackerman 2007; Madjidian et al. 2008; Rader et al. 2009).

MATERIALS AND METHODS

Study site and plant species

The study was conducted at the research farm "Oberer Hardthof" of the University of Giessen, Germany, which is surrounded by farmland and a small area of woodland. Four flower mixtures ("Lebensraum1", "Odin1", "Odin2", and "Veitshöchheimer Bienenweide") were sown on former arable land on 28 May 2013 in a block design, consisting of a row of eight blocks with four plots (4 m × 4 m) per block. Within each block, plots of the four flower mixtures were randomised and separated from each other by 3.5 m-wide strips of grassland. For determining pollination performances of flower visitors, we chose the three plant species *M. sylvestris, B. officinalis* and *O. viciifolia*, which were present in all (*M. sylvestris* and *O. viciifolia*) or in three (*B. officinalis*) of the four flower mixtures sown at the study site. Differences in pollinator performance between seed mixtures were not tested, because plant-community composition data did not allow for meaningful hypotheses.

Malva sylvestris, B. officinalis and *O. viciifolia* are frequently applied in flower mixtures, sown as flower strips in agricultural landscapes, and are cultivated as crops or for pharmaceutical purposes (Janick et al. 1989; Gasparetto et al. 2012; Hayot Carbonero et al. 2011). *Malva sylvestris* is native to Germany and the other two species are considered to be neophytes. The common mallow *M. sylvestris* (Malvaceae) generates actinomorphic, dish-shaped and upturned pinkish-purple flowers with at least 11 ovules per flower (Kumar et al. 2014). The central style divides into numerous filamentous stigmas and is surrounded by several anthers at its base (Kumar et al. 2014; Appendix I). Flowers of borage *B. officinalis* (Boraginaceae) are actinomorphic, dish-shaped and downward directed and change from pink to blue as they age. There are four ovules per flower (De Haro-Bailón & Del Rio 1998). The style ends in a terminal stigma and is surrounded by cone-like anthers at its base (Ghorbel & Nabli 1998; Appendix I). The common sainfoin *O. viciifolia* (Fabaceae) produces inflorescences with up to 120 white to pink, papilionaceous flowers (Goplen et al. 1991) with a single ovule per flower (Galloni et al. 2007). The stigma and the anthers are enclosed within the two fused and boat-shaped keel petals (Appendix I).

The three plant species produce pollen and large amounts of nectar. Nectaries are located at the base of the corollas and are indicated by prominent nectar guides in *M. sylvestris* and *O. viciifolia* (Westrich 1989). The plant species are considered self-compatible, but flowers show features to avoid self-pollination or self-fertilisation, i.e. spatial and temporal separation of mature stigmas and anthers (*M. sylvestris, B. officinalis*) (Montaner et al. 2000; Kumar et al. 2014) or presence of a stigmatic cuticle (*O. viciifolia*) (Galloni et al. 2007). Given these floral traits and observational data on flower visitors, the three plant species are assumed to rely mainly on insects such as bees (*M. sylvestris, B. officinalis, O. viciifolia*), hoverflies and butterflies (*M. sylvestris*) for pollen transfer between flowers of the same (geitonogamy) or other conspecific plants (allogamy) (Richards & Edwards 1988; Corbet et al. 1991; Goplen et al. 1991; Comba et al. 1999; Kumar et al. 2014).

Pollination efficiencies of flower visitors

We analysed the pollination efficiencies of flower visitors to *M. sylvestris, B. officinalis* and *O. viciifolia* directly by determining stigmatic pollen deposition per single visit. Sampling took place between 7:30 a.m. and 6:30 p.m. on warm and sunny days with low wind speeds from 3 June to 4 July 2014 (on 19 days for *M. sylvestris*, 15 days for *B. officinalis* and 13 days for *O. viciifolia*). Stigmas were sampled from all 32 plots (*M. sylvestris*, 1055 stigmas), from 16 plots representing all flower mixtures (*O. viciifolia*, 411 stigmas) and from 11 plots representing two flower mixtures (*B. officinalis*, 252 stigmas). Flowers of the three plant species were bagged before blooming (perforated polypropylene bags, holes 2 mm in diameter) and uncovered after flowers had fully opened and stigmas had turned receptive. We photographed the first flower visitor from an appropriate distance to prevent disturbances and removed the stigmas after the first visitor had left. Only flower-visitor groups with at least eight visited stigmas were further analysed. At the same time, receptive stigmas from bagged flowers were sampled as controls before any visit took place to account for pollen transfer due to self- or wind-pollination, animals smaller than 2 mm, which could enter the bags, or the experimental handling of the flowers. Sampled stigmas were stored in individual wells of plastic 96 cell-culture arrays in a freezer (-20 °C) until pollen analysis.

For counting attached pollen grains, visited and unvisited stigmas were unfrozen and mounted in glycerin-jelly slides stained with basic fuchsine following the protocol of Kearns

& Inouye (1993) to increase the contrast (Appendix I). Morphologically conspecific pollen grains attached to stigmas were counted by light microscopy (between ×40 and ×400). We identified the photographed flower visitors and grouped them according to size, coloration and/or taxonomy. Flower-visitor groups were considered to be efficient pollinators when pollen loads on visited stigmas exceeded those on unvisited stigmas. Pollinator efficiencies (i.e. mean numbers of pollen grains deposited per single visit) were calculated by subtracting the mean numbers of pollen grains found on unvisited stigmas, serving as controls, from mean numbers of pollen grains on stigmas visited by the flower-visitor groups for each plant species.

Visitation frequencies and pollinator effectiveness

For estimating pollinator effectiveness in terms of pollen deposition per hour, visitation frequencies of flower visitors to the three plant species were recorded during separate observations. Observations took place between 7:30 a.m. and 6:30 p.m. from 17 June to 6 July 2014 (on six days for *M. sylvestris*, 10 days for *B. officinalis* and eight days for *O. viciifolia*). *Malva sylvestris* was observed in 20 plots (representing all flower mixtures), *B. officinalis* was observed in seven plots (representing two flower mixtures) and *O. viciifolia* was observed in three plots (representing three flower mixtures). Per plot, six flowers of *M. sylvestris* and *B. officinalis* and one to four inflorescences of *O. viciifolia* (around 25 flowers in total) were observed for 15-minute periods (*M. sylvestris*) or 20-minute periods (*B. officinalis* and *O. viciifolia*) over the day. In total, receptive and unbagged flowers were observed for 201 15-minute periods (50.25 hours, *M. sylvestris*), 168 20-minute periods (56 hours, *B. officinalis*) and 113 20-minute periods (37.67 hours, *O. viciifolia*).

Per-hour visitation frequencies were calculated by dividing the total number of visits recorded for a flower-visitor group by the total number of observational hours and the number of flowers simultaneously observed. Pollinator effectiveness (i.e. mean numbers of pollen grains deposited per hour) was calculated only for flower-visitor groups being efficient pollinators per single visit. For this, the mean number of pollen grains deposited per single visit by each pollinator group was multiplied with its corresponding visitation frequency.

Statistical analysis

Wilcoxon rank-sum tests were used to test for differences in the numbers of pollen grains on visited and unvisited stigmas for each plant species (function "wilcox.test" in the stats package of R version 3.2.3; R Core Team 2015). To differentiate flower visitors from pollinators and rank flower-visitor groups according to their pollination efficiency, numbers of pollen grains on unvisited stigmas and on stigmas visited by the different flower-visitor groups were compared by Poisson generalized linear models (GLMs) with a correction for overdispersion (i.e. family was set to quasipoisson) for each plant species (function "glm" in the stats package of R version 3.2.3). To provide levels of uncertainty in pollinator effectiveness (a product of the mean pollen deposition and a constant visitation frequency), we calculated standard errors by multiplying the error of pollen deposition efficiencies with the corresponding visitation frequency.

RESULTS

Pollination efficiencies of flower visitors

Visited stigmas had significantly increased pollen grain numbers compared to unvisited stigmas for each plant species, the mean increase ranging from over two-fold (*B. officinalis*) to over three-fold (*O. viciifolia*) and over four-fold (*M. sylvestris*) (Tab. I). Stigmas sampled from the three plant species were visited mostly by bees, with other insects being less common.

For analysing pollen deposition, enough (eight or more) visited stigmas from *M. sylvestris* were obtained for four flower-visitor groups: *Apis mellifera*, the *Bombus lapidarius* complex (mainly *B. lapidarius*, but possibly also including the similarly coloured species *B. ruderarius*, *B. pratorum*, *B. soroeensis*), the *B. terrestris* complex (mainly *B. terrestris* or *B. lucorum*, but possibly also including the similarly coloured species *B. magnus*, *B. cryptarum*) and Halictidae (including only species of the genera *Halictus* and *Lasioglossum* smaller than 10 mm) (Tab. 2; Appendix II). Pollen deposition for each group of flower visitors was always significantly larger than pollen loads on unvisited stigmas (Tab. 2). Mean

TABLE I. The total numbers of visited and unvisited stigmas sampled from *Malva sylvestris*, *Borago officinalis* and *Onobrychis viciifolia* and the mean numbers of pollen grains on stigmas (± SE).

Plant species	Visitation type	No. of stigmas	Mean no. of pollen grains
M. sylvestris	Unvisited	304	57.4 ± 3.5
	Visited	751	246.3 ± 8.1***
B. officinalis	Unvisited	79	6.1 ± 0.9
	Visited	173	14.2 ± 1.6**
O. viciifolia	Unvisited	130	0.8 ± 0.1
	Visited	281	2.7 ± 0.2***

* P < 0.05; ** P < 0.01; *** P < 0.001.

P-values were calculated using Wilcoxon rank-sum tests and indicate differences between the mean numbers of pollen grains on visited compared to unvisited stigmas.

TABLE 2.   The numbers of stigmas sampled, mean numbers of pollen grains on stigmas after a single visit (± SE), pollinator efficiencies (i.e. mean numbers of pollen grains deposited per single visit, corrected for pollen loads on unvisited stigmas), visitation frequencies (visits per flower and hour and percentage of total visits in parenthesis) and pollinator effectiveness (i.e. mean numbers of pollen grains deposited per hour) of the analysed flower-visitor groups of *Malva sylvestris*, *Borago officinalis* and *Onobrychis viciifolia*. Please note that pollinator efficiencies and visitation frequencies are rounded, but pollinator effectiveness is calculated from exact values.

Plant species	Flower-visitor group	Stigmas sampled	Pollen per stigma (mean ± SE)	Pollinator efficiency	Visitation frequency	Pollinator effectiveness
M. sylvestris	*A. mellifera*	683	227.9 ± 7.1***	170.5	9.0 (88.4)	1,535.1 ± 63.9
	B. lapidarius complex	27	625.1 ± 76.7***	567.7	0.6 (5.9)	342.7 ± 46.3
	B. terrestris complex	11	669.8 ± 127.0***	612.4	0.2 (1.8)	113.8 ± 23.6
	Halictidae	13	169.7 ± 37.2***	112.3	0.1 (1.1)	7.8 ± 4.2
B. officinalis	*A. mellifera*	106	13.3 ± 1.7**	7.2	8.4 (87.3)	60.3 ± 14.3
	B. terrestris complex	22	19.6 ± 7.2***	13.4	0.7 (7.3)	9.4 ± 5.0
	B. lapidarius complex	9	26.3 ± 10.4***	20.2	0.1 (0.8)	1.5 ± 0.8
	B. sylvarum	9	23.7 ± 9.6***	17.6	0.1 (0.8)	1.3 ± 0.7
	Megachile	9	8.7 ± 2.9	-	0.1 (1.0)	-
	Halictidae	8	4.6 ± 1.4	-	0.1 (0.7)	-
O. viciifolia	*B. lapidarius* complex	187	3.0 ± 0.3***	2.2	1.4 (60.2)	3.1 ± 0.4
	B. sylvarum	11	4.3 ± 1.4***	3.5	0.2 (8.9)	0.7 ± 0.3
	Megachile	39	2.7 ± 0.8***	1.9	0.1 (5.4)	0.3 ± 0.1
	A. mellifera	9	2.0 ± 0.7	-	0.3 (11.4)	-
	Halictidae	21	0.9 ± 0.3	-	0.1 (2.2)	-

* P < 0.05; ** P < 0.01; *** P < 0.001.

P-values were calculated using Poisson generalized linear models (GLMs) with a correction for overdispersion (i.e. family was set to quasipoisson) and indicate differences between the mean numbers of pollen grains on stigmas visited by the analysed flower-visitor groups compared to unvisited stigmas.

Standard errors of pollinator effectiveness were calculated by multiplying the error of pollen deposition efficiencies with the corresponding visitation frequency.

pollen deposition ranged from 112.3 to 612.4 pollen grains per single visit after subtracting the mean number of pollen grains found on unvisited stigmas (Tab. 2, Fig. 1A). Both bumblebee groups showed significantly higher pollination efficiencies than *A. mellifera* and Halictidae (GLM: all P < 0.001).

For *B. officinalis*, enough visited stigmas were obtained for six flower-visitor groups: *A. mellifera*, *B. lapidarius* complex, *B. sylvarum*, *B. terrestris* complex, *Megachile* and Halictidae (Tab. 2; Appendix II). Pollen deposition was always significantly larger than pollen loads on unvisited stigmas, except for bees of the genus *Megachile* and the family Halictidae (Tab. 2). Mean pollen deposition ranged from 7.2 to 20.2 pollen grains per visit after subtracting the mean number of pollen grains found on unvisited stigmas (Tab. 2, Fig. 1B). Pollination efficiency of the *B. lapidarius* complex was significantly higher than the efficiency of *A. mellifera* (GLM: P < 0.05).

For *O. viciifolia*, enough visited stigmas were obtained for five flower-visitor groups: *A. mellifera*, *B. lapidarius* complex, *B. sylvarum*, *Megachile* and Halictidae (Tab. 2; Appendix II). Pollen deposition was always significantly

larger than pollen loads on unvisited stigmas, except for *A. mellifera* and Halictidae (Tab. 2). Mean pollen deposition ranged from 1.9 to 3.5 pollen grains per visit after subtracting the mean number of pollen grains found on unvisited stigmas of *O. viciifolia*, (Tab. 2, Fig. 1C). There was no significant difference in pollination efficiency between the *B. lapidarius* complex, *B. sylvarum* and *Megachile*.

Visitation frequencies and pollinator effectiveness

Virtually all insects visiting the three plant species during visitation surveys were honeybees, bumblebees or solitary bees (*M. sylvestris*: 97.6%, *B. officinalis*: 99.0%, *O. viciifolia*: 97.9%; Appendix III).

The four flower-visitor groups of *M. sylvestris* analysed for pollination efficiency were the most frequent flower visitors, with *A. mellifera* being by far the most frequent visitor overall (Tab. 2). Accordingly, *A. mellifera* turned out to be the most effective flower-visitor group per time unit (1,535.1 pollen grains/hour compared to the *B. lapidarius* complex: 342.7 grains/hour, the *B. terrestris* complex: 113.8

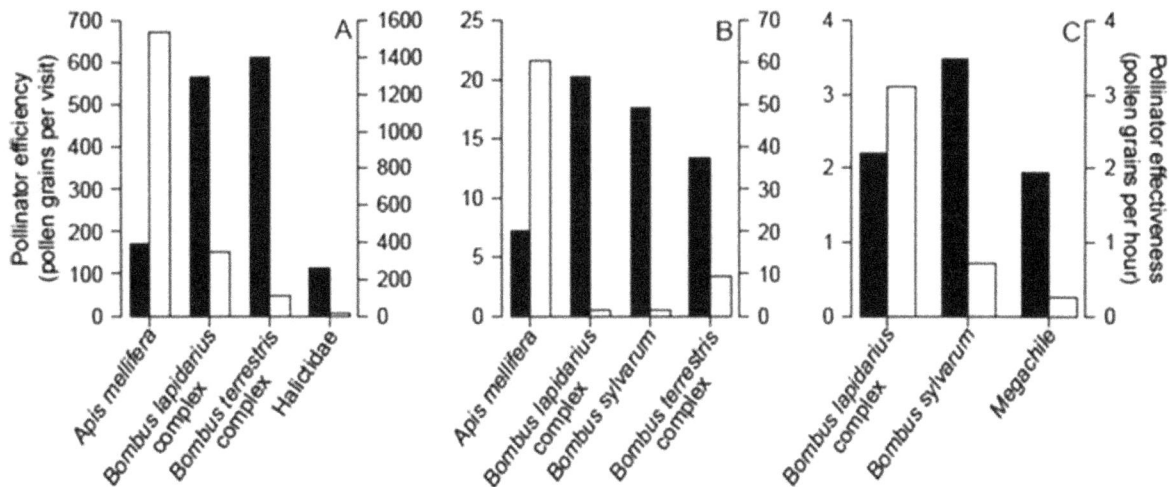

FIGURE I.   Pollinator efficiencies (black bars, i.e. mean numbers of pollen grains deposited per single visit) and pollinator effectiveness (white bars, i.e. mean numbers of pollen grains deposited per hour) of the flower-visitor groups exceeding mean number of pollen grains on unvisited stigmas of (A) *Malva sylvestris*, (B) *Borago officinalis* and (C) *Onobrychis viciifolia*.

grains/hour and Halictidae: 7.8 pollen grains/hour; Tab. 2, Fig. IA).

The six flower-visitor groups of *B. officinalis* analysed for pollination efficiency were among the seven most frequent flower visitors, again *A. mellifera* being the most frequent visitor (Tab. 2). *Apis mellifera* was also again the most effective flower-visitor group per time unit (60.3 pollen grains/hour compared to the *B. terrestris* complex: 9.4 pollen grains/hour, the *B. lapidarius* complex: 1.5 pollen grains/hour and the *B. sylvarum*: 1.3 pollen grains/hour; Tab. 2, Fig. IB).

The five flower-visitor groups of *O. viciifolia* analysed for pollination efficiency were also the most frequent flower visitors. The *B. lapidarius* complex was the most frequent flower-visitor group (Tab. 2) and was also most effective per time unit (3.1 pollen grains/hour compared to *B. sylvarum*: 0.7 pollen grains/hour and *Megachile*: 0.3 pollen grains/hour; Tab. 2, Fig. IC).

DISCUSSION

Pollination efficiencies of flower visitors

Overall, flower visitors deposited more pollen grains than wind or other vectors, supporting the pollinator dependencies of *M. sylvestris*, *B. officinalis* and *O. viciifolia* described in the literature (Richards & Edwards 1988; Corbet et al. 1991; Goplen et al. 1991; Comba et al. 1999; Kumar et al. 2014). We were able to identify four efficient pollinators of *M. sylvestris* (*A. mellifera*, *B. lapidarius* complex, *B. terrestris* complex and Halictidae), four of *B. officinalis* (*A. mellifera*, *B. lapidarius* complex, *B. terrestris* complex and *B. sylvarum*) and three of *O. viciifolia* (*B. lapidarius* complex, *B. sylvarum* and *Megachile*). Hence, assuming that within-flower autogamous pollination is minimised by floral traits of the three plant species, these bees were visiting other conspecific flowers before, thereby taking up pollen that stayed available for the pollination of subsequent flowers (Inouye et al. 1994; Ne'eman et al.

2010). On the other hand, two of the analysed flower-visitor groups of *B. officinalis* (*Megachile* and Halictidae) and two of *O. viciifolia* (*A. mellifera* and Halictidae) were not efficient pollinators. Thus, not all flower visitors can be classified as efficient pollinators, even when belonging to a group of potential pollinators (i.e. all flower-visitor groups were bees), or being efficient pollinators to other plant species (i.e. honeybees were efficient pollinators of both other plant species, Halictidae and *Megachile* of one other plant species each).

Variation in pollen deposition efficiency occurs due to differences in the degree to which flower and visitor traits match, defining a visitor's ability to take up and deposit pollen grains. Because possible factors determining pollen transfer, such as body size (Willmer & Finlayson 2014), hairiness, tongue length (Hobbs et al. 1961), floral constancy or preference (Waser 1986), pollen-transporting structures and nectar- and pollen-collecting behaviours (Inouye 1980; Thorp 1999; Michener 2007) are highly diverse in bees, generalisation about pollination performances is difficult (Fenster et al. 2004). Nevertheless, generalist bumblebee groups (Goulson & Darvill 2004) were the most efficient (for *M. sylvestris* and *B. officinalis*) or among the most efficient pollinators (for *O. viciifolia*) across the three plant species in our study. This supports that bumblebees can often be more efficient per single visit across plant species of different families than honeybees (Wilson & Thomson 1991; Willmer et al. 1994; Javorek et al. 2002; Ballantyne et al. 2015; Zhang et al. 2015).

Bumblebees are characterised by a large surface due to their large and hairy bodies, which probably increases the probability to touch anthers and stigmas and the amount of pollen transferred. Bumblebees' long tongues (Goulson et al. 2005) make them better pollinators of flowers with deep corollas (Hobbs et al. 1961) and probably induced positioning on the studied flowers favourable for contact with anthers and stigmas. Furthermore, like for other Fabaceae (Córdoba & Cocucci 2011), bumblebees were strong enough to open flowers of *O. viciifolia* and thus could reach the nectar hidden between the base of the banner and keel petals. By pushing their heads and tongues towards the nectar, keel petals folded down while stigma and anthers were released and touched the bumblebees' ventral side, enabling pollination. In *B. officinalis* dehiscence of anthers is introrse and pollen grains are held within the cone-like anthers (Corbet et al. 1988). This controls pollen removal by flower visitors, since large amounts of pollen are released only in response to sonication (Buchmann 1983; De Luca & Vallejo-Marín 2013). Bumblebees are well known to perform buzz pollination (Buchmann 1983; King & Buchmann 2003). They were observed to sonicate on *B. officinalis* in this study (see also Corbet et al. 1988), which probably increased the amount of pollen released and transferred to subsequent flowers in *B. officinalis*.

Other bees analysed in this study were generally smaller, less hairy and had shorter tongues, probably causing morphological and behavioural mismatches, which restricted pollen transfer compared to bumblebees in most cases. Bees of the family Halictidae seemed of inappropriate size to flowers of all three plant species, not touching stigmas while drinking nectar, but rather occasionally on their way to the nectaries or when foraging on pollen. Honeybees had to crawl underneath stigmas of *M. sylvestris* to reach the nectar or inserted the tongue laterally into flowers of *M. sylvestris* and *O. viciifolia*, reducing the contact with anthers and stigmas (see also Comba et al. 1999). Furthermore, honeybees were observed to intensively groom pollen from their bodies and discard it after visiting flowers of *M. sylvestris*, making the pollen unavailable for transfer (Inouye et al. 1994). In addition, sonication is not performed by bees of the genus *Apis* (Buchmann 1983; King & Buchmann 2003) and only very rarely within the Megachilidae (Neff & Simpson 1988), probably reducing their pollination efficiency on *B. officinalis*. In contrast to the literature (Richards & Edwards 1988; Goplen et al. 1991), *A. mellifera* was not a successful pollinator of *O. viciifolia* in this study, probably because closed flowers of *O. viciifolia* could not be opened easily, as has been shown for other Fabaceae (Córdoba & Cocucci 2011). Bees of the genus *Megachile*, however, have relatively long tongues and were strong enough to trigger the lever mechanism of *O. viciifolia*, as shown for other papilionate flowers (Córdoba & Cocucci 2011), explaining a pollen deposition efficiency similar to the ones of the bumblebee groups.

In summary, *A. mellifera* and Halictidae on *M. sylvestris* and *A. mellifera* on *B. officinalis* were pollinators of the respective plant species, although much less efficient than the bumblebee groups. According to the observed behaviours, the inefficient pollinators can be classified as (pollen or nectar) thieves or base workers instead (*sensu* Inouye 1980), feeding on floral resources without pollinating. All analysed flower-visitor groups of *M. sylvestris* were efficient pollinators, whereas two groups of *B. officinalis* and *O. viciifolia* were inefficient pollinators. This may reflect an increasing specialisation of the floral morphology from *M. sylvestris* to *B. officinalis* to *O. viciifolia*, which hinders pollination by less suitable flower visitors.

The efficient pollinators deposited all at least as many pollen grains on stigmas as ovules present per flower (see Material and Methods). Hence, all efficient pollinators can potentially provide pollination services for maximum seed set of the plant species studied in our system. Differences in foraging behaviour within a plant or within single flowers may still affect conspecific pollen grain quality with regard to viability or the degree of kinship, both affecting germination success (Snow et al. 1996). Similarly, low flower constancy increases transfer of heterospecific pollen (Waser 1986), which could lead to stigma clogging and prevent conspecific pollen to germinate. Thus, including fertilisation, e.g. seed set and germination rates, is an important next step to determine the required quantity and quality of conspecific pollen, and the effect of heterospecific pollen deposited by efficient pollinators (Garibaldi et al. 2014; Zhang et al. 2015).

Variation in the efficiency within flower-visitor groups may be explained by differences between species within groups (King et al. 2013) or between individuals within species (Jauker et al. 2016). This seems especially important for bumblebees, showing a wide intraspecific range of body sizes and tongue lengths (Goulson et al. 2002; Willmer & Finlayson 2014). Since only female bees collect pollen, morphological and behavioural differences between sexes may also lead to intraspecific differences in bees (Ne'eman et al. 2006). Future studies including analyses of pollinator performances at the lowest taxonomic rank possible and even within species will be necessary for further insights into pollination variability.

Visitation frequencies and pollinator effectiveness

Most visitors to flowers of *M. sylvestris*, *B. officinalis* and *O. viciifolia* were bees. The flower-visitor groups analysed for pollen deposition were among the most frequent visitors. *Apis mellifera* dominated the number of flower visits to *M. sylvestris* and *B. officinalis*, whereas bumblebees of the *B. lapidarius* complex were the most frequent visitors to flowers of *O. viciifolia*. Differences in dominant visitors between plant species suggest differing flower preferences and high flower constancy of *A. mellifera* (Hill et al. 1997) and bumblebees of the *B. lapidarius* complex (Chittka et al. 1997; Raine & Chittka 2005; Zych & Stpiczyńska 2012). It also exemplifies, that the overall most abundant pollinator species, the generalist honeybee, is not necessarily the most frequent visitor to all present plant species.

For *M. sylvestris* and *B. officinalis*, *A. mellifera* was the most effective pollinator, its high abundance compensating for the relatively low pollinator efficiency per single visit. For *O. viciifolia*, bumblebees of the *B. lapidarius* complex were the most effective pollinators, again based on higher abundances than the similarly efficient *B. sylvarum* and bees of the genus *Megachile* at the single-visit level. Hence, in this study the most frequent flower-visitor group of each plant species was most effective in depositing conspecific pollen per time unit, even though not showing the highest pollinator efficiency per single visit. In combination with previous studies, this suggests that the most frequent visitors are often the most important pollinators overall, even when being poor or equally efficient pollinators per single visit

(Olsen 1997; Vázquez et al. 2005; Madjidian et al. 2008; Rader et al. 2009; Zych et al. 2013). Even though pollen limitation was of minor importance in our study and several pollinator species can potentially cause maximum seed set, the results suggest that visitation rate could play a more important role in plant reproduction than pollination success on a per-interaction basis (Vázquez et al. 2005; Sahli & Conner 2006). Thus, pollinator abundance seems a similar important conservation measure for ecosystem functioning as species richness for biodiversity (Kleijn et al. 2015).

Visitation frequencies, however, did not explain pollinator effectiveness entirely. Flower visitors not contributing to pollination at all were similarly frequent as legitimate pollinators in both *B. officinalis* (*Megachile* and Halictidae) and *O. viciifolia* (*A. mellifera*). Furthermore, differences in visitation frequencies between flower-visitor groups are not proportional to the differences in pollinator effectiveness. For example, honeybees were 15 times more frequent on *M. sylvestris* than bumblebees of the *B. lapidarius* complex, but only five times more effective. Therefore, it seems more reliable and informative to use direct measurements when determining pollinator performances, instead of deducing effectiveness from visitation frequencies alone (Mayfield et al. 2001; Javorek et al. 2002; Ballantyne et al. 2015). Such information reveals differences in the density dependence of pollination success between pollinators and is ultimately necessary for estimating consequences of pollinator decline for plant population persistence. Revealing true pollinator performances will allow to construct more informative plant-pollinator networks (Ballantyne et al. 2015), find appropriate species for crop pollination (Westerkamp & Gottsberger 2000) and thus conserve and apply pollination service best (Garratt et al. 2014).

Conclusion

In conclusion, considerable differences in pollinator performance occurred even in closely related taxa. Not all bees were pollinators of the studied plants and pollinators differed in their pollen deposition efficiency per single visit. Although the most frequent flower visitors were most effective, pollinator effectiveness could not be explained by visitation frequencies alone. These findings emphasise the need to connect visitation frequencies to stigmatic pollen deposition to reveal true pollination performances of flower visitors.

ACKNOWLEDGEMENTS

We thank Daniela Warzecha for helping identify flower visitors and Verena Grob for advice on pollen identification. We further thank Verena Grob, Robert Garling and Daniela Warzecha for their assistance during the flower-visitor surveys.

APPENDICES

Additional supporting information may be found in the online version of this article:
APPENDIX I.   Flowers and stigmas with pollen of (A) *M. sylvestris*, (B) *B. officinalis* and (C) *O. viciifolia*.

APPENDIX II.   Analysed flower-visitor groups of (A) *M. sylvestris*, (B) *B. officinalis* and (C) *O. viciifolia*.

APPENDIX III.   All flower-visitor groups observed on (A) *M. sylvestris*, (B) *B. officinalis* and (C) *O. viciifolia*.

REFERENCES

Adler LS, Irwin RE (2006) Comparison of pollen transfer dynamics by multiple floral visitors: experiments with pollen and fluorescent dye. Annals of Botany 97:141-150.

Ballantyne G, Baldock KCR, Willmer PG (2015) Constructing more informative plant–pollinator networks: visitation and pollen deposition networks in a heathland plant community. Proceedings of the Royal Society of London B: Biological Sciences 282:20151130.

Buchmann SL (1983) Buzz pollination in angiosperms. In: Jones CE, Little RJ (eds) Handbook of experimental pollination biology. Van Nostrand Reinhold Company Inc., New York, pp 73-113.

Buchmann SL, Nabhan GP (1996) The forgotten pollinators. Island Press/Shearwater Books, Washington, DC.

Calderone SA, Lozier JD, Strange JP, Koch JB, Cordes N, Solter LF, Griswold TL (2012) Insect pollinated crops, insect pollinators and US agriculture: trend analysis of aggregate data for the period 1992–2009. PLoS ONE 7:e37235.

Chittka L, Gumbert A, Kunze J (1997) Foraging dynamics of bumble bees: correlates of movements within and between plant species. Behavioral Ecology 8:239-249.

Comba L, Corbet SA, Hunt L, Warren B (1999) Flowers, nectar and insect visits: evaluating British plant species for pollinator-friendly gardens. Annals of Botany 83:369-383.

Corbet SA, Chapman H, Saville N (1988) Vibratory pollen collection and flower form: bumble-bees on *Actinidia*, *Symphytum*, *Borago* and *Polygonatum*. Functional Ecology 2:147-155.

Corbet SA, Williams IH, Osborne JL (1991) Bees and the pollination of crops and wild flowers in the European Community. Bee World 72:47-59.

Córdoba SA, Cocucci AA (2011) Flower power: its association with bee power and floral functional morphology in papilionate legumes. Annals of Botany 108:919-931.

De Haro-Bailón A, Del Rio M (1998) Isolation of chemically induced mutants in borage (*Borago officinalis* L.). Journal of the American Oil Chemists' Society 75:281-283.

De Luca PA, Vallejo-Marín M (2013) What's the "buzz" about? The ecology and evolutionary significance of buzz-pollination. Current Opinion in Plant Biology 16:429-435.

Fenster CB, Armbruster WS, Wilson P, Dudash MR, Thomson JD (2004) Pollination syndromes and floral specialization. Annual Review of Ecology, Evolution, and Systematics 35:375-403.

Fumero-Cabán JJ, Meléndez-Ackerman EJ (2007) Relative pollination effectiveness of floral visitors of *Pitcairnia angustifolia* (Bromeliaceae). American Journal of Botany 94:419-424.

Gallai N, Salles J-M, Settele J, Vaissière BE (2009) Economic valuation of the vulnerability of world agriculture confronted with pollinator decline. Ecological Economics 68:810-821.

Galloni M, Podda L, Vivarelli D, Cristofolini G (2007) Pollen presentation, pollen-ovule ratios, and other reproductive traits in Mediterranean Legumes (Fam. Fabaceae - Subfam. Faboideae). Plant Systematics and Evolution 266:147-164.

Garibaldi LA, Steffan-Dewenter I, Winfree R, Aizen MA, Bommarco R, Cunningham SA, Kremen C, Carvalheiro LG, Harder LD, Afik O, Bartomeus I, Benjamin F, et al. (2014) Wild pollinators enhance fruit set of crops regardless of honey bee abundance. Science 339:1608-1611.

Garratt MPD, Coston DJ, Truslove CL, Lappage MG, Polce C, Dean R, Biesmeijer JC, Potts SG (2014) The identity of crop pollinators helps target conservation for improved ecosystem services. Biological Conservation 169:128-135.

Gasparetto JC, Martins CAF, Hayashi SS, Otuky MF, Pontarolo R (2012) Ethnobotanical and scientific aspects of *Malva sylvestris* L.: a millennial herbal medicine. Journal of Pharmacy and Pharmacology 64:172-189.

Ghorbel S, Nabli MA (1998) Pollen, pistil and their interrelations in *Borago officinalis* and *Heliotropium europaeum* (Boraginaceae). Grana 37:203-214.

Goplen BP, Richards KW and Moyer JR (1991) Sainfoin for western Canada. Agriculture Canada Publication 1470E:24.

Goulson D, Peat J, Stout JC, Tucker J, Darvill B, Derwent LC, Hughes WOH (2002) Can alloethism in workers of the bumblebee *Bombus terrestris* be explained in terms of foraging efficiency? Animal Behaviour 64:123-130.

Goulson D (2003) Effects of introduced bees on native ecosystems. Annual Review of Ecology, Evolution, and Systematics 34:1-26.

Goulson D, Darvill B (2004) Niche overlap and diet breadth in bumblebees; are rare species more specialized in their choice of flowers? Apidologie 35:55-63.

Goulson D, Hanley ME, Darvill B, Ellis JS, Knight ME (2005) Causes of rarity in bumblebees. Biological Conservation 122:1-8.

Hayot Carbonero C, Mueller-Harvey I, Brown TA, Smith L (2011) Sainfoin (*Onobrychis viciifolia*): a beneficial forage legume. Plant Genetic Resources 9:70-85.

Hill PSM, Wells PH, Wells H (1997) Spontaneous flower constancy and learning in honey bees as a function of colour. Animal Behavior 54:615-627.

Hobbs GA, Nummi WO, Virostek JF (1961) Food-gathering behaviour of honey, bumble, and leaf-cutter bees (Hymenoptera: Apoidea) in Alberta. The Canadian Entomologist 93:409-419.

Inouye DW (1980) The terminology of floral larceny. Ecology 61:1251-1253.

Inouye DW, Gill DE, Dudash MR, Fenster CB (1994) A model and lexicon for pollen fate. American Journal of Botany 81:1517-1530.

Janick J, Simon JE, Quinn J, Beaubaire N (1989) Borage: a source of gamma linolenic acid. In: Craker LE, Simon JE (eds) Herbs, spices, and medicinal plants: recent advances in botany, horticulture and pharmacology. Vol 4. Oryx Press, Phoenix, AZ, pp 145-168.

Jauker F, Speckmann M, Wolters V (2016) Intra-specific body size determines pollination effectiveness. Basic and Applied Ecology 17:714-719.

Javorek KE, Mackenzie SP, Vander Kloet SK (2002) Comparative pollination effectiveness among bees (Hymenoptera: Apoidea) on Lowbush Blueberry (Ericaceae: *Vaccinium angustifolium*). Annals of the Entomological Society of America 95:345-351.

Jędrzejewska-Szmek K, Zych M (2013) Flower-visitor and pollen transport networks in a large city: structure and properties. Arthropod-Plant Interactions 7:503-516.

Johnson SD, Steiner KE (2000) Generalization versus specialization in plant pollination systems. Trends in Ecology and Evolution 15:140-143.

Kearns CA, Inouye DW (1993) Techniques for pollination biologists. University Press of Colorado, Boulder.

King MJ, Buchmann SL (2003) Floral sonication by bees: mesosomal vibration by *Bombus* and *Xylocopa*, but not *Apis* (Hymenoptera: Apidae), ejects pollen from poricidal anthers. Journal of the Kansas Entomological Society 76:295-305.

King C, Ballantyne G, Willmer PG (2013) Why flower visitation is a poor proxy for pollination: measuring single-visit pollen deposition, with implications for pollination networks and conservation. Methods in Ecology and Evolution 4:811-818.

Kleijn D, Winfree R, Bartomeus I, Carvalheiro LG, Henry M, Isaacs R, Klein A-M, Kremen C, M'Gonigle LK, Rader R, Ricketts TH, Williams NM, et al. (2015) Delivery of crop pollination services is an insufficient argument for wild pollinator conservation. Nature Communications 6:7414.

Kumar G, Kadam GB, Saha TN, Girish KS, Tiwari AK, Kumar R (2014) Studies on floral biology of *Malva sylvestris* L. Indian Journal of Horticulture 71:295-297.

Madjidian JA, Morales CL, Smith HG (2008) Displacement of a native by an alien bumblebee: lower pollinator efficiency overcome by overwhelmingly higher visitation frequency. Oecologia 156:835-45.

Mayer C, Adler L, Armbruster WS, Dafni A, Eardley C, Huang, SQ, Kevan PG, Ollerton J, Packer L, Ssymank A, Stout JC, Potts SG (2011) Pollination ecology in the 21st century: key questions for future research. Journal of Pollination Ecology 3:8-23.

Mayfield M, Waser NM, Price M (2001) Exploring the 'Most effective pollinator principle' with complex flowers: bumblebees and *Ipomopsis aggregata*. Annals of Botany 88:591-96.

Michener CD (2007) The bees of the world. Johns Hopkins University Press, Baltimore.

Montaner C, Floris E, Alvarez JM (2000) Is self-compatibility the main breeding system in borage (*Borago officinalis* L.)? Theoretical and Applied Genetics 101:185-189.

Ne'eman G, Shavit O, Shaltiel L, Shmida A (2006) Foraging by male and female solitary bees with implications for pollination. Journal of Insect Behavior 19:383-401.

Ne'eman G, Jürgens A, Newstrom-Lloyd LE, Potts SG, Dafni A (2010) A framework for comparing pollinator performance: effectiveness and efficiency. Biological Reviews 85:435451.

Neff JL, Simpson BB (1988) Vibratile pollen-harvesting by *Megachile mendica* Cresson (Hymenoptera, Megachilidae). Journal of the Kansas Entomological Society 61:242-244.

Ollerton J, Tarrant S, Winfree R (2011) How many flowering plants are pollinated by animals? Oikos 120:321-326.

Olsen KM (1997) Pollination effectiveness and pollinator importance in a population of *Heterotheca subaxillaris* (Asteraceae). Oecologia 109:114-121.

Popic TJ, Wardle GM, Davila YC (2013) Flower-visitor networks only partially predict the function of pollen transport by bees. Austral Ecology 38:76-86.

R Core Team (2015) R: A language and environment for statistical computing. R Foundation for Statistical Computing, Vienna, Austria. URL https://www.R-project.org/.

Rader R, Howlett BG, Cunningham SA, Westcott DA, Newstrom-Lloyd LE, Walker MK, Teulon DAJ, Edwards W (2009) Alternative pollinator taxa are equally efficient but not as effective as the honeybee in a mass flowering crop. Journal of Applied Ecology 46:1080-1087.

Raine NE, Chittka L (2005) Comparison of flower constancy and foraging performance in three bumblebee species (Hymenoptera: Apidae: Bombus). Entomologia Generalis 28:81-89.

Richards KW, Edwards PD (1988) Density, diversity, and efficiency of pollinators of sainfoin, *Onobrychis viciaefolia* Scop. The Canadian Entomologist 120:1085-1100.

Sahli HF, Conner JK (2006) Characterizing ecological generalization in plant-pollination systems. Oecologia 148:365-72.

Slaa EJ, Sánchez CLA, Malagodi-Braga KS, Hofstede FE (2006) Stingless bees in applied pollination. Practice and perspectives. Apidologie 37:293-315.

Snow AA, Spira TP, Simpson R, Klips RA (1996) The ecology of geitonogamous pollination. In: Lloyd DG, Barrett SCH (eds) Floral biology: studies on floral evolution in animal-pollinated plants. Chapman and Hall, New York, pp 191-216.

Thorp RW (2000) The collection of pollen by bees. Plant Systematics and Evolution 222:211-223.

Vázquez DP, Morris WF, Jordano P (2005) Interaction frequency as a surrogate for the total effect of animal mutualists on plants. Ecology Letters 8:1088-1094.

Waser NM (1986) Flower constancy: definition, cause and measurement. The American Naturalist 127: 96-603.

Westerkamp C, Gottsberger G (2000) Diversity pays in crop pollination. Crop Science 40:1209-1222.

Westrich P (1989) Die Wildbienen Baden-Württembergs. Allgemeiner Teil. E Ulmer, Stuttgart.

Willmer PG, Bataw AAM, Hughes JP (1994) The superiority of bumblebees to honeybees as pollinators: insect visits to raspberry flowers. Ecological Entomology 19:271-284.

Willmer PG, Finlayson K (2014) Big bees do a better job: intraspecific size variation influences pollination effectiveness. Journal of Pollination Ecology 14:244-254.

Wilson P, Thomson JD (1991) Heterogeneity among floral visitors leads to discordance between removal and deposition of pollen. Ecology 72:1503-1507.

Zhang H, Huang J, Williams PH, Vaissière BE, Zhou Z, Gai Q, Dong J, An J (2015) Managed bumblebees outperform honeybees in increasing peach fruit set in China: different limiting processes with different pollinators. PLoS ONE 10:e0121143.

Zych M, Goldstein J, Roguz K, Stpiczyńska M (2013) The most effective pollinator revisited: pollen dynamics in a spring-flowering herb. Arthropod-Plant Interactions 7:315-322.

Zych M, Stpiczyńska M (2012) Neither protogynous nor obligatory out-crossed: pollination biology and breeding system of the European Red List *Fritillaria meleagris* L. (Liliaceae). Plant Biology 14:285-294.

Pollen transfer efficiency of *Apocynum cannabinum* (Apocynaceae)

Tatyana Livshultz*, Sonja Hochleitner, Elizabeth Lakata

Department of Biodiversity Earth and Environmental Science and Academy of Natural Sciences, Drexel University, 1900 Benjamin Franklin Parkway, Philadelphia, PA 19103-1101, U.S.A.

Abstract—Pollen transfer efficiency (PTE), the percentage of removed pollen delivered to conspecific stigmas, has been implicated in the morphological evolution, population dynamics, and lineage diversification of flowering plants. Pollinia, the aggregated contents of pollen sacs, present in Apocynaceae subfamilies Asclepiadoideae (milkweeds), Secamonoideae, and Periplocoideae and orchids (Orchidaceae), are the pre-eminent example of a plant trait that elevates PTE (to ca. 25%). However, comparison of species with pollinia to "average" flowers (PTE ca. 1%) may over-estimate the gains from pollinia. We hypothesize that elevated PTE evolved in Apocynaceae prior to pollinia. We measured PTE and pollen to ovule ratio, a possible correlate of PTE, in *Apocynum cannabinum*, a milkweed relative with pollen tetrads (instead of pollinia) and simple bands of style head adhesive (instead of complex pollinium-carrying translators), comparing them to reports of other species collated from the literature. PTE of *A. cannabinum* is 7.9%, in the 24th percentile of reports for 36 milkweed species, but more than twice the highest PTE reported for a species with monads (3.4%). The bands of style head adhesive are functionally equivalent to the translators of milkweeds. The pollen to ovule ratio of *A. cannabinum,* at 19.8, is in the 94th percentile of ratios reported for milkweeds (mean 9.6). Our results are consistent with the hypothesis that floral novelties of Apocynaceae that evolved prior to pollinia also promote aggregated pollen transport and elevated PTE.

Keywords: floral function; pollen ovule ratio; pollen transfer efficiency; pollinium; pollination; tetrad

Introduction

Pollen transfer efficiency (PTE, the percentage of pollen removed from a flower that is subsequently deposited on conspecific stigmas) is one of the primary components of male reproductive fitness in plants, since only those pollen grains that are delivered to conspecific stigmas may father the next generation (Harder 2000; Harder & Johnson 2008; Thomson 2006). All else being equal, the more efficient pollen donors in a population will have greater siring success. Thus, to the extent that plant traits influence pollen transfer efficiency, selection for male fitness via PTE can shape the evolution of plant traits (Kobayashi et al. 1997; Ren & Tang 2010; Thomson 2006). Low PTE can contribute to pollen-limitation, and thus limit female reproductive output (seed production) at the population level (Harder & Aizen 2010). Since low female reproductive output can contribute to the extinction of populations and species (Groom 1998), differences in PTE among species may produce different species-specific probabilities of population persistence over ecological time scales and different rates of lineage accumulation over evolutionary time scales (Armbruster & Muchhala 2009). Thus, plant traits that result in significantly greater PTE could potentially be key innovations that facilitate species accumulation (Armbruster & Muchhala 2009; Livshultz et al. 2011). Given that plants at low population densities are more likely to experience pollen limitation, and are at higher risk of reproductive failure due to mate-finding Allee effects (Gascoigne et al. 2009; Ghazoul 2005a; Ghazoul 2005b), both selection on individuals within populations and differential extinction of populations may produce a pattern of higher PTE in plant species that typically live at low population densities than in species that live at high densities (Livshultz et al. 2011). The importance of pollination success for plant demography remains to be determined empirically, however, since other factors such as maternal resources, seedling mortality, and clonality may be of greater importance in population persistence.

The observed pollen transfer efficiency in animal-pollinated plants is an emergent property of the interaction among the pollinator community (Thomson 2006; Wilson & Thomson 1991; Young et al. 2007), demographic variables such as population size and density (Ghazoul 2005b; Kunin 1997), the larger plant community which may reduce or facilitate pollination of the target species (Flanagan et al. 2011; Ghazoul 2006), and plant traits such as floral morphology and reward (Castellanos et al. 2003; Kobayashi et al. 1997; Ren & Tang 2010). Pollinia, formed by the aggregation of the pollen contents of anther sacs into single masses, are the pre-eminent example of a plant trait that contributes to elevated PTE (Cruden 2000; Harder 2000; Harder & Johnson 2008; Johnson et al. 2005; Livshultz et al. 2011). Species with pollinia have pollen transfer efficiencies more than an order of magnitude higher than those measured in plants with monads (Harder & Johnson 2008). Pollinia have evolved in only two families of flowering plants,

---
*Corresponding author: tl534@drexel.edu

Orchidaceae and Apocynaceae, but other forms of pollen aggregation (tetrads, polyads, massulae, viscin threads) have evolved at least 39 times among flowering plants (Harder & Johnson 2008). The effect of these alternate forms of pollen aggregation on PTE is largely unknown, since pollen transfer efficiency has been reported for only a single species with pollen tetrads (Harder & Johnson 2008; Wilson 1995). Furthermore, the rarity of pollinia among flowering plants and the fact that the vast majority of species disperse their pollen as monads, suggest that the fitness benefits of aggregated pollen dispersal outweigh the costs only under certain, relatively rare, circumstances (Harder & Johnson 2008; Harder & Thomson 1989).

Apocynaceae are an ideal group for the investigation of the effect of pollen aggregation and other modifications of floral morphology on PTE in a phylogenetic context (Livshultz et al. 2011). Within Apocynaceae, pollinia have evolved at least three times independently: once in the common ancestor of subfamilies Secamonoideae and Asclepiadoideae (the milkweeds) (Livshultz et al. 2007; Straub et al. 2014), and at least twice within subfamily Periplocoideae (Ionta & Judd 2007). Pollen tetrads have likely evolved at least five times independently within the family: at least once in subfamily Periplocoideae (Livshultz et al. 2007; Straub et al. 2014), and at least once within each of the tribes Apocyneae (Livshultz et al. 2007), Alyxieae (van der Ham et al. 2001), Melodineae (Van De Ven & Van Der Ham 2006), and Tabernaemontaneae (van der Weide & van der Ham 2012).

In the milkweed flower, pollinia are one of four functionally integrated floral structures that together constitute the highly efficient pollen transfer mechanism. The others are the gynostegium, lignified anther guide-rails, and translators (Fishbein 2001; Livshultz et al. 2007). The gynostegium consists of the adnate androecium and gynoecium (Fig. IA, B). The anther guide-rails are formed by the adpressed lignified margins of adjacent anthers (Fig. IB, ag). Together these two structures function to trap and guide the appendages of flower visitors first to the stigma (Fig. IB, st) and then to the pollen for export (Fig. IB, ps). They are hypothesized to increase the precision of pollen transfer (Fallen 1986) and may also increase PTE. The gynostegium and anther guide-rails evolved prior to pollinia in the most recent common ancestor of the APSA clade (Livshultz et al. 2007), the lineage within which milkweeds evolved, and function together with solitary pollen grains in ca. 700 species classified within the paraphyletic subfamily Apocynoideae (Endress et al. 2014; Livshultz et al. 2007). Lignified anther guide-rails also evolved independently in tribe Tabernaemontaneae where they function without a gynostegium (Simões et al. 2010; Simões et al. 2007). Translators are structures comprised of hardened, molded style-head secretion that function to attach pollen to a pollinator. Each flower has five translators, each removed as a unit with a load of attached pollen. Translators likely evolved independently in milkweeds and in Periplocoideae (Straub et al. 2014), and possibly also in the apocynoid genera *Apocynum* and *Forsteronia* (Leggett 1872; Nilsson et al. 1993), although no functional evidence for the presence of translators in either of these genera has been presented until now.

Comparative study of Apocynaceae species has the potential to elucidate the individual contribution of each of the four structures (gynostegium, guide-rails, translators, pollinia) that constitute the milkweed pollination mechanism to overall PTE. It may also provide an appropriate context for understanding the selective pressures that drove the evolution of the extremely efficient pollen transfer mechanism of milkweeds (Livshultz et al. 2011).

Low pollen to ovule ratio, a trait that has been correlated with high PTE in xenogamous species (Cruden 2000; Cruden & Jensen 1979; Erbar & Langlotz 2005; Harder & Johnson 2008), has been reported in Apocynaceae species from across the phylogeny [range 20-315 outside milkweeds (11 species with monads or tetrads), 2.7-22 within milkweeds (46 species with pollinia)] (Cruden 1977; Darrault & Schlindwein 2005; de Moura et al. 2011; Erbar & Langlotz 2005; Herrera 1991; Lin & Bernardello 1999; Raju & Ramana 2009; Raju et al. 2005; Tanaka et al. 2006; Wyatt et al. 2000). This points to the possibility that efficiency gains from pollinia may not be as large as they appear when milkweeds are compared to "average" xenogamous or facultatively xenogamous flowers, which have mean pollen to ovule ratios of 5859 and 797, respectively (Cruden 1977). The plesiomorphic flower of Apocynaceae, with style-head secretion that glues pollen to floral visitors (Fallen 1986), a basal collar on the style-head that traps pollen (Simões et al. 2007), and tight packing of the anthers and style-head within a narrow-tubed salverform corolla (Fallen 1986), may already possess an elevated PTE as compared to the average non-Apocynaceous xenogamous flower.

As a first step toward the comparative study of the evolution of floral function in Apocynaceae, we here report the PTE of *Apocynum cannabinum* L. This is the first measurement of PTE in a species of Apocynaceae outside subfamily Asclepiadoideae, and only the second report of PTE in an animal pollinated plant with pollen dispersed in tetrads (Harder & Johnson 2008; Wilson 1995). We also present the first evidence that the adhesive bands in *Apocynum* flowers function as translators and report which orders of insects remove them.

MATERIALS AND METHODS

Study taxon

The genus *Apocynum* belongs to Apocynaceae tribe Apocyneae, a well-supported monophyletic lineage of 23 genera (Endress et al. 2014; Livshultz 2010; Livshultz et al. 2007; Middleton & Livshultz 2012). Within Apocyneae, the species of *Apocynum* are highly unusual in their morphology, biogeography, and ecology (Livshultz et al. 2011). All four *Apocynum* species are erect perennial herbs or shrubs with pollen dispersed in tetrads (Woodson 1930). Floral morphology of *Apocynum* has been extensively studied and illustrated (Lipow & Wyatt 1999; Nilsson et al. 1993; Omlor 1996; Rosatti 1989; Safwat 1962; Sennblad et al. 1998; Woodson 1930). Translators (or "glands" or "plates") have been reported in three *Apocynum* species based on morphological criteria (Demeter 1922; Leggett 1872; Nilsson

FIGURE I. *Apocynum cannabinum*. A. Inflorescence with flowers at anthesis showing the gynostegium (g). B. Gynostegium with two anthers removed to show pollen clump (p) on the stigma (st); the anthers (a) are adnate to the collar (c) of the style-head via adaxial pads of hairs, "retinacles" (r); the pollen sacs (ps) which hold pollen for export are thus isolated from the stigma (st) inside a chamber formed by the five anthers and the style-head apex (sa); lignified basal anther appendages, "anther guide-rails" (ag), trap and guide insect appendages as they reach for the nectaries (n), each located below a pair of anther guide-rails. C. Style-head with deposited pollen clump (p) on the stigma below the style-head collar (c); insect appendage (i) still attached to the pollen clump via an adhesive band (ab). D. Close-up of pollen clump in C with attached adhesive band (ab) and insect appendage (i). E. pollen clump with fragment of stigma (st) squashed and viewed with transmitted light; pollen tetrads (t) stained pink with Calberla's fluid; pollen tubes (pt) are whitish filaments; the adhesive band (ab) is a yellowish rectangle. All scale bars are 100 μm. **a** anther, **ab** adhesive band, **ag** anther guiderail, **as** anther sac, **c** style-head collar, **g** gynostegium, **i** insect appendage, **n** nectary, **o** ovary, **p** pollen clump, **ps** pollen sac, **pt** pollen tube, **r** retinacle, **sa** style-head apex (non-receptive), **st** stigma, **t** pollen tetrad.

et al. 1993; Omlor 1996; Safwat 1962), but never confirmed by functional analysis. Species of *Apocynum* are components of herb-dominated communities in temperate Asia, Europe, and North America (Rosatti 1989). In contrast, all other species of Apocyneae are woody lianas with pollen monads occurring in tropical to sub-tropical forests of Asia and Australasia (Livshultz et al. 2007; Middleton 2007).

Translators have never been reported in any species of Apocyneae outside *Apocynum*.

Apocynum cannabinum L. is a common component of early successional herbaceous communities across temperate North America (Rosatti 1989). A significant agricultural weed in some regions (Schultz & Burnside 1979), it forms large clonal colonies via underground roots (Johnson et al.

1998). *Apocynum cannabinum* is an obligate out-crosser with a late-acting self-incompatibility system, similar to that of *Asclepias* and other milkweed genera (Lipow & Wyatt 1999).

The minute flowers, ca. 3-4 mm long, are produced in dichasial inflorescences (Fig. 1A). Each flower has five greenish sepals, and a white-to-greenish urceolate corolla, with five short corolla lobes that diverge at anthesis (Fig. 1A). The five nectaries are situated around the ovary, opposite the corolla lobes and alternating with the stamens (Fig. 1B, n), each nectary positioned below a minute corolline corona lobe located in the lower half of the corolla tube. The five anthers form a tightly closed cone around the style-head (the apex of the gynoecium) via appression of their lignified margins (Fig. 1A, B). The ovoid style-head has a narrow collar (Fig. 1B, c), situated ca. 1/3 of the distance from its apex to its base, and sits atop the two half-inferior ovaries (Fig. 1B, o). The gynostegium is formed by adhesion of each anther (Fig. 1B, a) via an adaxial patch of hairs (Fig. 1B, r) in the shape of an inverted V (termed a "retinacle") to the style-head collar (Fig. 1B, c). The two pollen sacs (Fig. 1B, ps) are on the adaxial side of the anther, above the retinacle. The five closely packed anthers and the distal surface of the style-head, the "style-head apex" (Fig. 1B, sa), form a closed chamber that retains the pollen, which remains in tetrads at maturity. We confirm that the vast majority of pollen tetrads are retained within the pollen sacs at and after anthesis (Lipow & Wyatt 1999), and are not shed onto the apex of the style-head as has been reported (Rosatti 1989). The location of the pollen sacs, inside a closed pollen chamber distal to the retinacle and style-head collar, separate from the stigmatic surface (Fig. 1B, st) of the style head, which is situated below the collar (Fallen 1986; Lipow & Wyatt 1999), prevents both pollen loss from the anthers (e.g. via wind agitation) and autonomous pollen transfer to the stigma. The only way pollen can exit the flower is by way of removal by an animal visitor, and the only way pollen tetrads can move to the stigma is via the action of an animal pollinator. A rectangular band of yellowish style head adhesive (Fig 1C, D, E, ab), with an extremely viscous and gummy texture at anthesis, is formed in each of the five alterni-staminal zones of the style-head apex, distal to the style-head collar. At anthesis, the bands of style-head adhesive are often found attached to the anthers, below the proximal pollen sacs of two adjacent anthers, spanning the narrow space in between them, and unattached to the style-head. These adhesive bands have been termed "translators" (Nilsson et al. 1993), but will be referred to as "adhesive bands" in this paper since a "translator" is defined by function (removal as a single unit with a load of pollen) and not by structure.

Pollen transfer in *Apocynum* (and most other species in the paraphyletic subfamily Apocynoideae) has been hypothesized to occur via trapping and guidance of insect appendages by the appressed lignified anther guide-rails (Fig. 1B, ag) as the insect reaches toward the base of the flower for nectar (Fig. 1B, n). Once the insect appendage is trapped between the bases of the anther guide-rails, the insect pulls up to escape and is guided first to the stigma, depositing any attached pollen (from previous floral visits), then to the adhesive band, and finally to the pollen in the pollen sacs (Fallen 1986; Rosatti 1989; Waddington 1976). Species of five insect orders have been recorded as floral visitors of *A.*

cannabinum, but only butterflies (10 species) have been reported as carrying pollen attached to their proboscises (Waddington 1976). Our observations of appendages characteristic of Hymenoptera attached to pollen clumps deposited on stigmas, indicate that other insect orders are also likely legitimate pollinators (Fig. 1C, D, i). Some insects cannot escape once trapped by the anther guide-rails (Bailey 1874).

Floral development

Late stage floral development was followed on June 22-30, 2010 in one population, Lemon Hill, Philadelphia. Six buds from each of fourteen haphazardly selected inflorescences (84 buds total) were marked with an oil paint marker on the pedicel, and their development was followed on a daily basis until they abscised, the corolla withered, and/or they began developing into fruits. Developing fruits were confirmed on July 2, 2010. Censuses were conducted daily between 6 and 10 AM.

To determine whether flowers that abscised in the closed hydrated state were more likely to have been pollinated than those that abscised after the corolla withered, we made bulk collections of closed hydrated flowers and closed withered flowers, and scored each flower for the presence or absence of pollen clumps on the stigma.

Sampling for estimation of population pollen transfer efficiency

Inflorescences were collected from three populations, referred to as populations A, B, and C, in Burlington County, New Jersey on July 16, 2009. Two of these populations were located in the vegetation margin between a paved road and a planted agricultural field; the third was located in the vegetation margin between a paved road and a mown lawn. Insects, including Hymenoptera and Lepidoptera, were observed visiting flowers in all three populations on the day of collection. One to three inflorescences were collected from each flowering stem in the population and preserved in 70% ethanol (added to each sample within 24 hours of collection) and stored at 4°C.

Inflorescences from 6 flowering stems were selected haphazardly from each population, and all post-anthesis flowers with turgid, unwithered corollas (Fig. 1A) were placed in a 50 mL Falcon tube with 70% ethanol; each sample was therefore limited to flowers that had closed 0-3 days prior to the collection date (Fig. 2). Five post-anthesis flowers were selected haphazardly from each sample for quantification of pollen removal and deposition. Flowers that showed evidence of florivory were excluded. In total, pollen removal and deposition were quantified for 90 flowers (3 populations × 6 flowering stems/population × 5 flowers/flowering stem).

Pollen and ovule counts

We used two ultra-fine forceps to disperse the pollen in a drop of Calberla's fluid (Dafni 1992) on a microscope slide; basic fuchsin stained the pollen pink. After the dissection, a drop of fluid was applied to the tips of each forceps and released into the drop on the slide to remove any pollen clinging to the forceps. A cover slip was added and the slide

day 0 → day 1 → day 2-4/5 → day 5/6-7 → day 8 → day 10
bud   open,   closed,   closed,   closed,   developing
         corolla lobes corolla turgid corolla withered corolla withered, fruit,
         diverging   **   calyx base enlarged corolla
                                                                             abscised

↓   ↓
flowers   flowers
abscised   abscised

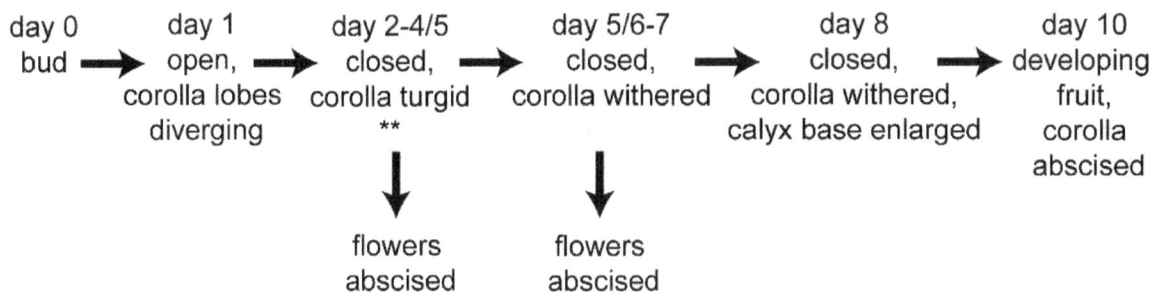

FIGURE 2. The most common developmental patterns from bud to flower abscission or fruit initiation in 78 flowers censused daily; six (6) herbivore damaged flowers were excluded from analysis. Three flowers initiated fruit; one flower had neither abscised nor initiated a fruit on day 10, the last day of the census. ** Flowers in this stage were sampled from bulk population collections to quantify population rates of pollen removal, deposition, and pollen transfer efficiency (See Tab. 1, Fig. 4).

scanned systematically under 100× magnification with a compound microscope, counting all tetrads using a thumb tally counter. Ovules were counted by dissecting out the placenta of one ovary and using two ultra-fine forceps to tear it up and release the ovules into a drop of water on a microscope slide. A cover slip was added, the slide scanned systematically under 50× magnification with a compound microscope, and all ovules were counted.

Pre-anthesis pollen content of flowers from each of the same 18 sampled flowering stems was estimated by counting all the tetrads from one undehisced anther from one bud and multiplying by 5, the number of anthers. The largest bud with undehisced anthers (the one chronologically closest to the post-anthesis flowers) was selected for counting.

Pollen to ovule ratio was calculated for 11 of the 18 flowering stems by also counting all the ovules in one of the two ovaries from the same bud and multiplying by 2. The number of tetrads (four pollen grains in each) was multiplied by 4 to obtain the number of pollen grains, which was then divided by the number of ovules in the flower.

Pollen removal was estimated by subtracting the pollen remaining in a post-anthesis flower from the calculated pre-anthesis pollen content of a bud from that same flowering stem. The amount of pollen remaining in each post-anthesis flower was assessed in two ways: first by scoring, and then by counting. Scoring was accomplished under a dissecting microscope (20-30×) by carefully dissecting out the five anthers and laying them, pollen sacs up, on a slide. The pollen content of each of the 10 pollen sacs (2 per anther) was estimated as 0 (no pollen remaining), 0.33 (some pollen remaining), 0.67 (some pollen removed) and 1 (no pollen removed). Late flower buds (pre-anthesis but with anther sacs already dehisced) were used as the reference point for the four scores. The estimated pollen contents of the 10 pollen sacs were summed and divided by two to obtain the number of full anther equivalents of pollen remaining in the flower; this estimate was multiplied by the estimated pre-anthesis pollen content to obtain the pollen remaining. In a few flowers, clumps of tightly packed pollen were found outside the pollen sacs but inside the chamber formed by the attachment of the anthers to the style-head. Each of these clumps was broken up, and counted and added to the total of pollen remaining.

Pollen remaining was both estimated and counted for 18 flowers, one from each of the sampled flowering stems. Linear regression showed that the estimate closely predicted the count (Fig. 3A). Thus, pollen remaining was scored but not counted for the remaining 62 flowers. The linear regression equation derived from the 18 flowers that were both scored and counted was used to calculate pollen remaining for all 90 flowers. Pollen clumps found inside the anther chamber were always counted and the count added to the estimate.

Pollen deposition. For each of the 90 post-anthesis flowers, the number of pollen clumps on the stigma (Fig. 1B-D, p) was counted, the presence of pollen tubes noted, and the number of pollen tetrads in all clumps counted (Fig. 1E). Solitary pollen tetrads on the stigma were not counted since their presence was likely an artefact of the process of dissecting the flower. Pollen tubes were never observed from any solitary pollen tetrad on the stigma even though both compatible and incompatible pollen germinates on the stigmas of *Apocynum cannabinum* (Lipow & Wyatt 1999).

Pollen transfer efficiency (PTE) was calculated per flower, following the design of Harder & Johnson (2008), and defined as PTE = (pollen deposited on stigma)/(pre-anthesis pollen content - post-anthesis pollen content). For each population, and for the species, PTE was calculated by pooling pollen export and import for all flowers using the formula:

$$PTE = \sum(\text{pollen deposited on stigma})/\sum(\text{pre-anthesis pollen content - post-anthesis pollen content}).$$

Adhesive band removal and deposition. The number of missing adhesive bands (from 0 to 5) was scored for each of the 90 flowers scored for pollen removal and deposition. The proportion of pollen clumps deposited with adhesive bands was not recorded for these 90 flowers. Instead, post-anthesis flowers from one population were dissected until 20 flowers with deposited pollen clumps were observed. For each of these 20 flowers, the pollen clumps and attached stigmatic regions were dissected out with minimal disturbance, placed on slides with Calberla's fluid (Dafni 1992), topped with a coverslip, and examined with 100× magnification for presence of a yellowish mass of adhesive attached to the clump of tetrads (Fig. 1E, ab).

Adhesive band removal was also scored for virgin flowers presented to single insect visitors in the mid-morning to early afternoon of July 2 and 4, 2010. Inflorescences at the Lemon Hill population were covered with fine mesh cloth for 48 hours, and then uncovered and observed until one insect probed for nectar; some insects revisited individual flowers multiple times before moving on to a different inflorescence. After the insect departed the inflorescence, all open flowers were immediately collected into 70% alcohol. The flowers were transferred to 100% alcohol for ca. 10 minutes prior to dissection to harden the adhesive bands so that they would not stretch and break when the anthers were pulled apart, as we observed when flowers preserved in 70% ethanol were dissected. The number of bands remaining in each flower was scored, and any partial or broken bands noted. Voucher specimens of plants and insect visitors to *A. cannabinum* are deposited in the collections of the Academy of Natural Sciences of Drexel University.

Statistical analyses

We used the Pearson χ^2 test of goodness of fit to test if flowers that abscised in the closed hydrated state were more likely to have been pollinated than those that abscised after the corolla withered, linear regression to estimate the correlation between pollen estimates and pollen counts, and Shapiro-Wilk W to test for normality of distributions. For normally distributed data (pollen tetrads removed), we used 1-way ANOVA and Student's T-tests to test for statistical significance. For data that are not normally distributed (pollen tetrads and pollen clumps deposited, adhesive bands removed, pollen transfer efficiency), we used the Kruskal-Wallis χ^2 test. All statistical analyses were conducted with the Analyse-it Standard Edition plug-in (Analyse-it Software, Ltd.) for Microsoft Excel (Microsoft Corporation).

Comparative data from other species

Pollen to ovule ratios of Apocynaceae species (Appendix I) were taken from Erbar & Langlotz (2005) citing the work of Ali & Ali (1989); Christ et al. (2001); Cruden (1977); Lohne et al. (2004); Torres & Galetto (1999); Wyatt et al. (2000) and supplemented with counts from de Araujo et al. (2011); Darrault & Schlindwein (2005); de Moura et al. (2011); Herrera (1991); Lin & Bernardello (1999); Raju & Ramana (2009); Raju et al. (2005). Pollen transfer efficiencies (Fig. 4, Appendix II) were taken from Harder & Johnson (2008) and articles cited by them (Aizen & Raffaele 1996; Broyles & Wyatt 1995; Freitas & Paxton 1998; Galen & Stanton 1989; Harder & Thomson 1989; Hiei & Suzuki 2001; Kunze & Liede 1991; Lipow & Wyatt 1998; Ollerton et al. 2003; Pauw 1998; Snow & Roubik 1987; Tanaka et al. 2006; Vieira & Shepherd 2002; Webb & Bawa 1983; Wilson 1995; Wyatt 1976; Young & Stanton 1990) and additional articles located by searches in Google Scholar on the phrases "pollen transfer efficiency" and "pollination efficiency" (Castellanos et al. 2003; Conner et al. 1995; Ren & Tang 2010; Shuttleworth & Johnson 2006; Shuttleworth & Johnson 2008; Shuttleworth & Johnson 2009; Thostesen & Olesen 1996; Wolff et al. 2008; Young et al. 2007). Studies were included if they reported both removal and deposition of pollen grains. When pollen transfer efficiencies were reported

for multiple populations of a species, the populations with the lower values were excluded if evidence was presented that the pollination environment for that population was less than optimal [e.g. due to absence of the most effective pollinator (Shuttleworth & Johnson 2008) or due to pollinator behaviour (Young et al. 2007)]. The pollen transfer efficiency for the species was then calculated as the mean of all the other populations. For *Impatiens capensis* Meerb. (Balsaminaceae), the pollen transfer efficiency measured by Young et al. (2007), 0.64% for nectar collecting *Apis mellifera*, almost an order of magnitude higher than that measured by Wilson & Thomson (1991), 0.088% for nectar collecting *Bombus impatiens*, was included.

RESULTS

Flower development

Seventy-eight of the 84 buds followed developed without obvious damage. Six that showed evidence of chewing or premature wilting were excluded from analysis. The most common developmental sequences are summarized in Fig. 2. In 73 of the 78 intact flowers (93%), the corolla lobes were reflexed on the first day of anthesis (Fig. 2, day 1). In five (5) flowers (7%), anthesis was not observed; they were buds on day 0, and the lobes were already closed when censused on day 1. Of the 73 flowers observed at anthesis on day 1, 61 (83%) were closed on day 2, while 12 (17%) were still partially open, the lobes parallel or not quite closed. Two of these 12 were still not quite closed on day 3 but closed on day 4. Thus, anthesis lasts <24 hours for 7% (5 of 78) flowers, ca. 24 to <48 hours for 78% (61 of 78) of flowers, >24 but <96 hours for 15% (12 of 78) flowers.

Once the lobes closed, the corolla remained turgid and well-hydrated for two days (2 flowers, 3%), three days (46 flowers, 59%), or four days (30 flowers, 38%) (Fig. 2, day 2-4/5). During these days when the lobes were closed but the corolla still turgid, many flowers began to nod due to geotropic curvature of the pedicel, corollas changed colour from white/cream to yellowish and finally brownish, and pedicels changed from pale green to greenish-yellow. Finally, on day 4 or 5, the whole flower was gone (59%, 46 of 78 flowers), presumably abscised in the 24 hours between censuses (Fig. 2, 46 flowers abscised), or the corolla withered, became dry and shrivelled (Fig. 2, day 5/6-7). Once the corolla shrivelled, either the flower was easily dislodged with a touch within one or two days (36%, 28 flowers) (Fig. 2, 28 flowers abscised), or fruit initiation was detected within four days of withering (4%, 3 flowers) (Fig. 2, day 8), or neither abscission nor fruit initiation were detectable after six days with a withered corolla (1%, 1 flower).

The three flowers that initiated fruit were each at anthesis on day 1, closed and nodding with turgid corollas on days 2-4, and with withered corollas on day 5. Unlike the flowers that abscised, however, the pedicels were noticeably green on day 5. The flowers remained in this state, nodding, withered corolla, green pedicel, until day 8, when the base of the calyx was noticeably enlarged (Fig. 2, day 8). On day 10, nine days after anthesis, a pair of initiating fruits was visible as the

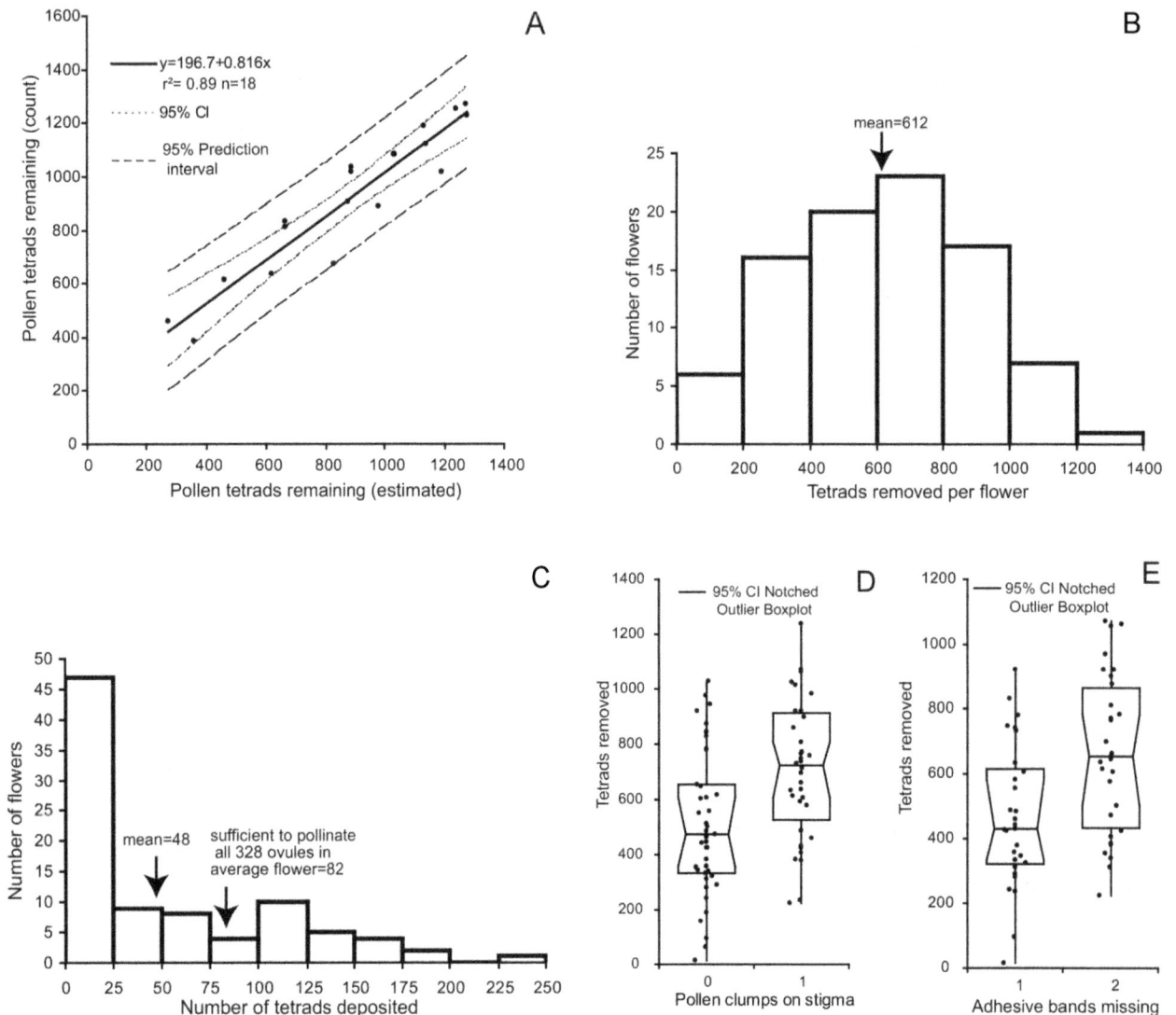

FIGURE 3.   Pollen removal and deposition in 90 flowers of *Apocynum cannabinum* pooled from three populations (Table I). A. Linear regression of count of pollen tetrads remaining in 18 post-anthesis flowers as a function of estimates of pollen tetrads remaining calculated by scoring the pollen content of each pollen sac and multiplying by the pollen content of a pre-anthesis bud (See "Pollen Removal" in Materials and Methods). B. Distribution of tetrad removal per flower. C. Distribution of tetrad deposition per flower. D. Comparison of tetrads removed per flower in flowers with 0 versus I pollen clump deposited on the stigma. E. Comparison of tetrads removed per flower in flowers with I versus 2 adhesive bands missing.

enlarging ovaries pushed the withered corolla away from the receptacle (Fig. 2, day I0).

We collected 22 flowers with a hydrated corolla and 25 flowers with a withered corolla that all fell with a gentle touch. Of the 22 hydrated flowers, 14 were pollinated, seven were not, and one was damaged by herbivory and not counted. Of the 25 withered flowers, 20 were pollinated, four were not pollinated, and one was excluded due to herbivory. The pollination rate for flowers with hydrated versus withered corollas was not significantly different (Pearson $\chi^2 = 1.68$, $DF = I$, $P = 0.19$).

Pollen and ovules

Buds contained an average of I536 ± 28 *S.E.* ($N = 18$ buds) tetrads or 6I44 ± II2 *S.E.* monads ($N = 18$ buds) and 328 ± 19 *S.E.* ($N = II$ buds) ovules. The pollen to ovule ratio (monads per ovule) is 19.8 ± 1.3 *S.E.* ($N = II$ buds).

Pollen removal. Estimates and counts of tetrads remaining in post-anthesis flowers were highly correlated ($r^2 = 0.89$, $N = I8$ flowers) (Fig. 3A). The regression equation $y = I96.7 + 0.8I6x$ was used to estimate pollen remaining.

Pollen removal from the I0 pollen sacs within a flower was typically very heterogeneous. Both full and empty sacs were found within a single flower. The two adjacent sacs of two adjacent anthers were usually similar, i.e. either both full or both empty, while the two pollen sacs within a single anther were often different, i.e. one full and one empty.

The average post-anthesis flower had 6I2 ± 274 *S.D.* tetrads removed (Fig. 3B). The distribution was not significantly different from normal (Shapiro-Wilk $W = 0.98$, $DF = 89$, $P = 0.30$). The average number of tetrads removed per flower varied among populations (Tab. I) from 508 ± 269 *S.D.* (population A) to 744 ± 295 *S.D.* (population B).

TABLE I.   Comparison of three populations of *Apocynum cannabinum* for pollen removal, deposition, transfer efficiency, and adhesive band removal. Means or distributions that are not significantly different ($P > 0.05$) are designated with the same superscript letter. Six (6) flowering stems and 5 flowers per flowering stem were sampled per population. Significance of differences between populations was tested with one-way ANOVA (tetrads removed) or with the non-parametric Kruskal-Wallis test (all other parameters).

Parameter	Population		
	A, $N = 30$ flowers	B, $N = 30$ flowers	C, $N = 30$ flowers
Tetrads removed (mean per flower)	508 ± 269 *S.D.*[a]	744 ± 295 *S.D.*[b]	584 ± 201 *S.D.*[ab]
Pollen clumps deposited (mean per flower)	0.40 ± 0.56 *S.D.*[a]	1.0 ± 0.79 *S.D.*[b]	0.60 ± 0.68 *S.D.*[ab]
Tetrads deposited (mean per flower)	25 ± 45 *S.D.*[a]	75 ± 67 *S.D.*[b]	44 ± 56 *S.D.*[ab]
Pollen transfer efficiency (mean % per flower)	5.6 ± 13.3 *S.D.*[a]	11.6 ± 13.7 *S.D.*[b]	8.4 ± 12.9 *S.D.*[ab]
Population pollen transfer efficiency (%)	4.9	10.1	7.6
Adhesive bands removed (mean per flower)	1.4 ± 1.2 *S.D.*[a]	2.2 ± 1.3 *S.D.*[b]	2.1 ± 0.86 *S.D.*[b]

Populations A and B were significantly different from each other (1-way ANOVA, $F = 6.56$, $DF = 2$, $P = 0.0022$, difference between means $= 236$ tetrads, 95% CI 74 to 399, $P < 0.05$), but neither was significantly different from population C (Tab. 1).

Pollen deposition. Pollen was deposited below the style head collar in distinct clumps producing a tangle of pollen tubes (Fig. 1B-D). Forty-three (48%) of 90 post-anthesis flowers had no deposited pollen clumps, 36 (40%) had one deposited pollen clump, 10 (11%) had two deposited pollen clumps, and one (1%) had three deposited pollen clumps. The average post-anthesis flower had 48 ± 60 *S.D.* tetrads deposited, but the distribution is highly left skewed (Shapiro-Wilk $W = 0.80$, $DF = 89$, $P < 0.0001$) (Fig. 3C). Flowers with one pollen clump deposited had significantly more tetrads removed (709 ± 249 *S.D.*) than flowers with no deposited pollen clumps (511 ± 259 *S.D.*) (t $= 3.45$, $DF = 75.5$, 1-tailed $P = 0.0005$) (Fig. 3D).

The average number of clumps and tetrads deposited per flower varied among populations (Tab. 1): 0.40 ± 0.56 *S.D.* clumps and 25 ± 45 *S.D.* tetrads (population A) to 1.0 ± 0.79 *S.D.* clumps and 75 ± 67 *S.D.* tetrads (population B). Populations A and B were significantly different from each other for both number of clumps (Kruskal-Wallis test $\chi^2 = 10.49$, $DF = 2$, $P = 0.0053$, mean rank difference $= 19$, $P = 0.0041$) and number of tetrads deposited (Kruskal-Wallis test $\chi^2 = 10.81$, $DF = 2$, $P = 0.0045$, mean rank difference $= 21$, $P = 0.0028$), but neither was significantly different from population C (Tab. 1).

Pollen transfer efficiency per flower varied from 5.6 ± 13.3% *S.D.* (population A) to 11.6 ± 13.7% *S.D.* (population B). It was significantly different between these two populations (Kruskal-Wallis test $\chi^2 = 6.40$, $DF = 2$, $P = 0.041$, mean rank difference $= 16$, $P = 0.035$) but neither was significantly different from population C (Tab. 1). Population pollen transfer efficiency varied from 4.9%

(population A) to 10.1% (population B) (Tab. 1). Species-level pollen transfer efficiency is 7.9% when all 90 flowers are pooled, 7.5% when the three populations are averaged. The species-level pollen transfer efficiency falls in the 24th percentile of pollen transfer efficiencies reported for 36 species of milkweeds in subfamily Asclepiadoideae (mean $= 26$%, median $= 20$%, Fig. 4).

Adhesive band removal varied from zero (seven flowers, 8%) to five (three flowers, 3%) bands missing per flower, with most flowers missing either one (29 flowers, 32%) or two adhesive bands (31 flowers, 34%). Significantly more tetrads had been removed from flowers with two missing adhesive bands (660 ± 241 *S.D.*) than from flowers with one band missing (465 ± 219 *S.D.*) (t $= 3.28$, $DF = 58$, 1-tailed $P = 0.0009$) (Fig. 3E).

Population A had significantly fewer bands removed per flower than both populations B and C (Kruskal-Wallis test $\chi^2 = 10.91$, $DF = 2$, $P = 0.0043$; A versus B mean rank difference $= 17.9$, $P = 0.0135$; A versus C, mean rank difference $= 19.1$, $P = 0.0075$) while populations B and C are not significantly different (Tab. 1).

Of 78 virgin flowers visited by bumblebees (*Bombus* sp.), 11 had adhesive bands missing (10 with four of five bands remaining, one with only one band remaining), as did two of 22 virgin flowers visited by small bees (one with three and one with four bands remaining). Two flowers visited by butterflies (*Pieris rapae*) did not have missing adhesive bands. One flower visited by a wasp had three intact bands remaining along with fragments of broken bands in the other positions.

Adhesive band deposition. Of 20 pollen clumps dissected out from stigmas, 19 had yellowish adhesive bands attached (Fig. 1B, ab). In the 20th flower, the adhesive band along with a part of the pollen clump was not attached to the stigma but stuck in the hairs of the retinacle and on the stamen filaments, below the attachment of the retinacles to the style head collar (Fig. 1A). Among the pollinated flowers, four were discovered

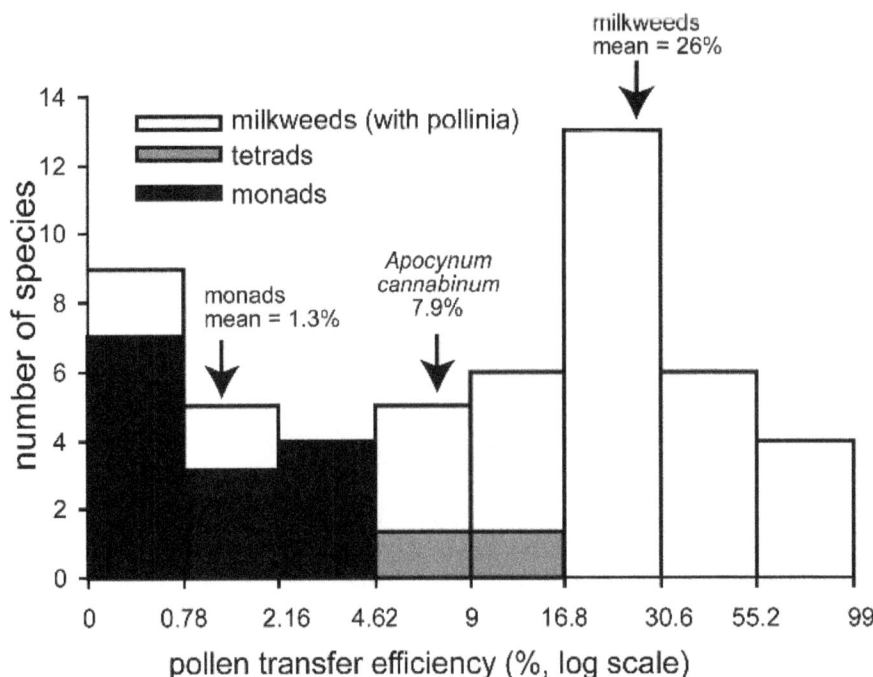

FIGURE 4. Distribution of reported pollen transfer efficiencies in 36 species with pollinia (all milkweeds in Apocynaceae subfamily Asclepiadoideae), 14 species of animal-pollinated angiosperms with pollen in monads (none in Apocynaceae), and two species with pollen in tetrads [*Drosera tracyi* (Droseraceae) and *Apocynum cannabinum* (Apocynaceae)]. The x-axis is scaled logarithmically (base 10). Source data in Appendix II.

with insect appendages still attached to the pollen clumps, the adhesive bands sandwiched between the appendage and the pollen (Fig. 1C, D).

DISCUSSION

Selection for increased pollen transfer efficiency (PTE) may have been an important force in shaping the evolution of flowers. Milkweeds and orchids (Orchidaceae), the two plant lineages with pollinia, have average PTE around 25% (Harder & Johnson 2008), versus 1.3% in species with monads (Fig. 4). However, such broad comparisons can confound efficiency gains due to pollinia with those resulting from other floral novelties. Flowers of *Apocynum cannabinum* (Apocynaceae), like those of milkweeds, have gynostegia and lignified anther guide-rails, but they produce pollen tetrads rather than pollinia and simple bands of style head adhesive instead of complex pollinium-carrying translators. To provide a phylogenetically appropriate comparison to quantify efficiency gains from the evolution of pollinia, we measured PTE in *A. cannabinum*. We also tested the frequently proposed hypothesis (Nilsson et al. 1993) that the adhesive bands in flowers of *Apocynum cannabinum* are translators functionally similar to those found in flowers of milkweeds and Periplocoideae. We compiled reports of pollen to ovule ratios, a correlate of PTE (Harder & Johnson 2008), from Apocynaceae to investigate when elevated PTE may have evolved in the family.

Pollen transfer efficiency of *Apocynum cannabinum* is 7.9%. This is in the 24th percentile of pollen transfer efficiencies reported for 36 species of milkweeds in subfamily Asclepiadoideae (mean $= 26\%$, median $= 20\%$, Fig. 4). It is more than twice as high as any pollen transfer efficiency reported for animal pollinated angiosperms with monads (mean $= 1.3\%$, median $= 0.64\%$, range $= 0.07$-3.4%, $N =$

14 species), but less than that reported for *Drosera tracyi* (13%), the only other species with tetrads where pollen transfer efficiency has been measured (Fig. 4, Appendix II) (Harder & Johnson 2008; Wilson 1995).

These results are consistent with the following hypotheses: 1) Apocynaceae flowers without pollinia have elevated pollen transfer efficiency conferred by other floral novelties such as the gynostegium and style head adhesive; 2) evolution of pollinia in milkweeds resulted in even higher pollen transfer efficiency (Fig. 4); and 3) intermediate levels of pollen aggregation (tetrads) elevate pollen transfer efficiency (Fig. 4). Any statistical test of these statements requires much more data on pollen transfer efficiency in other species of Apocynaceae and in other angiosperm families. Outside the orchids and milkweeds, where pollinia greatly simplify the task, pollen transfer efficiency has rarely been measured in a way that permits comparison across species (Harder & Johnson 2008), despite the fact that selection for pollen transfer efficiency may drive selection on many plant traits and may be an important factor in plant demography and population dynamics. Of particular interest for Apocynaceae, is comparison of flowers of *Apocynum*, which have been proposed as a model for the early ancestors of milkweeds (Livshultz et al. 2011), to other species in the APSA clade with pollen in monads and with undifferentiated style head adhesive, proposed as appropriate models for the ancestral flowers from which early milkweeds diverged.

Translators

The pollen translators of milkweeds and Periplocoideae develop via restricted secretion and hardening of the style-head adhesive present in all Apocynaceae (Kunze 1993; Kunze 1994; Omlor 1996; Safwat 1962; Schick 1982). Each of the five translators is removed as a discrete unit along with the contents (pollinia or tetrads) of the adjacent pollen sacs of two adjacent anthers. In milkweeds, the translator is usually broken and remains attached to the insect when the pollinium is deposited, although pollinia can be deposited with all or part of the translator still attached (Kunze 1991). To our knowledge, the mode of pollen deposition has never been reported for any species of Periplocoideae.

The five adhesive bands of *Apocynum* (Fig. 1C, D), also illustrated in Nilsson et al. (1993); Omlor (1996); Safwat (1962), function in fundamentally the same way as translators of milkweeds but with less precision. The adhesive bands are removed as discrete units. Virgin flowers visited by insects usually had zero (87 flowers) or one (11 flowers), rarely up to four (two flowers) adhesive bands removed. Fragmentation and possible partial removal of adhesive bands was detected in only one flower, visited by a wasp, bringing into question whether wasps are legitimate pollinators of *Apocynum*.

The two adjacent pollen sacs of two adjacent anthers are frequently either both full or both empty in post-anthesis flowers, similar to the situation in milkweeds where pollinia are either removed or not. Removal of two adhesive bands from a flower is linked to significantly more pollen removal than removal of only one adhesive band (Fig. 3E). On average, removal of a second adhesive band resulted in removal of 195 additional tetrads, 660 ± 241 *S.D.* versus 465 ± 219 *S.D.* (Fig. 3E), about 63% of the pollen content of an average anther (307 tetrads) with considerable scatter around the mean (Fig. 3E). Overall, however, pollen removal per flower fits a continuous, normal distribution (Fig. 3B). This indicates that both removal of pollen without an adhesive band and removal of an adhesive band with less than the full pollen contents of two adjacent anther sacs occur frequently, making this pollen transfer mechanism less precise than that of milkweeds where removal of one translator always results in removal of two (Asclepiadoideae) or four (Secamonoideae) pollinia.

The pattern of pollen deposition (Fig. 3C) is likewise similar to but less precise than that of milkweeds where each flower is either unpollinated or receives at least one pollinium, each with sufficient pollen to fertilize all ovules in the flower (Wyatt et al. 2000). Almost half of *Apocynum* flowers (43, 48%) received no pollen clumps while 24 flowers (27%) received 82 or more tetrads (Fig. 3C), sufficient to pollinate all ovules in an average flower, 328 ± 19 *S.E.* ovules. The largest single deposited pollen clump we discovered had 158 tetrads; and butterflies have been reported to carry clumps of up to 123 tetrads (Waddington 1976).

Nineteen of 20 pollen clumps dissected out from stigmas were deposited with adhesive bands attached (Fig. 1E). Given the structural simplicity of these bands in *Apocynum*, it was not possible to determine whether some of the adhesive remained on the pollinator or not. When an insect appendage was left behind with the pollen clump (Fig 1C, D), the adhesive band was clearly visible attaching the pollen clump to the pollinator (Fig. 1D, ab).

Confirmation that simple bands of style-head adhesive function as translators in *Apocynum cannabinum* makes it more likely that similar structures in *Forsteronia affinis*

(Mesechiteae) (Nilsson et al. 1993) are also translators. This would imply multiple independent origins of simple translators. A careful survey of species from across the Apocynaceae to better document the occurrence of these minute and easily overlooked structures is required, along with additional functional studies.

The pollen to ovule ratio of *Apocynum cannabinum*, 19.8 ± 1.3 *S.E.*, is lower than that reported for any species of Apocynaceae with pollen in monads (range 31-204, $N = 9$ species) but in the 94th percentile of ratios reported for milkweeds of subfamily Asclepiadoideae (mean = 9.6 ± 0.7 *S.E.*, range = -2.7-21.9, $N = 46$ species), consistent with the angiosperm-wide trend of lower pollen to ovule ratios in species with more highly aggregated pollen dispersal (Harder & Johnson 2008). Interestingly, both the highest [315 for *Periploca aphylla* (Wyatt et al. 2000)] and among the lowest [20 for *Decalepis hamiltonii* (Raju & Ramana 2009)] ratios in Apocynaceae are reported for species of Periplocoideae. *Periploca* has translators that carry loose tetrads (Verhoeven & Venter 2001) while *Decalepis* has translators that carry four pollinia of aggregated tetrads (Raju & Ramana 2009; Verhoeven & Venter 1998). A possible interpretation of this is that the pollination mechanism of *Periploca* may result in less aggregated pollen deposition (i.e. a few tetrads deposited in each of many flowers) than that of other Apocynaceae and thus select for greater pollen production via increased pollen competition (Harder & Johnson 2008; Queller 1984). A corollary of this hypothesis is that multiple paternity should be more frequent in fruits of *Periploca* than of other Apocynaceae. Unfortunately, pollen deposition and paternity has never been studied in any species of Periplocoideae. Further studies are also necessary to understand why Apocynaceae overall have such low pollen to ovule ratios and to what extent these are linked to aggregated pollen deposition and/or elevated pollen transfer efficiency.

Population variation

All measures of pollination success including pollen removal and deposition, translator removal, and pollen transfer efficiency were significantly higher in population B than in population A with population C intermediate (Tab. 1). The population level pollen transfer efficiency varied two fold between population A (4.9%) and B (10.1%). Large differences among populations in pollen transfer efficiency (1.9% versus 13-16%) were detected in the South African milkweed *Xysmalobium undulatum* and attributed to the low abundance of the more effective pollinator (pompilid wasps) at the site with low pollen transfer efficiency (Shuttleworth & Johnson 2008). Even the same species of pollinator can produce significantly different pollen transfer efficiencies with different behaviours (Young et al. 2007). Likewise, population density and the larger plant community can affect pollen transfer efficiency (Ghazoul 2005a; Ghazoul 2005b; Ghazoul 2006).

In spite of the diversity of factors other than floral morphology that impact on pollen transfer efficiency, the pollen transfer efficiency measured for population A (4.9%) was still higher than the highest measured for any angiosperm with monads, 3.4% for *Campsis grandiflora* (Bignoniaceae) pollinated by vespid wasps (Ren & Tang 2010). Thus, while pollen transfer efficiency is an emergent property of the interaction of floral morphology with many ecological factors, the role of floral morphology (which is likely much more constant among populations than the other parameters) can be elucidated by studying multiple populations of a species and/or the same population at multiple time points. Furthermore, even cursory studies of multiple species with a shared floral trait can produce a distribution that shows how the trait performs on average across a diversity of habitats, pollinator assemblages, population densities, etc. (Fig. 4).

Pollinators of Apocynum cannabinum.

Species of Hymenoptera, Diptera, Coleoptera, and Hemiptera have been recorded as floral visitors to *Apocynum cannabinum* (syn. *Apocynum sibiricum* Jacq.), but only species of Lepidoptera have been previously reported as carrying pollen (Waddington 1976). Our observations, including the deposition of pollen clumps with attached appendages that could not have come from butterflies (Fig. 1C, D) and removal of translators by bumblebees and small bees after visits to virgin flowers, suggest that *Apocynum cannabinum* has a more generalized pollination system than previously reported, although the importance of each of these different groups of insects to the male and female fitness of *A. cannabinum* remains to be determined.

Flower development and sampling for pollen transfer efficiency

Understanding late stage flower development (from anthesis to fruit initiation) and how it is modified by pollination is important for unbiased sampling of flowers for estimation of population level pollen transfer efficiency. In milkweeds, pollen transfer efficiency has typically been estimated by scoring the standing population of open flowers for pollinia removal and deposition (Ollerton et al. 2003; Shuttleworth & Johnson 2008; Wyatt 1976). This provides an unbiased sample of pollinated and unpollinated flowers only if anthesis of flowers is neither abbreviated nor extended by pollination; something that has not been reported in any study that measured pollen transfer efficiency in milkweeds. Our preliminary studies of several Apocynaceae species show that anthesis may be either extended or truncated by pollination (Livshultz, unpublished), potentially a widespread phenomenon across flowering plants (Fung & Thomson 2017).

Because anthesis of *Apocynum cannabinum* flowers is brief (ca. 24 to <48 hours, Fig. 2), we sampled recently closed flowers (still turgid) to estimate pollen transfer efficiency since the ratio of pollen deposition to pollen removal is likely to increase through the life-span of a one-day flower. The vast majority (97%) of flowers whose development was followed in our census were in this closed, turgid state for three to four days (Fig. 2, day 2-4/5), including both flowers that ultimately abscised without change of corolla hydration and flowers whose corollas ultimately withered prior to either abscission or fruit initiation (Fig. 2). Turgid flowers that abscised with a gentle touch were not significantly more likely to be pollinated than flowers with withered corollas that abscised with a gentle touch. We thus have no evidence that the standing population of closed turgid flowers was either depleted or enriched for pollinated flowers.

Conclusions

The measured pollen transfer efficiency of *Apocynum cannabinum*, 7.9%, fits neatly with predictions from floral morphology. It is less efficient than the average Apocynaceae species with pollinia (milkweeds, 26%), but more efficient than an average flower with monads (1.25%) (Fig. 4). The elevated efficiency of *Apocynum cannabinum* is likely due to the combined function of floral structures shared by *Apocynum* and milkweeds: gynostegium, lignified anther guide-rails, aggregated pollen, and translators (Fig. 1). These structures occur in various combinations among species of Apocynaceae, making them an ideal group for comparative study of functional floral morphology.

ACKNOWLEDGEMENTS

Thanks to Marina Potapova for assistance with S.E.M, Ling Ren for assistance with light microscopy, and Julian Golec for assistance with virgin flower presentation experiments. Gretchen Ionta, Suzanne Koptur, and Daniel Schott provided valuable comments on the manuscript.

APPENDICES

Additional supporting information may be found in the online version of this article:

APPENDIX I.   Literature sources for pollen: ovule ratios of Apocynaceae.

APPENDIX II.   Literature sources for pollen transfer efficiencies (PTE) in Fig. 4.

REFERENCES

Aizen MA, Raffaele E (1996) Nectar production and pollination in *Alstroemeria aurea*: responses to level and pattern of flowering shoot defoliation. Oikos 76:312-322.

Ali T, Ali SI (1989) Pollination biology of *Calotropis procera* subsp. *hamiltonii* (Asclepiadaceae). Phyton-Annales Rei Botanicae 29:175-188.

Armbruster WS, Muchhala N (2009) Associations between floral specialization and species diversity: cause, effect, or correlation? Evolutionary Ecology 23:159-179.

Bailey WW (1874) *Apocynum*. Bulletin of the Torrey Botanical Club 5:9-10.

Broyles SB, Wyatt R (1995) A reexamination of the pollen-donation hypothesis in an experimental population of *Asclepias exaltata*. Evolution 49:89-99.

Castellanos MC, Wilson P, Thomson JD (2003) Pollen transfer by hummingbirds and bumblebees, and the divergence of pollination modes in *Penstemon*. Evolution 57:2742-2752.

Christ K-D, Dieterle A, Gottsberger G (2001) Pollinators, pollen ovule ratio and the extent of cross-versus self-fertilization in the groundlayer of a spring wildflower community in a central european forest. Phytomorphology 51:529-540.

Conner JK, Davis R, Rush S (1995) The effect of wild radish floral morphology on pollination efficiency by four taxa of pollinators. Oecologia 104:234-245.

Cruden RW (1977) Pollen-ovule ratios - conservative indicator of breeding systems in flowering plants. Evolution 31:32-46.

Cruden RW (2000) Pollen grains: why so many? Plant Systematics and Evolution 222:143-165.

Cruden RW, Jensen KG (1979) Viscin threads, pollination efficiency and low pollen-ovule ratios. American Journal of Botany 66:875-879.

Dafni A (1992) Pollination Ecology a Practical Approach. Oxford University Press, Oxford, United Kingdom.

Darrault RO, Schlindwein C (2005) Limited fruit production in *Hancornia speciosa* (Apocynaceae) and pollination by nocturnal and diurnal insects. Biotropica 37:381-388.

de Araujo LDA, Quirino ZGM, Machado IC (2011) Fenologia reprodutiva, biologia floral e polinização de *Allamanda blanchetii*, uma Apocynaceae endêmica da Caatinga. Revista Brasileira de Botânica 34:211-222.

de Moura TN, Webber AC, Torres LNM (2011) Floral biology and a pollinator effectiveness test of the diurnal floral visitors of *Tabernaemontana undulata* Vahl. (Apocynaceae) in the understory of Amazon rainforest, Brazil. Acta Botanica Brasilica 25:380-386.

Demeter K (1922) Vergleichende Asclepiadeenstudien. Flora 115:130-176.

Endress ME, Liede-Schumann S, Meve U (2014) An updated classification for Apocynaceae. Phytotaxa 159:175–194.

Erbar C, Langlotz M (2005) Pollen to ovule ratios: standard or variation - a compilation. Botanische Jahrbuecher fuer Systematik Pflanzengeschichte und Pflanzengeographie 126:71-132.

Fallen ME (1986) Floral structure in the Apocynaceae: morphology, functional, and evolutionary aspects. Botanische Jahrbuecher fuer Systematik, Pflanzengeschichte, und Pflanzengeographie 106:245-286.

Fishbein M (2001) Evolutionary innovation and diversification in the flowers of Asclepiadaceae. Annals of the Missouri Botanical Garden 88:603-623.

Flanagan RJ, Mitchell RJ, Karron JD (2011) Effects of multiple competitors for pollination on bumblebee foraging patterns and *Mimulus ringens* reproductive success. Oikos 120:200-207.

Freitas BM, Paxton RJ (1998) A comparison of two pollinators: the introduced honey bee *Apis mellifera* and an indigenous bee *Centris tarsata* on cashew *Anacardium occidentale* in its native range of NE Brazil. Journal of Applied Ecology 35:109-121.

Galen C, Stanton ML (1989) Bumble bee pollination and floral morphology: Factors influencing pollen dispersal in the Alpine Sky Pilot, *Polemonium viscosum* (Polemoniaceae). American Journal of Botany 76:419-426.

Gascoigne J, Berec L, Gregory S, Courchamp F (2009) Dangerously few liaisons: a review of mate-finding Allee effects. Population Ecology 51:355-372.

Ghazoul J (2005a) Implications of plant spatial distribution for pollination and seed production. In: Burslem D, Pinard M, Hartley S (eds) Biotic interactions in the tropics: their role in the maintenance of species diversity. Cambridge University Press, Cambridge, UK, pp 241-266.

Ghazoul J (2005b) Pollen and seed dispersal among dispersed plants. Biological Reviews 80:413-443. doi: doi:10.1017/S1464793105006731.

Ghazoul J (2006) Floral diversity and the facilitation of pollination. Journal of Ecology 94:295-304.

Groom MJ (1998) Allee effects limit population viability of an annual plant. The American Naturalist 151:487-496. doi:10.1086/286135.

Harder LD (2000) Pollen dispersal and the floral diversity of Monocotyledons. In: Wilson K.L. MD (ed) Monocots: systematics and evolution. CSIRO Publishing, Melbourne, pp 243–257.

Harder LD, Aizen MA (2010) Floral adaptation and diversification under pollen limitation. Philosophical Transactions of the Royal Society B: Biological Sciences 365:529-543. doi: 10.1098/rstb.2009.0226.

Harder LD, Johnson SD (2008) Function and evolution of aggregated pollen in angiosperms. International Journal of Plant Sciences 169:59-78.

Harder LD, Thomson JD (1989) Evolutionary options for maximizing pollen dispersal of animal-pollinated plants. The American Naturalist 133:323-344.

Herrera J (1991) The reproductive biology of a riparian Mediterranean shrub, *Nerium oleander* L. (Apocynaceae). Botanical Journal of the Linnean Society 106:147-172.

Hiei K, Suzuki K (2001) Visitation frequency of *Melampyrum roseum* var. *japonicum* (Scrophulariaceae) by three bumblebee species and its relation to pollination efficiency. Canadian Journal of Botany 79:1167–1174. doi: doi:10.1139/cjb-79-10-1167.

Ionta GM, Judd WS (2007) Phylogenetic relationships in Periplocoideae (Apocynaceae s.l.) and insights into the origin of pollinia. Annals of the Missouri Botanical Garden 94:360-375.

Johnson SA, Bruederle LP, Tomback DF (1998) A mating system conundrum: hybridization in *Apocynum* (Apocynaceae). American Journal of Botany 85:1316-1323.

Johnson SD, Neal PR, Harder LD (2005) Pollen fates and the limits on male reproductive success in an orchid population. Biological Journal of the Linnean Society 86:175-190.

Kobayashi S, Inoue K, Kato M (1997) Evidence of pollen transfer efficiency as the natural selection factor favoring a large corolla of *Campanula punctata* pollinated by *Bombus diversus*. Oecologia 111:535-542.

Kunin WE (1997) Population size and density effects in pollination: pollinator foraging and plant reproductive success in experimental arrays of *Brassica kaber*. Journal of Ecology 85:225-234.

Kunze H (1991) Structure and function in asclepiad pollination. Plant Systematics and Evolution 176:227-253.

Kunze H (1993) Evolution of the translator in Periplocaceae and Asclepiadaceae. Plant Systematics and Evolution 185:99-122.

Kunze H (1994) Ontogeny of the translator in Asclepiadaceae S-Str. Plant Systematics and Evolution 193:223-242.

Kunze H, Liede S (1991) Observations on pollination in *Sarcostemma* (Asclepiadaceae). Plant Systematics and Evolution 178:95-106.

Leggett WH (1872) *Apocynum*. Bulletin of the Torrey Botanical Club 3:53-55.

Lin S, Bernardello G (1999) Flower structure and reproductive biology in *Aspidosperma quebracho-blanco* (Apocynaceae), a tree pollinated by deceit. International Journal of Plant Sciences 160:869-878.

Lipow SR, Wyatt R (1998) Reproductive biology and breeding system of *Gonolobus suberosus* (Asclepiadaceae). Journal of the Torrey Botanical Society 125:183-193.

Lipow SR, Wyatt R (1999) Floral morphology and late-acting self-incompatibility in *Apocynum cannabinum* (Apocynaceae). Plant Systematics and Evolution 219:99-109.

Livshultz T (2010) The phylogenetic position of milkweeds (Apocynaceae subfamilies Secamonoideae and Asclepiadoideae): Evidence from the nucleus and chloroplast. Taxon 59:1016-1030.

Livshultz T, Mead JV, Goyder DJ, Brannin M (2011) Climate niches of milkweeds with plesiomorphic traits (Secamonoideae; Apocynaceae) and the milkweed sister group link ancient African climates and floral evolution. American Journal of Botany 98:1978-1988. doi: 10.3732/ajb.1100202.

Livshultz T, Middleton DJ, Endress ME, Williams JK (2007) Phylogeny of Apocynoideae and the APSA clade (Apocynaceae s.l.). Annals of the Missouri Botanical Garden 94:324-359.

Lohne C, Machado IC, Porembski S, Erbar C, Leins P (2004) Pollination biology of a *Mandevilla* species (Apocynaceae), characteristic of NE-Brazilian inselberg vegetation. Botanische Jahrbuecher fuer Systematik Pflanzengeschichte und Pflanzengeographie 125:229-243.

Middleton DJ (2007) Flora Malesiana Series I: Apocynaceae (subfamilies Rauvolfioideae and Apocynoideae)- volume 18. Foundation Flora Malesiana, Leiden.

Middleton DJ, Livshultz T (2012) *Streptoechites* gen. nov., a new genus of Asian Apocynaceae. Adansonia 34:365-375.

Nilsson S, Endress ME, Grafstrom E (1993) On the relationship of the Apocynaceae and Periplocaceae. Grana Supplement 2:3-20.

Ollerton J, Johnson SD, Cranmer L, Kellie S (2003) The pollination ecology of an assemblage of grassland asclepiads in South Africa. Annals of Botany 92:807-834.

Omlor R (1996) Do *Menabea venenata* and *Secamonopsis madagascariensis* represent missing links between Periplocaceae, Secamonoideae and Marsdenieae (Asclepiadaceae)? Kew Bulletin 51:695-715.

Pauw A (1998) Pollen transfer on bird's tongues. Nature 394:731–732.

Queller DC (1984) Pollen-ovule ratios and hermaphrodite sexual allocation strategies. Evolution 38:1148-1151.

Raju AJS, Ramana KV (2009) Pollination and seedling ecology of *Decalepis hamiltonii* Wight & Arn. (Periplocaceae), a commercially important, endemic and endangered species. Journal of Threatened Taxa 1:497-506.

Raju AJS, Zafar R, Rao SP (2005) Floral device for obligate selfing by remote insect activity and anemochory in *Wrightia tinctoria* (Roxb.) R.Br. (Apocynaceae). Current Science 88:1378-1380.

Ren M-X, Tang J-Y (2010) Anther fusion enhances pollen removal in *Campsis grandiflora*, a hermaphroditic flower with didynamous stamens. International Journal of Plant Sciences 171:275-282. doi: doi:10.1086/650157.

Rosatti TJ (1989) The genera of suborder Apocynineae (Apocynaceae and Asclepiadaceae) in the Southeastern United States. Journal of the Arnold Arboretum 70:443-514.

Safwat FM (1962) The floral morphology of *Secamone* and the evolution of the pollinating apparatus in Asclepiadaceae. Annals of the Missouri Botanical Garden 49:95-129.

Schick B (1982) Zur Morphologie, Entwicklung, Feinstruktur und Funktion des Translators von *Periploca* L. (Asclepiadaceae). Tropische und Subtropische Pflanzenwelt 40:7-45.

Schultz ME, Burnside OC (1979) Distribution, competition, and phenology of hemp dogbane (*Apocynum cannabinum*) in Nebraska. Weed Science 27:565-570.

Sennblad B, Endress ME, Bremer B (1998) Morphology and molecular data in phylogenetic fraternity: The tribe Wrightieae (Apocynaceae) revisited. American Journal of Botany 85:1143-1158.

Shuttleworth A, Johnson SD (2006) Specialized pollination by large spider-hunting wasps and self-incompatibility in the African milkweed *Pachycarpus asperifolius*. International Journal of Plant Sciences 167:1177-1186.

Shuttleworth A, Johnson SD (2008) Bimodal pollination by wasps and beetles in the African milkweed *Xysmalobium undulatum*. Biotropica 40:568–574.

Shuttleworth A, Johnson SD (2009) Palp-faction: An African milkweed dismembers its wasp pollinators. Environmental Entomology 38:741-747.

Simões AO, Endress ME, Conti E (2010) Systematics and character evolution of Tabernaemontaneae (Apocynaceae, Rauvolfioideae) based on molecular and morphological evidence. Taxon 59:772-790.

Simões AO, Livshultz T, Conti E, Endress ME (2007) Phylogeny and systematics of the Rauvolfioideae (Apocynaceae) based on molecular and morphological evidence. Annals of the Missouri Botanical Garden 94:268-297.

Snow AA, Roubik DW (1987) Pollen deposition and removal by bees visiting two tree species in Panama. Biotropica 19:57-63.

Straub SCK, Moore MJ, Soltis PS, Soltis DE, Liston A, Livshultz T (2014) Phylogenetic signal detection from an ancient rapid radiation: Effects of noise reduction, long-branch attraction, and model selection in crown clade Apocynaceae. Molecular Phylogenetics and Evolution 80:169-185. doi: 10.1016/j.ympev.2014.07.020.

Tanaka H et al. (2006) Andromonoecious sex expression of flowers and pollinia delivery by insects in a Japanese milkweed *Metaplexis japonica* (Asclepiadaceae), with special reference to its floral morphology. Plant Species Biology 21:193-199.

Thomson JD (2006) Tactics for male reproductive success in plants: contrasting insights of sex allocation theory and pollen presentation theory. Integrative and Comparative Biology 46:390-397.

Thostesen AM, Olesen JM (1996) Pollen removal and deposition by specialist and generalist bumblebees in *Aconitum septentrionale*. Oikos 77:77-84.

Torres C, Galetto L (1999) Factors constraining fruit set in *Mandevilla pentlandiana* (Apocynaceae). Botanical Journal of the Linnean Society 129:187-205.

Van De Ven EA, Van Der Ham R (2006) Pollen of Melodinus (Apocynaceae): Monads and tetrads. Grana 45:1-8.

van der Ham R, Zimmermann YM, Nilsson S, Igersheim A (2001) Pollen morphology and phylogeny of the Alyxieae (Apocynaceae). Grana 40:169-191.

van der Weide JC, van der Ham RWJM (2012) Pollen morphology and phylogeny of the tribe Tabernaemontaneae (Apocynaceae, subfamily Rauvolfioideae). Taxon 61:131-145.

Verhoeven RL, Venter HJT (1998) Pollinium structure in Periplocoideae (Apocynaceae). Grana 37:1-14.

Verhoeven RL, Venter HJT (2001) Pollen morphology of the Periplocoideae, Secamonoideae, and Asclepiadoideae (Apocynaceae). Annals of the Missouri Botanical Garden 88:569-582.

Vieira M, Shepherd G (2002) Removal and insertion of pollinia in flowers of *Oxypetalum* (Asclepiadaceae) in southeastern Brazil. Rev. Biol. Trop. 50:37–43.

Waddington KD (1976) Pollination of *Apocynum sibiricum* (Apocynaceae) by Lepidoptera. The Southwestern Naturalist 21:31-36.

Webb CJ, Bawa KS (1983) Pollen dispersal by hummingbirds and butterflies: A comparative study of two lowland tropical plants. Evolution 37:1258-1270.

Wilson P (1995) Variation in the intensity of pollination in *Drosera tracyi* - selection is strongest when resources are intermediate. Evolutionary Ecology 9:382-396.

Wilson P, Thomson JD (1991) Heterogeneity among floral visitors leads to discordance between removal and deposition of pollen. Ecology 72:1503-1507.

Wolff D, Meve U, Liede-Schumann S (2008) Pollination ecology of Ecuadorian Asclepiadoideae (Apocynaceae): How generalized are morphologically specialized flowers? Basic and Applied Ecology 9:24-34.

Woodson RE, Jr. (1930) Studies in the Apocynaceae. I. A critical study of the Apocynoideae (with special reference to the genus *Apocynum*). Annals of the Missouri Botanical Garden 17:1-172+174-212.

Wyatt R (1976) Pollination and fruit-set in *Asclepias* - reappraisal. American Journal of Botany 63:845-851.

Wyatt R, Broyles SB, Lipow SR (2000) Pollen-ovule ratios in milkweeds (Asclepiadaceae): an exception that probes the rule. Systematic Botany 25:171-180.

Young HJ, Dunning DW, Hasseln KWv (2007) Foraging behavior affects pollen removal and deposition in *Impatiens capensis* (Balsaminaceae). American Journal of Botany 94:1267-1271. doi: 10.3732/ajb.94.7.1267.

Young HJ, Stanton ML (1990) Influences of floral variation on pollen removal and seed production in wild radish. Ecology 71:536-547

PERMISSIONS

LIST OF CONTRIBUTORS

Victoria Wojcik
Pollinator Partnership, 423 Washington Street 5th Floor, San Francisco, CA 94111
University of California, Berkeley, Department of Environmental Science, Policy, and Management, Berkeley, CA 94720

Dara A. Stanley
Botany and Plant Science, School of Natural Sciences and Ryan Institute, National University of Ireland, Galway, Ireland

Mark Otieno
Department of Agricultural Resource Management, Embu University College, Embu, Kenya

Karin Steijven
Department of Animal Ecology and Tropical Biology, Biocentre - University of Würzburg, Am Hubland, 97074 Würzburg, Germany
Department of Bee Health, Van Hall Larenstein – University of Applied Sciences, Agora 1, 8901 BV Leeuwarden, the Netherlands

Emma Sandler Berlin
Department of Biology, Lund University, Sölvegatan 37 223, 62 Lund, Sweden

Tiina Piiroinen
Faculty of Science and Forestry, Department of Environmental and Biological Sciences, University of Eastern Finland, FI-80101 Joensuu, Finland

Pat Willmer
School of Biology, Harold Mitchell Building, University of St Andrews, St Andrews, Fife, KY16 9TH, UK

Clive Nuttman
Tropical Biology Association, The David Attenborough Building, Pembroke Street, Cambridge CB2 3QZ, UK

Favio Gerardo Vossler
Laboratorio de Actuopalinología, CICyTTP–CONICET / FCyT-UADER, Dr. Materi y España, E3105BWA, Diamante, Entre Ríos, Argentina

Pushpa Raj Acharya, Sunthorn Sotthibandhu and Sara Bumrungsri
Department of Biology, Prince of Songkla University, Hat Yai Thailand

Paul A Racey
Centre for Ecology and Conservation, University of Exeter in Cornwall, UK

Leif L. Richardson
Department of Biological Sciences, Dartmouth College, Hanover NH, 03755 USA
Gund Institute for Ecological Economics, University of Vermont, Burlington, VT 05405 USA

Rebecca E. Irwin
Department of Biological Sciences, Dartmouth College, Hanover NH, 03755 USA
Department of Applied Ecology, North Carolina State University, Raleigh, NC 27695 USA

Laurent Penet, Benoit Marion and Anne Bonis
UMR CNRS 6553, ECOBIO, Université de Rennes 1, Campus de Beaulieu, 35042 Rennes Cedex, France

Jenny A. Hazlehurst and Jordan Karubian
Department of Ecology and Evolutionary Biology, Tulane University, 400 Lindy Boggs Center, New Orleans LA 70118

Boris Tinoco and Santiago Cárdenas
Escuela de Biologia, Ecología y Gestión, Universidad del Azuay, Av. 24 de Mayo 7-77 y Hernán Malo, Cuenca, Ecuador

Petra Wester
Institute of Sensory Ecology, Heinrich-Heine-University, Universitätsstr. 1, 40225 Düsseldorf, Germany

Peter Bernhardt, Dowen Jocson, Justin Zweck and Gerardo R. Camilo
Department of Biology, Saint Louis University, St. Louis, MO, USA 63103

Retha Edens-Meier
School of Education, Saint Louis University, St. Louis, MO, USA 63103

Zong-Xin Ren
Key Laboratory for Plant Diversity and Biogeography of East Asia, Kunming Institute of Botany, Chinese Academy of Sciences, 132 Lanhei Road, Kunming, Yunnan 650201, P. R. China

Michael Arduser
325 Atalanta Ave., Webster Groves, MO, USA 63119

Shawn Krosnick
Department of Biology, Tennessee Technological University, Cookeville, TN, 38505, U.S.A

Tim Schroeder
Department of Biochemistry and Chemistry, Southern Arkansas University, 100 East University Street, Magnolia, AR, 71753, U.S.A

Majesta Miles
Department of Biology, Southern Arkansas University, 100 East University Street, Magnolia, AR, 71753, U.S.A

Samson King
Department of Neurobiology and Anatomy, Wake Forest School of Medicine, Winston-Salem, NC, 27517, U.S.A

Peter B. Sørensen, Christian F. Damgaard, Beate Strandberg and Marianne B. Pedersen
Aarhus University, Bioscience, Vejlsøvej 25, 8600 Silkeborg, Denmark

Yoko L. Dupont, Jens Mogens Olsen and Melanie Hagen
Aarhus University, Bioscience, NyMunkegade 114, 8000 Aarhus C, Denmark

Luisa G. Carvalheiro
Institute of Integrative and Comparative Biology, University of Leeds, Leeds LS2 9JT, UK
NCB-Naturalis, postbus 9517, 2300 RA, Leiden, The Netherlands

Jacobus C. Biesmeijer
NCB-Naturalis, postbus 9517, 2300 RA, Leiden, The Netherlands

Simon G. Potts
University of Reading, School of Agriculture, Policy and Development, UK

Andre Sanfiorenzo and Manuel Sanfiorenzo
Department of Agriculture Technology University of Puerto Rico-Utuado, Puerto Rico

Lisette Waits
Department of Fish and Wildlife Sciences, University of Idaho, 83844-1136, United States

Ronald Vargas
Scientific department, Organization for Tropical Studies (OTS) La Selva, Puerto Viejo de Sarapiquí, Costa Rica

Bryan Finegan
Forests, Biodiversity and Climate Change Program, Tropical Agricultural Research and Higher Education Center (CATIE), Turrialba 30501, Costa Rica

Jennie F. Husby and Carri J. LeRoy
The Evergreen State College, 2700 Evergreen Parkway NW, Olympia, Washington 98505, USA

Cheryl Fimbel
Center for Natural Lands Management, 120 E, Union #215, Olympia, Washington 98501, USA

James D. Thomson
Department of Ecology and Evolutionary Biology, University of Toronto, 25 Harbord Street, Toronto, Ontario M5S 3G5, Canada

Takashi T. Makino
Department of Ecology and Evolutionary Biology, University of Toronto, 25 Harbord Street, Toronto, Ontario M5S 3G5, Canada
Department of Biology, Faculty of Science, Yamagata University, 1-4-12 Kojirakawa, Yamagata, 990-8560, Japan

Amanda Soares Miranda
Núcleo em Ecologia e Desenvolvimento Ambiental de Macaé. Universidade Federal do Rio de Janeiro. Av. São José Barreto, 764. São José do Barreto, Macaé/ RJ – Brazil; CEP: 27965-045

Milene Faria Vieira
Departamento de Biologia Vegetal. Universidade Federal de Viçosa. Viçosa/MG - Brazil; CEP: 36570-900

Jeff Ollerton, Shawn Connerty, Julia Lock, Leah Parker and Ian Wilson
Landscape and Biodiversity Research Group, School of Science and Technology, University of Northampton, Avenue Campus, Northampton, NN2 6JD, UK

Stella Watts
Landscape and Biodiversity Research Group, School of Science and Technology, University of Northampton, Avenue Campus, Northampton, NN2 6JD, UK
Current address: Laboratory of Pollination Ecology, Institute of Evolution, University of Haifa, Haifa 31905, Israel

Sheila K. Schueller
School of Natural Resources and Environment, University of Michigan 440 Church Street Ann Arbor, MI 48109-1115, USA

Andrea A. Cocucci
Instituto Multidisciplinario de Biología Vegetal (IMBIV). Conicet-Universidad Nacional de Córdoba. Casilla de Correo 495. 5000, Córdoba. Argentina

Julieta Nattero
Instituto Multidisciplinario de Biología Vegetal (IMBIV). Conicet-Universidad Nacional de Córdoba. Casilla de Correo 495. 5000, Córdoba. Argentina Cátedra de Introducción a la Biología, Facultad de Ciencias Exactas, Físicas y Naturales, Universidad Nacional de Córdoba. X5000JJC, Córdoba, Argentina

Ido Izhaki
Department of Evolutionary and Environmental Biology, Faculty of Science and Science Education, University of Haifa, 31905 Haifa, Israel

Anton Pauw
Dept of Botany and Zoology, Stellenbosch Univ., Matieland, 7602, South Africa

Sjirk Geerts
Dept of Botany and Zoology, Stellenbosch Univ., Matieland, 7602, South Africa
South African National Biodiversity Institute, Kirstenbosch National Botanical Gardens, Claremont, South Africa

Jane C. Stout
Trinity Centre for Biodiversity Research and School of Natural Sciences, Trinity College Dublin, Dublin 2, Republic of Ireland

Alison J. Parker and James D. Thomson
University of Toronto, Department of Ecology and Evolutionary Biology, 25 Harbord Street, Toronto, Ontario M5S 3G5 Canada

Laura A. D. Doubleday
Graduate Programs in Organismic and Evolutionary Biology and Entomology, University of Massachusetts, Amherst, Massachusetts, 01003 USA

Christopher G. Eckert
Department of Biology, Queen's University, Kingston, Ontario, K7L 3N6 Canada

Diana B. Robson
Manitoba Museum, 190 Rupert Avenue, Winnipeg, MB Canada R3B 0N2

Cary Hamel
Nature Conservancy of Manitoba, 611 Corydon Ave., Winnipeg, MB, Canada, R3L 0P3

Rebekah Neufeld
Nature Conservancy of Manitoba, 207-1570 18th Street, Brandon, MB, Canada R7A 5C5

Hannah F. Fung
Department of Ecology and Evolutionary Biology, University of Toronto, 25 Harbord Street, Toronto, Ontario, M5S 3G5, Canada

James D. Thomson
Department of Ecology and Evolutionary Biology, University of Toronto, 25 Harbord Street, Toronto, Ontario, M5S 3G5, Canada
Rocky Mountain Biological Laboratory, Crested Butte, CO 81224-0519, USA

Juan Carlos Ruiz-Guajardo
Department of Evolution and Ecology, University of California at Davis, Davis CA 95618 USA
Mpala Research Centre, Laikipia, Kenya

Andrew Schnabel
Department of Biological Sciences, Indiana University South Bend, South Bend IN 46634 USA

Britnie McCallum
Belk College of Business, University of North Carolina, at Charlotte, Charlotte NC 28223 USA

Adriana Otero Arnaiz
Office of Agricultural Affairs, USDA, U.S. Embassy, Mexico City

Katherine C. R. Baldock
School of Biological Sciences, University of Bristol, Bristol BS8 1TQ UK

Graham N. Stone
Institute of Evolutionary Biology, University of Edinburgh, Edinburgh EH9 3JT UK

Anna Gorenflo
Applied Entomology, Department of Insect Biotechnology, Justus Liebig University Giessen, Heinrich-Buff-Ring 26-32, 35392 Giessen, Germany

Tim Diekötter
Department of Landscape Ecology, Institute for Natural Resource Conservation, Kiel University, Olshausenstrasse 75, 24118 Kiel, Germany

Mark van Kleunen
Ecology, Department of Biology, University of Konstanz, Universitätsstrasse 10, 78457 Konstanz, Germany
Zhejiang Provincial Key Laboratory of Plant Evolutionary Ecology and Conservation, Taizhou University, Taizhou 318000, China

Volkmar Wolters and Frank Jauker
Department of Animal Ecology, Justus Liebig University Giessen, Heinrich-Buff-Ring 26-32, 35392 Giessen, Germany

Tatyana Livshultz, Sonja Hochleitner and Elizabeth Lakata
Department of Biodiversity Earth and Environmental Science and Academy of Natural Sciences, Drexel University, 1900 Benjamin Franklin Parkway, Philadelphia, PA 19103-1101, USA

Index